Introduction to Modern EW Systems

For a listing of recent titles in the
Artech House Radar Library,
turn to the back of this book.

Introduction to Modern EW Systems

Andrea De Martino

ARTECH HOUSE
BOSTON | LONDON
artechhouse.com

Library of Congress Cataloging-in-Publication Data
A catalog record for this book is available from the U.S. Library of Congress.

British Library Cataloguing in Publication Data
A catalog record for this book is available from the British Library.

ISBN-13: 978-1-60807-207-1

Cover design by Vicki Kane

10 9 8 7 6 5 4 3 2 1

Contents

Foreword

A definition of the main subject of this book, electronic warfare, is "a set of measures and actions performed by the conflicting sides to detect and electronically attack enemy electronic systems for the control of forces and weapons, as well as to electronically defend one's own electronic systems and other targets." Electronic warfare (EW) was first applied in an extensive manner during the Second World War, mainly to defeat new electronic instruments such as radar and identification friend or foe (IFF) proving to be very effective, especially in night operations. A well-known example is the passive, most secret anti-radar technique then called Window (by Allied forces) or Düppel (by German forces), and now called Chaff, first—and very successfully—used on 24 June 1943 during the night bombing raids on Hamburg to disturb the main sensor of the German air defense, which was the 54-cm Würzburg radar.

EW is aimed to negate or at least reduce the effectiveness of a wide class of enemy systems, not only radar but also infrared sensors/seekers, radio communications and navigation equipment. Therefore, today's EW systems are essential to protect all military platforms when operating in hostile environments.

A vast topic such as EW can be treated in many ways that can be roughly categorized either as "top-down" or as "bottom-up" approaches, depending on the main starting point: operations and system architectures, the former, and technology, the latter. The former choice has been rather natural to the author, Dr. Ing. Andrea A. De Martino, an internationally well-known scientist and systems engineer with more than two decades of experience in radar systems analysis, evaluation, and design; as well as two decades of experience in the EW domain. He is presently Chief Technology Officer of one of the main (third in the world's rank) companies in this area, the Italy-based Elettronica S.p.A.

Dr. De Martino's book starts with operational aspects and scenarios, analyzes their relevant emitters and their signals, continues with the topics of emitter detection and classification/identification as well as the very demanding aspect of precision emitter location and ends with the different electronic measures used against the hostile emitter/platform. Some basic, needed elements are suitably dealt with in four Appendixes, i.e.: signal detection, estimation theory and methods, array antennas, analog modulation; a fifth appendix treats the effects of jamming in frequency shift keying radio.

Such an organization makes the book of interest to both operational and technical people, probably with more emphasis on this latter class of readers. As a matter of fact, there are significant aspects of this book that provide real added value compared to others published on the same topic until about ten years ago. The first is the genuine system engineering approach, as discussed before. The second is the updated treatment of the most recent (and powerful) methods, techniques and technologies for EW systems (such as: phased array antennas for both reception and jamming, digital RF memories, signal acquisition by direct RF /IF sampling, fast digital signal processors, and advanced detection and estimation techniques for emitters identification and localization, including the low-probability-of-intercept techniques). The third and last benefit is the thoroughness of the text, with detailed and deep discussions of items not frequently treated in most EW books. These topics include the intercept and jamming of radio communications, of infrared systems, of laser systems, as well as the electronic actions against the main weapons of the so-called asymmetric warfare, i.e. against the improvised explosive devices (IED), and portable weapons (man pads, rocket propelled grenades).

To conclude, I'm pleased to welcome this significant and very useful addition to the technical literature, and expect it will benefit EW engineers and users, as well the broader group of operators and designers in the electronic defense, and attack arena.

Gaspare Galati
Professor Radar Theory and Techniques
Tor Vergata University
Rome, Italy
July 2012

Introduction

Modern warfare is largely based on electronic systems (such as sensors, actuators, communications equipment, data processors and displays) which constitute the bulk of the information operation system and of the weapon systems.

Information is essential to successfully accomplish any military mission, both defensive as well as offensive, and to ensure the survivability of one's own human resources and assets/weapons.

Information operation is performed through the acquisition, transmission, storage of information and data in order to understand the enemy's intentions, and to make the right decisions about subsequent actions involving the use of weapons and human resources. The above cited process is represented by the so-called Observe-Orient-Decide-Act (OODA) loop that is required to be performed as quickly as possible to gain superiority with respect to the enemy.

Weapon systems effectiveness is completely dependent on the identification of the target/threat and the provision of its position and kinematics data with high accuracies, all provided by the electronic eystems.

Within the above systems is electronic warfare (EW) playing an important role in modern warfare as the passive components of those systems can enhance the acquired information about the emitter threat by adding its identity, achieving an excellent situation awareness, while the active components of those systems can dispense countermeasures according to an optimized strategy, both for the protection of one's own military human resources and assets as well as for offensive actions against enemy defenses.

This book focuses on modern EW Systems and is composed of six chapters.

Chapter 1 describes both the symmetric and asymmetric conflict scenario. In the first scenario all the threats to own assets (i.e. the weapon systems of ground, air, space and sea surface platforms, such as missiles and guns) are driven by electronic sensor systems (such as radars and/or IR/EO sensors). In the second scenario the threats are constituted by terrorist troops/snipers and small weapons such as short range IR missiles and improvised explosive devices (IED). In both scenarios the EW systems have the tasks of threat detection and their countering, although it will be shown that they are quite different in terms of frequency band and type.

Chapter 2 deals with the evolution of the signal emitters (to address how modern systems such as Radar, RF and Laser guided weapons, and communications equipment differ from those in the past) and IR/EO sensors, which both constitute threats of the symmetric conflict scenario and also partially of the asymmetric conflict scenario.

In Chapter 3 the main challenges of the Passive EW RF Band Systems, called electronic support measure (ESM) are discussed through a description of the requirements and the increased performances of the novel passive EW equipment architectures, based on high speed digital processing solutions with respect to the recent equipment architectures

Chapter 4 deals with the modern RF direction finding (DF) and emitter location techniques and technologies, as employed in ESM systems.

Chapters 5and 6 deal respectively with the modern RF and IR active countermeasure equipment requirements, architecture and employed technologies (such as digital RF memories [DRFM] and solid state active wide band phased arrays) and with the main generic (i.e. effective against many types of similar threats) countermeasure techniques against both RF active equipment (radars, RF missile seeker heads and communication equipment) as well as IR/EO missile seeker heads.

The appendixes provide brief introductory concepts about signal detection in sensor receivers (Appendix 1), estimation theory to assess sensor measurement accuracies (Appendix 2), antennas and phased arrays employed in RF equipment (Appendix 3), analog modulation methods employed in communication system (Appendix 4) and evaluation of the bit error rate (BER) increase for noise and CW tone jamming in communication systems employing BFSK modulation(Appendix 5).

Acknowledgments

In writing this book I received much support from friends, such as Professor G. Galati, who revised the manuscript and provided useful comments, and from colleagues of my firm Elettronica S.p.A. of Roma (Italy) who need devoted thanks. Among the latter, special thanks are due to V. Rossi and L. Girardi, who provided all the drawings for the book. Further colleagues to be thanked for their contributions are R. Ardoino, D. Benvenuti, F. Conte, L. Dinoi, G. Foglia, G. Lubello, A. Pantano, G. Panci, D. Pistoia, and A. Zaccaron. I apologize if I am forgetting someone.

Special thanks are also due to my firm Elettronica and to the Italian Defense Companies Selex Galileo, Selex SI, and MBDA who provided and permitted the reproduction of pictures of their equipment in this book.

Finally, very grateful thanks are due to my family members for their patience during the whole period of writing this book.

CHAPTER 1

Introduction to Electronic Warfare Scenarios

1.1 Definitions and EW Role in the Military Field

The formal military terminology [1] defines electronic warfare (EW) as a military action whose objective is control of the electromagnetic spectrum (EMS). This objective is achieved through offensive electronic attack (EA), defensive electronic protection (EP), intelligence gathering and threat recognition electronic warfare support (ES) actions. These current definitions give each function a broader scope with respect to terminology used previously, which was, respectively, electronic countermeasures (ECM), electronic counter-countermeasures (ECCM), and electronic warfare support measures (ESM).

The new EA function definition includes, in addition to the previous ECM function, the use of direct energy weapons (DEWs), antiradiation missiles (ARM), and electromagnetic and nuclear pulses (EMP and NEMP) to destroy enemy electronic equipment.

The new EP function definition includes, in addition to the previous ECCM function, the use of electromagnetic emissions control (EMCON), the electromagnetic hardening of electronic equipment, EW frequency conflict removal, and communications security (COMSEC) actions.

The new ES function definition is not very different from the previous ESM definition as both include near-real-time threat recognition in support of immediate decisions relevant to EA, EP, weapon avoidance, and targeting actions.

The classification of the EW functions is shown in Figure 1.1.

EW is one of the military actions of command and control (C2) warfare, whose objective is to obtain C2 superiority by breaking the connectivity of the adversary C2, thus leading to success in military operations. C2 systems are, in fact, the links through which the commanders' decisions, based on the information and intelligence that flow through those systems, are communicated to battlefield forces.

A C2 system is a network of nodes, each consisting of a number of subsystems, such as sensors, navigation systems, command and data fusion centers, and communication links. A critical node is an element whose disruption or destruction

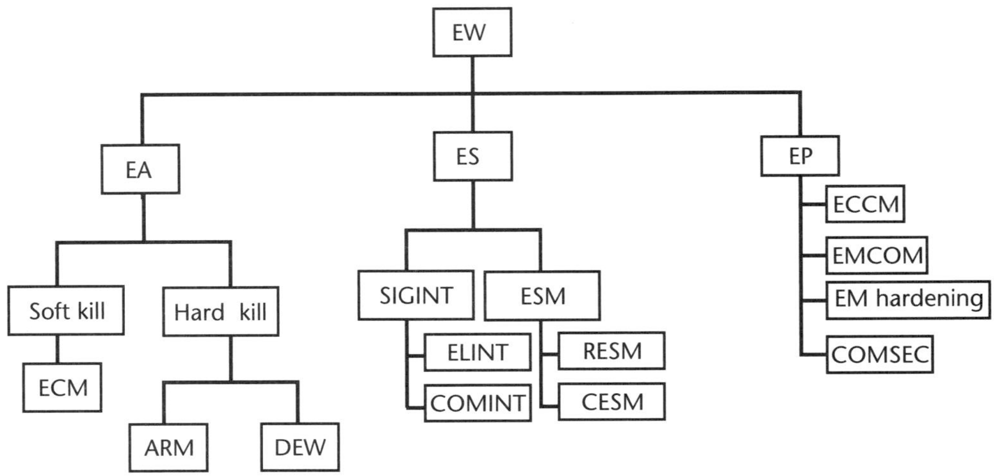

Figure 1.1 Classification of EW functions.

immediately degrades the ability of a C2 system to effectively conduct combat operations.

C2 warfare (C2W) consists of five military actions, which are integrated in both C2 protection and C2 attack actions:

- Operation security (OPSEC);
- EW;
- Psychological operations (PSYOPS);
- Military deception;
- Physical destruction.

Usually OPSEC and military deception are most effective during the so-called "competitive phase," which is the introductory phase to a military conflict, but all five actions are essential to war.

OPSEC is the process of denying the enemy any information about our own capabilities and intentions by controlling and hiding information about our military planning. PSYOPS is a planned process of conveying influential messages to enemy groups in order to convince them to perform actions that are beneficial to own forces.

Deception operations are meant to degrade the accuracy of hostile intelligence gathering, surveillance, target acquisition, and reconnaissance (ISTAR) so that the adversary will draw misleading conclusions about our forces' capabilities and disposition.

EW supports deception operations with electromagnetic deception on intelligence collection, communications, and radar systems.

Physical destruction of a critical node is the most desired and effective option. It involves the use of weapons and of targeting and battle damage assessment (BDA) functions, which are usually accomplished by electronic and optical imagery equipment. EA also participates in the destruction process by the use of ARMs against radar and communication sources and by DEWs in the disruption or burnout of adversary electronic equipment.

In conclusion, in the military field (and namely in the C2W framework), EW plays an important role in that EA supports C2W attacks for combat operations, EP contributes to C2W protection against enemy electronic attacks, and ES provides information that can be used to plan our own C2 attacks and to assess battle damage of the involved sites, thus providing feedback on the effectiveness of the C2W plan.

Section 1.2 gives a brief description of the principles of operation of the main weapons systems that are driven by electronic sensors and, thus, potentially prone to EW actions. We then discuss EW applications during symmetric conflicts (Section 1.3) and asymmetric conflicts (Section 1.4).

1.2 Main Weapons Systems of Interest to EW

The armed forces of each country and its alliances are usually comprised of at least three main groups: army, navy, and air force (marines are considered here to be a navy army), each tasked with different missions with respect to the defense of the country/alliance territory and to any war actions in enemy foreign territories. In the execution of those missions, the armed forces have at their disposal a number of different platforms employing weapons systems, such as munitions, guns, and missiles, whose task is the destruction of enemy assets.

Destruction of enemy assets is preferably accomplished from a distance in order to prevent any harm to our own assets: The larger the distance, the safer our own assets, because the probability of effective retaliation by the enemy is reduced. Modern weapons are being conceived to reach targets at ever longer ranges and to cover such distances at ever higher speeds, so as to reduce the time for an effective enemy defensive reaction, thus increasing the probability of success of the weapon. The provision that the weapons systems have sufficient aiming accuracy at long distances is ensured by electromagnetic sensors (with electronic equipment) that detect and accurately track targets at long distances, thus allowing for accurate bombing and accurate guidance of artillery fire or missiles toward the target. A pictorial representation of the various air attack to ground assets is reported in Figure 1.2 [2].

The air force mission is to ensure air/space superiority and air defense over the territories of interest. The accomplishment of those tasks is performed by an integrated air defense system (IADS) comprised of a number of assets, such as the following:

- An air defense system, composed of ground and airborne early warning radars, C2 centers for surface-to-air missiles (SAM) and antiaircraft artillery (AAA) for defense of important sites and airports;
- Air superiority (or counter-air) fighter aircraft (A/C);
- Intruder/strike A/C.

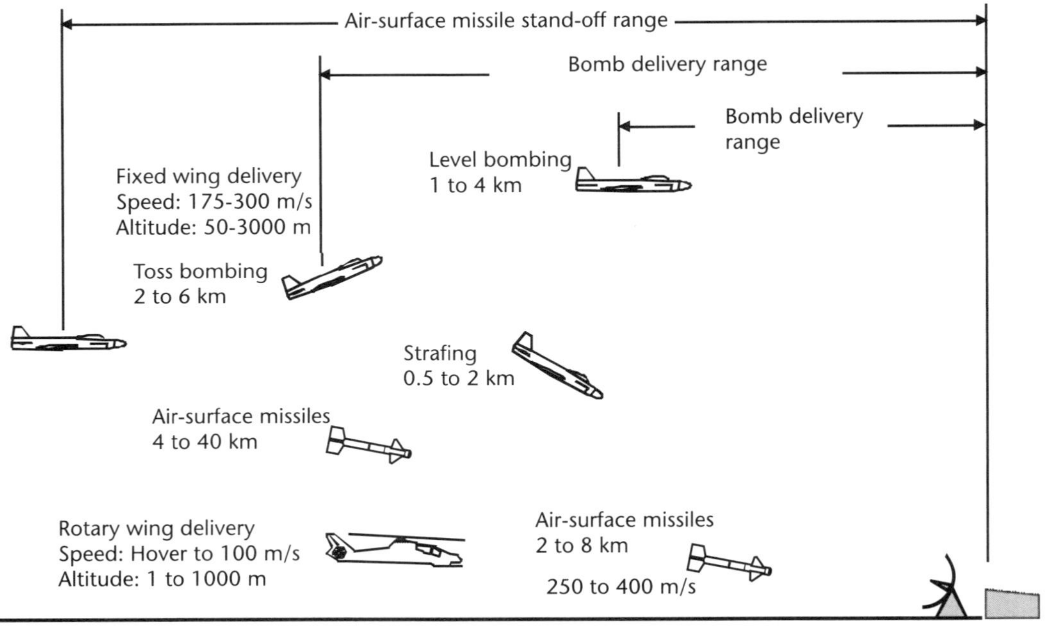

Figure 1.2 Pictorial representation of the various air attack to ground assets.

The objective of EW is to control the EMS in order to gain superiority for our own armed forces and to operate against enemy sensors, both by locating them and by performing actions aimed at reducing their performance. This is done by amplifying the enemy's problems and limits in order to dilute the effectiveness (i.e., the kill probability) of their associated weapons systems.

In the brief notes that follow, the operating principles of the main weapons systems are discussed. A detailed discussion about their operation and the limits of their driving sensors is deferred to Chapter 2.

The main weapons systems employed by the armed forces are of the following types:

1. Early warning surveillance systems, either ground/shipboard or airborne. Early warning systems consist of long-range (in excess of 400 km) surveillance radars mainly operating in the lower radar RF band (from 100 to 3,500 MHz).
2. Missiles systems of various types:
   - Antiaircraft SAMs;
   - Antiaircraft AAMs;
   - Antiship or antiground value target missiles, either launched from the air (ASM) or from the surface (SSM).

   All the above types of missiles can be guided either by radar or by an infrared (IR) seeker in accordance with the stand-off range (i.e., the range of the flight from the launch platform; IR-guided missiles usually have shorter ranges than the radar-guided ones). Antiship missiles are mostly radar guided, but some have hybrid guidance (radar + IR). Additional types of missiles are:

- Antitank systems, mostly wire or IR laser guided, but some guided by millimeter-wave (MMW) radars,
- Antiradiation missiles (ARM);
- Ballistic missiles (types of attack capabilities are shown in Figure 1.3).

Typical ranges of the various types of missiles are reported in Figure 1.4.

3. Artillery systems, such as radar-guided AAA.
4. Precision guided munitions against tanks or buildings, consisting of projectiles guided by optical/IR trackers and laser range finders.

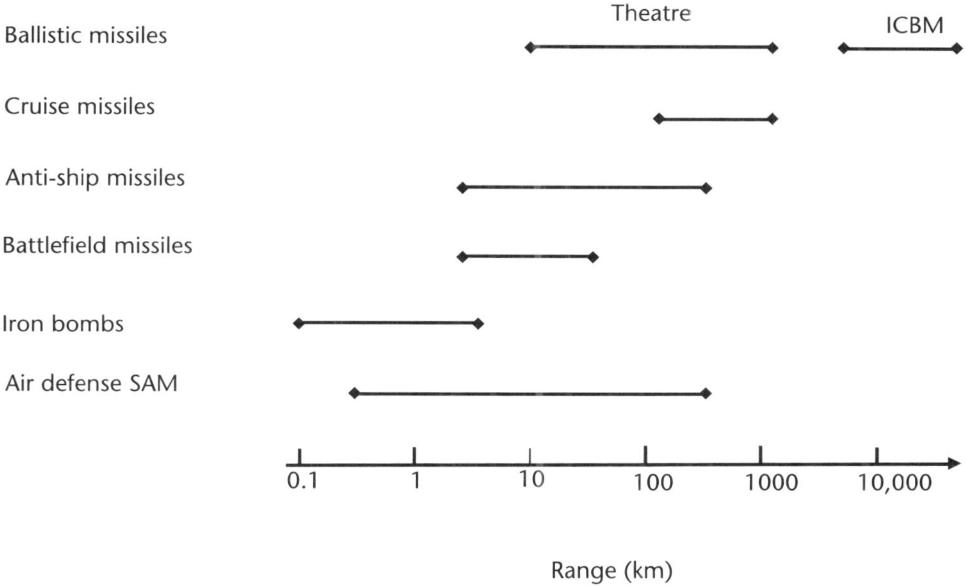

Figure 1.3 Various types of missile attacks.

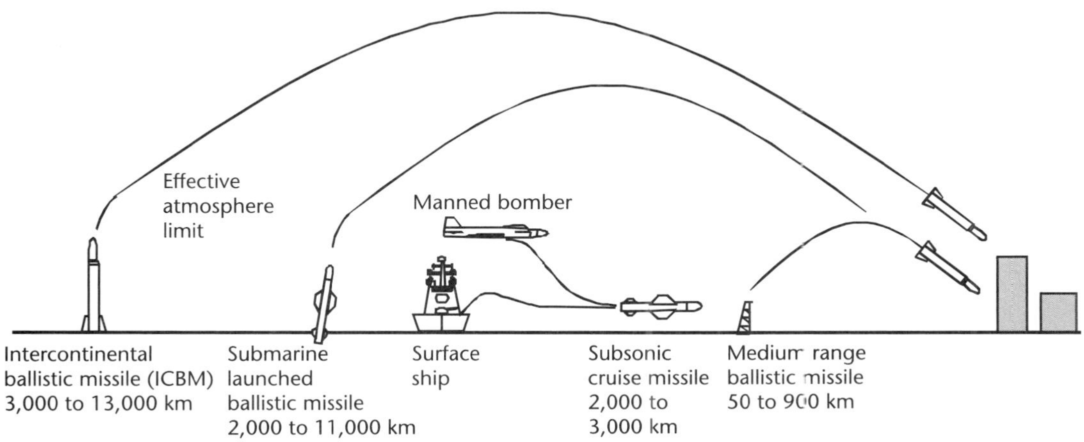

Figure 1.4 Typical stand-off ranges of land and maritime attack weapons compared to the range span of air defense systems.

5. Communication networks, which are the backbone of all military operations, providing combat situation information to the C2 and handling commands from it.

1.2.1 Artillery Systems

Artillery systems usually consist of guns of various calibers that shoot projectiles against fixed or mobile targets and are driven by commands from a fire control system (FCS). The FCS computes the projectile trajectory to the target on the basis of target parameters, projectile and gun parameters, and atmospheric conditions.

For mobile targets, usually aircraft, the AAA fire control system is provided with tracking radar that, on receipt of a 2-D data (dimensions: range and azimuth) designation from surveillance radar, starts an acquisition search at the elevation of the designated target. As soon as the target is acquired, the tracking radar continuously provides the FCS with accurate 3-D (range, azimuth, and elevation) target coordinates, which allow it to determine the kinematics data of the target. With these data and from an extrapolation of the target's trajectory, we can determine the interception point of the fired projectile with the target as shown in Figure 1.5.

A number of shots must be fired in order to achieve the required kill probability. During the shooting period, the tracking radar measures the miss distances of

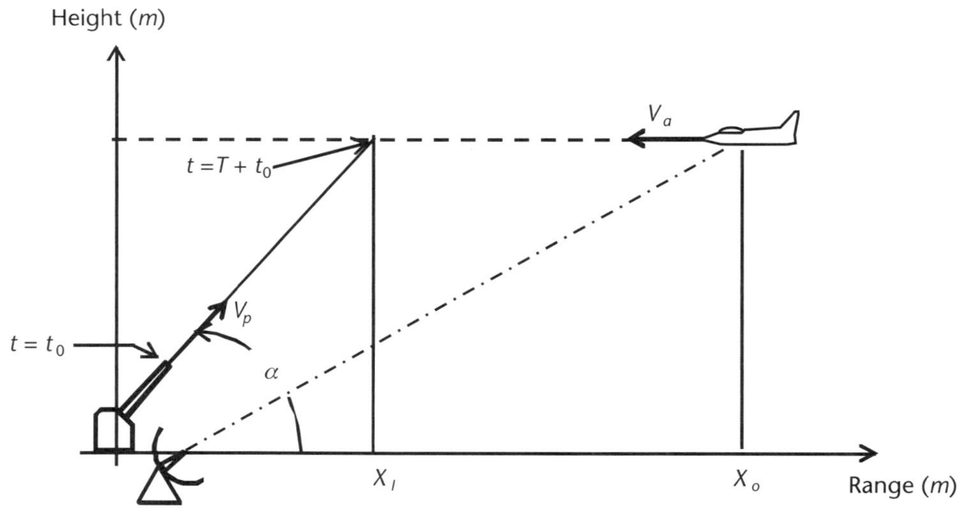

V_a = Target velocity (m/s) T = flight time of the projectile (s)

V_p = average velocity of the projectile during the flight time (m/s)

α = elevation angle (rad)

X_I = Intercept point (m)

X_0 = target position at firing time t_0(m)

$$X_I = X_o \frac{V_a \cos \alpha}{V_a + V_p \cos \alpha}$$

Figure 1.5 Intercept point on the target trajectory of an AAA engagement.

the shot projectiles from the aircraft in order to enable the FCS to correct the gun's aim and the computed projectile's trajectory. In the case of a successful hit, the tracking radar also provides the kill assessment of the target engagement.

The kill probability of the AAA is mainly dependent on the measurement accuracy of the target kinematics and trajectory data as provided by the tracking radar and somewhat dependent on the lethal radius of the projectile.

1.2.2 Missile Systems

Artillery systems are very effective at short ranges, when target kinematics and trajectory measurement are quite accurate (especially if the target is not maneuvering) and the flight time of the projectile is also short. For long ranges, which require a longer projectile flight time, the target gains the opportunity to invalidate the computed interception point by performing a moderate maneuver.

The increases intercept point range is achieved by missiles that belong to two classes: those that attack fixed locations using inertial or global positioning system (GPS) guidance up to the target [ballistic missiles and terrain contour matching (TERCOM)], and those that are guided to a moving target even if the target is maneuvering during the engagement period (i.e., guided missiles).

The latter class of missiles has, in turn, two subclasses, which differ in terms of the type of guidance employed: *command guidance* or *seeker guidance*. Indeed, a missile can be guided to its target by being commanded either by a remote sensor (radar, IR, or EO head), which tracks the target and the missile, or by a sensor (called a *seeker*) internal to the missile.

A guided missile is constituted [3] by an airframe usually consisting of the following equipment:

- Either a receiver that demodulates the command data transmitted by the remote sensor or
- a seeker (radar, IR, or EO head), enclosed in an ogival radome/IR dome, which tracks the target and generates the command signals to the guidance system.
- A warhead containing the explosive material used to damage the target. The explosive material is usually ignited by a proximity fuse, which allows the target to be hit anywhere within the warhead's lethal radius.
- A guidance system or autopilot that converts the command signals from the seeker into the positions of the control fins, hence guiding the missile toward the interception point with the target.
- An engine, fueled by propellant, which provides the thrust to the missile and feeds the power supply needed by both the seeker and the guidance system.

A schematic representation of a missile and of its seeker/guidance operation is shown in Figure 1.6.

Missile weapons systems have search radar that reports to a C2 in case of ground or shipboard systems, or to the main computer in the airborne case, which evaluates the threat and designates the tracking radar and associated missile

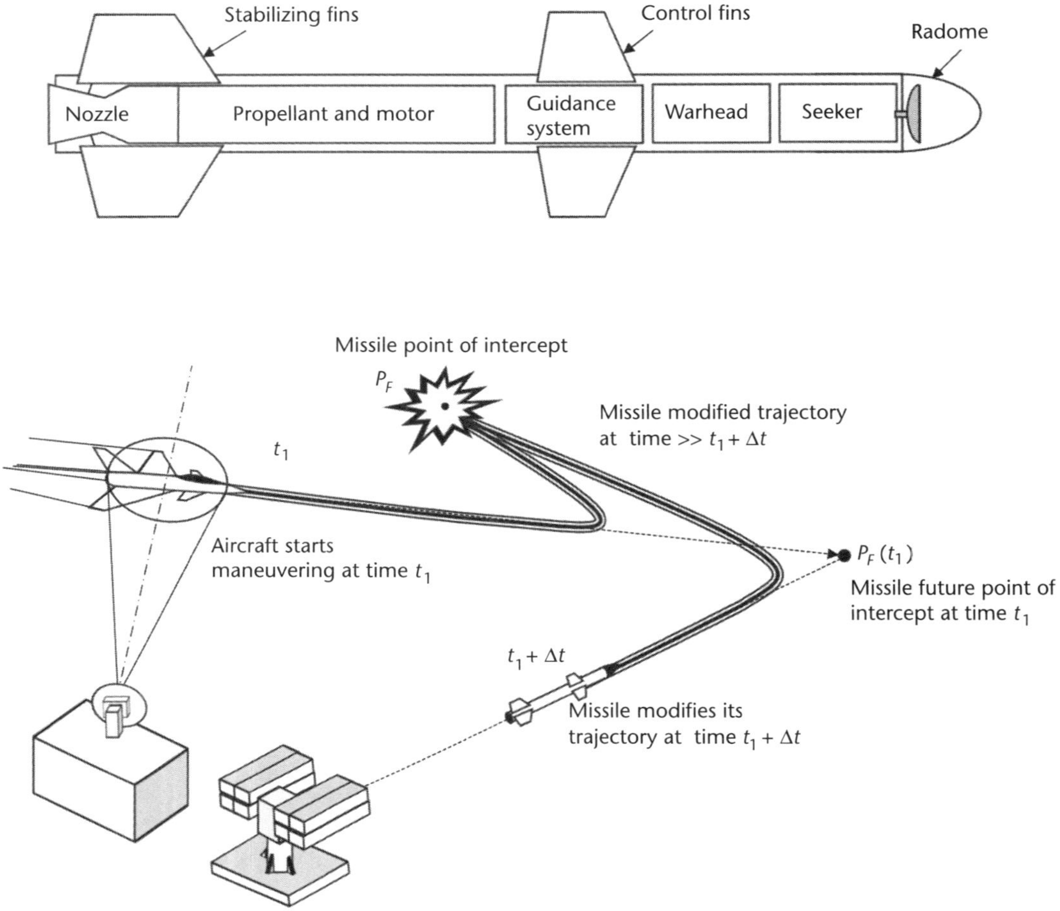

Figure 1.6 Schematic representation of a missile configuration and of its seeker/guidance operation.

launcher of the weapons systems, which is usually able to launch multiple missiles in a quick sequence.

Missile systems are of various types, from the very-long-range ballistic missiles and long-range cruise missile systems designed to attack fixed locations in enemy territory, to the local-area medium- to long-range (50- to 150-km) missile systems designed to defend a relatively wide area (both at ground and at sea) and to the medium- to short-range (5- to 50-km) battlefield missile systems designed to defend/attack high-value targets such as ships, airfields, and military ground assets.

Missiles can employ many different guidance systems, from short-range command and beam-riding guidance to medium- to long-range semiactive and active homing missiles. Very-long-range and long-range missiles usually employ active homing only in the terminal part of their flight to the target; during the rest of their course they are guided by an inertial or satellite navigation system or by recorded terrain data.

Command missiles are guided by commands transmitted from the command link of the ground/shipboard missile weapons systems, which is provided with two tracking radars: a target tracking radar (TTR) and a missile tracking radar (MTR). The missile is usually provided with a beacon in order to ease the tracking of the

MTR. The two tracking radars are managed by a common C2 center that evaluates the data received from them and computes the commands to be sent to the missile. The use of two independent tracking radars ensures that the best possible trajectory-to-target interception is selected for the missile. Alternative command guidance is constituted by a single tracker for both the target and the missile. In this case, called command-to-line of sight (CLOS), the missile has to stay always within the radar beam aimed at the target.

Because the angular target tracking accuracy σ_θ is a fraction of the radar beamwidth θ_B (usually $\sigma_\theta = 0.01$ to $0.05\ \theta_B$), the miss distance m_d from the target of the missile, which is supposed to be aligned with the radar boresight, increases with the distance R of the target from the tracking radar, that is, $m_d = R\sigma_\theta$. For this reason command missiles are used for short ranges. The advantages of these systems are that the missile is a simple one and that the available high effective radiated power (ERP) of the tracking radars employed on ground/ships provides a good ECCM feature.

Beam-riding missiles have an on-board receiver that senses whether or not the missile is aligned with the tracking radar boresight and provides that data to the missile guidance system, which commands the missile to stay aligned with it. As shown in Figure 1.7, the missile is forced to follow a trajectory that requires strong accelerations in the terminal phase of the engagement even in the absence of target maneuvers. Therefore, this type of missile guidance is mostly exploited against slow nonmaneuvering targets, for which only weak accelerations are required.

Semiactive homing missiles are provided with an RF seeker that has a receiver capable of passively tracking in angles the scattered signal from a target when this is illuminated by a powerful continuous wave (CW) or interrupted continuous wave (ICW) tracking radar. The advantage of this type of missile is the absence of an onboard transmitter, which is usually quite expensive. This type of missile, similar to the active pulse Doppler (PD) missile, has the ability to discriminate between the target and clutter by using Doppler frequency signal processing. The angle tracking signals are processed within a very narrow bandwidth (on the order of 1 kHz, thus also achieving a high receiver sensitivity), which is positioned at the Doppler frequency f_d relevant to the closing speed v_c (i.e., $f_d = 2v_c/\lambda = 2(v_t + v_m)/\lambda$, where

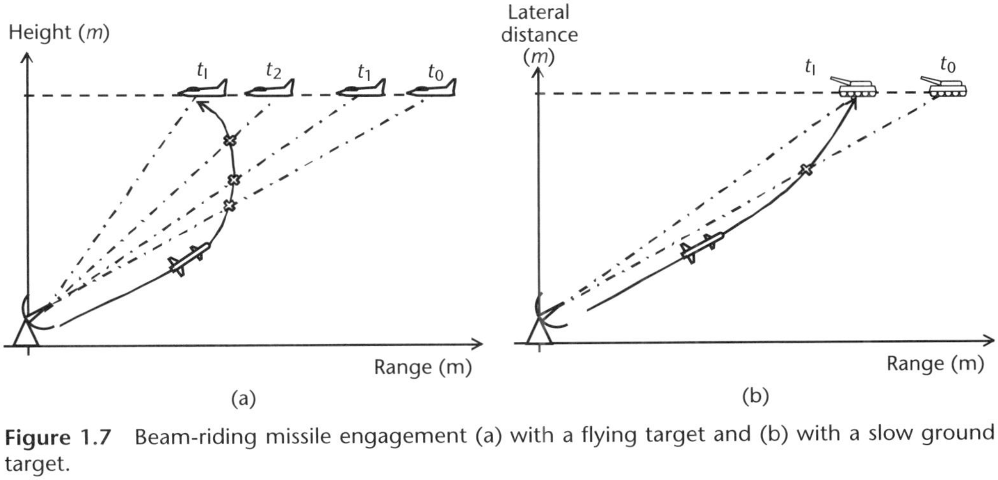

Figure 1.7 Beam-riding missile engagement (a) with a flying target and (b) with a slow ground target.

λ is the transmitted wavelength and v_t and v_m are, respectively, the projections of the target and missile speeds along the target–missile line, as shown in Figure 1.8.

The missile angular tracking accuracy in the terminal phase of the engagement no longer depends on the distance from the tracking/illuminator radar because the seeker is now close to the target. Because there is no more need for the missile to stay within the illuminator beam, the missile can exploit the proportional guidance, which is depicted in Figure 1.9.

The missile is launched toward the predicted interception point, while the seeker antenna tracks the target. With reference to Figure 1.7, in which the missile engagement with the target is reported and the angle parameters are defined, we can show that the correction commands to the missile velocity vector (i.e., only a lateral acceleration a_{lm}, as the missile is flying at constant velocity v_m in modulus) are applied only through its rotation rate $d\gamma/dt$, which is made proportional to the measured rotation rate of the seeker boresight $d\alpha/dt$ in accordance with the selected navigation constant (dimensionless) $N = k(v_t/v_m)$, where v_t is the target velocity and k is a constant (usually assumed to be between 2 and 4):

$$\dot{\gamma} = \frac{d\gamma}{dt} = N\frac{d\alpha}{dt} = N\dot{\alpha} = k\frac{v_t}{v_m}\dot{\alpha} \tag{1.1}$$

Because the lateral acceleration can be expressed as from the above equation, we then have:

$$a_{lm} = v_m\dot{\gamma} = v_m k\frac{v_t}{v_m}\dot{\alpha} = kv_t\dot{\alpha} = kalt \tag{1.2}$$

which means that the lateral acceleration impressed to the missile is proportional to the target lateral acceleration a_{lt}. With this type of guidance, no lateral acceleration is impressed on the missile in order to intercept a target flying on a constant course. So the full available acceleration of the missile can be spared and impressed on the missile only in the case of an evasive maneuver by the target. This ability provides the missile with a longer intercept range and available residual acceleration in the terminal phase of the engagement with the target.

Semiactive missile systems are very effective and constitute the majority of the medium- to long-range SAM population. The major disadvantage of this type of guidance is that constant target illumination is required during the entire time of flight of the missile, which can be dangerous in the case of an air-to-air engagement because the launching platform has to keep on the approach to the target, thus becoming subject to possible retaliation from enemy aircraft.

1.2.3 Active Homing Missiles

Active homing missiles are provided with a seeker that is a complete tracking radar; that is, it is equipped with a transmitter that can engage its target autonomously without any further assistance from the launch platform. For this reason it is called a "fire-and-forget" missile.

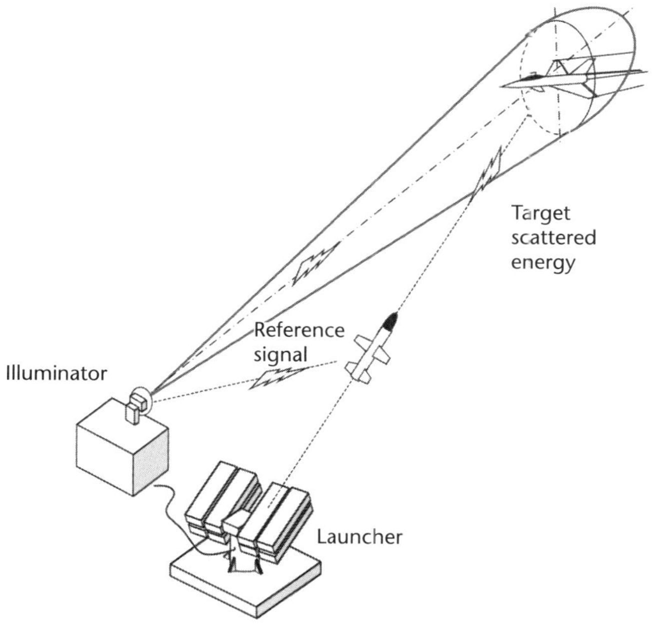

Figure 1.8 Principle of operation of a semiactive missile.

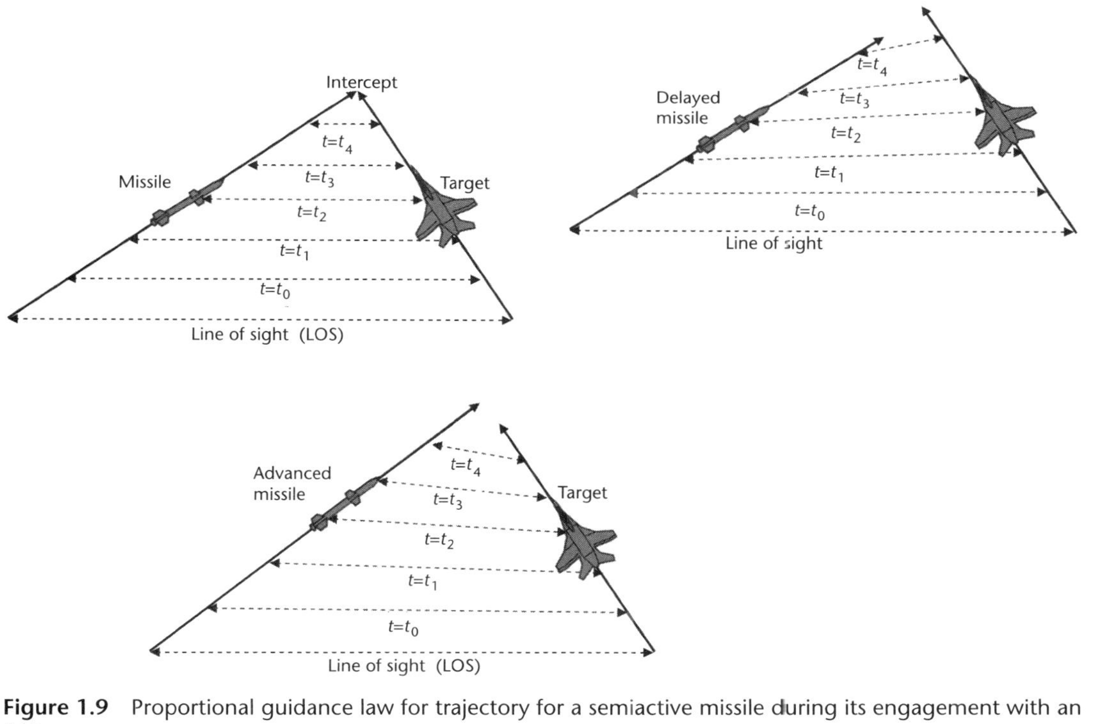

Figure 1.9 Proportional guidance law for trajectory for a semiactive missile during its engagement with an A/C.

Medium- to long-range active homing missiles are provided with a dual guidance system: an inertial or command guidance in the first phase of the flight toward the target, which is then switched to active homing proportional guidance on the target on arrival at some distance from the target.

Due to the presence of a transmitter of this type, the missile is quite expensive.

1.2.4 Track via Missile Systems

Track via missile (TVM) systems [4] employ a mix of command and semiactive missile guidance principles in order to achieve improved medium- to long-range performance at a reasonable cost.

They require a ground illuminator and a semiactive radar sensor on board the missile. However, the angular tracking data are not processed within the missile; they are instead sent via a downlink to a powerful central processor at the launch platform on the ground. Here the trajectory data of both the target and missile are processed, and accurate guidance commands are sent back to the missile via an uplink.

Modern systems use multifunction radar based on phased array technology (Patriot, PAAMS, etc.), which provides them with the ability to engage many targets simultaneously, regardless of which type of missile is use: semiactive and TVM (with the use of ICW illuminators), active seeker, or command guidance. A single missile system provided with a phased array fire control radar may launch and track simultaneously in time-sharing mode many missiles against several different targets. In the case of semiactive and TVM guidance, the phased array fire control radar illuminates the targets in ICW mode, thus providing both tracking of the target and the missile as well as target illumination. This type of weapons systems is shown in Figure 1.10.

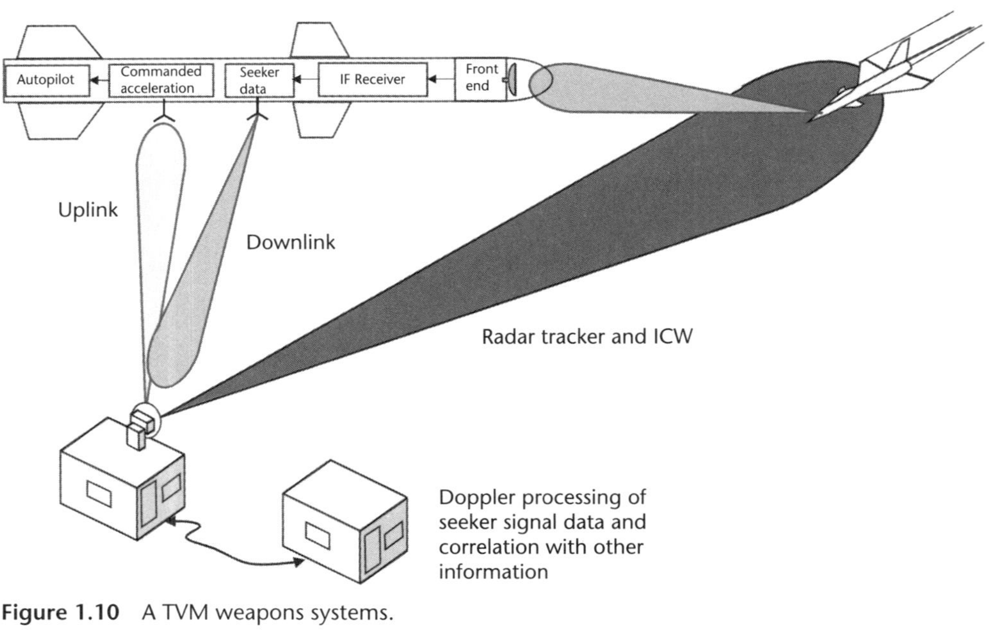

Figure 1.10 A TVM weapons systems.

1.2.5 Passive IR-Guided Missiles

Passive IR-guided missiles are provided with an IR seeker that autonomously tracks the heat signature of the target and provides commands to the proportional navigation guidance system during its approach to the target. Thus they also are of the "fire-and-forget" type.

These types of missiles are extensively used both as medium-range AAMs, due to the relatively lower propagation attenuation at high altitude and their high maneuverability (because they are lightweight), and also as short-range SAMs, due to the higher propagation attenuation at low altitude. They are most often portable and shoulder-launched missiles (in fact, they are called MANPADs) for the defense of ground troops and armored vehicles (and, unfortunately, in current periods, also by terrorists).

Descriptions of the various types of IR seekers are discussed in Section 2.6 of Chapter 2.

1.2.6 Sea-Skimming Missiles

This type of active homing missile is dedicated to ship attacks and flies at a very low altitude over the sea surface as if it is skimming the waves. It can be launched from long distances either from aircraft or ships/littoral sites and is usually provided with a dual guidance system (inertial along the cruise phase and active homing on the target during the terminal phase) in the horizontal plane and with an altimeter to keep a constant height in the vertical plane. A pictorial representation of maritime missile attacks is reported in Figure 1.11.

Ship targets present both a problem and an advantage for these missiles. The problem is constituted by the large angular glint fluctuations (see Section 2.3.8

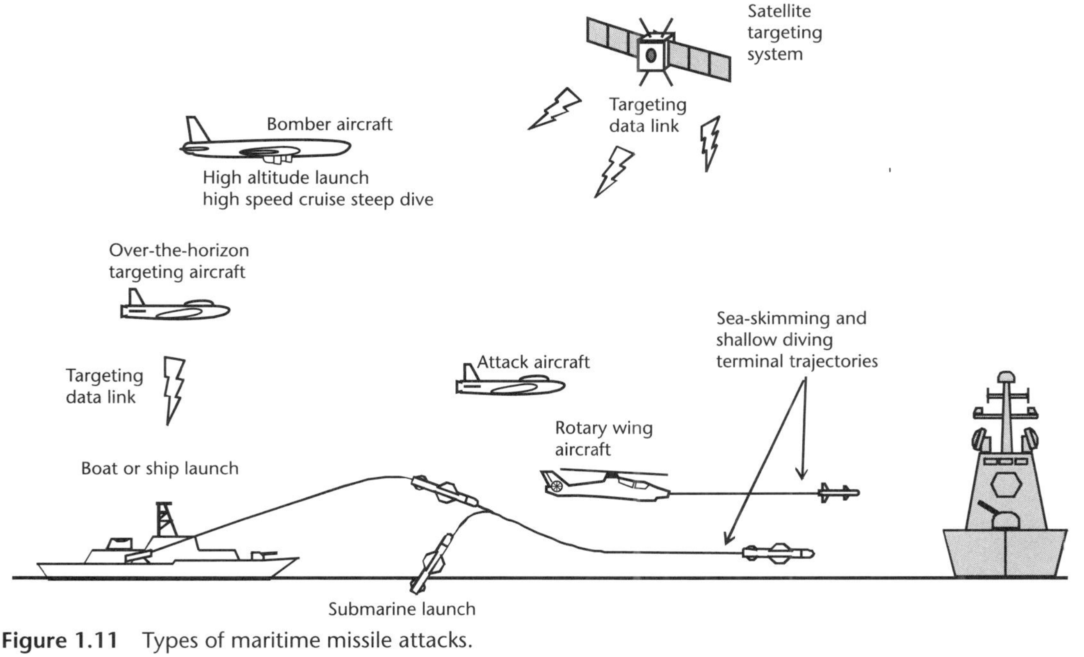

Figure 1.11 Types of maritime missile attacks.

of Chapter 2) produced by the extended length of the ship that require robust angular data filtering to smooth them. The advantage is constituted by the large ship's radar cross section (RCS), which provides large target returns that exceed the sea clutter ones, thus allowing the missile seeker to operate in superclutter conditions [i.e., with a high signal-to-noise power ratio (SNR) and signal-to-clutter (S/C) power ratio on a single radar pulse].These conditions permit the use of large carrier frequency pulse-to-pulse agile radar waveforms with very short pulse widths, which constitutes a big challenge to the ECM ship defenses (as discussed in Section 6.2 of Chapter 6).

1.2.7 Antiradiation Missiles

An antiradiation missile (ARM) is provided with a passive seeker that is similar to a small wideband superheterodyne ESM receiver (see Section 3.4 of Chapter 3), which is capable of acquiring and tracking in angles the signal transmitted by a selected surface radar or communication source and to home on it with high precision. ARMs are mostly of the ASM type and are installed on board aircraft dedicated to the suppression of enemy air defense (SEAD) missions.

The ARM seeker is cued by the airborne ESM system of the A/C, which, upon interception, identification, and location of the victim radar, provides the ARM with all of the emitter data and verifies that the ARM is locked on to that emitter.

Once launched, the ARM seeker, through the use of highly accurate phase direction-finding (DF) antennas, computes from the received emitter signals (in a similar way as for the active homing missiles) the angular commands to be sent to the guidance system in order to home on the victim radar.

The ARM is a typical example of a stand-off "fire-and-forget" weapon because it can be launched from long distances without any further assistance from the launching platform. Many ARMs have been developed that cover the complete radar bands from 0.5 to 18 GHz. The only protective actions that the victim radar can use are to stop transmitting (this does not, however, ensure survival as the missile can home on it on the basis of memorized coordinates) or to activate nearby decoy transmitters that emit waveforms similar to those of the victim radar.

Table 1.1 lists various aircraft characteristics, and Tables 1.2 through 1.5 give the reader insight into the various missiles characteristics.

Table 1.1 Representative Aircraft Characteristics

Aircraft Type	*Wing Span (m)*	*Wing Area (m^2)*	*Number of Engines*	*Combat Radius (km)*
F/A- 18	11.43	37.16	2	1,480
F-16	9.45	27.87	1	925
A-10	17.53	47.01	2	463
Mig 27	14.25	27.25	1	700
Saab AJ-37	10.60	46.00	1	1,000
Mirage 2000	9.13	41.00	1	1,480

Table 1.2 Battlefield ASM Characteristics

Missile	*Origin*	*Targets*	*Launch Platform*	*Type of Guidance*	*Maximum Range (km)*	*Speed (m/s)*
HOT 2	France/ Germany	Tanks	Helicopter	Optical track-wire guided	4	250
AGM-114 (Hellfire)	USA	Tanks/vehicles	Helicopter/fixed wing	Semiactive laser	7	380
AS 30	France	Hard targets	Fixed wing	Semiactive laser	11	510
AGM-65 (Maverick)	USA	Tanks, bridges, ships	Fixed wing	TV; semiactive laser; imaging IR	20–40	–

Table 1.3 Antiship ASM Characteristics

Missile	*Origin*	*Targets*	*Launch Platform*	*Type of Guidance*	*Maximum Range (km)*	*Speed (m/s)*
SeaSkua	UK	Ships	Helicopter/ fixed wing	Semiactive radar + altimeter	25	316
AM 39	France	Ships	Helicopter/ fixed wing	Active radar	Low: 50 High: 70	316
AGM-84A (Harpoon)	USA	Ships	Fixed wing	Active radar + altimeter	100	255
AS-6 (Kingfish)	Russia	Ships/land	Fixed wing	Inertial +Active radar or ARM	Low: 250 High: 700	Low: 408 High: 850

Table 1.4 ASM-ARM Characteristics

Missile	*Origin*	*Targets*	*Launch Platform*	*Type of Guidance*	*Maximum Range (km)*	*Speed (m/s)*
AST-1228 ALARM	UK	Radars	Fixed wing/ helicopter	Antiradar homing	20	680
AGM-88A	USA	Radars	Fixed wing	Antiradar homing	80	680
Armat	France	Radars	Fixed wing	Antiradar homing	120	380

Table 1.5 ASM (Strategic) Characteristics

Missile	*Origin*	*Targets*	*Launch Platform*	*Type of Guidance*	*Maximum Range (km)*	*Speed (m/s)*
AS-14/15	Russia	Strategic	Fixed wing	Inertial + TERCOM	1,200	306
AGM-86A/B (ALCM)	USA	Strategic	Fixed wing	Inertial + TERCOM	2,500	224

1.3 EW in Symmetric Conflicts

Symmetric conflicts occur when the two adversaries' armed forces are both militarily structured, well organized, and provided with weapons systems such as the ones described in the previous sections. In such conflicts the concepts expressed in Section 1.1 about C2W apply.

In this section we expand on the description of the EW (ES, EA and EP) actions that play an important role in C2W.

C2W depends on all sources: timely intelligence about the adversary's capabilities, objectives, and operational concepts. Intelligence sources are the means used to provide such information. Signal intelligence (SIGINT), an ES function, is one of those sources. SIGINT systems collect, analyze, identify and locate emitter signals throughout the entire communications and radar band. Usually they are composed of two dedicated systems: COMINT and ELINT.

Communications intelligence (COMINT) systems are concerned with the activity detection, collection, classification, identification, and DF of communications systems (both voice and digital), data links, satellite communications, and cellular phones. Location of emitters for real-time targeting is usually performed through signal time differential of arrival (TDOA) and differential Doppler frequency (DD) techniques applied in a multiple aircraft systems [5].

Electronic intelligence (ELINT) systems measure direction and time of arrival (DOA and TOA) along with all of the radar waveform signature parameters, such as carrier frequency, pulse width, signal bandwidth, and pulse repetition interval. These data are used both to upgrade the database on the adversary's radars and their associated weapons systems, called ELINT parameters limits (EPLs), and to provide, after comparison with the data in the database, the electronic order of battle (EOB), that is, the disposition of the adversary's weapons systems on his territory. The EPLs are provided to our own ESM equipment in order to allow emitter identification whenever they come in contact with them. Emitter location is accomplished through the methods described in Chapter 4, which include triangulation, trilateration, with time difference on arrival (TDOA), and frequency difference on arrival (FDOA).

ES is largely present in the C2W in the form of radar warning receivers or ESM on board aircraft to warn them of the presence and direction of radar threats.

EA is usually performed by aircraft penetrating the adversary's territory for SEAD purposes or by ships and ground assets during self-protection or mutual protection actions against an air attack.

The adversary's territory is usually protected by an IADS made up of surveillance, acquisition, and tracking radars, the latter providing the guidance to the weapons systems (AAA and missiles) as described in the previous section. A typical EW scenario during symmetric conflicts and its relevance to multirole A/C is shown in Figure 1.12.

EA performed by aircraft penetrating the adversary's territory against search radars is performed either with jamming, with the goal of preventing or delaying their target designation to the associated weapons systems, or with ARMs, with the goal of destroying the radar. Common jamming modes are discussed next.

Active self-protection jamming (SPJ) is performed either by noise jamming, with the purpose of denying range, velocity, and angle information about its platform, or by the generation of a large number of false targets, with the purpose of producing confusion to both the radar operator and the automatic detection and tracking processing system.

Chaff is launched to effect jamming. Chaff consists of a large volume cloud composed of thousands of elemental radar reflectors that provide a very large amount of volume radar reflection with respect to the echo of the protected aircraft, thus preventing target detection. Chaff effects may persist for a relatively

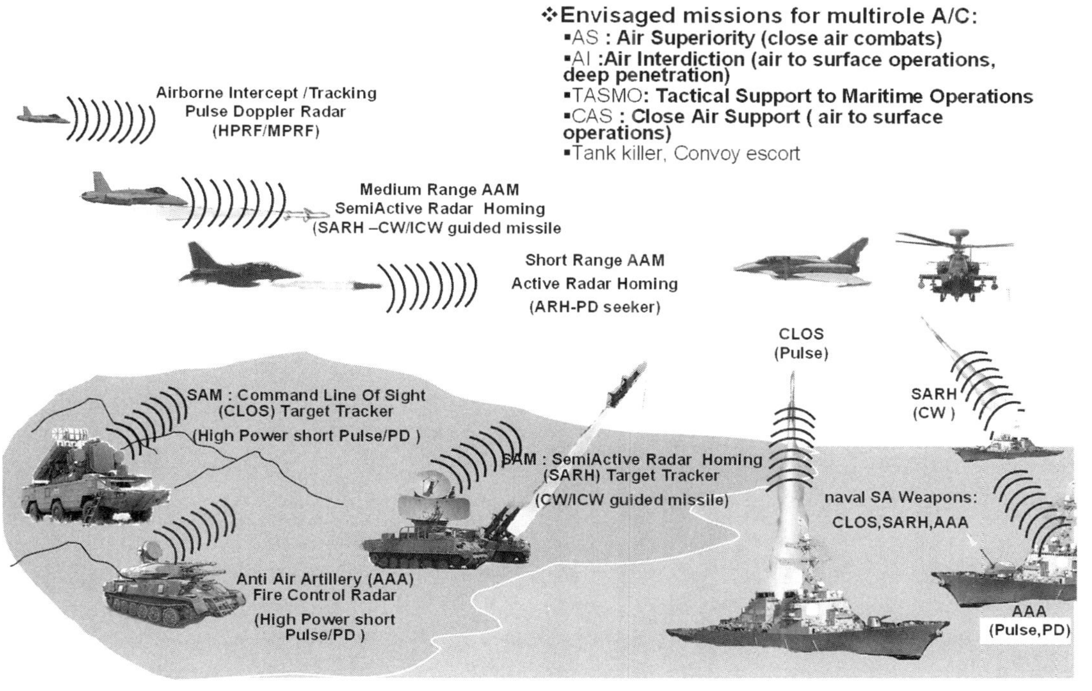

Figure 1.12 EW scenario in symmetric conflicts and its relevance to multirole A/C.

long time (tens of minutes). Because chaff is relatively inexpensive, it is one of the most used EA techniques. A Typical SPJ operation is shown in Figure 1.13.

Support jamming (also called mutual protection) includes stand-off jamming (SOJ); escort jamming (EJ), which usually operates in the search radar antenna's sidelobe region, as shown in Figure 1.14 or stand-in jamming (SIJ), which has the same purpose as the SPJ but applies to the entire raid of intruders.

Support jamming can employ noise jamming, deceptive electronic countermeasures, or deployment of corridors of chaff. SOJ missions are conducted by an

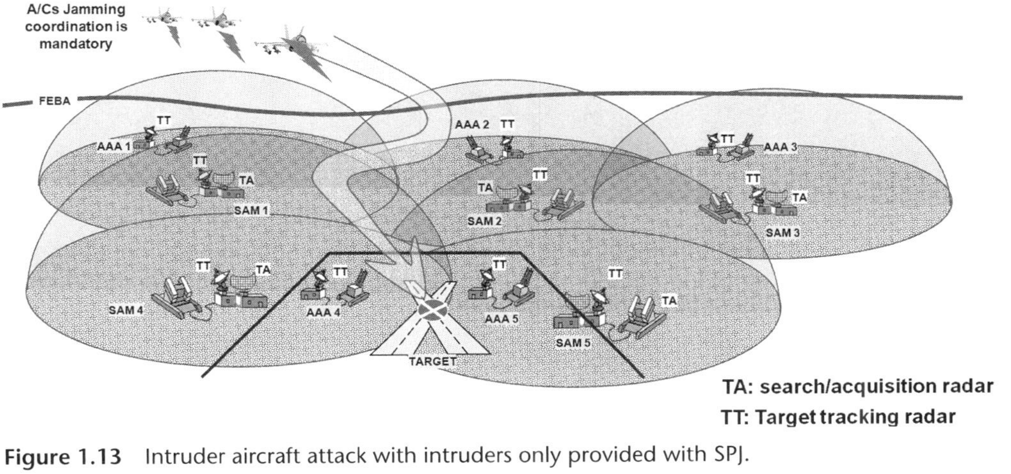

Figure 1.13 Intruder aircraft attack with intruders only provided with SPJ.

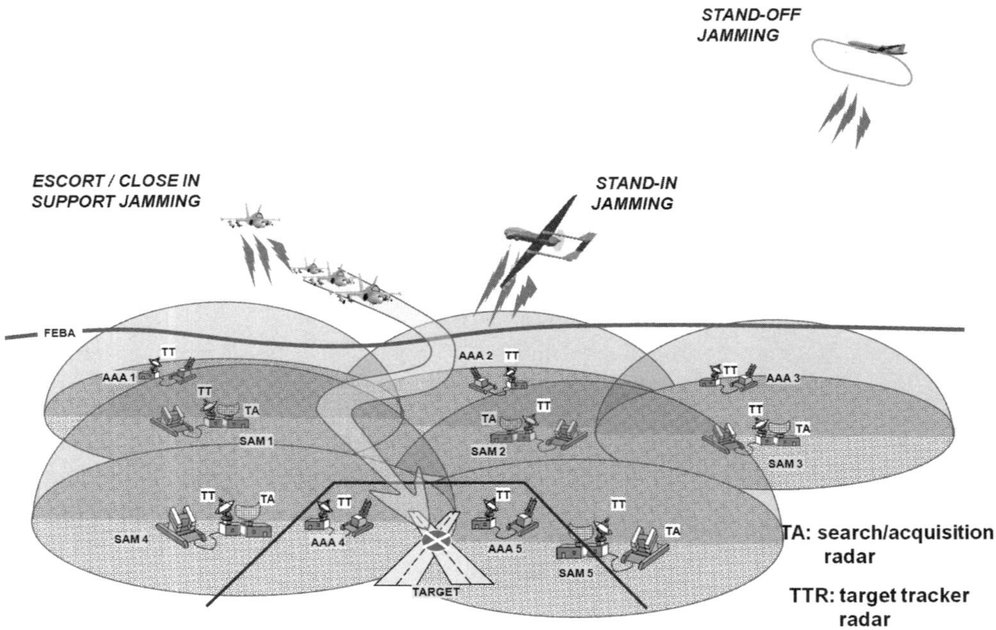

Figure 1.14 Intruder aircraft attack with intruders supported by SOJ, EJ, or SIJ.

aircraft loitering outside the lethal zones of hostile weapons systems in order to hide the radar echoes of the friendly intruder aircraft, thus preventing their exposure to the hostile fire of air defense systems triggered by the designation from the search radar.

EJ missions are conducted by aircraft that accompany the intruders as they penetrate the enemy's territory. As the EJ aircraft is penetrating the combat zone, it is vulnerable to enemy weapons and needs to be provided with self-protection jamming capabilities like those of the intruders. EJ aircraft are generally used when the intruder aircraft does not have a sufficient payload to protect itself.

EJ missions are usually conducted by multiple EJ aircraft, widely spaced among them, in order to provide confusion in the enemy's defense systems, which are trying to determine the intruder attack direction through jamming strobes. The confusion that is generated helps in delaying the air defense system response, thus increasing the probability of success of the intruder attack.

SIJ missions are conducted by unmanned combat aerial vehicles (UCAVs), which come nearer to the victim radars and exploit a lower jamming power to perform the same mission as the EJs.

EA performed by aircraft penetrating the adversary's territory is used primarily against target tracking radars (TTRs),which are used to guide the weapons systems (both defensive as well offensive) by locking on to the target and providing accurate direction (i.e., azimuth and elevation), range, and velocity data. The velocity data are used to predict the future position of the target and to point the weapon to the lead direction (as shown earlier in Figure 1.5). EA against TTRs is performed in a manner similar to that used for the search radars with jamming and ARM launches.

SPJ ECM systems are employed to counter the tracking radar through the use of deception techniques (DECM) that attempt to cause the tracking radar to break lock. Noise jamming is rarely employed against missiles, because most seeker heads can track the ECM radiation in angles (track-on jam), which is sufficient for many types of missiles (e.g., the semiactive missile) to lock on the target.

Illustrations of intruder aircraft attacks on an enemy's highly defended territory are shown in Figures 1.13 and 1.14 for, respectively, attack with intruders only provided with SPJ and with intruders supported by SOJ, EJ, or SIJ.

EA is also performed by ships when attacked by air raids or by sea-skimming missiles. In the first case noise and deception jamming are performed in order to deny range and position information, thus preventing either the launch of ASMs or an accurate bomb release. In the latter case a combination of noise jamming, launch of chaff, and deception jamming is performed in conjunction with the ship's close-in weapons systems (CIWS) operation to divert the missile from the ship or to destroy it.

EA is performed in a similar way for a ship (with the exception of chaff launches) in the defense of ground assets against air raids with launch of RF-guided ASMs or ARMs.

EA is also performed against communications networks in C2W operations. The primary objective in military communications jamming is to disrupt the network communication links so as to separate the C2 from its asset nodes. EA in the communication bands is conducted against combat net radios, tactical data links, and military satellite communications.

Combat radio nets, which provide short- and medium-range voice communications in the 30- to 88-MHz VHF band and operate with digital messages in the frequency hopping (FH) mode, can be somewhat jammed with broadband noise jamming or frequency-follower jamming.

Tactical data links are used to provide connectivity among distributed ground, sea, and air mobile nodes. They are conceived to be nodeless, which means that no communication center is needed because all users transmit to everyone else and operate in very fast FH mode, thus preventing frequency-follower jamming. Only tone jamming can have some effect, as will be discussed in Chapter 6.

Satellite communication systems (SATCOMs) provide worldwide connectivity at high data rates among forces that are quite distant from each other. Old SATCOMs operate in the UHF band (250 to 400 MHz). More recent SATCOMs operate in the SHF band (7 to 8 GHz) or EHF band (20 to 44 GHz).

UHF communications have the advantage of using a simple antenna that can be easily installed in mobile platforms such as aircraft, small ships, and small land vehicles. But because of the large antenna beamwidth they can be easily disturbed by ground jammers. SHF and EHF communications are harder to jam, and only tone jamming can provide some effect.

EP is strongly present in all radar and communication receivers and signal processors with a number of very effective ECCMs, as will be described in Chapters 2 and 6, thus rendering the contest between EA and EP a sort of chess match where each move of a player is met with a countermove by the other player.

1.4 EW in Asymmetric Conflicts

Asymmetric warfare occurs when the two opponents have highly unequal military capabilities and the weaker one uses methods of attack such as sabotage, terrorism, or insurgency that avoid contact with military forces in areas where the opponent is stronger. Conceptually, it is similar to guerrilla warfare or to Special Forces operations. The weapon usually employed by the weaker opponent in the above-cited actions are light gunnery and improvised explosive devices (IEDs).

In the asymmetric conflicts EW is strictly limited to the communications bands, and it is used only by the stronger opponent and employed in slightly different modes in the two areas into which a territory can be divided: rural areas and urban areas.

In rural areas EW essentially performs detection of hostile communications and communication intelligence from an adversary's messages using CESM and COMINT equipment (which is the same as that employed for symmetric conflicts), EW also involves countering a radio-controlled IED (RCIED) with some appropriate counter-RCIED jammer (CECM).

RCIEDs are triggered by an RF signal to initiate their explosive charge. They offer the user the advantage of flexibility because there is no physical proximity between him and the device. RCIEDs are used in ambushes of military convoys and patrols as they traverse roads. The RCIED is usually planted in locations where the convoy/patrol cannot go off-road to avoid the device. The RCIED user is located in an observation point where he can see the approach of the convoy/patrol to the device and is able to send the triggering message to the planted device at the right time. The radio link used is usually a short-range one, with the exception of the use of mobile phone systems, such as GSM, where existing networks are available. In the latter case the trigger command may be sent from up to a few kilometers away.

The following EW actions are applied to defend against RCIEDs:

- Detection of an adversary's communications in the interested area (which may be performed by stand-off systems such as UAVs patrolling the area for reconnaissance and intelligence collection in the days prior to the passage of the convoy),
- Preemptive initiation of the explosive device before its intended use by using a high-power jammer,
- Force protection jamming around the convoy/patrol to create an RF shield.

The counter-RCIED methods and jammer devices are discussed in Section 6.3.4 of Chapter 6.

EW actions in urban areas employ CESM and COMINT equipment in the same way as for the rural area to provide, respectively, targeting for the EA equipment of the adversary's communication links and the usual communication intelligence on the adversary's information exchanged through those links. EW actions, however, differ in the type of CECM equipment employed because in this case they are part of an offensive information operation, called electronic isolation, where the CECM action to deny the adversary's exchange of information through the radio links (by injecting noise or some appropriate signals into the target receivers) is accompanied by a computer network attack and is extended to prevent all adversary's communication with civilians, such as television. radio, mobile phones, wireless connectivity (Wi-Fi or Wi-max), and satellite communications.

An important consideration to be made regarding the use of EW attacks in urban areas is the collateral effects on friendly and neutral services. So these attacks need to be carefully planned.

Needless to say, in addition to the above–mentioned electronic isolation actions, urban combat requires the use of counter-RCIED jammers by every combat vehicle and dismounted patrol. The type of jammer required is the responsive type (also called a *signal-initiated jammer*) in order to allow our own radio links to be able to operate unaffected.

References

[1] Schleher, C., *Electronic Warfare in the Information Age*, Norwood, MA: Artech House, 1999.

[2] Macfadzean, R., *Surface-Based Air Defense System Analysis*, Norwood, MA: Artech House, 1992.

[3] Neri, F., *Introduction to Electronic Defense Systems, 2nd Ed.*, Raleigh, NC: SciTech Publishing, 2006.

[4] Carey, D., and Evans, W., "The Patriot Radar in Tactical Air Defense," *Microwave Journal*, Vol. 31, May 1987, pp. 325–332.

[5] Wiley, R., *ELINT: The Interception of Radar Signals*, Norwood, MA: Artech House, 1985.

CHAPTER 2

Evolution of Signal Emitters and Sensors

2.1 Introduction

The objective of EW systems is the protection of military troops, assets, and platforms (which we will call generically *targets*) from the electromagnetic (EM) threats present in the environment. EM threats are realized by means of weapons systems provided with sensors that can guide gun projectiles, launched munitions, and missiles from long distances.

Sensors that operate in the EM spectrum can transform the radiated energy impinging on them into electrical signals that are subsequently processed by electronic devices. Sensors are used for detection, tracking, recognition, and identification of targets. Some sensors provide their own source of target illumination, such as microwave, millimeter-wave, and laser radars. Such sensors are called *active sensors*. Other sensors depend on natural conditions for target illumination, such as electro-optical (EO) and infrared (IR) systems, which use, respectively, the reflected light energy from the target or the radiated heat energy of the natural background and of human-made engines and objects. These types of sensors are called *passive sensors*. Active sensors are equipped with both a transmitting and a receiving unit; passive sensors consist only of a receiving unit.

This chapter merely provides an introduction to modern available sensor technologies and techniques at the depth necessary to understand the requirements of the EW equipment described in the following chapters that is relevant to the interception, classification, and jamming of signals and receivers.

In Section 2.2 we discuss sensors features that are relevant to their operational wavelength: their dimensions and signal propagation characteristics, especially within the atmosphere. Three of the book's appendices also deal with the subject of sensors: Appendix A covers the signal detection capabilities of receivers, which can be impaired by the receiver noise produced by any electronic device and by other environmental disturbances; Appendix B provides information about the accuracy of sensors as derived from the concepts of estimation theory; and Appendix D provides a discussion of relevant analog modulation methods used in communications.

In Section 2.3 we tackle the principal active sensor in the EW domain: the radar. First we consider the capabilities offered by the tremendous evolution of the waveform generation and novel signal processing techniques with respect to maximum target range detection, range, range rate (i.e., radial velocity), and angular measurement accuracies, and then we review the some important types of radars.

Other active sensors that are of interest to EW systems are the communication systems (which are considered in Section 2.4) and the navigation systems (which are considered in Section 2.5). Both systems have enormously enhanced operational capabilities thanks to the evolution of signal processing techniques and technologies.

Finally, in Sections 2.6 and 2.7 we discuss, respectively, the most important passive sensors such as IR sensors and missile seeker heads and, in continuity with the IR band of the EM spectrum, the description of the capabilities of active laser systems.

2.2 Sensor Electromagnetic Spectrum and Atmospheric Propagation

Sensors can be classified in accordance with their wavelength because their component units (such as transmitters, receivers, apertures, and detectors) are sized proportional to the wavelength.

Table 2.1 compares microwave, millimeter wave (MMW), and EO/IR sensors characteristics.

An important component of any sensor is its aperture, where the energy enters (reception) and/or leaves (transmission) the sensor. Microwave and MMW sensor apertures are antennas, whereas EO/IR sensor apertures are optical lenses and mirrors.

The directivity G_D of antennas and lenses is defined as the ratio of the full spherical solid angle (4π sterad) to the solid angle in which the transmitted or received energy is primarily concentrated . For a rectangular aperture of solid angle with –3-dB beamwidths θ_B and ϕ_B, we have:

$$G_D = \frac{4\pi}{\theta_B \varphi_B} \tag{2.1}$$

which for circular apertures and lenses becomes:

Table 2.1 Sensor Characteristics

Characteristics	*Microwave*	*Millimeter Wave*	*EO/IR*
Wavelength	cm (10^{-2} m)	mm (10^{-3} m)	μm (10^{-6} m)
Component size	Large	Small	Small
Aperture size	Large	Small	Small
Angular resolution	Degrees (10^{-2} rad)	mrad (10^{-3} rad)	μrad (10^{-6} rad)
Prevalent mode of operation	Active	Active	Passive and active
Type of sensor	Radar	Radar	IR, thermal imager, laser radar
Performance in adverse environmental conditions (weather, dust, smokes)	Minor degradation	Degradation at long ranges	Degraded

$$G_D = \left(4\pi/\theta_B^2\right) \tag{2.2}$$

where θ_B is either the –3-dB beamwidth of the antenna or the lens diffraction limit.

From antenna theory (see [1] or Appendix C) it can be shown that for a circular aperture of diameter D:

$$G_D = \frac{4\pi A}{\lambda^2} = \frac{4\pi \cdot \pi D^2/4}{\lambda^2} = \left(\frac{\pi D}{\lambda}\right)^2 \tag{2.3}$$

By equating (2.2) to (2.3) we have:

$$\theta_B = k\,\lambda/D \tag{2.4}$$

where $k = \sqrt{4/\pi} = 1.13$.

As an example, for a microwave radar antenna operating at $\lambda = 3$ cm with diameter $D = 1\text{m} = 100$ cm, we have $\theta_B = 1.13 \times 3/100 = 0.035\text{rad} = 35\text{mrad} = 2°$ and $G_D=4\pi/(0.35)^2 = 10314 = 40.1$ dB, while, for a CO_2 laser operating at $\lambda = 10.6$ μm and a lens with diameter $D = 10$ cm, we have

$$\theta_B = 1.13 \cdot 10.6/10 \cdot 10^{-4} \text{ rad} = 0.119 \text{ mrad} = (6.9 \cdot 10^{-3})°$$

From the preceding numerical examples, we can deduce that the EO/IR sensors, because of their very short wavelengths, have the highest resolution characteristics, which are useful for imaging and target identification. In contrast, the microwave and MMW radars, due to their large beamwidths, are useful for volume search, tracking, guidance, and cueing of EO/IR sensors.

The theoretical values expressed in (2.1) to (2.4) are always in excess of actual values and have to be corrected by an efficiency factor η on the order of 0.6 to 0.7 depending on the physical structure and EM properties of antennas and lenses. As an example the gain of a microwave antenna is expressed as $G = \eta G_D$, where the antenna efficiency factor is assumed to be $\eta = 0.6$ to 0.65 for a good antenna design.

The frequencies of the EW sensors are reported in Table 2.2 for the radar microwave and MMW bands and in Table 2.3 for the IR spectrum.

An important consideration in the design of any sensor system is that of atmospheric effects such as rain, fog, and dust, which can degrade their performance, as well as the atmospheric attenuation vs. frequency. Figures 2.1, 2.2, and 2.3 illustrate, respectively, the one-way atmospheric attenuation (dB/km) vs. frequency, the attenuation due to rain and fog, and the atmospheric transmission coefficient at IR wavelength for high-frequency sensors.

From the above figures it is evident that visible and IR sensors performances are significantly reduced under adverse atmospheric conditions and also MMW radar are strongly affected.

Table 2.2 EW Sensor and Microwave/MMW Radar Frequency Band Denominations and Values

New EW Band Denomination	*Frequency Range (GHz)*	*Old Radar Band Denomination*	*Frequency Range (GHz)*
A	<0.25	VHF	0.03–0.3
B	0.25–0.5	UHF	0.3–1
C	0.5–1	—	—
D	1–2	L	1–2
E	2–3	S	2–4
F	3–4	—	—
G	4–6	C	4–8
H	6–8	—	—
I	8–10	X	8–12
J	10–20	Ku	12–18
K	20–40	K	18–27
L	40–60	Ka	27–40
M	60–100	V	40–75
—	—	W	75–110

Table 2.3 IR Spectrum

Region	*Wavelength Range (μm)*
Ultraviolet	0.1–0.38
Visible	0.38–0.76
Very near IR	0.76–1.0
Near IR	1–3
Middle IR	3–8
Longwave IR	8–14
Far IR	14–50

2.3 Radar Principles and Types

The name radar was assigned early on (prior to World War II) as a contraction of the words *radio detection and ranging*. The basic radar principle was developed by many nations, including the United Kingdom, the United States, Germany, Italy, France, and Russia. It was developed primarily for the frequency bands from 25 to 500 MHz [3]. This development accelerated during WW II. The U.K. development of the magnetron led to major advances in microwave radar that took place in parallel at the Massachusetts Institute of Technology's Radiation Laboratory and elsewhere in the United States and United Kingdom after 1941.

Modern radars can usually extract much more information from the target than just the range. They can extract angle, radial velocity, pointwise or distributed extension, aspect changes, jet engine vs. propeller modulation, and so forth. However the measurement of range (i.e., the distance to a target) is still the most important radar function. The basic radar principle is shown in Figure 2.4.

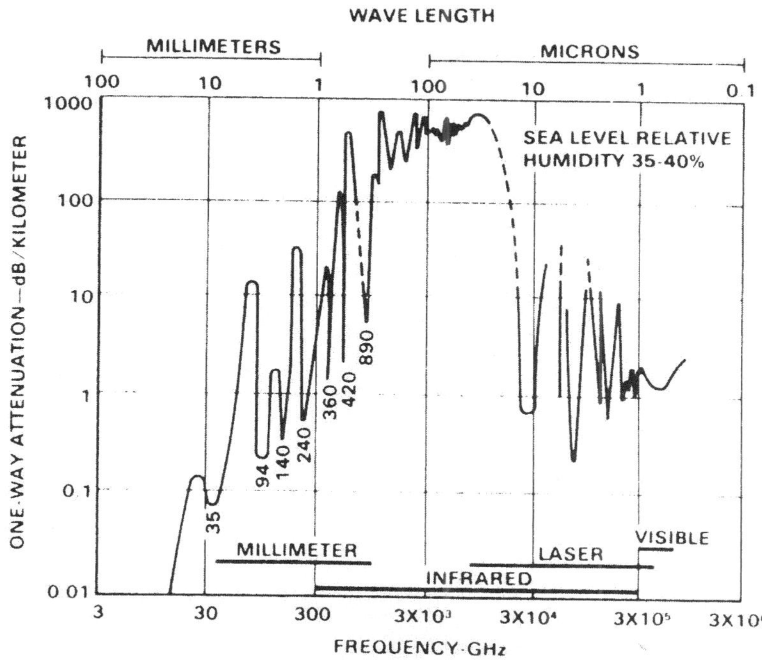

Figure 2.1 One-way atmospheric attenuation (dB/km) vs. frequency. (Reproduced with permission from S. Hovanessian [2].)

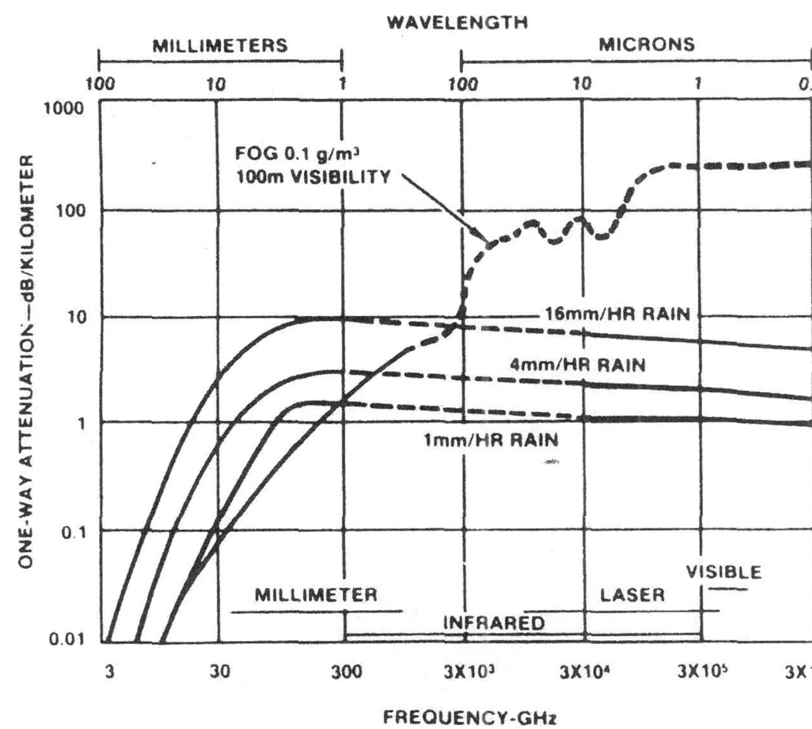

Figure 2.2 Attenuation due to rain and fog. (Reproduced with permission from S. Hovanessian [2].)

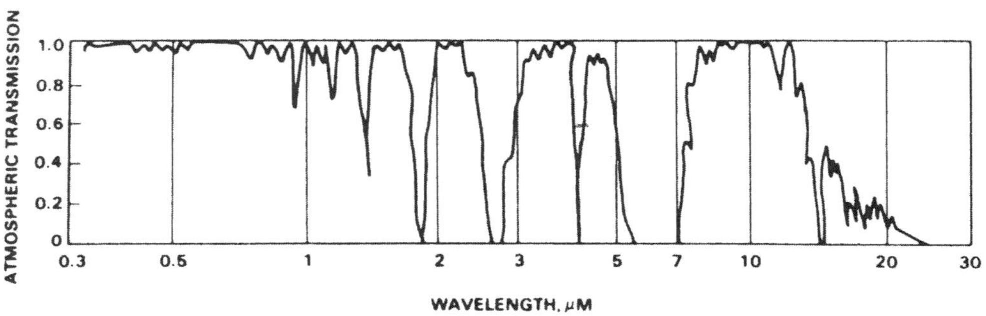

Figure 2.3 Atmospheric transmission coefficient at IR. (Reproduced with permission from S. Hovanessian [2].)

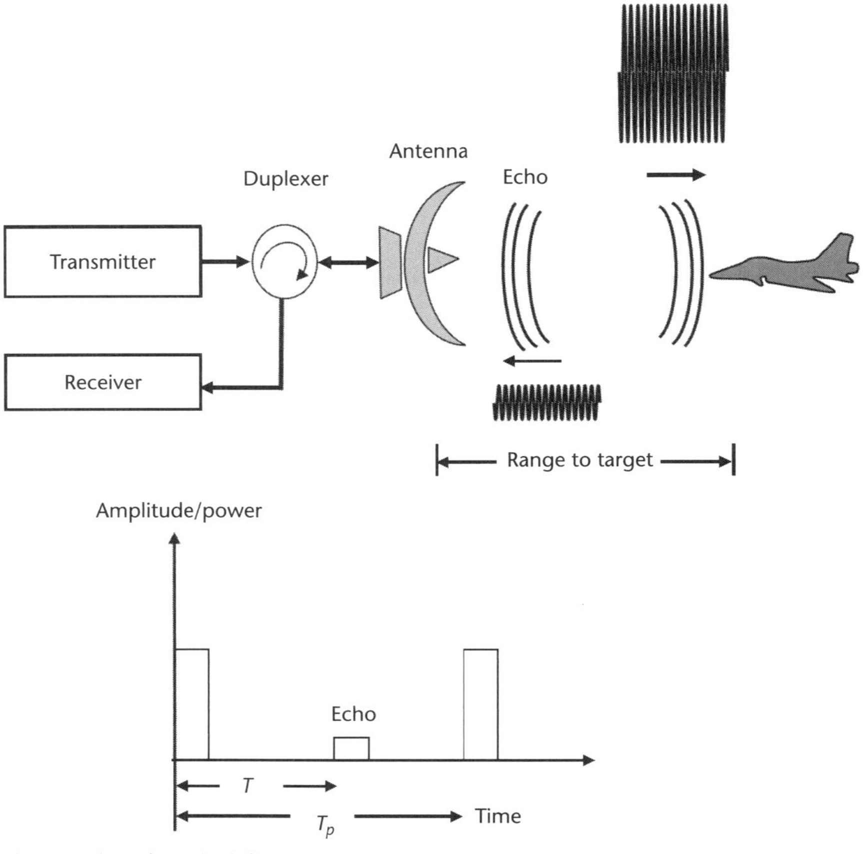

Figure 2.4 Basic radar principle.

A transmitter (Tx) generates a RF signal, usually a train of narrow pulses modulating a sine wave carrier, that is radiated into space by an antenna. A portion of the transmitted energy is intercepted by the target and scattered in many directions, including back toward the radar antenna. The backscattered signal is fed to the receiver (Rx), which processes it to detect the presence of the target and to determine its location—range and direction. The direction is provided by the narrow beamwidth of the antenna. Range R is obtained by measuring the delay Δt from the transmitted pulse to the received echo from the target, which is the time taken by the radar signal to travel at the speed of light $c = 3 \times 10^8$ m/s to the target and back to the radar:

$$\Delta t = \frac{2R}{c} \text{ and thus } R = \frac{c\Delta t}{2} \tag{2.5}$$

Each microsecond (μs) of round-trip travel corresponds to a distance of 150m.

Once a pulse is radiated into space by a radar, sufficient time must elapse to allow for all echo signals to return to the radar before the next pulse is transmitted. The rate at which the pulses are transmitted is therefore determined by the longest range at which targets are expected. If the period T_p between pulses is too short, an echo signal from a far target may arrive after the transmission of the next pulse and thus can be associated in error with an echo of the latter pulse than to the former one. These echoes are called second-time-around echoes (or even multiple-time-around echoes if they are relevant to earlier pulses) and provide ambiguous range measurements because the echoes would appear at much shorter distances than the actual ones. The range beyond which targets appear as second-time-around echoes is called the *maximum unambiguous range* R_{un} and is expressed as:

$$R_{un} = \frac{cT_p}{2} = \frac{c}{2f_p} \tag{2.6}$$

where T_p is the pulse repetition interval (PRI) and $f_p = 1/T_p$ is the pulse repetition frequency (PRF).

The simplest and typical radar pulse waveform of Figure 2.4 is characterized by a high peak power $P_{t,}$ a pulse width (PW) τ, and a small duty cycle d, this last defined as the ratio of the average transmitted power P_{av} to the peak power $P_{t,}$ that is, $d = P_{av}/ P_t = \tau/T_p$.

A short-duration pulse waveform is attractive because the transmitter is not radiating when the weak echo signal is being received and because it provides a good resolution of the measured distance; that is, it can discriminate between two targets that are not very far ($>\tau$) between them. A short-duration pulse waveform, however, requires a very high peak power P_t in order for the radar receiver to be able to obtain sufficient energy to detect small targets at long ranges. Different radar waveforms will be considered after the introduction of the radar equation and the structure of a typical radar.

2.3.1 Radar Equation

The radar equation relates the detection range of a radar to the characteristics of its architecture (i.e., transmitter, receiver, antenna, and waveform), as depicted in Figure 2.5, and of the target and the environment. As such the equation is a fundamental tool in radar system design.

If the transmitter peak power P_t(W) is radiated by an isotropic antenna, the power density PD_i at a distance R(m) from the radar is:

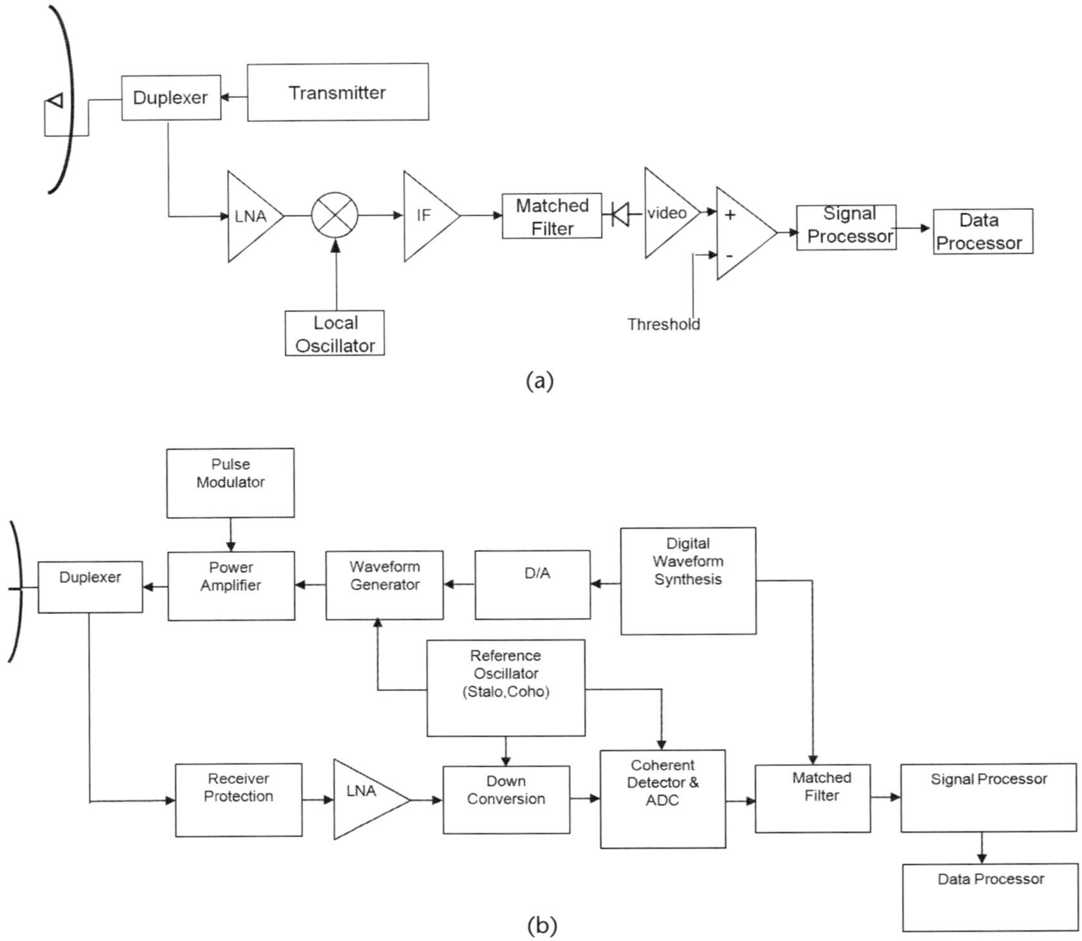

Figure 2.5 (a) Simplified pulse radar (vintage) block diagram. (b) Modern radar simplified block diagram, where STALO = stable local oscillator and COHO = coherent oscillator.

$$PD_i = \frac{P_t}{4\pi R^2} \quad \left(\mathrm{W/m^2}\right) \tag{2.7}$$

However, because radars transmit through a directive antenna of gain G_t, the power density at a distance R along the direction of the narrow antenna beamwidth is:

$$PD_i = \frac{P_t G_t}{4\pi R^2} \quad \left(\mathrm{W/m^2}\right) \tag{2.8}$$

The product $P_t G_t$ is the effective radiated power (ERP) in watts.

The target intercepts a portion of the above incident energy and scatters it in various directions. Only the backscattered power density in the direction of the radar is useful for target detection. This reradiated power density PD_r at the radar is determined by an equivalent (fictional) reflecting area σ, the RCS, which is not related to the physical size of the target, but rather to the target's shape. Thus:

$$PD_r = \frac{P_t G_t}{4\pi R^2} \frac{\sigma}{4\pi R^2} \tag{2.9}$$

The radar antenna captures a portion of the PD_r in accordance with the effective area A_e of the receiving antenna. The effective area is related both to the physical area size A by the relationship $A_e = \eta A$, where $\eta < 1$ is the antenna aperture efficiency as well as to the receiving antenna gain G_r by the relationship $G_r = 4\pi A_e/\lambda^2$. Usually, in monostatic radars (i.e., radars with colocated transmitter and receiver) the same antenna is used both for transmission and reception, making the antenna a reciprocal device and, thus, $G_t = G_r = G$. Therefore, the received power at the radar receiver input is:

$$P_r' = PD_r \cdot A_e = \frac{P_t G_t}{4\pi R^2} \frac{\sigma}{4\pi R^2} \frac{G_r \lambda^2}{4\pi} = \frac{P_t G^2 \lambda^2 \sigma}{(4\pi)^3 R^4} \tag{2.10}$$

The actual received power P_r is less than P_r' by a factor $L_{atm} > 1$, which expresses the signal power loss due to the two-way path length atmospheric attenuation:

$$P_r = P_r'/L_{atm} \tag{2.11}$$

The detection of the target return power, which is impaired by the receiver noise power N (proportional to the bandwidth B_r of the radar waveform), is performed by setting a threshold (see Appendix A) well above this value in order to significantly reduce the probability of occurrence of false detections (called false alarms and hence the probability is called probability of false alarms, P_{fa}) produced by noise spikes Usually P_{fa} is required to be $<10^{-6}$.The target return power must largely overcome the above threshold in order to ensure a high detection probability P_d (usually >0.9) .To achieve the above-mentioned detection conditions (P_d,P_{fa}), the ratio of the target power to the noise power [which is called the signal-to-noise power ratio and indicated with (SNR)] must be larger than the minimum SNR_0 value that satisfies the detection conditions, that is, $SNR > SNR_0$. The target return power therefore must have a minimum power $S_{min} = SNR_0 \cdot N$ in order for the target to be reliably detected.

The maximum range R_{max} of a radar is the distance beyond which the target cannot be detected at the required detection conditions (P_d, P_{fa}). This occurs when the received power P_r is equal to the minimum detectable signal power, S_{min}, which depends, as mentioned earlier, on the radar waveform and the structure of the receiver (which both determine the receiver noise power) and the required SNR_0 for detection conditions. Thus R_{max} is expressed from (2.10) and (2.11) by substituting P_r with S_{min}:

$$R_{max} = \left[\frac{P_t G^2 \lambda^2 \sigma}{(4\pi)^3 S_{min} L_{atm}} \right]^{1/4} \tag{2.12}$$

As will be discussed in Chapters 5 and 6 one of the principal EW countermeasure techniques against radars is the transmission of noise-like waveforms that are meant to increase the noise power at the receiver to a level that significantly reduces the maximum range of target detection.

2.3.2 Radar Structure

The operation of a traditional (vintage) pulse radar can be described with the aid of the simplified block diagram of Figure 2.5(a).The structure of modern radar thereafter evolved as represented in the block diagram of Figure 2.5(b). Radar system evolution is discussed in the following paragraphs.

The radar signal is produced at low power by a waveform generator (WFG) and is fed to the power amplifier, which constitutes the radar transmitter. In the past, low duty waveform power oscillators, such as magnetrons, were used as transmitters. Currently, because modern radars require larger duty coherent waveforms, a power amplifier is preferred. The power amplifier can be a crossed-field amplifier, a klystron, a travelling-wave-tube (TWT) or a solid-state (SS, i.e., transistor) amplifier. In most power amplifiers, except for the SS ones, and in power oscillators, a modulator turns the Tx on and off in synchronism with the input pulses from the WFG.

The output of the Tx is delivered via a duplexer to the antenna, where it is radiated into space. The duplexer acts as a switch and allows for a single antenna to be used on a time-shared basis between transmission and reception. The duplexer is generally a gaseous device that produces a short circuit (an arc discharge) at the input of the receiver when the transmitter is operating, so that the high power flows to the antenna and not to the receiver (thus protecting it). On reception the duplexer directs the echo signal to the receiver and not to the transmitter.

The high-gain (directive) antenna can consist of mechanically steered parabolic reflectors, mechanically steered planar arrays, or electronically steered phased arrays (see Appendix C).

The rotation of a surveillance radar antenna through 360° in azimuth is an example of an antenna scan. Typical scan rates range from 5 to 6 revolutions per minute (rpm) for long-range surveillance radar up to 30 to 60 rpm for medium- and short-range surveillance radars.

The receiver is almost always superheterodyne (SHR). The RF input is amplified by a low-noise amplifier (LNA) to achieve high sensitivity and then down-converted to an intermediate frequency (IF), usually 30 or 60 MHz in past radars, where it is further amplified by an IF amplifier. The signal bandwidth of a SHR is determined by the bandwidth of its IF stage, which in the case of radar equipment is the one of the matched filters (see later discussion), that is, the filter that maximizes the output peak SNR. For the widely used rectangular pulse shape, the matched filter is approximated by conventional receiver filters with a bandwidth $B_{IF} \approx 1/\tau$, that is, the inverse of the pulse width.

The IF amplifier is followed in legacy radars by an envelope detector, which is traditionally called a *demodulator*, whose task is to extract the signal modulation from the carrier. The chain composed of IF amplifier, demodulator, and video amplifier acts as an envelope detector (see Appendix A) to pass the pulse modulation (envelope) and remove the IF carrier frequency.

At the output of the Rx, a decision is made about whether or not a target is present. The decision is based on the amplitude of the receiver video output with respect to a predetermined threshold, set in order to reduce the false alarm crossing rate due to noise to a tolerable level. For a case in which, instead of receiver noise, external noise (such as unintentional interference, deliberate noise jamming, or clutter echoes) becomes the dominant cause of false alarms, the threshold has to be varied adaptively to maintain a constant false alarm rate by means of a constant false alarm rate (CFAR) receiver.

A radar usually receives many echo pulses from a target. Summing these pulses to achieve a larger SNR_{out} before the detection decision is called *integration of pulses*. The integrator is part of the signal processor, whose function is to pass the desired echo signal, but reject the unwanted signals, noise, or clutter before the detection decision is made.

Modern radars are also provided with a data processor that performs automatic target tracking by using previously measured locations of the target to establish its track (i.e., its past trajectory) and forecast its location at the next scan of the radar.

In modern radars matched filtering, detection, signal, and data processors are implemented with digital techniques.

2.3.2.1 Radar Clutter Signal Processing

The basic principles of radar discussed in the previous section illustrated only the detection capabilities of targets in the presence of noise. This situation occurs only with targets flying at high altitudes and at distances where the antenna beamwidth is not affected by echoes from the natural environment. In the vast majority of situations, however, radar target detection is severely impaired by the natural environment such as land, sea, and weather echoes, which are called *clutter*, since they can "clutter" the radar display and prevent the detection of desired targets (aircraft, ships, or moving vehicles). Clutter echoes can be many orders of magnitude larger than target echoes and when they both are present in the same resolution cell the target is undetectable, unless proper signal processing is applied.

The most powerful method for detecting a moving target in the presence of clutter is by taking advantage of the differential Doppler shift and in general the different spectrum of the target echo from the clutter echo. CW radar target detection processing is essentially based on differential Doppler shift filtering. Pulse radars that employ the differential Doppler shift for detecting are the moving target indicator (MTI), Doppler filter banks and moving target detection (MTD), airborne MTI (AMTI), and pulse Doppler (PD) radars. These types of radars are briefly discussed after we introduce the concept of coherent radar detection.

The Doppler frequency shift f_d consists of the frequency change with respect to the transmitted frequency impressed to the echo of a moving target by its relative radial speed v with respect to the radar. Its evaluation can be deduced from the total phase change ϕ with time in the two-way propagation path from the radar to a target in motion, initially placed at a distance R.

Total phase change ϕ can be expressed as:

$$\phi = (2\pi/\lambda)\cdot 2R = 4\pi R/\lambda = 4\pi R f_t/c \tag{2.13}$$

where $\lambda = c/f_t$, and f_t is the transmitted frequency. Its change with time is the derivative of the preceding expression:

$$\frac{d\phi}{dt} = \frac{4\pi}{\lambda}\frac{dR}{dt} = \frac{4\pi}{\lambda}v_r = 2\pi f_d \tag{2.14}$$

where $v_r = v\cos\theta$ is the radial component of the target speed relative to the radar as shown in Figure 2.6.

The Doppler frequency shift f_d can therefore be expressed as:

$$f_d = 2v_r/\lambda = 2f_t v_r/c \tag{2.15}$$

Note that the preceding expression is valid for two-way propagation, which is typical of a monostatic radar. In the case of one-way propagation (i.e., reception from a passive receiver) the Doppler frequency shift f_d with respect to the transmitted frequency f_t is half of the above value.

Note that an approaching target (i.e., $dR/dt < 0$) produces a negative Doppler shift, whereas a receding target produces a positive one.

To exploit the Doppler shift of the moving targets, the radar receiver must be coherent (i.e., it needs to have a stable transmitted frequency generation) and use a *phase detector* instead of an amplitude detector (as discussed in the previous paragraphs and in Appendix A) following the IF stage (which is designed as a matched filter). The structure of the radar is thus modified as shown in Figure 2.7

It can easily be shown from the above scheme that the echo signal from a moving target ($f_d \neq 0$) produces time-varying output, whereas the echo from stationary clutter ($f_d \approx 0$) produces a constant output. This consideration is applied in radar Doppler signal processing, for which the earliest and simplest development was the MTI.

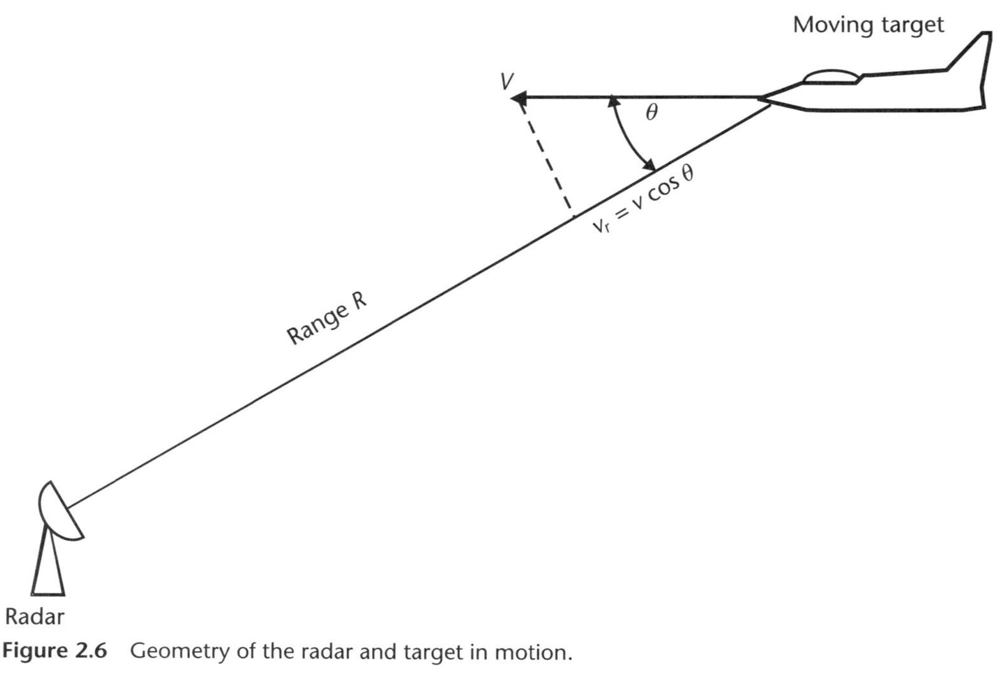

Figure 2.6 Geometry of the radar and target in motion.

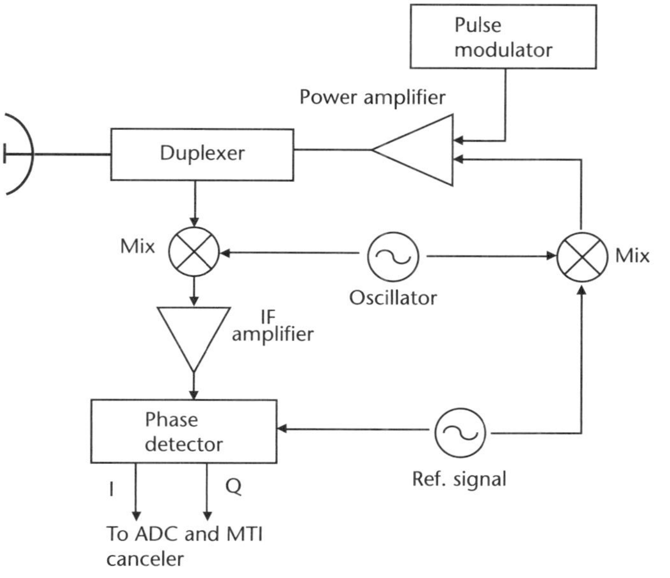

Figure 2.7 Coherent MTI radar block diagram.

MTI

Consider the output of the coherent receiver shown in Figure 2.7 to be applied to a single delay-line canceller as shown in Figure 2.8, which represents the modern digital version of the canceller. We then have:

$$y_i(t) = A\sin(2\pi f_d t - \phi_0)$$
$$y_{i-1}(t) = A\sin(2\pi f_d (t - T_p) - \phi_0)$$

and the difference

$$x(t) = y_i(t) - y_{i-1}(t) = 2A\sin(\pi f_d T_p)\cos(2\pi f_d (t - T_p / 2) - \phi_0) \tag{2.16}$$

That is, the frequency response of the single delay-line canceller (i.e., the ratio of output amplitude to input amplitude) is:

$$H(f) = 2\sin(\pi f T_p) \tag{2.17}$$

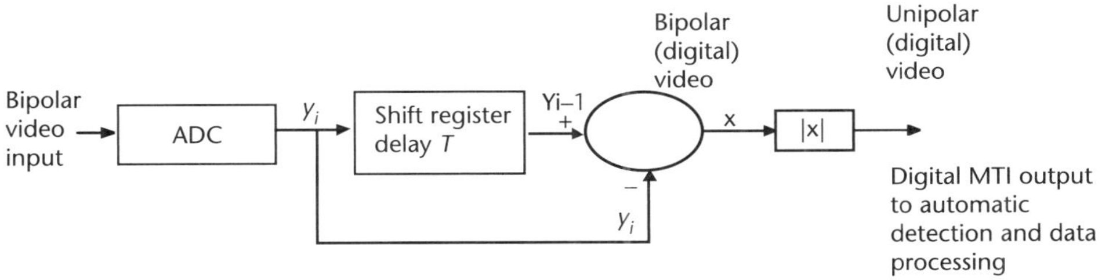

Figure 2.8 Block diagram of a digital MTI with single delay-line canceller.

and its modulus vs. frequency is shown in Figure 2.9.

A detailed discussion about MTI processing is beyond the scope of this book and we recommend that interested readers refer to [3]. Suffice it to say here that, as can be noted from Figure 2.9, the single delay-line canceller actually cancels the clutter response at zero Doppler shift, but unfortunately also the echoes of the targets moving with a Doppler shift $f_d = k/T_p$ (i.e., at multiples of the PRF = $1/T_p$), for which the corresponding velocities $v_r = k\lambda/2T_p$ are called *blind speeds*. From this last expression note that, in order to increase the velocity range, unless operating at long wavelengths, the usable methods consist of using high PRFs (and this is the case for the PD radar) or a number of different PRFs.

The rationale of this second method is that the blind speeds that occur with one PRF are generally different from those occurring at a different PRF. So targets that are strongly attenuated with one PRF may easily be detected with a different PRF. Usually to increase the value of the first blind speed, a number of PRFs are employed, with the ratios between them being expressed with large prime numbers.

The PRFs can be either changed from batches to batches of pulses or from pulse to pulse (this case is called *staggered* PRF), which has been popular for air traffic control radar (ATCR) and surveillance radars. Staggered PRFs provide an important waveform signature for the radar identification process operated by ESM systems.

As an example [3], consider the increase of the first blind velocity obtained by selecting the staggered PRFs as $k_1\text{PRF}_1 = k_2\text{PRF}_2 = ... = k_r\text{PRF}_r$, where the numbers $\{k_i\}$ are relatively prime integers and among them (as an example, 25, 30, 27, 31) is $v_1 = 28.25\ v_{av}$.

With the evolution of high-speed digital technology, MTI processing is currently performed in complete digital fashion, as shown in Figure 2.10 with the following advantages:

- Compensation for "blind phases," by using in-phase and quadrature (I and Q) processing,
- Greater dynamic range,
- Better achievement of the required filter characteristics.

Doppler Filter Banks

A Doppler filter bank is a set of narrowband contiguous filters for detecting targets, as shown for the case of N = 8 in Figure 2.11. With respect to MTI filters, it has the advantages of:

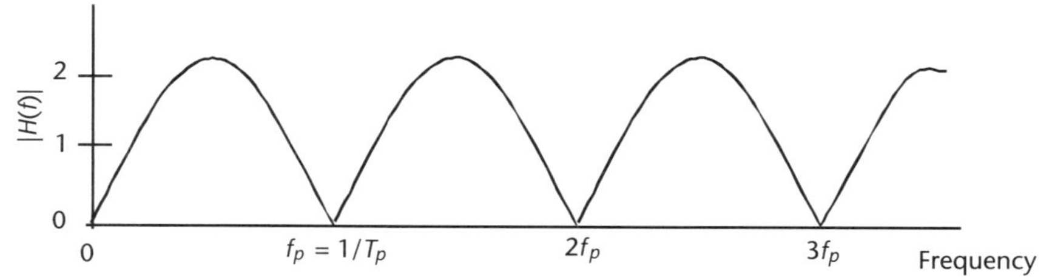

Figure 2.9 Modulus of the single delay-line canceller vs. frequency.

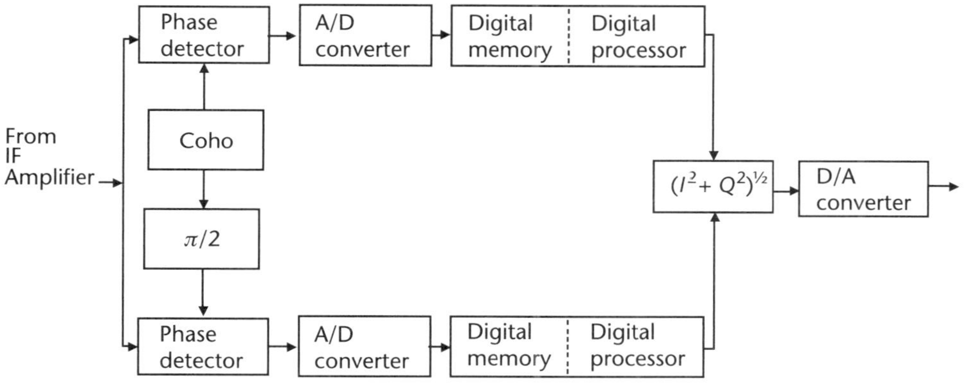

Figure 2.10 Block diagram of a digital MTI processor.

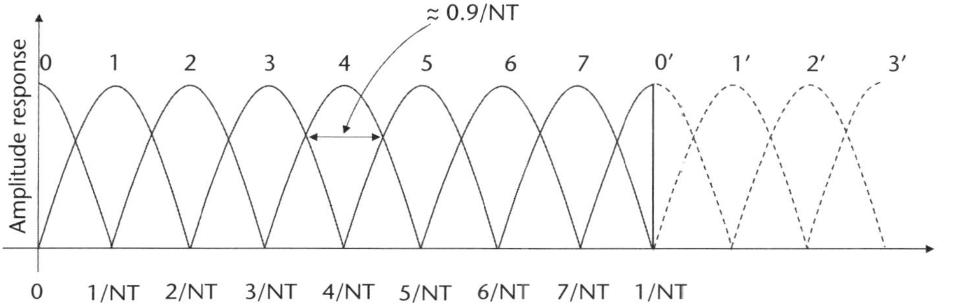

Figure 2.11 Doppler filter bank.

- Separation of undesired moving clutter (rainstorms or birds) with non-zero Doppler shifts,
- Raw measure of target radial velocity, the ambiguity of which can be resolved by changing the PRF,
- Provision of coherent integration at the output of each filter.

With the evolution of high-speed digital technology, the filter bank is now implemented via fast Fourier transform (FFT).

2.3.3 Radar Signal Processing Fundamentals

Radars must not only detect the presence of a target in the environment but also extract a number of pieces of information about it, such as location (range, azimuth, and possibly elevation), radial velocity, and even target recognition.

Moving target detection and extraction of target information is impaired by the presence of noise, echoes from the natural environment such as land, sea, and weather clutter and by intentional (jamming) and nonintentional interference. Methods for the detection of the desired echo signal and rejection of undesired noise, clutter, and other types of interference are called radar signal processing, and they are discussed in the following sections after introduction of one basic theoretical radar signal process—the matched filter—and one important radar waveform design tool—the ambiguity function.

2.3.3.1 Matched Filter

The matched filter has the task of maximizing the output peak SNR of the radar receiver so as to maximize the detectability of a target. Let us consider, as shown in Figure 2.12, a signal with finite energy E whose complex envelope is $s(t)$ and its Fourier transform $S(f)$ at the input of a filter with transfer function $H(f)$[1]. These assumptions are formally expressed as:

$$E = \int_{-\infty}^{+\infty} |s(t)|^2 \, dt \tag{2.18}$$

and

$$S(f) = \int_{-\infty}^{+\infty} s(t) \cdot \exp(-j2\pi ft) \cdot dt$$

The instantaneous power of the output signal $s_0(t)$ is given by the expression

$$|s_0(t)|^2 = \left| \int_{-\infty}^{+\infty} S(f)H(f) \cdot \exp(j2\pi ft) \cdot df \right|^2 \tag{2.19}$$

In the same way we can evaluate the output N of a white-noise filter (i.e., a uniform amplitude spectrum) having one-sided spectral power density N_0 (and $N_0/2$ for the two-sided version):

$$N = \frac{N_0}{2} \int_{-\infty}^{+\infty} |H(f)|^2 \, df \tag{2.20}$$

$$S(f) = \int_{-\infty}^{+\infty} s(t) \cdot \exp(-j2\pi ft) \cdot dt$$

$$H(f) = g \cdot S^*(f) \exp(-j2\pi f t_0)$$

Figure 2.12 Matched filter concept.

1. In the following discussion we will adopt for notation elegance the two-sided frequency domain, although we will return later to only positive (real) frequencies when dealing with practical equipment.

The factor ½ that appears before the integral takes into account that while the limits of the integral extend from $-\infty$ to $+\infty$, the noise power density is defined only over positive values of the frequency f.

We are interested in maximizing the ratio Γ, defined as:

$$\Gamma = \frac{|s_0(t)|^2}{N} = \frac{\left|\int_{-\infty}^{+\infty} S(f)H(f)\exp(j2\pi f t_0)\cdot df\right|^2}{\frac{N_0}{2}\int_{-\infty}^{+\infty}|H(f)|^2\cdot df} \tag{2.21}$$

where we have indicated with t_0 the time at which the maximum value of $|s_0(t)|^2$ occurs.

Schwartz's inequality states that if P and Q are two complex functions of the real variable x, then

$$\int P^* P\cdot dx \int Q^* Q\cdot dx \geq \left|\int P^* Q\cdot dx\right|^2 \tag{2.22}$$

In our case, by letting $P^* = S(f)\exp(j2\pi f t_0)$ and we have:

$$\Gamma_{max} = \frac{\left|\int S(f)H(f)\exp(j2\pi f t_0)\cdot df\right|^2}{\frac{N_0}{2}\int |H(f)|^2\cdot df} \leq \frac{\int |S(f)|^2\, df \cdot \int |H(f)|^2\, df}{\frac{N_0}{2}\int |H(f)|^2\, df} = \frac{\int |S(f)|^2\, df}{\frac{N_0}{2}} \tag{2.23}$$

and taking into account Parseval's theorem, which states that

$$\int |S(f)|^2\, df = \int |s_0(t)|^2\, dt = E \tag{2.24}$$

we have:

$$\Gamma_{max} \leq \frac{2E}{N_0} \tag{2.25}$$

The equality sign in (2.22), and hence also in (2.25), is reached when $P = gQ$ where g is a real constant. Thus, if

$$H(f) = g\cdot S^*(f)\exp(-j2\pi f t_0) \tag{2.26}$$

then (2.26) defines the optimum filter transfer function in the frequency domain. Its expression $h(t)$ in the time domain can be obtained by the inverse Fourier transform of $H(f)$:

$$h(t) = g\int_{-\infty}^{+\infty} S^*(f)\exp(-j2\pi f(t_0 - t))df = g\left(\int_{-\infty}^{+\infty} S(f)\exp(j2\pi f(t_0 - t))df^*\right) \tag{2.27}$$

That is, by considering (2.19)

$$h(t) = g \cdot s^*(t_0 - t) \tag{2.28}$$

which shows that the impulse response of the optimum filter is (except for an amplitude constant g) a time-reversed, conjugated, and delayed version of the input signal $s(t)$. That is, it is *matched* to the input signal. With such a filter the ratio Γ is maximized and we have:

$$\Gamma_{max} = \frac{2E}{N_0} \tag{2.29}$$

An important consideration follows from the linear filter theory, which states that the output $y(t)$ of a filter, with impulse response $h(t)$, to an input $x(t) = s(t)+n(t)$ is expressed as the *convolution* of the input and the impulse response, that is:

$$y(t) = \int_{-\infty}^{+\infty} h(\tau) \cdot x(t-\tau) d\tau = x(t) * h(t) \tag{2.30}$$

In our case, by considering the matched filter $h_m(t) = s^*(t_0 - t)$, we have:

$$y_m(t) = \int_{-\infty}^{+\infty} s^*(t_0 - \tau) \cdot x(t-\tau) d\tau = \int_{+\infty}^{-\infty} s^*(\lambda) \cdot x(\lambda + t - t_0) d\lambda \tag{2.31}$$

having used in the last integral the variable $\lambda = t_0 - \tau$.

The last integral in (2.31) has the same form of the cross-correlation between the transmitted signal $s(t)$ and the received signal $x(t)$, thus we can conclude that the output of the matched filter is the cross-correlation between the transmitted and the received signal.

From (2.28) and (2.31) it is clear that, for a pulsed radar signal $s(t)$ of duration T, the matched filter is a *causal* one [with a waveform as shown in Figure 2.13(b)] and the output of the matched filter is the cross-correlation of the two waveforms, as shown in Figure 2.13(c).

When the SNR is large, $x(t) \approx s(t)$ and the output signal from the matched filter $y_m(t)$ is approximated by the autocorrelation function of the transmitted signal $s(t)$.

The above considerations show that it is possible to implement the matched filter as a correlation receiver in which the input signal $x(t)$ is multiplied by a delayed replica of the transmitted signal $s(t - t_0)$, where t_0 is an estimate of the time delay of the target echo signal. Because the cross-correlation receiver and the matched receiver are mathematically equivalent, the choice as to which to use in a particular radar application is determined by which is most practical to implement.

In practice, the matched filter cannot always be perfectly implemented. In Table 2.4, derived from Skolnik [3], some practical matched filter approximations are reported, together with their efficiency (i.e., the SNR loss with respect to the true matched filter).

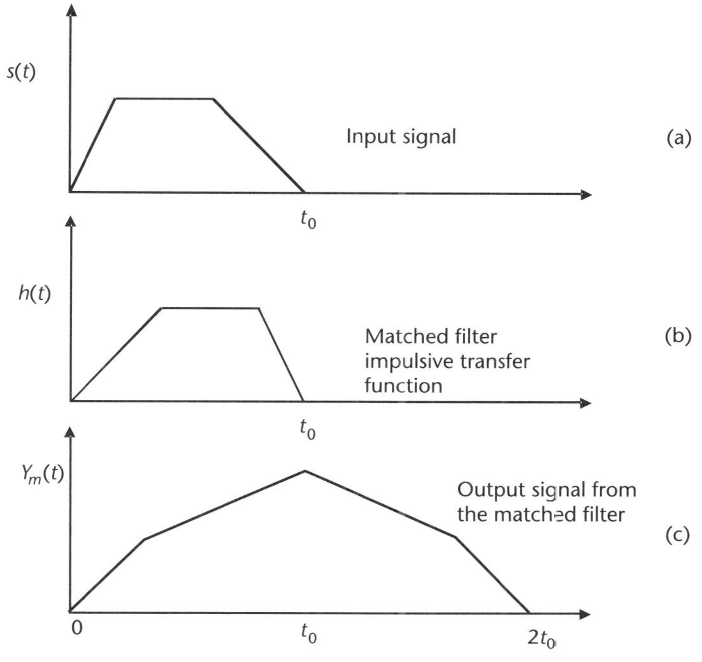

Figure 2.13 Matched filter time response.

Table 2.4 Efficiency of Practical Approximations of Matched Filters*

Input Signal Shape	*Filter Implementation*	*Optimum* $B\tau$	*SNR Loss (dB)*
Rectangular pulse	Single tuned	0.4	0.88
Rectangular pulse	Double tuned (Butterworth)	0.81	0.46
Rectangular pulse	Gaussian	0.74	0.51
Gaussian pulse	Gaussian	0.44	0 (matched)

* Where B is the filter half-power bandwidth and τ is the duration of the pulse. Derived from Skolnik [3].

The matched filter receiver (or its approximation for large SNR, the autocorrelation receiver) makes the detectability of signals a function only of the received signal energy E and the input noise spectral density $N_0 = kT_0F$. Thus, the maximum detection range of the radar is not dependent on the shape of the transmitted signal or the receiver bandwidth, which can therefore be selected to optimize the extraction of the information relevant to the target, as shown in the following sections.

2.3.3.2 The Ambiguity Function

The previous section demonstrated that for large SNR the matched filter output can be considered as the autocorrelation function of the transmitted signal, that is, the noise can be neglected. However, when delayed, the received signal has a Doppler shift, its carrier frequency is different from the transmitted one, and the matched filter output can no longer be considered to be the autocorrelation function of the transmitted signal. In this case the matched filter output, always for large SNR, must be considered as the cross-correlation of the received Doppler-shifted signal and the transmitted signal.

From the shape of the matched filter output as a function of both time and Doppler frequency, it is possible to understand many properties of the transmitted radar waveform with regard to range resolution, ambiguities in range and radial velocity, range and velocity measurement accuracies, and the response to clutter (an environmental echo at the same speed of the radar).

As seen from (2.31) the output of the matched filter [always neglecting the noise contribution to the received signal, that is, $x(t) \approx s_r(t)$] is:

$$y_m(t_0) = \int_{-\infty}^{+\infty} s_r(t)s^*(t - t_0)dt \tag{2.32}$$

where t_0 is the estimate of the time delay of the echo, which has to be considered as a variable.

By using the complex notation, the transmitted signal can be expressed as

$$s(t) = u(t)\exp(j2\pi f_0 t) \tag{2.33}$$

where $u(t)$ is the complex modulation function, whose modulus is the envelope of the real signal and f_0 is the carrier frequency. Therefore, the received signal is assumed to be the same as the transmitted one, except for a Doppler frequency shift f_d and the time echo delay t_1:

$$s_r(t) = u(t - t_1)\exp(j2\pi(f_0 + f_d)(t - t_1)) \tag{2.34}$$

Therefore, from (2.32) we have

$$y_m(t_0) = \int_{-\infty}^{+\infty} u(t - t_1)u^*(t - t_0)\exp\left[j2\pi((f_0 + f_d)(t - t_1) - f_0(t - t_1)\right]dt \tag{2.35}$$

For the sake of understanding the properties of this equation, let us consider $t_1 = 0$ and $f_0 = 0$ and define $\tau = -t_0$ and hence

$$y_m(\tau, f_d) = \int_{-\infty}^{+\infty} u(t)u^*(t + \tau)\exp(j2\pi f_d t)dt \tag{2.36}$$

A positive τ indicates a target around the true time delay t_1 and a positive Doppler frequency f_d indicates an approaching target. The squared magnitude of $y_m(\tau,f_d)$, that is, $|y_m(\tau, f_d)|^2$ is called the *ambiguity function*[2] and it is usually represented as:

2. The name ambiguity function given by Woodward to (2.29) can be misleading since this function describes more about the properties of a waveform than just the ambiguities it produces. Woodward was interested

$$|\chi(\tau, f_d)|^2 = |y_m(\tau, f_d)|^2 = \left| \int_{-\infty}^{+\infty} u(t)u^*(t+\tau)\exp(j2\pi f_d t)\cdot dt \right|^2 \qquad (2.37)$$

Its three-dimensional plot as a function of time delay τ and Doppler frequency f_d is called the *ambiguity diagram*. The properties of the ambiguity function $|\chi(\tau, f_d)|^2$ are as follows:

1. Maximum value $|\chi(\tau, f_d)|^2_{\max} = |\chi(0,0)|^2 = 1$, that is, the maximum occurs at the origin, which is the true location of the target at $f_d = 0$ and 1 is the energy of the echo signal;
2. Symmetry $|\chi(-\tau, -f_d)|^2 = |\chi(\tau, f_d)|^2$;
3. Behavior on the time-delay axis $|\chi(\tau, 0)|^2 = \left| \int_{-\infty}^{+\infty} u(t)u^*(t+\tau)dt \right|^2$, that is, the square of the autocorrelation function
4. Behavior on the Doppler frequency axis $|\chi(0, f_d)|^2 = \left| \int_{-\infty}^{+\infty} |u(t)|^2 \exp(j2\pi f_d t)dt \right|^2$ is the square of the inverse Fourier transform of $|u(t)|^2$;
5. Total volume $\int_{-\infty}^{+\infty}\int_{-\infty}^{+\infty} d\tau df_d |\chi(\tau, f_d)|^2 = 1$.

The "ideal" ambiguity diagram would consist of a Dirac's impulse [as shown in Figure 2.14(a)], which has no ambiguities both in range and Doppler frequency. Its infinitesimal thickness at the origin would permit a determination of both time delay and target radial velocity to the desired accuracy, to resolve two very closely spaced targets and reject all clutter (except at the origin). However the Dirac's impulse does not satisfy properties 1 and 5 above, which state that both the maximum of the ambiguity diagram and the total volume should be 1.

An approximation of a good ambiguity diagram is shown in Figure 2.14, in which the waveform has a single peak and no ambiguities in the (τ, f_d) space. However, the single peak might be so wide along either time or Doppler frequency axes

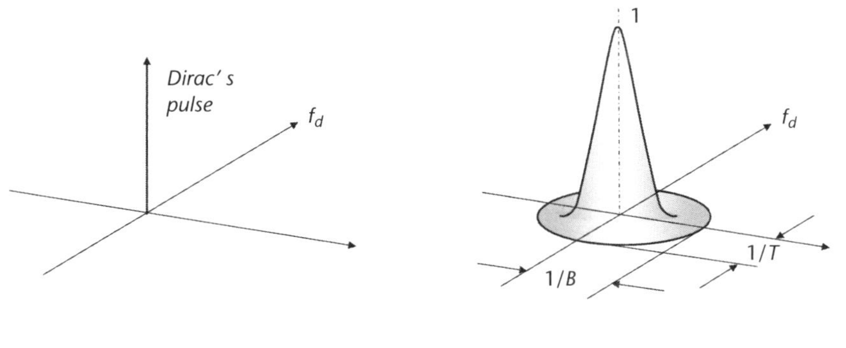

Figure 2.14 (a) Ideal and (b) approximated ideal ambiguity diagram.

---

in the spread of the matched filter response, which introduces an ambiguity as to the exact range when a Doppler shift is present [11].

or both that it might provide poor measurement accuracy and resolution in one or both domains.

In what follows only the projection on the plane (τ, f_d) of the ambiguity diagrams of some practical waveforms will be reported. So the interested reader is referred to [4] for deeper insight into the ambiguity diagrams of most radar waveforms.

Figure 2.15(a) shows a two-dimensional projection on the plane (τ, f_d) of the ambiguity diagram for a single pulse of sine wave of width T. Figures 2.15(b) and (c), respectively, show the shape of the ambiguity function along the time and Doppler frequency axis. In part (a) the darkened shaded area indicates the region where the ambiguity function is large (greater than –6 dB), the lightly shaded area indicates the region where it is small (less than –6 dB), and no shading indicates the area where the ambiguity function is zero. As can be seen in Figure 2.15(a), the simple pulse presents the limitations that if range resolution is improved by reducing width T, the Doppler frequency accuracy is worsened and vice versa if width T is increased. The short pulse waveform is *Doppler tolerant* in that, as can be noted in Figure 2.15(b), the response of a matched filter to a moderate Doppler frequency shift does not change too much with respect to the response to zero Doppler frequency.

Figure 2.16 shows the two-dimensional projection on the plane (τ, f_d) of the ambiguity diagram of a linearly frequency-modulated (LFM) rectangular pulse of width T over a bandwidth B.

Let us evaluate the ambiguity function of a LFM pulse of bandwidth B and duration T, which can be expressed as:

$$s(t) = \frac{1}{\sqrt{T}} rect\left(\frac{t}{T}\right) \exp\left(j\pi \frac{B}{T} t^2\right)$$

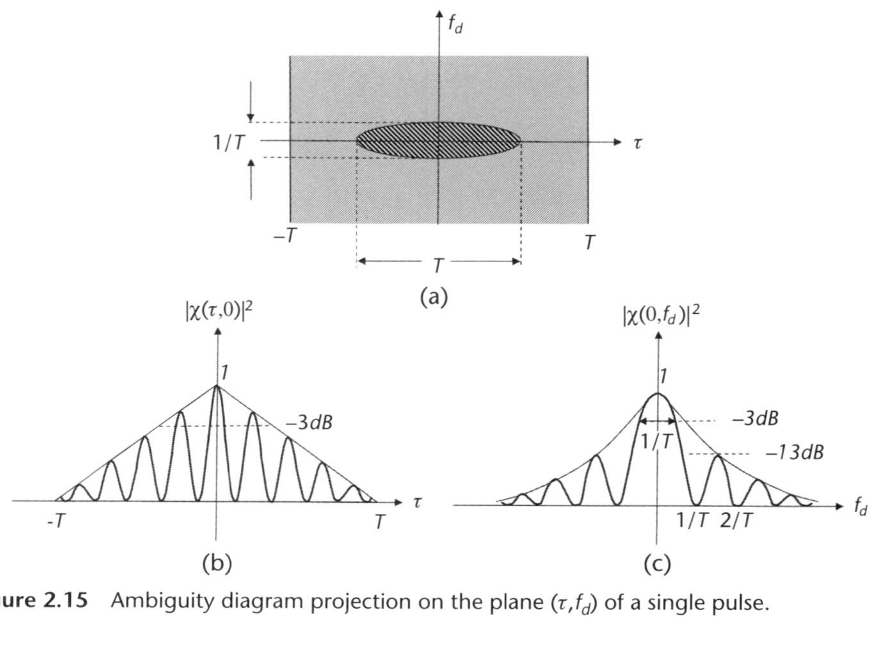

Figure 2.15 Ambiguity diagram projection on the plane (τ, f_d) of a single pulse.

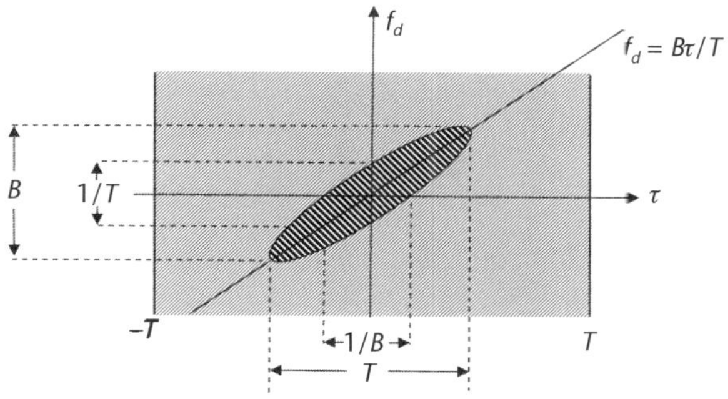

Figure 2.16 Ambiguity diagram for a single LFM pulse of width τ and bandwidth B.

By letting $k = B/T$ we have:

$$\begin{aligned}
|\chi(\tau, f_d)|^2 &= \left|\int_{-\infty}^{+\infty} s(t)s^*(t+\tau)\exp(-j2\pi f_d t)dt\right|^2 = \left|\frac{1}{T}\int_0^{T-\tau}\exp(j\pi k t^2 - j\pi k(t+\tau)^2 - j2\pi f_d t)\cdot dt\right|^2 \\
&= \frac{1}{T^2}\left|\exp(-j\pi k\tau^2)\right|^2 \cdot \left|\int_0^{T-\tau}\exp(-j2\pi(k\tau - f_d)t)\cdot dt\right|^2 = \frac{1}{T^2}\left|\frac{T-\tau}{\alpha}\right|^2\left|\int_0^{-j2\alpha}\exp(z)dz\right|^2 \\
&= \left(\frac{T-\tau}{T}\right)^2 \cdot \left(\frac{\sin^2\alpha}{\alpha^2}\right)
\end{aligned}$$

where

$$\alpha = \pi(T-\tau)(k\tau - f_d) \tag{2.38}$$

Note from this equation that along the time axis ($f_d = 0$) the first zero crossing is at $\alpha\tau = \pi$, that is, $(T - \tau)k\tau = 1$, which becomes $\tau = 1/kT = 1/B$ for $\tau << T$; and along the Doppler frequency axis ($\tau = 0$) the first zero crossing is at $\alpha_f = \pi$, that is, $\pi T f_d = \pi$ or $f_d = 1/T$. Further note that for $\tau = T/2$ we have $f_d = B/2$ and thus the total frequency excursion is $f_d = B$.

As can be noted from Figure 2.16, the time-delay measurement accuracy is proportional to $1/B$ and the Doppler frequency accuracy is proportional to $1/T$. since the pulse width T and the bandwidth B can be chosen independently, the two above-mentioned accuracies can both be obtained.

Because the selected radar waveform affects target detection, range and Doppler frequency measurement accuracy and resolution, ambiguities and clutter rejection, its ambiguity diagram is very useful in the radar design process to assess qualitatively the achievement of the above-mentioned capabilities. A noise waveform, such as pseudorandom variations of phase or frequency, has an ambiguity diagram called a *thumbtack*, as shown in Figure 2.17. This waveform has the advantage that the time-delay and frequency measurement accuracies are independently determined, respectively, by the bandwidth of the modulation and the duration of the pulse.

The shape of the thumbtack shows that there are no ambiguities, and it resembles the approximation of the ideal ambiguity diagram shown in Figure 2.14(a) with the exception of the plateau on which the main response rests. This plateau

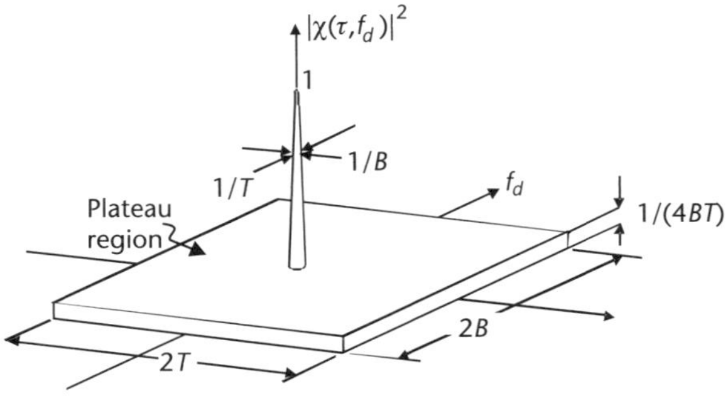

Figure 2.17 Thumbtack ambiguity diagram.

extends over a dimension $2T$ along the time axis and $2B$ along the frequency axis. When the product BT is large, the volume under the peak response is small and almost all of the volume is in the plateau, which hence has a height h = (volume/plateau area) = $1/(4BT)$.

2.3.3.3 Information from Radar Signals

Knowing that a target is present (detection) in the environment is of no use if more information on the target is not obtained. A radar obtains information about a target by comparing the received echo signal with the signal that was transmitted. From a single received echo, the radar obtains range, radial velocity, azimuth, sometimes the elevation of the target, and, in the case of a high-resolution radar, can assess whether it is a point target or a distributed target (with respect to the dimension of the radar resolution cell).

After several observations of a moving target over a period of time, the target trajectory or track is obtained, thus estimating also the two components of the target's velocity (radial and tangential). The value of any measurement resides in its accuracy, which is the estimate of the error between the unbiased measurement of a parameter and its true value (see Appendix B). In the following paragraphs we discuss the radar's measurements accuracy, taking into account that receiver noise is the fundamental limitation to accurate radar measurements and, hence, because the measurements are samples of a statistical parameter estimation the accuracy will correspond to the root-mean-square (rms) error value.

The expressions of the theoretical accuracies of the radar measurements (assumed unbiased) are valid for large values of SNR, since it has already been shown that radar detection conditions (P_d and P_{fa}) require large SNR and detection is a precondition for the measurement.

A further assumption is that the measurement error associated with a particular parameter is independent of the errors in any other parameter and the accuracy is limited only by receiver noise. With the above assumptions the theoretical rms error ΔM of a radar measurement M has the following expression:

$$\delta M = \frac{kM}{\sqrt{2E/N_0}} = \frac{kM}{\sqrt{SNR}} \tag{2.39}$$

where k is a constant, E is the received signal energy, and N_0 is the noise power per unit bandwidth (two sided).

Theoretical radar accuracies can be derived by a variety of statistical methods and also by simple geometrical relationships among the signal, the noise, and the parameter to be measured. The latter is the method followed by Skolnik [3], which affirms that this method, for a rectangular pulse, gives results similar to those for the more rigorous statistical methods.

The derivation of the expressions are beyond the scope of this book. In the following we report on only the expressions and make some comments about their implications. We recommend to the interested reader the books from Skolnik [3] or Galati and Crescimbeni [5] for gaining deeper insight into the matter.

Range Accuracy

Range is measured by the round-trip time delay $T_R = 2R/c$. Thus the range accuracy is given by $\Delta R = c\Delta T_R/2$, where ΔT_R is the accuracy of the time delay. This accuracy can be evaluated through the shift in time of the leading edge (affected by the receiver noise) of the detected video pulse (which has a trapezoidal shape) at the threshold crossing . Thus $\Delta T_R = t_r /(2S/N)^{1/2}$, where t_r is the pulse rise time and S is the signal power. Because the rise time of the video pulse is limited by the spectral bandwidth B of the rectangular-shaped IF filter, we have $t_r \approx 1/B$. Remembering that $S = E/\tau$, where τ is the pulse width and $N = N_0B$, we have:

$$\delta T_R = \left(\tau/(2BE/N_0)\right)^{1/2}$$

If the same consideration is applied to the trailing edge and assuming that the noise samples that affect the pulse rise and fall time are independent, the above rms accuracy can be improved by a factor $\sqrt{2}$, thus giving

$$\delta T_R = \left(\tau/(4BE/N_0)\right)^{1/2} \tag{2.40}$$

The other more rigorous statistical methods cited earlier give the following expression for the time-delay accuracy:

$$\delta T_R = \frac{1}{\beta(2E/N_0)^{1/2}} \tag{2.41}$$

where β is the so-called *effective bandwidth* or *rms bandwidth*, defined as

$$\beta^2 = \frac{\int_{-\infty}^{+\infty} (2\pi f)^2 |S(f)|^2 \, df}{\int_{-\infty}^{+\infty} |S(f)|^2 \, df} = \frac{1}{E}\int_{-\infty}^{+\infty} (2\pi f)^2 |S(f)|^2 \, df \tag{2.42}$$

The effective bandwidth β is such that $(\beta/2\pi)^2$ is the normalized second moment of $|S(f)|^2$ about its mean, which is assumed to be at $f = 0$, where $S(f)$ is the video spectrum with both negative and positive frequencies. The effective bandwidth is different from all other bandwidths encountered in electronic engineering because it is not related to either the noise bandwidth or the half-power bandwidth. In fact, the more the spectral energy is concentrated at the two ends of the bandwidth, the larger is β and the more accurate is the measurement of the time delay T_R. For a trapezoidal pulse of finite bandwidth B_s ($\pm B_s/2$ around $f = 0$), it can be shown that for large $B_s\tau_r$:

$$\beta^2 \approx \frac{2B_s}{\tau_r} \approx \frac{2}{\tau_r \cdot t_r} \tag{2.43}$$

where it was assumed that $t_r \approx 1/B_s$ (i.e., the total bandwidth instead of half-power bandwidth $B \approx 1/\tau$). By substituting the above value of β in (2.41), we have:

$$\delta T_R = \left(\tau_r/(4B_S\, E/N_0)\right)^{1/2} = \left(\tau_r t_r/(4E/N_0)\right)^{1/2} \tag{2.44}$$

Accuracy of Doppler Frequency and Radial Velocity

In a similar way as done for the range accuracy, the rms error in the measurement of frequency can be expressed as:

$$\delta f_d = \frac{1}{\alpha(2E/N_0)^{1/2}} \tag{2.45}$$

where α is called the *effective time duration* of the signal $s(t)$ and is defined, in a manner similar to that for β as:

$$\alpha^2 = \frac{\int_{-\infty}^{+\infty} (2\pi t^2)\cdot s^2(t)dt}{\int_{-\infty}^{+\infty} s^2(t)dt} \tag{2.46}$$

So $(\alpha/2\pi)^2$ is the normalized second moment of $s^2(t)$ about the mean epoch, assumed to be $t = 0$. The value of α for various simple waveforms is reported by Skolnik [3].

The rms error in the radial velocity is straightforwardly derived from the expression $v_r = cf_d/2f$ and is:

$$\delta v_r = \frac{c}{2f}\delta f_d = \frac{c}{2\alpha f(2E/N_0)^{1/2}}$$

The error expressions discussed above apply to a single observation. When n independent measurements are made, the resultant rms accuracy is reduced by a factor $n^{1/2}$.

From the above expressions relevant to the effective bandwidth β and effective duration α relevant to the employed waveform $s(t)$, note that by applying the Schwartz's inequality to the product of α and β we obtain

$$\alpha\beta \geq \pi \tag{2.47}$$

This equation, which is a consequence of the Fourier transform relationship between a time waveform and its spectrum, states that there is no theoretical restriction on the accuracy with which a radar can simultaneously locate the position of a target and determine its velocity. In fact, if we take the product of the two accuracies $\Delta T_R \Delta f_d$ we have:

$$\delta T_R \delta f_d = \frac{1}{\alpha\beta(2E / N_0)} \leq \frac{1}{\pi(2E / N_0)} \tag{2.48}$$

Thus, by designing a radar to yield a sufficiently large ratio E/N_0 or, with fixed E/N_0 and by selecting a waveform with a large $\alpha\beta$ product, both of the preceding accuracies, as small as desired, can be achieved. Note that a large $\alpha\beta$ product requires waveforms with long time durations and wide spectral widths.

By expressing (2.48) in terms of ΔR and Δv_r, we obtain

$$\delta R \delta v_r \leq \frac{c\lambda}{4\pi(2E / N_0)} \tag{2.49}$$

This equation states that the shorter the wavelength λ and, more important, the greater the product $\alpha\beta$ (i.e., the compression ratio), the better will be the accuracy that can be achieved in simultaneous measurement of range and radial velocity.

Angular Accuracy

In a similar way to the time waveform, there is a Fourier relationship between the illumination $A(x,y)$ of an antenna aperture and its complex radiation pattern $g(\theta,\phi)$. For simplicity we will show only the one-dimensional case. The one-dimensional electric-field strength radiation pattern in one plane is expressed as follows, where D is the dimension of the aperture and λ is the wavelength:

$$g(\theta) = \int_{-D/2}^{+D/2} A(x)\exp\left(j2\pi \frac{x}{\lambda}\sin\theta \right) dx$$

Hence, the effective aperture width γ is expressed as:

$$\gamma^2 = \frac{\int_{-\infty}^{+\infty} \left(2\pi \frac{x}{\lambda}\right)^2 |A(x)|^2 \, dx}{\int_{-\infty}^{+\infty} |A(x)|^2 \, dx} \tag{2.50}$$

and the rms error for angle measurement can be obtained as follows:

$$\delta\theta = \frac{1}{\gamma(2E/N_0)^{1/2}} \tag{2.51}$$

For uniform amplitude illumination across the aperture, the theoretical angle measurement error is:

$$\delta\theta = \frac{\sqrt{3}\lambda}{\pi D(2E/N_0)^{1/2}} = \frac{0.628\theta_B}{(2E/N_0)^{1/2}}$$

where $\theta_B = 0.88\lambda/D$ is the half-power beamwidth of the aperture for uniform amplitude illumination.

2.3.4 Automatic Detection

Automatic detection is digital processing that is run on the radar video signal. The radar video signal performs the detection process, which is presented both to the operator and to subsequent data processors associated with the radar. Automatic detection involves the following steps:

- Quantization of the radar coverage into range and angle resolution cells;
- Sampling in time the radar signal at a rate of two samples per range resolution cell;
- Analog-to-digital conversion of the above samples with an adequate number of bits in accordance with the selected signal processing to remove as much noise, clutter echoes, and interference as practicable (see below) before the detection decision is made;
- Integration of the available data samples at each resolution cell in order to improve SNR_{out} and achieve the required detection conditions in terms of P_D and P_{FA};
- Perform CFAR processing in order to keep the false alarm rate in the presence of clutter and interference residues from the signal processing, with threshold detection to present the target echoes to the subsequent automatic detection and tracking (ADT) processing (see Section 2.3.7).

The first three steps are common to all digital signal processing and data presentations and as such they do not need to be discussed here. In what follows we discuss mainly the improvement of the SNR due to pulse signal integration and the achievement of a constant false alarm rate in the presence of clutter and interference residues that either Doppler (see next section) and/or ECCM (see Chapter 5) processing had been unable to remove.

2.3.4.1 Integrator

As mentioned in the previous sections, a point target when illuminated by a radar beam, either steady or scanning, provides the radar with a number n of successive echoes at the PRF rate, where n is limited either by the data processing bandwidth of the receiver (in the case of a steady beam) or by the so-called time-on-target (TOT, or observation time within the scanning beam). The TOT is determined by the beamwidth θ_B (rad) and the scanning speed of the beam ω (rad/s) as follows (and as also shown in Figure 2.18):

$$TOT = \frac{\theta_B}{\omega}$$

and, hence,

$$n_{scan} = PRF \odot TOT = \frac{\theta_B}{\omega \cdot PRI}$$

where PRI = 1/PRF is the pulse repetition interval.

As an example, for a scanning surveillance radar antenna rotating at a speed ω = 6 rpm = 0.63 rad/s = 36°/s with beamwidth θ_B = 2° = 35 mr, and PRF = 0.5 kHz, we have:

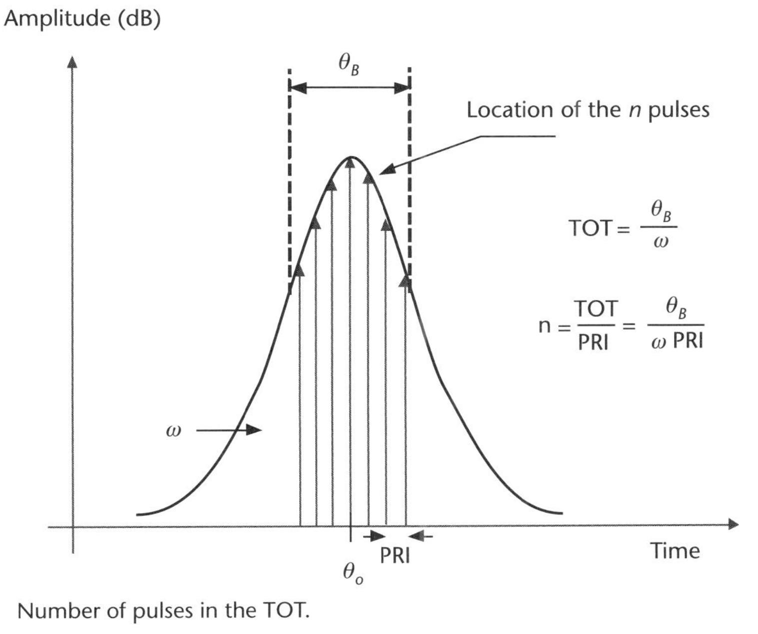

Figure 2.18 Number of pulses in the TOT.

TOT = 55.5 ms and n_{scan} = 28 pulses

As discussed in Appendix A, the integration of the n available echo pulses can be performed either coherently or incoherently. Coherent integration is more efficient, but until advances in digital processing became available, continuous incoherent integration of the last n pulses belonging to the same range resolution cell will be performed by the so-called moving window integrator (MWI), whose block diagram is shown in Figure 2.19.

In a current digital processor it might be possible to apply weights to the tap outputs, based on the two-way gain of the antenna pattern. This would help to improve SNR_{out}. In practice, uniform weighting is applied giving a loss in the SNR_{out} of about 0.5 dB (with respect to the optimum weighting) [3].

The MWI operates according to the double-threshold detector logic or *m out of n* detection logic, which means that it is sufficient to achieve m primary detections within the n available pulses to declare the detection of a target pulse. The primary threshold Th_1 can be set at a low probability of false alarm (e.g., $p_{fa} = 10^{-2}$) to obtain a primary pulse detection with a low SNR_{in}. The output of $m \geq n/2$ detections provided by the crossing of the second threshold $Th_2 = m$ provides a target detection (at the usual number $n > 10$ within the beamwidth) at probability conditions[3] $P_D \geq 90\%$ and $P_{FA} \leq 10^{-6}$ with an SNR_{in} = 5 dB. (Remember that for the same probability conditions the single pulse $SNR_{in} \geq 13$ dB.)

The angular estimate can be obtained by taking either the angular value of the maximum of the running sum (and then correcting it for the bias $n/2$ or $\theta_B/2$) or the

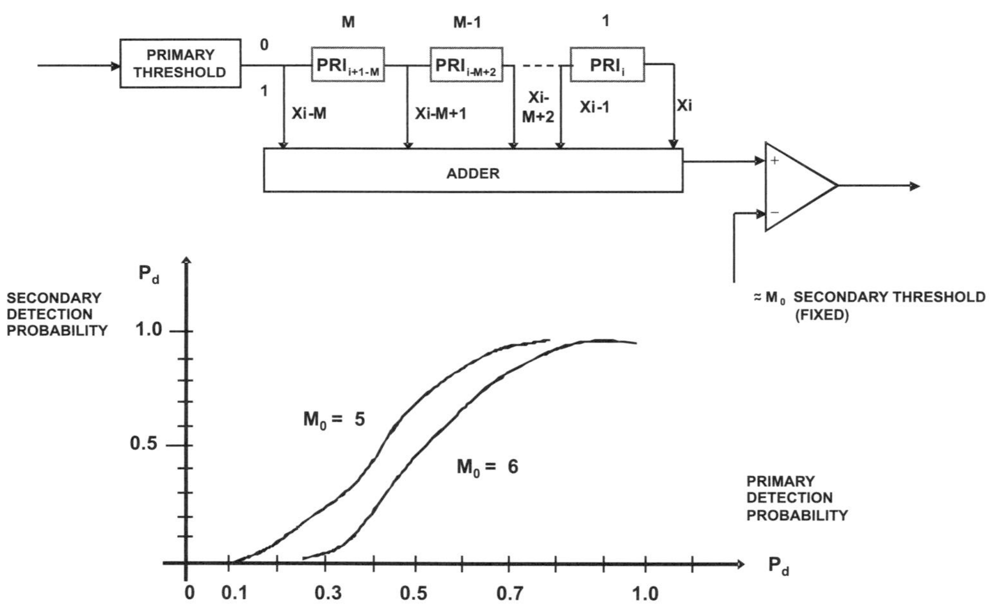

Figure 2.19 (a) Moving window integrator schematic block diagram. (b) Operational characteristics, that is, detection probability P_D after secondary threshold vs. single pulse detection probability P_d.

3. The output P_{FA} can be evaluated according to Bernouilli's probability expression as $P_{FA} = \sum_{m}^{n} (a_{n,k}) p_{fa}^{k} \cdot (1 - p_{fa})^{n-k}$ where $a_{n,k} = \frac{n!}{k!(n-k)!}$.

mean angular value between the first and the last crossing of the primary threshold. The standard deviation (angular rms error) of these estimators is about 20% higher than the optimal estimate specified by the Cramer-Rao bound (see Appendix B and relevant references).

The MWI is a relatively simple robust automatic detector that is less sensitive with respect to the other detectors to the effects of a single interference or clutter pulse, regardless of how large it is, because it converts any detection only into a 1 in the sum.

2.3.4.2 CFAR

In many situations a detection threshold that is set according to the receiver noise level can be easily crossed by clutter echoes or hostile noise jamming, which usually have much larger response levels than the receiver internal noise. This produces many harmful false alarms to the ADT function. In the ADT function, false alarms might strongly degrade the computer capacity to recognize and discard the false targets; therefore, it is necessary to add a CFAR processing step at the output of the radar receiver, before a target detection is declared, in order to keep a constant false alarm rate. CFAR processing does not provide any improvement in the signal-to-clutter ratio (SCR); instead, it produces some loss in the SNR.

The usual form of CFAR processing is constituted by the *cell-averaging* (CA) scheme reported in Figure 2.20. Two tapped delay lines sample the echo signals in range into a number of M reference cells (each of the same size as the range resolution) located on both sides of the range cell of interest (the test cell). The output of the test cell is compared with an adaptive threshold derived by the averaged sum (multiplied by a factor K) of the output of the reference cells. If the radar output noise or clutter is described by a Rayleigh probability density function (pdf), the factor K can be determined by means of classical detection theory (see Appendix A). When the clutter statistic is described by some other pdf, the factor K has to be estimated by simulation.

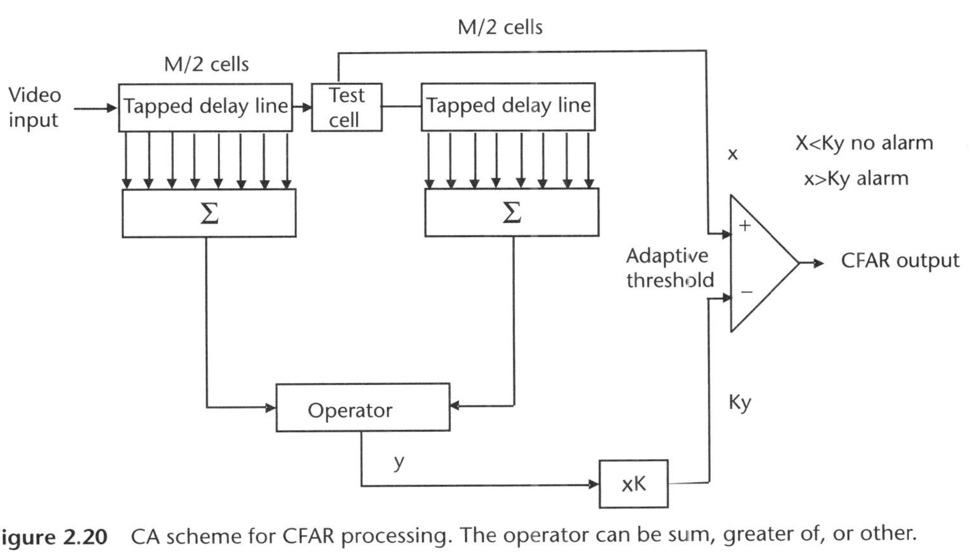

Figure 2.20 CA scheme for CFAR processing. The operator can be sum, greater of, or other.

The greater the number M of reference cells, the better the estimate of the background clutter or jamming noise and the less loss in detectability (or SNR loss) of the target.

The scheme shown in Figure 2.20 illustrates the situation in which the edges of a patch of clutter are exceeding the threshold (which is evaluated taking into account that most of the reference cells contain the clutter). The threshold crossing of clutter edges is reduced by determining a further threshold that is obtained by summing separately only the leading and lagging reference cells and by using the greater of the two sums as the CFAR processing threshold. This method is called *G-greater of the two sums* or GO-CFAR and introduces a further SNR loss of about 0.3 dB.

A further CFAR method is the so-called *clutter map* (CM), where the radar coverage is subdivided into range-azimuth cells and the measured clutter echo of each cell is stored in it and used as the detection threshold for that cell. Since the clutter echo can change with time, the value of the clutter echo in each cell is updated periodically upon average on a number (on the order 10) of successive scans. The CM-CFAR has the advantage with respect to CA and GO-CFAR that it is not affected by nonhomogeneous clutter and to the clutter edges.

A much used CFAR processing technique that is quite effective against impulse-like and broadband noise jamming is the *Dicke-Fix* or *hard-limiter* processing. It consists of the cascade of a broadband IF filter (of bandwidth B_L) followed by a hard-limiter (set to a level about 10 times below the receiver noise) and a narrowband matched filter (of bandwidth $B_{IF} << B_L$) as shown in Figure 2.21. The method is used with pulse compression waveforms, where the matched filter provides the pulse compression gain needed for target detection. This method cannot be effectively used with MTI radar processing because it would strongly reduce its obtainable improvement factor (see Section 2.3.2.1). Moreover, the dynamic range downstream of the pulse compression is equal to the compression ratio regardless of the dynamic range upstream from the pulse compression.

2.3.4.3 Space-Time Adaptive Processing

When greater clutter attenuation is required in phased array radars on board mobile platforms, then adaptivity in both antenna pattern and Doppler processing can be applied. Indeed in a phased array antenna with N elements, it is possible to adjust, through a control loop, the amplitude and phase at each element in order to place one or more antenna pattern nulls in the directions of external noise, jamming, or clutter for the purpose of reducing or eliminating unwanted signals. The adaptive phased array antenna is thus acting as a spatial filter.

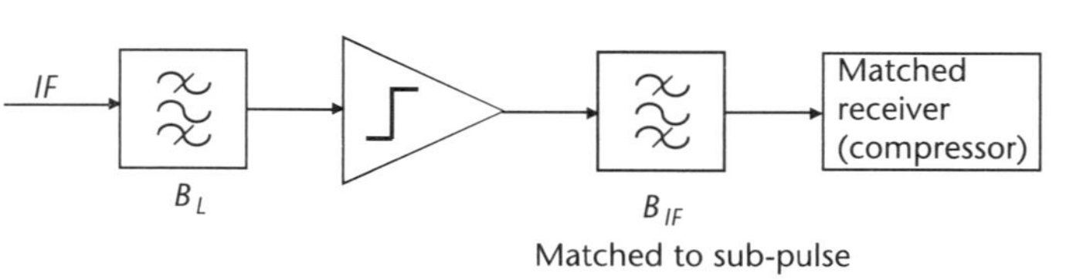

Figure 2.21 Hard-limiter CFAR.

After the adaptive antenna has minimized the clutter echo received from the antenna sidelobes, an adaptive Doppler filter can operate on the main beam by adjusting, through a control loop, its weights in order to shape its frequency response so as to maximize the S/C ratio at its output on the basis of the input clutter spectrum. This process of selecting weights corresponds as an example in placing the zeroes of the adaptive MTI filter [6] at the Doppler frequencies of the regions occupied by the clutter spectrum.

The combination of the two adaptive techniques is called space-time adaptive processing (STAP) and it provides the following advantages for a radar:

- Reduction of clutter in the main beam through the adaptive Doppler filtering,
- Reduction of clutter and jamming from the sidelobes through the adaptive antenna,
- Adaptive selection of antenna weights to provide platform motion compensation.

Of course, the larger the number of antenna elements, the number of pulses within a coherent processing interval (CPI), and the number of zeroes of the adaptive Doppler filter, the better is the performance of interference reduction. In practice, however, there is a limit on the number N of adaptive elements and K pulses. In fact, it is unusual to have a fully adaptive phased array (the elements are usually hundreds or thousands) so the elements are grouped in order to have $M << N$ subarrays and only their combined outputs are connected to the adaptive control loop. The number of pulses is limited by the beam dwell time in a direction and the selected PRF to resolve range ambiguities. A notional schematic block diagram of STAP is shown in Figure 2.22.

2.3.5 Pulse Compression

Simple short pulse waveforms were used in the early radars in conjunction with the high peak power transmitters to achieve target detection at long ranges with moderate range accuracy. But high peak power transmitters have a serious technical limitation because at high powers voltage breakdown (arc discharge) may occur, especially at high frequency, where the waveguide dimensions are small, and at high altitudes, where the air density is lower.

The need to transform the radar from a simple target detection system to a source of accurate information on the targets in order to assess their dangerousness, such as their radial velocity, their route, and their separation in a highly dense target environment (thus involving also good radar resolution), required a reduction in the pulse width (or, in other terms, an increase in the signal bandwidth) to achieve good range resolution.

Because the detection capabilities are dependent on the energy of the signal (i.e., the product of the peak power times the duration of the signal), high peak power can be reduced by increasing the duration T of the signal(long pulses), but modulating it either in frequency or phase so as to increase the signal bandwidth B and

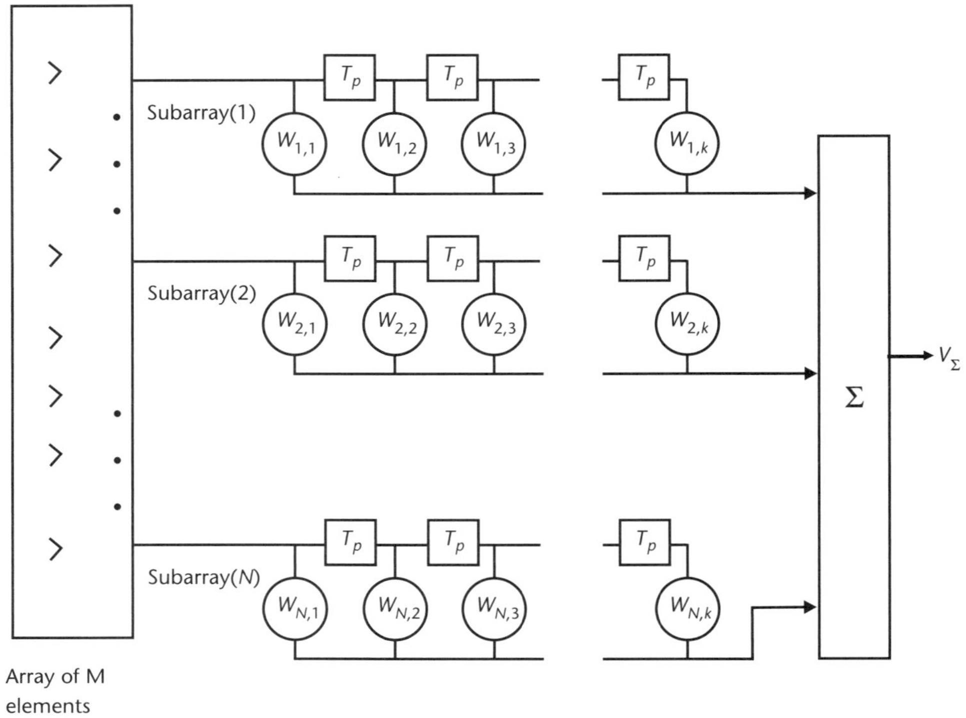

Figure 2.22 STAP block diagram.

hence the range resolution[4]. The long modulated pulse width is then compressed in the matched filter to achieve a short pulse width $\tau = 1/B$. Pulse compression thus allows a radar to achieve the energy of a long pulse and simultaneously the resolution of a short pulse, without requiring high peak power. The pulse compression ratio (PCR) is defined as the ratio of the long pulse width T to the compressed pulse width τ (also called a subpulse), that is, PCR $= T/\tau =$ TB

In the past, two classes of pulse compression waveforms were used: the linear frequency modulation (LFM; also known as "chirp") and the binary phase-shift code or key (BPSK). In more recent times, further classes of pulse compression waveforms were added, such as:

- Polyphase codes, whose phase quantizations are rad, with $n = 1, 2, \ldots$;
- Costas code, in which the frequencies of each subpulse are selected in accordance with a rule;
- Nonlinear FM (NLFM);
- Complementary codes.

These compression waveforms are discussed briefly in the following subsections.

4. Amplitude modulation, which also increases the signal bandwidth, is not generally used because good transmitter efficiency requires constant saturated power output.

2.3.5.1 Linear FM

The FM waveform is generated at low power and then amplified by a power amplifier to be transmitted. The transmitted waveform consists of a rectangular pulse of constant amplitude A and duration T. The carrier frequency either increases linearly within the pulse from f_1 to f_2 (up-chirp) or decreases linearly from f_2 to f_1 (down-chirp). On reception the LFM signal is passed through the pulse compression filter, which is a delay line whose velocity of propagation is proportional to the frequency (up-chirp) or inversely proportional to the frequency (down-chirp). In the case of an up-chirp, it speeds up the higher frequencies at the trailing edge of the pulse relative to the lower frequencies at the leading edge of the pulse and hence compresses the signal to a width , where $B = f_2 - f_1$. The pulse compression filter is a matched filter; hence, its output envelope is the autocorrelation function of the input. Its analytic expression (derived earlier) is of the type $\sin(\pi B\tau)/(\pi B\tau)$. The peak power of the pulse is increased by the PCR, where PCR = $T/$ after passage through the filter. The LFM transmitted waveform and the output from the matched filter are shown in Figure 2.23.

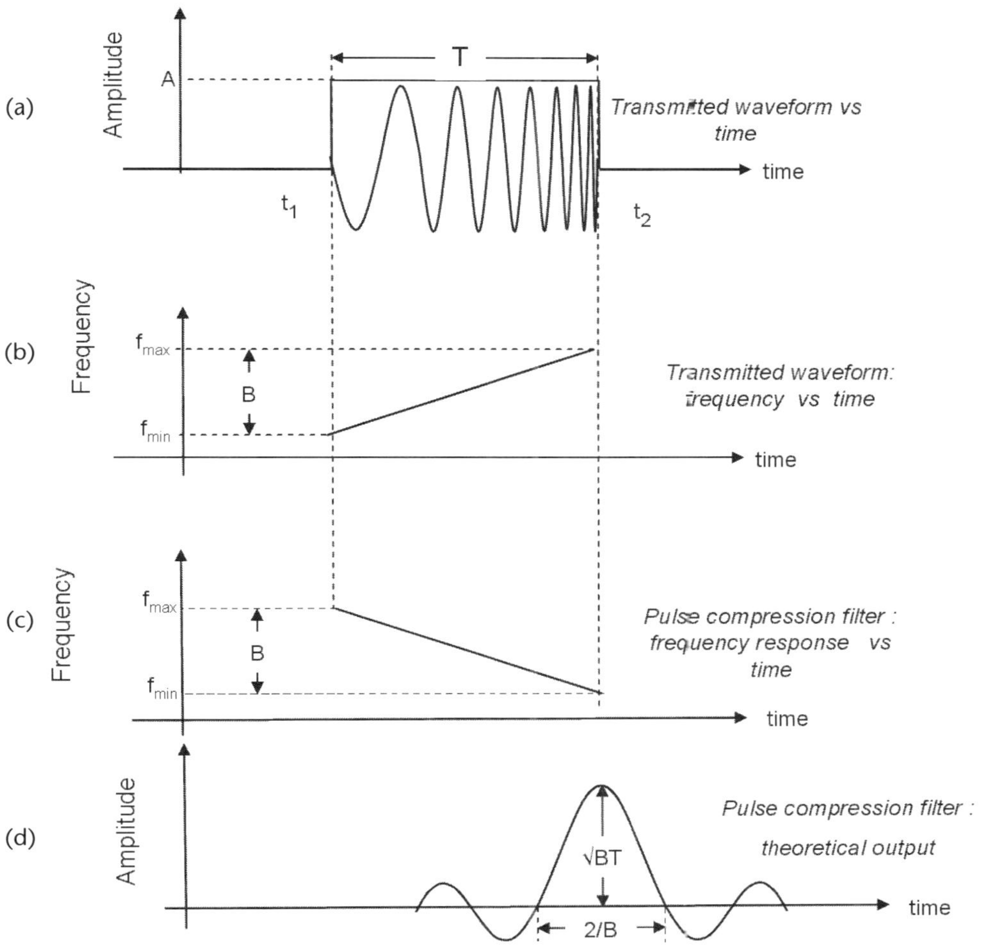

Figure 2.23 LFM transmitted waveform and the output from the matched filter.

The ambiguity diagram for a LFM pulse compression waveform shows that a large Doppler shift in the echo signal can indicate a range that is not the true range. This effect is known as the *range-Doppler coupling* . In many cases, the range error due to the Doppler shift is small and can be tolerated. If the range is large, the effect of the Doppler shift error can be compensated for by averaging the two range indications obtained with an up-chirp and a down-chirp waveform of the same bandwidth.

The envelope out of the matched filter of a LFM waveform has relatively high peak time sidelobes of –13.2 dB adjacent to the main response. This is usually not acceptable because it may be mistaken for further targets or mask some nearby weaker target. The amplitude of the time sidelobes can be reduced on reception by applying an amplitude weighting within the matched filter. The filter is thus mismatched and provides some loss to the SNR. Table 2.5 reports the first peak time sidelobe level and the relevant SNR loss and widening of the mainlobe response due to some weighting functions.

2.3.5.2 Binary Phase-Code Pulse Compression

An increase in signal bandwidth can be achieved by dividing a long pulse of duration T in N subpulses of duration and by changing the phase of each subpulse (the change of phase with time is a frequency). The simplest form of phase change is binary phase coding in which the phase of each subpulse can be selected as either 0 or π rad according to some specified criterion. If the selection of the phases $0,\pi$ is made at random, the waveform approximates a noise-modulated signal, which has a thumbtack-like ambiguity diagram.

The output of the matched filter is a compressed pulse of width and amplitude N times greater than that of the long pulse. The PCR is thus PCR = $N = T/\tau$, and the enlarged bandwidth is $B = 1/\tau$.

The matched filter output (the autocorrelation function) is extended from $-T$ to $+T$ with respect to the peak response. Time sidelobes are always present, and if the selection of the phases is made at random the expected (power) maximum sidelobe (SLL) is about $2/N$ below the peak of the compressed output. So random selection of phases is not the favored criterion.

A better criterion is the selection of codes that provide equal (minimum) time sidelobes and these are the Barker codes, which are listed in Table 2.6. As can be noted from Table 2.6, the Barker codes have a maximum length of 13 and the sidelobe (power) level is $1/N^2$ that of the peak signal. The Barker code matched filter

Table 2.5 Properties of Some Weighting Functions to Reduce Time Sidelobes

Weighting Function	*Peak Sidelobe (dB)*	*Loss (dB)*	*Main Beamwidth (relative)*
Uniform	–13.2	0	1.0
$\cos^2(\pi f/B)$	–31.7	1.76	1.65
Hamming: $0.08 + 0.92\cos^2(\pi f/B)$	–42.8	1.34	1.5

Table 2.6 Barker Codes (+ = phase 0; – = phase π)

Code Length	*Code Elements*	*Sidelobe Level (dB)*
2	+– ; ++	–6.0
3	++–	–9.5
4	++–+ ; +++–	–12.0
5	+++–+	–14.0
7	+++––+–	–16.9
11	+++–––+––+–	–20.8
13	+++++––++–+–+	–22.3

is constituted by a tapped delay line as shown in Figure 2.24 for the Barker code of length $N = 7$.

Larger PCRs than those provided by the Barker codes can be obtained by an N-stage binary shift register with feedback and modulo 2 adder such as the one depicted in Figure 2.25. The actual sequence obtained depends on the feedback connections and the initial loading of the shift register. An N-stage binary device like the one depicted in Figure 2.25 has a total of 2^N different possible states, with the exception of the state in which all the bits are 0. So it can generate a sequence of $2^N - 1$ bits, called the *maximal length sequence* or *pseudo-noise* (PN) *sequence*, after which the sequence will be repeated.

There can be more than one maximal length sequence depending on the feedback connections. With the device of Figure 2.25, up to eight different maximal length sequences of length 31 can be generated. With the proper code, the highest (power) sidelobe level can be about 12 to 18 dB (or $1/2N$ to $2/N$, with average $1/N$) below the peak response (which achieves a PCR of $N^2 = 30$ dB).

2.3.5.3 Polyphase Codes

When more than two phase quantizations are applied to the subpulses of a long pulse the waveforms are called *polyphase codes*. In general, these codes produce lower sidelobe levels and are more tolerant to Doppler frequency shift than the binary phase codes.

An example of polyphase code is the Frank code of length M^2, which is generated by an $M \times M$ square matrix. Each element of the matrix is defined as:

$$\phi_{n,k} = 2\pi(n-1)(k-1)/M \tag{2.52}$$

where $n = 1, 2, \ldots, M$ is the index of the row and $k = 1, 2, \ldots, M$ is the index of the column of the matrix. For example, if $M = 4$ and, hence, $2\pi/M = \pi/2$ rad, we have the following matrix:

$$\mathbf{F}_4 = \begin{bmatrix} 0 & 0 & 0 & 0 \\ 0 & 1 & 2 & 3 \\ 0 & 2 & 4 & 6 \\ 0 & 3 & 6 & 9 \end{bmatrix} \cdot \pi/2\pi$$

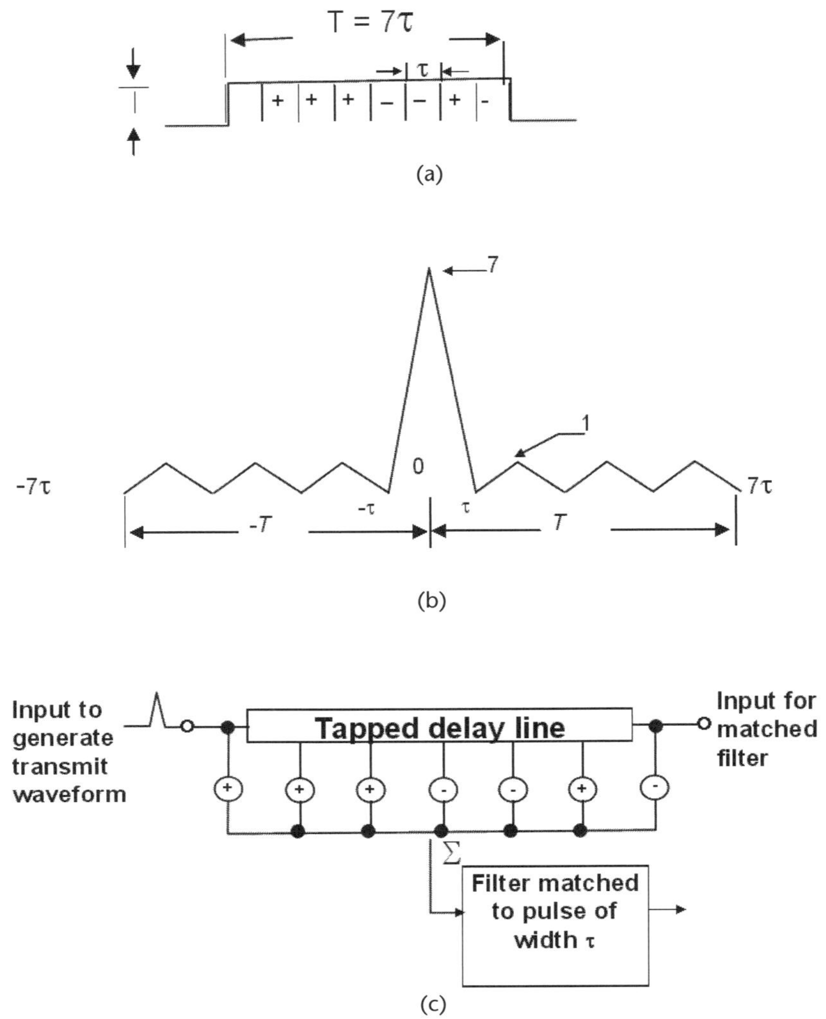

Figure 2.24 Barker code of length $N = 7$: (a) Barker code, (b) matched filter output, and (c) matched filter block diagram.

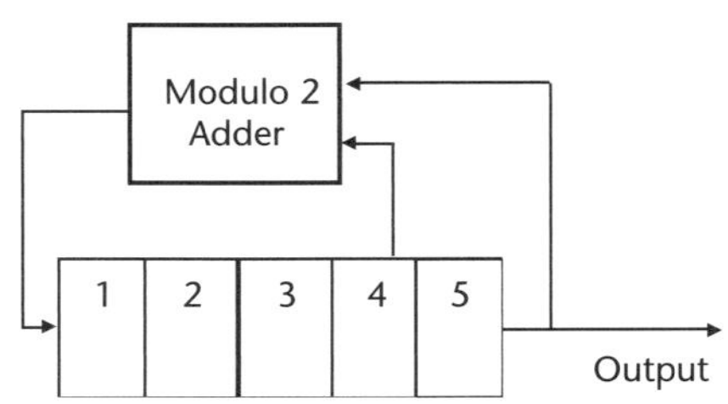

Figure 2.25 Five-stage binary shift register with feedback and modulo 2 adder.

Then the rows of the matrix are concatenated in order to achieve the following sequence of 16 phase elements modulo 2π to be transmitted:

$$[0 \quad 0 \quad 0 \quad 0][0 \quad \pi/2 \quad \pi \quad 3\pi/2][0 \quad \pi \quad 0 \quad \pi][0 \quad 3\pi/2 \quad \pi \quad \pi/2]$$

The Frank codes can be retained as approximating a stepped-LFM waveform, as shown in Figure 2.26.

The ambiguity diagram for a polyphase code is similar to that of a LFM waveform, but it can show a loss of about 3 to 4 dB in the signal at Doppler frequencies that are odd-integer multiples of π rad per subpulse.

The highest time sidelobe of a polyphase code, relative to the peak of the compressed pulse, is $1/(\pi^2 M^2)$. In the example given above ($M^2 = 16 = 12$ dB), the peak sidelobe is about –22 dB below the peak response.

Further polyphase codes are the so-called *P codes*, presented by Lewis and colleagues [7],which are also stepped approximations to LFM (P_1 and P_2 codes) or directly LFM (P_3 and P_4 codes).

Lewis [8] has also shown that the time sidelobes of the polyphase codes can be significantly reduced (by about 10 dB) by introducing at the output of the pulse compression filter a two-sample sliding window subtractor (in the case of Frank and P_1 codes) or a two-sample sliding window adder (in the case of P_3 and P_4 codes).

With this additional processing, the time sidelobes of a polyphase code of length N are uniform and are at a level where SLL = $4/N^2$ relative to the peak. The width of the compressed pulse, however, is doubled (i.e., width = 2τ) and the PCR is N/2.For example in the case of a Frank code of $N = M^2 = 100$ elements, the PCR = $N/2 = 50 = 17$ dB and the SLL = $4/N^2 = -28$ dB below the peak output. This is the SLL that would be achieved if a Barker code of 50 elements existed.

The importance of polyphase coding is that such codes increase substantially the code length, thus achieving large PCRs, while decreasing the time sidelobe level of the autocorrelation function (i.e., the matched filter output).

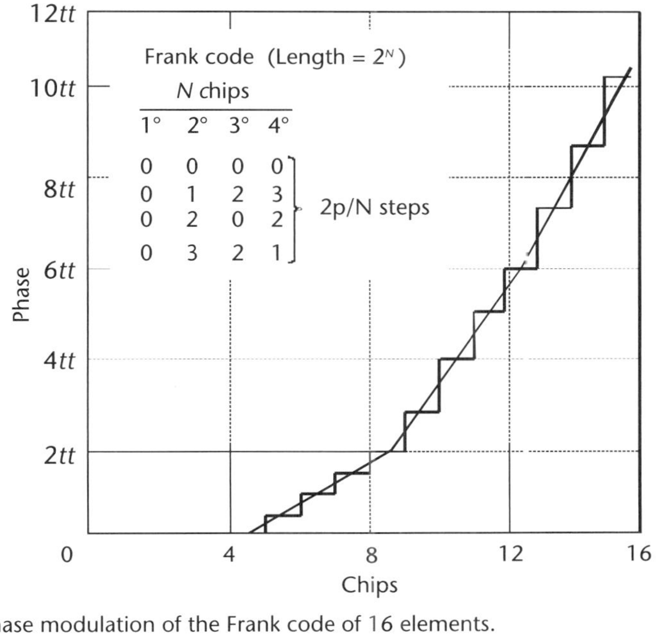

Figure 2.26 Phase modulation of the Frank code of 16 elements.

2.3.5.4 Costas Codes

The waveform, a frequency-hopping code, is generated by dividing the long pulse width T into a series of M contiguous subpulses of width τ, each at a frequency selected within a bandwidth B, as shown in Figure 2.27. The frequencies are separated by the reciprocal of the subpulse width (i.e., $\Delta f = 1/\tau = M/T$) and the bandwidth $B \geq M\Delta f$. Because the frequencies are selected at random, the ambiguity diagram has the thumbtack shape. The PCR is $BT = M\Delta f \cdot T = M^2$, but it is reached with only $M = (BT)^{1/2}$ subpulses instead of the BT subpulses required for binary phase-coded pulses.

Not all combinations of M frequencies produce an ambiguity diagram with low sidelobes in time and frequency domains. Costas [6] has suggested a procedure for selecting the order of frequencies to achieve a nearly thumbtack ambiguity diagram with a sidelobe level of $1/M^2$ relative to the peak power level in those regions of the ambiguity diagram away from the central peak and a SLL up to $4/M^2$ near the central peak.

The waveforms based on the Costas criterion are described in Chapter 6 of [3]. We recommend the interested reader consult that reference for deeper insight into these codes.

2.3.5.5 Further Coding

Nonlinear Frequency Modulation

NLFM offers the advantage over LFM of producing low time sidelobes, because the nonlinear rate of change of frequency has the same effect of amplitude weighting of the spectrum. In fact, if less time is spent over some part of the spectrum, this is equivalent to reducing the amplitude of that part of the spectrum. Further, there is no significant widening of the compressed pulse. When a symmetrical NLFM is used, the ambiguity diagram is that of a thumbtack, but as such is not Doppler tolerant. A nonsymmetrical NLFM waveform presents the classical ridge in the

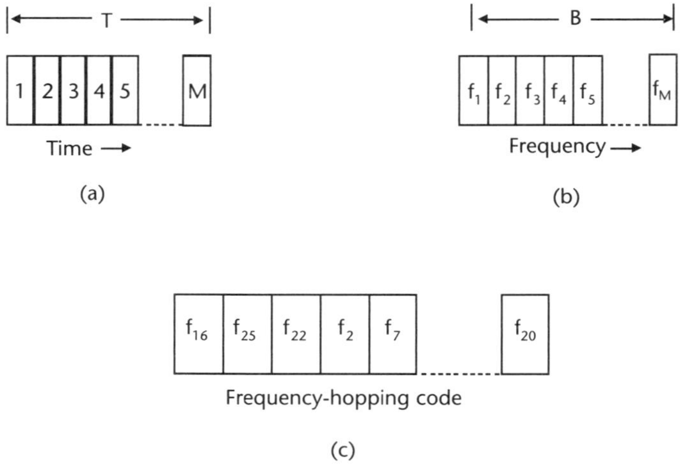

Figure 2.27 Costas code: (a) waveform time representation, (b) waveform frequency band, and (c) an example of transmitted waveform.

ambiguity diagram and has some of the range-Doppler coupling characteristics of the LFM.

Complementary Codes (Golay Codes)

Complementary codes are binary codes widely used in LPI radars, especially in air-to-ground waveforms [9]. They are defined as a pair of equal length codes, called A and $\tilde{A}$, which have the property that the time sidelobes of the autocorrelation function of one code are equal in amplitude but opposite in sign with respect to the code's complement. Therefore, if the autocorrelation functions of two complementary codes of length N are added, the resulting peak is $2N$ and the time sidelobes are identically zero, as shown in Figure 2.28.

By alternately transmitting the two codes and summing bin by bin, an infinite peak-to-sidelobe ratio can be achieved. In practice, implementation errors limit this performance to a high value better than 30 dB. Furthermore, good performance is also obtained for very long codes.

Golay [10] found complementary codes of length $N = 2^n$, where n is an integer, and Welti [2] found complementary binary "D-codes" of length $N = 2^n$ that are formed from shorter complementary codes. More details on these codes can be found in [9].

The large variety of complementary codes used in LPI radars compels the ESM interceptors either to have a large number of parallel decoders or to use a square law detector across the entire radar operating bandwidth. The latter gives a detection loss of $N^{1/2}$ with respect to the pulse compression of the emitter. This loss can be substantial when code lengths are larger than 10^4 (i.e., loss > 20 dB).

Complementary codes are not much used because they have two main weaknesses: (1) The uncancelled time sidelobes are present at f_d = PRF/2 and (2) they are not Doppler tolerant, like all binary and polyphase codes.

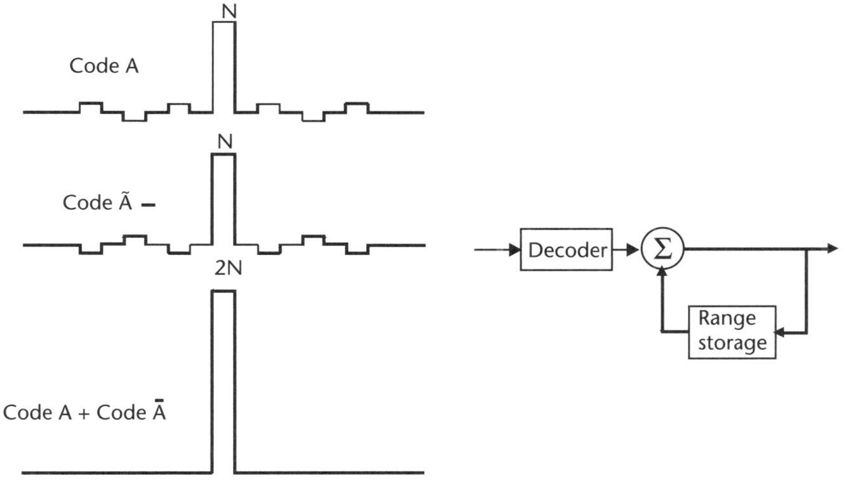

Figure 2.28 Property of the autocorrelation functions of complementary codes.

2.3.5.6 Combined or Compound Barker Codes

Longer binary codes can be created as products of Barker codes. The inner Barker code of chip τ can be phase modulated by an outer Barker code, whose chip T is equal to 13τ. Thus, we can achieve a code length of 169 chips with a PCR of 169. The code and its autocorrelation function are shown in Figure 2.29. As can be noted from the autocorrelation function, the central peak is surrounded on each side by 12 peaks of height 13, of which 6 are in the central region. These time sidelobes are quite high and it is necessary to introduce a sidelobe suppression filter [9], even at the expense of some loss in SNR (Table 5.2 of [9] reports a SNR loss of –1.1 dB).

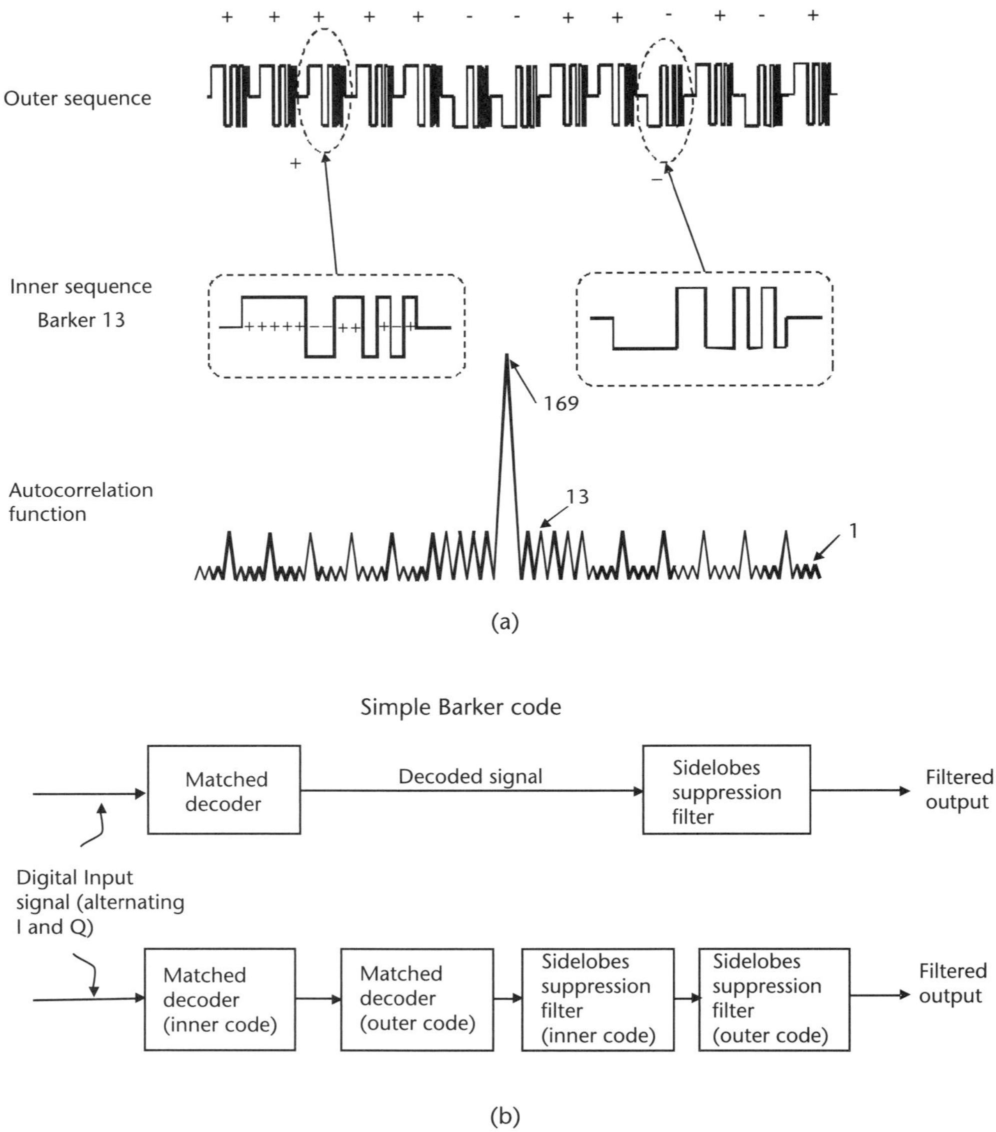

Figure 2.29 Compound Barker code of length 169 and its autocorrelation function.

2.3.6 Surveillance Radars

2.3.6.1 Automatic Detection and Tracking with Surveillance Radars

Tracking with air-surveillance 2D radar with a mechanically rotating antenna is performed at a much lower data rate (i.e., the inverse of revisit time) between observations of the same target than a single target tracker. The latter has revisit times of less than a tenth of second, while the former can have revisit times of from one to many seconds (this is inversely reflected in the measurement tracking accuracy). The advantage of the surveillance radar is that it can track up to many hundred targets within its spatial coverage. Such a large amount of tracked targets requires automatic methods for target detection and coordinate extraction for track initiation and tracking (i.e., track update and filtering) and this is what ADT does.

ADT is possible if the surveillance radar is able to eliminate clutter echoes, so the radar, in addition to Doppler processing (described in Section 2.3.3), must employ CFAR techniques to completely eliminate clutter echoes, even if CFAR techniques introduce some inconvenience (such as some limited detection loss and the suppression of weaker targets in the vicinity of a stronger one).

Automatic Detection

The coverage area (radius = maximum detection range) is quantized in range (range quantization is the equivalent pulse width range $c\tau/2$, called a *range bin*) and in azimuth (usually the beamwidth). At each range-azimuth quantization cell, the pulses received during the TOT are integrated and a detection decision is made. CFAR is incorporated before the decision process. The pulse integration is performed in some sort of automatic integrator, such as the MWI, which examines continuously the last n pulses and announces the presence of a target if at least *m out of n* (secondary threshold) of the pulses exceed the primary threshold. MWI is thus determining an azimuth measurement, whose accuracy depends on the number of pulses available and their SNR. A rough rule of thumb is that the accuracy of azimuth measurement is about $\theta_B/5$ at high SNR (see the discussion of MWIs in Section 2.3.2).

Track Initiation

In principle, a track can be initiated from the above range-azimuth information on two successive scans of the radar. (The first detection initiates the process and the second one allows the computation of velocity and prediction of the third range-azimuth detection.) In practice, however, to prevent false track initiation in the presence of a large number of echoes within the spatial coverage, target information is needed from three or more successive scans. The time elapsed after a detection to initiate a track can be too long for a quick defensive reaction and usually military radar have a look-back beam directed to the angle of the original detection. This can be achieved easily with a rotating phased array radar, or by providing a mechanically rotating radar with an additional look-back beam in order to get a quick second look of the detected target and to initiate a track.

Track Association

When a new target is detected, an attempt is made to associate it with an existing track. This is done by establishing for each track a predicted small area (or gate) within which the target should appear at the next scan. To take into account a possible target maneuver initiated within the scan period, a larger area is also established. When an echo is not found within the nonmaneuvering gate, the larger gate is then searched. Sometimes, due to signal fading, a nonmaneuvering target may not appear in the predicted position, while a further target is appearing in the larger (maneuvering) gate. To avoid the problem caused by a target fade and a false indication in the larger gate, the track can be divided into two tracks (this is known as *bifurcation* of the track): One is the old track with no detection in the smaller gate and the other is the new track based on the target found in the larger gate. After receiving the target detection on the next scan of the radar, a decision is made as to which of the two tracks should be dropped.

Track Smoothing (Filtering)

Two track smoothing methods are available: the α-β tracker and the Kalman filter. The former method is very effective for a linear target's motion (constant velocity vector) and was used in the past because of its relative low computational complexity. The latter filter, although more complex computationally, can inherently provide for maneuvering targets and is the most used currently thanks to the enormous computing capability of modern processors.

The discussion of the above two methods is beyond the scope of this book; we recommend the interested reader refer to [11].

2.3.7 LPI Radars

The low probability of intercept (LPI) requirement for a radar was generated in response to the increased capability of modern intercept receivers to detect and locate a radar emitter. The interception of a radar transmission can quickly lead to it either being jammed or being destroyed by means of precision guided munitions (PGMs) or ARMs [12].

The classical radar concept of using simple high-peak-power pulsed waveforms to achieve long detection ranges and good range resolution therefore turned into the modern radar concept of using:

- Low sidelobe antennas,
- Low-peak-power, CW type of transmission,
- Power management (in the case of stealth applications),
- A wideband modulation technique that achieves good range resolution,
- Frequency agility waveforms with large compression ratios,
- Infrequent antenna scan modulation or use of antennas with a broad non-scanning Tx beam combined with a stationary set of high-gain Rx beams.

All the above features are conceived to make it difficult for an intercept receiver to achieve a detection range advance factor (RAF; see Chapter 3), as well as to classify the radar type and to identify the parameters of its waveform.

The change of waveforms exploits to the maximum degree the dependence of the detection capability of a radar on the energy collected by the receiver as backscattered by the target RCS. Energy is the product of the average transmitted power and the collection (or observation) TOT.

The typical duty cycle (or the ratio of average to peak transmitted power) of a classical coherent pulsed radar is 1% (or less) to 10%, while the typical duty cycle of modulated CW signals as used by modern LPI radars is 100%. This allows the LPI radar to use a substantially lower transmitting power to maintain the same detection capabilities of the classical coherent pulsed radar. LPI radars, therefore, transmit large–bandwidth, periodically modulated CW signals with incrementally variable transmitting periods.

The typical LPI waveforms are of the following types:

- LFM and NLFM,
- Phase modulation (PSK),
- Frequency hopping (FSK),
- Combined PSK/FSK,
- Pseudorandom noise (PRN) modulation.

All these types of modulation have been discussed in previous sections because they are also used in coherent pulsed radars. In the following discussion we will first highlight, according to [13], the differences in the use of the above waveforms in an LPI radar and then discuss, as an example, the FMCW radar.

2.3.7.1 Phase-Coded CW Radar

The block diagram of a phase-coded CW radar is shown in Figure 2.30. The transmitted signal is generated by a direct digital synthesizer (DDS). Various phase modulations can be used from time to time.

The target echo signal is amplified by a LNA, downconverted using a local oscillator (LO), and digitized with an analog-to-digital converter (ADC). The

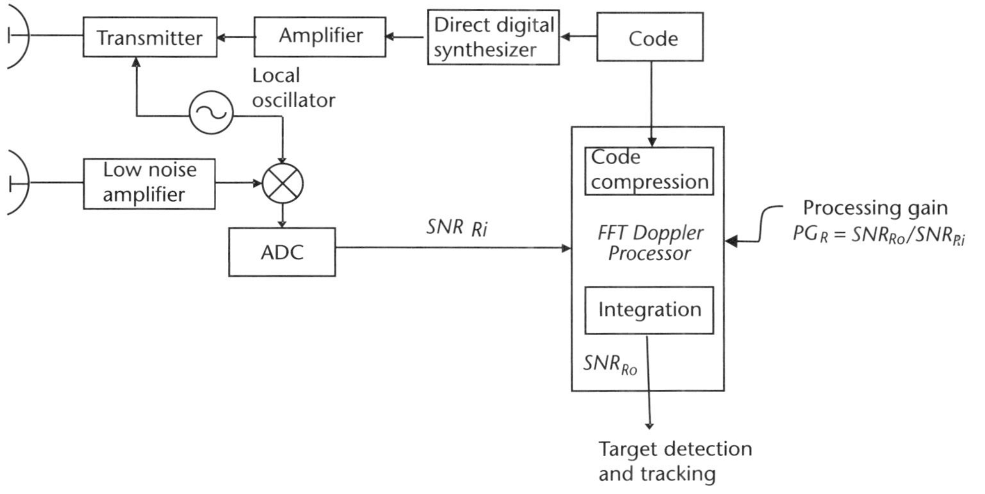

Figure 2.30 Block diagram of a phase-coded CW radar.

digitized samples are then processed by a digital compressor, which cross-correlates the transmitted code with the received signal.

The phase-coded CW waveform of duration T is constituted by a number Nc of subcodes of period t_b, so the cross-correlation achieves a compression ratio (also called processing gain)

$PCR = T/t_b = Nc$. Further noncoherent integration of N codes is also performed by the digital processor to increase the processing gain by a factor of $>N/\sqrt{2}$.The phase-coded CW waveform is shown in Figure 2.31, where $P > N$ indicates a larger number of periods than the noncoherently integrated ones.

All of the LPI waveforms are further coded on transmission by a spreading code, called the reference signal, which is used to avoid detection and identification by noncooperative intercept receivers. The reference signal can take the form of an LFM, a FSK, or a PSK.

Due to the physical constraints, such as the illumination time and the size of the receiver's correlation processor, which is constituted by FFT processors (for the FMCW waveforms) and finite duration coherent correlators (for PSK waveforms), only a portion of the continuously transmitted CW waveform is processed to extract the target information.

The LPI receiver correlates (or compresses) the received signal using the stored reference signal, which has a duration NT in order to perform target detection. So the correlation receiver is the "matched receiver" for portions of the received signal aligned to the reference signal (i.e., same duration as the latter) as shown in Figure 2.32.

The LPI receiver can be modeled as a coherent correlation processor of finite duration NT as shown in Figure 2.32. The echo signal is received by the correlation receiver, which performs a cross-correlation between the received signal and the stored reference signal, which is the complex conjugate of N periods of the transmitted signal with $N < P$. Before correlation the echo signal is first processed by a filter matched to the subcode of duration t_b, which demodulates the subcode and provides a 0 or 1 bit relevant to the binary phase code of the waveform. The demodulated signal is the piped through a tapped delay line with delays $D = t_b$. The signal in the tapped delay line is multiplied by the reference signal, and summed separately for each of the N code periods, thus performing the cross-correlation.

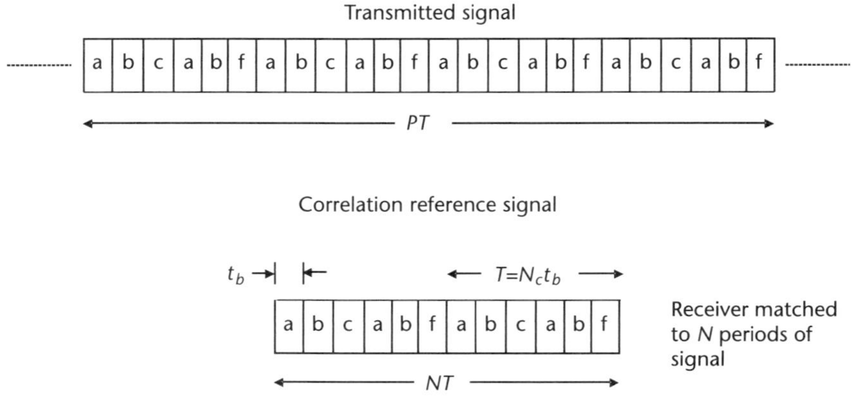

Figure 2.31 CW transmitted waveform and receiver reference signal.

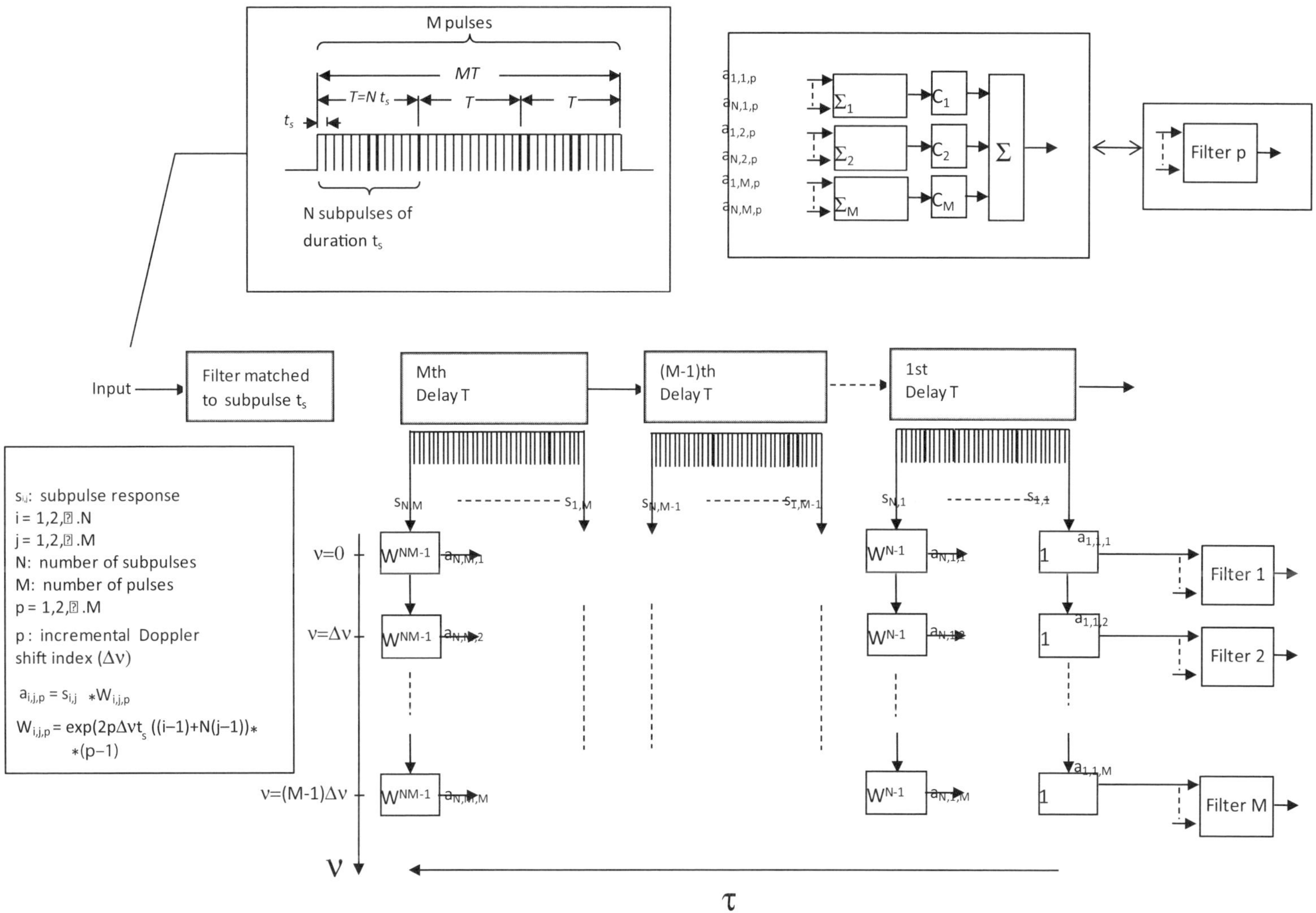

Figure 2.32 LPI receiver coherent correlation processor of finite duration *NT*.

The output of the sum block can then be weighted by coefficients C_1 to C_N. If the coefficients of the first stage are all equal to 1, then the output represents the autocorrelation function, which is the response of the receiver for zero Doppler shift ($f_d = 0$).

The responses of each of the successive $M - 1$ stages of the correlation processor, which constitute a Doppler filter bank, are the responses of the receiver to Doppler shifts $k\Delta f_{d,}$(k = 0, 1, 2, ..., M – 1) as each bit of the echo signal is progressively multiplied at each stage by a weight $q^{i,}$ where $i = 0, 1, 2, \ldots, NM - 1$ and $q = \exp(j2\pi\Delta f_d t_b)$ and then passed to the corresponding element of the successive stage.

In phase-coded CW radar systems, the echo signals with Doppler shift do not correlate perfectly because the Doppler shift changes the phases of the code across its period and this effect has to be analyzed through the ambiguity function. Pace [13] listed the properties of the ambiguity function of the usable LPI waveforms by analyzing for each of them the (1) periodic autocorrelation function (PACF) and (2) periodic ambiguity function (PAF).

Further discussion of these properties is beyond the scope of this book, therefore we recommend the interested reader to refer to [13] for deeper insight into this topic.

2.3.7.2 FMCW Radar

The two essential parameters required from a radar are the detection range (with defined statistical conditions Pd and P_{fa}, usually $Pd \geq 90\%$ and $P_{fa} \leq 10^{-6}$) on a target with a defined RCS and the range resolution. The further target location parameters, the azimuth and elevation angles, are provided by the beamwidth of the receiving antenna.

For a CW radar that uses the same receiving antenna as a pulsed radar to provide the required parameters at the same level of a pulsed radar, it is necessary to transmit in sequence (or simultaneously) at least two CW waveforms at different carrier frequencies f_1 and f_2, separated by $\Delta f = f_1 - f_2$. The phase shifts of the echo signal with respect to the transmitted one at the two frequencies from a target at distance R are, respectively:

$$\phi_1 = \frac{4\pi R f_1}{c} + \varepsilon \text{ and } \phi_2 = \frac{4\pi R f_2}{c} \tag{2.53}$$

Therefore, the range is measured as

$$R = \frac{c(\phi_1 - \phi_2)}{4\pi(f_1 - f_2)} = \frac{c \cdot \Delta\phi}{4\pi \cdot \Delta f} \tag{2.54}$$

Because the phase difference $\Delta\phi$ becomes ambiguous when $\Delta\phi \geq 2\pi$, the selected frequency difference Δf has to be selected in order to get the desired unambiguous detection range R_{un}:

$$R_{un} = \frac{2\pi \cdot c}{4\pi \cdot \Delta f} = \frac{c}{2\Delta f} \tag{2.55}$$

As an example if $\Delta f = 10$ kHz, we have $R_{un} = 15$ km.

If instead of two frequencies a number $n + 1$ frequencies are used, spaced between with Δf (the so-called stepped frequency waveform), where $f_i = f_0 + i\,\Delta f$, with i = 0, 1, 2, …., n, thus covering a bandwidth $\Delta F = f_0 + n\,\Delta f - f_0 = n\,\Delta f$, we can have a range resolution ΔR:

$$\Delta R = \frac{c}{2n\Delta f} = \frac{c}{2\Delta F} \tag{2.56}$$

So, as an example by selecting $\Delta f = 10$ kHz and $\Delta F = 15$ MHz, that is, $n =$ 1,500, we can have $R_{un} = 15$ km and $\Delta R = 10$m.

Another way to achieve a large bandwidth and the accompanying benefits of good range resolution and unambiguity is to use the LFM (such as a sawtooth) CW waveform (FMCW) as depicted in Figure 2.33, which is characterized by a sweep of duration T_m and a frequency excursion ΔF.

In this case the phase of the echo signal is linearly changing with the time:

$$\phi(t) = 4\pi R f_c(t)/c = 4\pi R(f_0 + \Delta F t/(cT_m))/c \tag{2.57}$$

where f_0 is the carrier and ΔF is the LFM swept bandwidth and T_m is the period of the sawtooth and, therefore, produces a frequency as follows:

$$f_b = \frac{1}{2\pi}\frac{d\phi(t)}{dt} = \frac{2R\Delta F}{cT_m} \tag{2.58}$$

which is called beat frequency and is proportional to the range R. Of course, the targets of interest for a radar are not only steady targets for which the measurement of the echo delay is sufficient, but also mobile targets, for which echoes are modulated by the Doppler frequency.

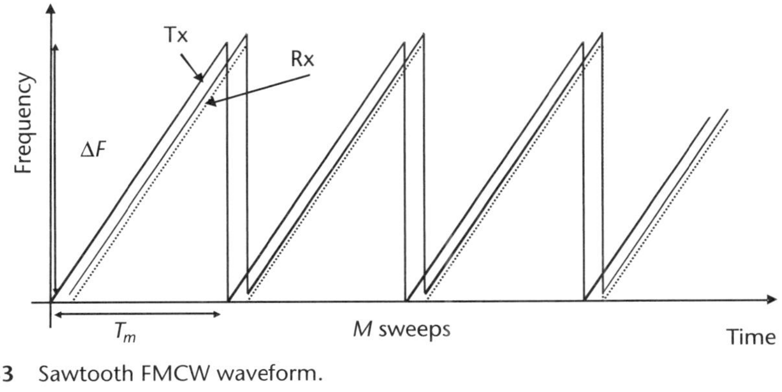

Figure 2.33 Sawtooth FMCW waveform.

In a case where the target is provided with a radial velocity v_r toward the radar, the phase shift of the echo signal with respect to the transmitted one is:

$$\phi(t) = \frac{4\pi(R - v_r t)f_c(t)}{c} \tag{2.59}$$

and hence the beat frequency f_b' becomes:

$$f_b' = \frac{1}{2\pi}\frac{d\phi(t)}{dt} = \left[f_b - \frac{2v_r f_0}{c} - \frac{4v_r \Delta F t}{c^2 T_m} \right] \approx f_b - f_d \tag{2.60}$$

that is, the steady target beat frequency less the usual Doppler frequency shift $f_d = 2vrf_0/c$, because the third term within parentheses in negligible. The Doppler shift unambiguous interval is $f_{d\max} = 1/T_m$ and there is no range migration. Its resolution over M sweeps is $\Delta f_d = 1/(MT_m)$.

Another useful FMCW waveform is the triangular waveform shown in Figure 2.34, in which the two beat frequencies for mobile targets are

$$f_{b,up} = \frac{2R\Delta F}{cT_m} - \frac{2v_r f_0}{c} \text{ and } f_{b,down} = \frac{2R\Delta F}{cT_m} + \frac{2v_r f_0}{c} \tag{2.61}$$

By summing and subtracting the two expressions, we obtain:

$$\frac{f_{b,up} + f_{b,down}}{2} = \frac{2R\Delta F}{cT_m} \text{ i.e., } R = \frac{cT_m}{\Delta F} \cdot \frac{f_{b,up} + f_{b,down}}{2}$$

$$\frac{f_{b,down} - f_{b,up}}{2} = \frac{2v_r f_0}{c} = f_d \tag{2.62}$$

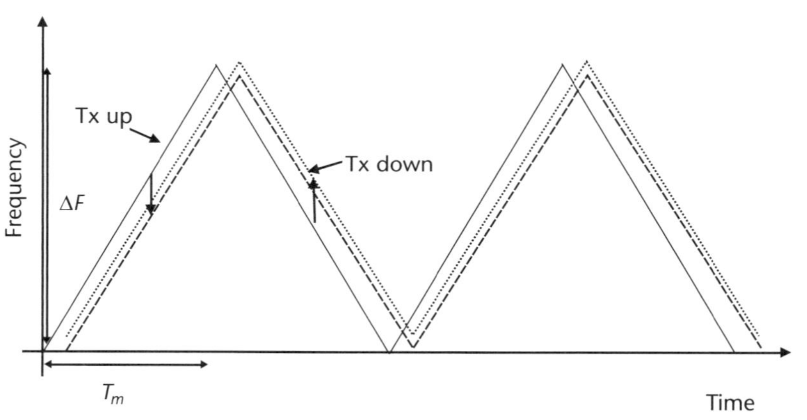

Figure 2.34 Triangular FMCW waveform.

thus providing a range-independent measure of the radial velocity.

To evaluate the angular coordinate of the target, a number of solutions can be adopted depending on whether two separate antennas are employed for transmission and reception (in case the power coupling is too high and can impair reception) or only one antenna is used for transmission and reception. In the first case the Tx antenna could be (1) either a directive (pencil beam) or a (2) floodlight antenna. In this case the Rx antenna should be of the same type and aligned to the Tx one when scanning over its required spatial coverage. In this second case, the Rx antenna could be either a directive (pencil beam) antenna or a couple of broad beam antennas providing the measurement angle through the monopulse technique (see Section 2.8) based on either amplitude or interferometry (of course in the latter case taking into account that the distance between the two antenna phase centers is $\leq\lambda/2$) and that the receivers should be amplitude and phase matched. The block diagram of the FMCW radar is shown in Figure 2.35.

The processing gain of a radar is defined as PGR = SNR_{out}/SNR_{in}. In the case of an FMCW radar, the PGR = ΔFT_m. The simplest FMCW radar receiver is the homodyne receiver, whose block diagram is depicted in Figure 2.36, in which a single antenna is used for Tx and Rx.

The appealing concept of FMCW radar had in the past a number of technological problems that have since been solved, such as the following:

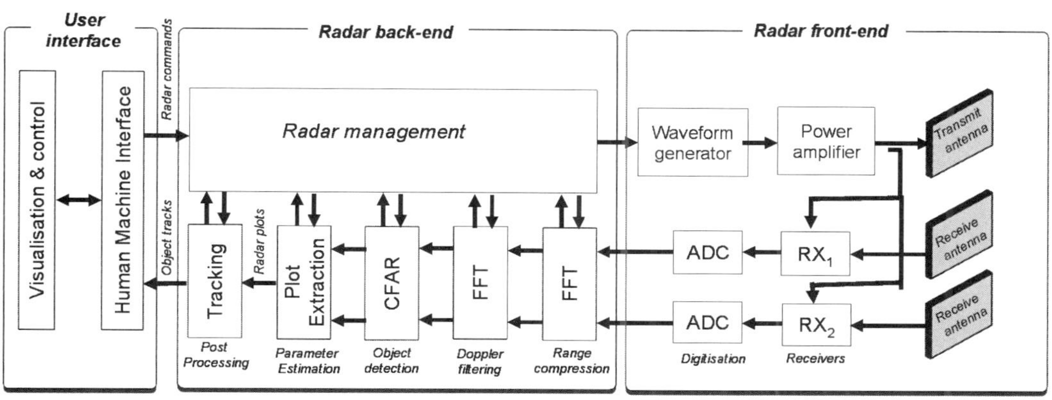

Figure 2.35 Block diagram of the FMCW radar.

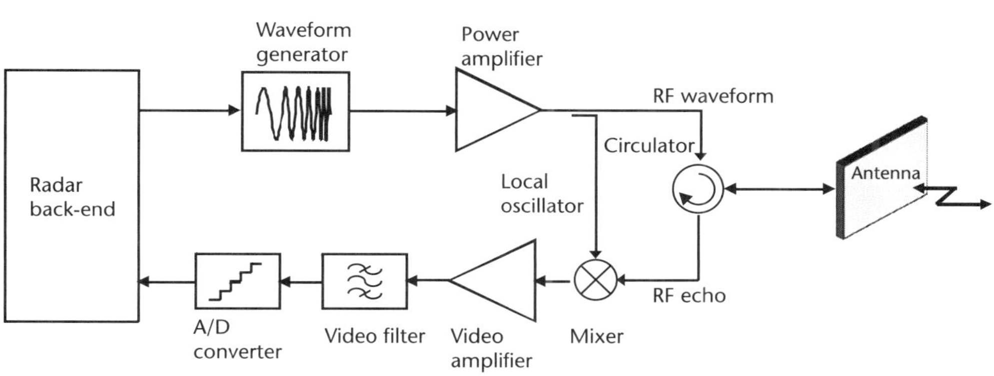

Figure 2.36 Block diagram of a homodyne receiver.

- Nonlinearity of frequency sweeps generated by voltage controlled oscillators (VCOs) which produced a nonconstant beat frequency for a target at constant range and therefore also a degraded range resolution. The new technology of a DDS-based open-loop and closed-loop frequency generator solved this problem.
- Single-antenna homodyne receivers were affected by power leakages (i.e., reflected power at the antenna) at the circulator and mixer stages, which reduced the receiver sensitivity. Currently a reflective power canceller, whose block diagram is shown in Figure 2.37, is available that provides better than 30-dB cancellation.
- A separate Tx and Rx antenna with increased isolation (up to 60 dB) so as to allow the introduction of a LNA in the Rx path.

To complete the discussion on FMCW radar, let us make a performance comparison between a coherent pulse radar and a FMCW radar based on two distinctive parameters: detection range and range resolution.

The detection range equations of the two radars in terms of single-pulse SNR_{out} are as follows:

$$SNR_{pulse} = P_p \tau \cdot \frac{G^2 \lambda^2 \sigma}{(4\pi)^3 LFkT_0 R^4} \text{ and } SNR_{FMCW} = P_{av} T_m \cdot \frac{G^2 \lambda^2 \sigma}{(4\pi)^3 LFkT_0 R^4}$$

Then let us assume that the second factor is equal in both expressions. The equalization of $SNR_{pulse} = SNR_{FMCW}$ is therefore simply expressing that the detection capability is only dependent on the transmitted energy by the two radars; that is:

$$P_p \tau = P_{av} T_m \tag{2.63}$$

Let us now equate the two range resolutions, which are

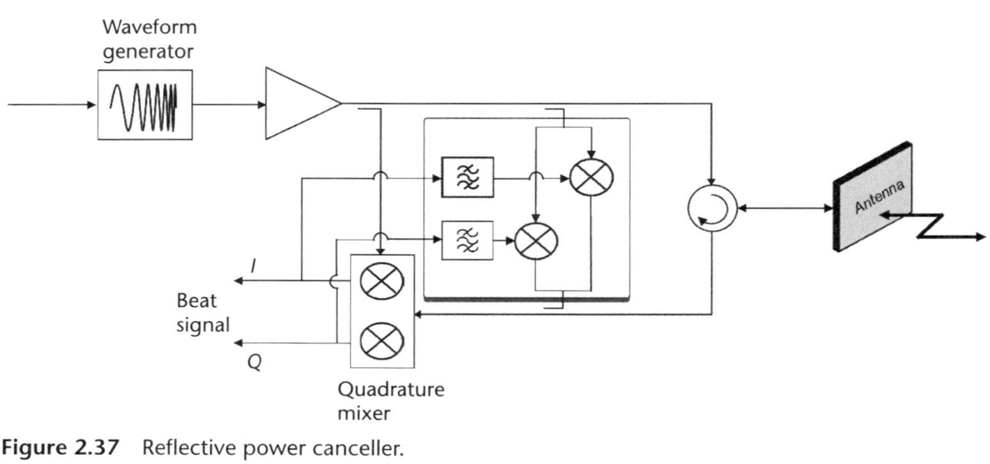

Figure 2.37 Reflective power canceller.

$$\Delta R_{pulse} = c\tau/2 \text{ and } \Delta R_{FMCW} = c/2\Delta F$$

and we have

$$\tau = 1/\Delta F$$

By substituting τ in (2.63) we have the ratio of the pulse radar peak power to the average (or CW) power of the FMCW radar :

$$\frac{P_p}{P_{av}} = \Delta F \cdot T_m \tag{2.64}$$

Because the two parameters ΔF and T_m can be selected independently in the FMCW waveform, we can select a normal value such as $T_m = 1$ ms and $\Delta F = 20$ MHz (to achieve a range resolution of 7.5m with $\tau = 0.05$ μs). The get the product $T_m\Delta F = 2 \times 10^{-3} \times 10^7 = 2 \times 104 = 43$ dB. This means that if we assume that the pulse radar needs a peak power $Pp = 100$ kW = 80 dBm to achieve a given detection range R, the FMCW radar needs only an average power $Pav = 37$ dBm = 5W.

The big advantage of the FMCW radar is with respect to it intercept receiver, because if we assume that the pulse radar is intercepted at a distance $R0$, the FMCW will be intercepted at a distance $R'/R0 = (P_{av}/P_p)^{1/2} = -21.5$ dB $= 0.707 \times 10^{-2}$.

Table 2.7 compares the features of the FMCW and pulse/pulse Doppler radars.

2.3.8 Pulse Doppler Radars

In Section 2.3 when we discussed MTI Doppler processing, which is largely used in ground surveillance radars operating at relatively low frequencies, we showed that the PRF, which limits the nonambiguous Doppler shift range, was mainly selected to avoid range ambiguities and hence tolerates some blind speeds (which were then removed through the PRF stagger). Also when discussing the AMTI, we discussed the need to not only compensate for the clutter not being any greater at

Table 2.7 Comparison of the Features of FMCW and Pulse/Pulse Doppler Radars

FMCW Radar	*Pulse/Pulse Doppler Radar*
Simultaneous Tx and Rx	Sequential Tx and Rx
No eclipsing (blind range intervals) at *n* PRI	Eclipsing at *n* PRI
High impact of Tx signal stability on detection performance	Low impact
High impact of large short-range scatterers	Little impact
Short detection range	Long detection range
Low peak power	High peak power
Short intercept range	Very long intercept range
Low interference with on-board RF systems	High interference
Homodyne transceiver	Heterodyne transceiver
Small dimension and low power consumption	Large dimension and high power consumption
High range resolution with low sampling rate	High range resolution with high sampling rate

zero Doppler shift but also to compensate for the widening of the clutter spectra due to the antenna scanning during the platform motion.

For the case of high-speed aircraft, such as a fighter A/C, where the need for a compact antenna compels the use of high frequencies (around 9 GHz), we can immediately see that the PRF must be high in order not to reduce the Doppler shift range too much. The increased PRF immediately bears range ambiguities. Thus, there are three different types of pulse radars that use Doppler shift processing:

- The MTI radar operating at low PRF with no range ambiguities and many Doppler ambiguities;
- The high PRF (HPRF) PD with just the opposite: no Doppler ambiguities and many range ambiguities;
- The medium PRF (MPRF) PD with some range and Doppler ambiguities.

In the past there were also differences in the structures of the three radar types, but current technology allows the use of a similar structure in which the differences reside in the transmitted waveforms and in the higher clutter power content in the PD radars (because of the foldover in the range of clutter echoes when range-ambiguous PRFs are employed).

2.3.8.1 HPRF PD

Let us consider the geometry of an airborne radar when the A/C is flying straight and level (i.e., constant speed v_a and altitude h_a) as shown in Figure 2.38. As the radar transmits a train of pulses of duration τ at the PRF = f_p, it has a line spectrum (with envelope of the type sin $\pi f\tau/\pi f\tau$ centered around the carrier frequency f_0) with the separation between the spectral lines equal to the PRF. The clutter spectrum

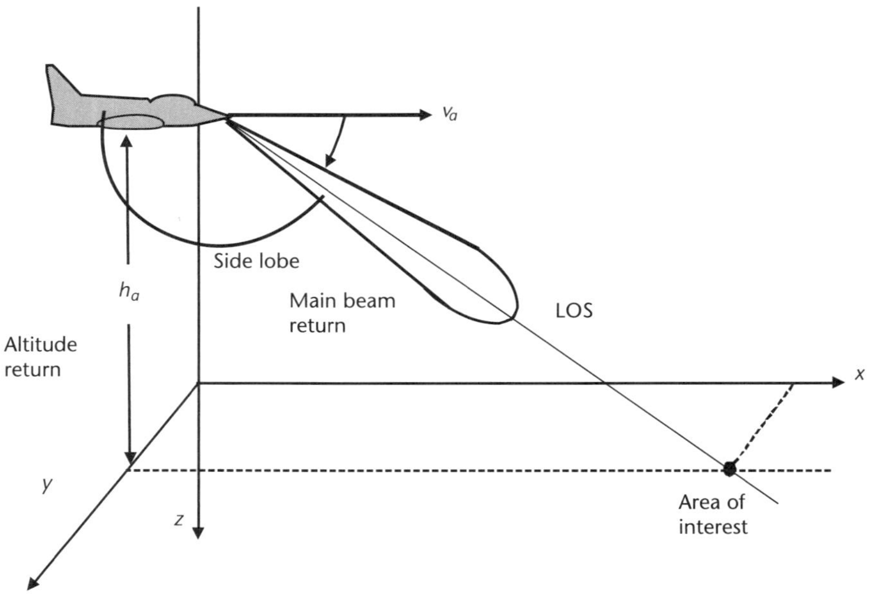

Figure 2.38 Geometry of airborne radar.

around the carrier frequency is shown in Figure 2.39. At the carrier frequency f_0 there is a large return due to echoes received directly from below (altitude return), which is at zero relative speed to the aircraft. Because of the aliasing, the altitude line is repeated, (as well as any other contribution is at frequencies), $f_0 \pm nf_p$ (with n integer).

The antenna sidelobes illuminate the clutter over a large number of incidence angles (from 0° to 90°), so that clutter echoes can extend almost for $\pm 2v_r/\lambda$ around the carrier frequency. For convenience , the shape of the sidelobe clutter is shown as being nearly uniform, but in reality it is not uniform.

The sidelobe clutter in a HPRF PD is large because there are many range-ambiguous pulses simultaneously illuminating the clutter. To detect A/C targets within the sidelobe clutter, a bank of narrowband Doppler filters with adaptive thresholds can be used. To keep the sidelobe clutter sufficiently limited, the airborne radar antenna must have exceptionally low antenna sidelobes (on the order of better than –40 dB). The clutter echo seen by the main antenna beam can be large in amplitude and it lies somewhere within the sidelobe clutter region. Because the antenna scans in angles, the main beam center frequency varies.

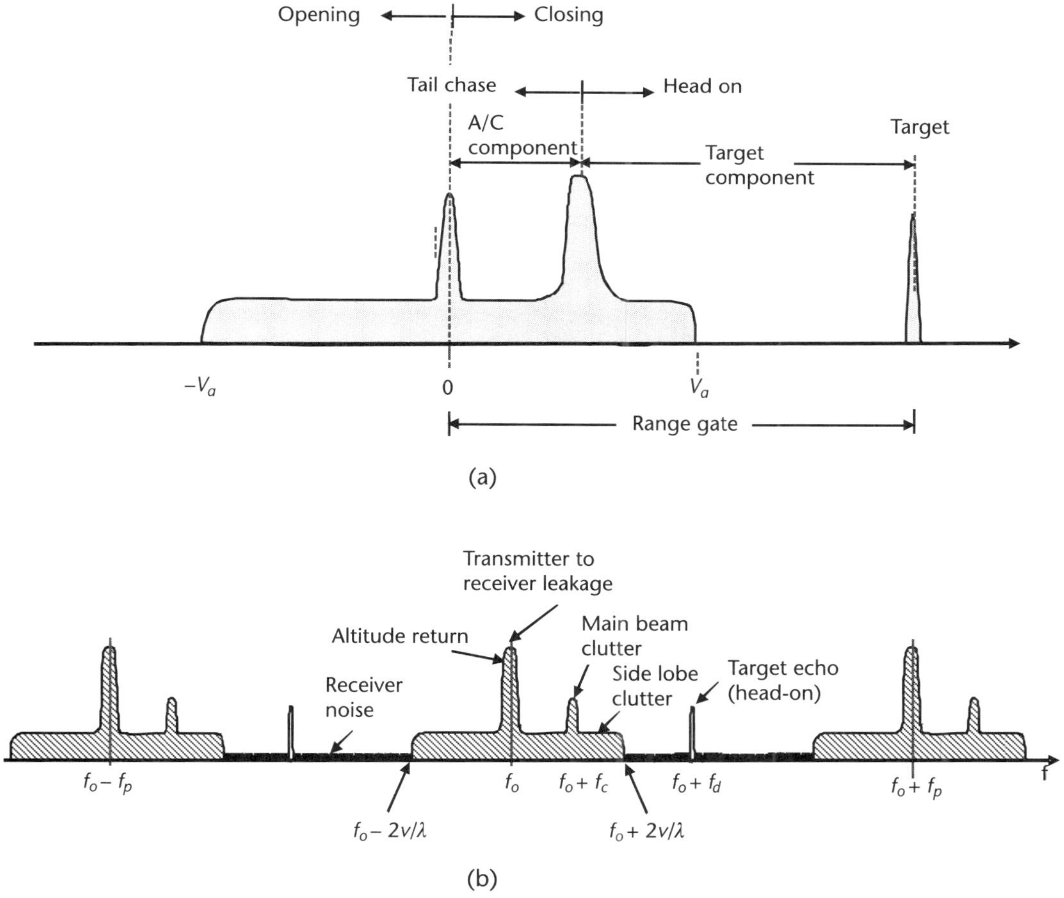

Figure 2.39 (a) CW radar clutter spectrum. (b) Enlarged view of clutter spectrum around the carrier frequency of a HPRF airborne radar. (Reproduced with permission from G. Morris [14].)

The presence of a clutter-free region is an important advantage of HPRF radars, because this region corresponds to viewing nose-on approaching targets (especially if detection of high-speed closing targets at long ranges is desired). If the relative velocity of the target is low (such as when the targets are viewed tail-on or when the target is on a crossing trajectory), the echo can fall within the sidelobe clutter region and its detectability is much less than that of a high-speed target in the clutter-free region.

The PRF of an X-band (around 9 GHz) HPRF PD radar for military use ranges from 150 to 300 kHz. With such a high PRF, the pulse width PW is usually short and in the order of 0.3 to 0.5 μs. By assuming a PRF of 200 kHz (i.e., T_p = 1/PRF = 5 μs) we have a duty $d = \text{PW}/T_p = 10\%$. The PRF is selected on the basis of the maximum ground speed of the target and radar platform. In fact, by inspection of Figure 2.39 if we want the target be in the clutter-free region its Doppler shift should be at least $2(v_A + v_T)/\lambda$ far from f_0 and $2v_A/\lambda$ far from $f_0 + f_p$; thus $f_p > 4v_A/\lambda + 2v_T/\lambda$, where v_A is the radar platform speed and v_T is the target speed.

Eclipsing Loss

Because the pulse Doppler radar cannot receive when it is transmitting, the high duty cycle can result in an eclipsing loss if the echo signal arrives when a pulse is being radiated and the receiver is gated off. This effect produces holes in the range coverage, which are by their nature temporary because the relative motion of the target and platform will make the target appear after a short time (on the order of few hundreds milliseconds) at a shorter range. Eclipsing losses longer than 1 sec are not acceptable during tracking of a target; therefore, several PRFs are provided to reduce the effects of the eclipses.

2.3.8.2 MPRF PD

The MPRF mode has become one of the most important modes for modern airborne PD radar, primarily because of the capability it provides to detect slow closing rate targets from high-speed platforms.

The equations for unambiguous velocity ($v_{un} = \lambda\text{PRF}/2$) and unambiguous range ($R_{un} = c/2\text{PRF}$) are both functions of the PRF and therefore cannot be independently chosen. However, they are related at the X-band (λ = 3.3 cm) by the relation $v_{un}\, R_{un} = \lambda c/4 = 2.5$ km^2/s, which represents a hyperbola in the plane v_{un}, R_{un} and is shown in Figure 2.40.

Note from the curve of Figure 2.40 that between the two operating points that define, respectively, the maximum value of PRF that permits LPRF operations ($\text{PRF}_L \leq 2$ kHz) and the minimum PRF that permits HPRF operations ($\text{PRF}_H \geq 150$ kHz) there is the domain of MPRF operations, which are characterized by ambiguities in both range and velocity. The method for resolving both ambiguities employs the range-Doppler space relevant to a PRF and the use several PRFs in a dwell and it is drawn from [14], to whom we address the interested reader for deeper insight into the matter.

The criterion for resolving range and Doppler ambiguities of a target detected in a clear region of the range-Doppler space is that the target must be detected on at least two PRFs to resolve the range ambiguity. Usually the radars are designed to

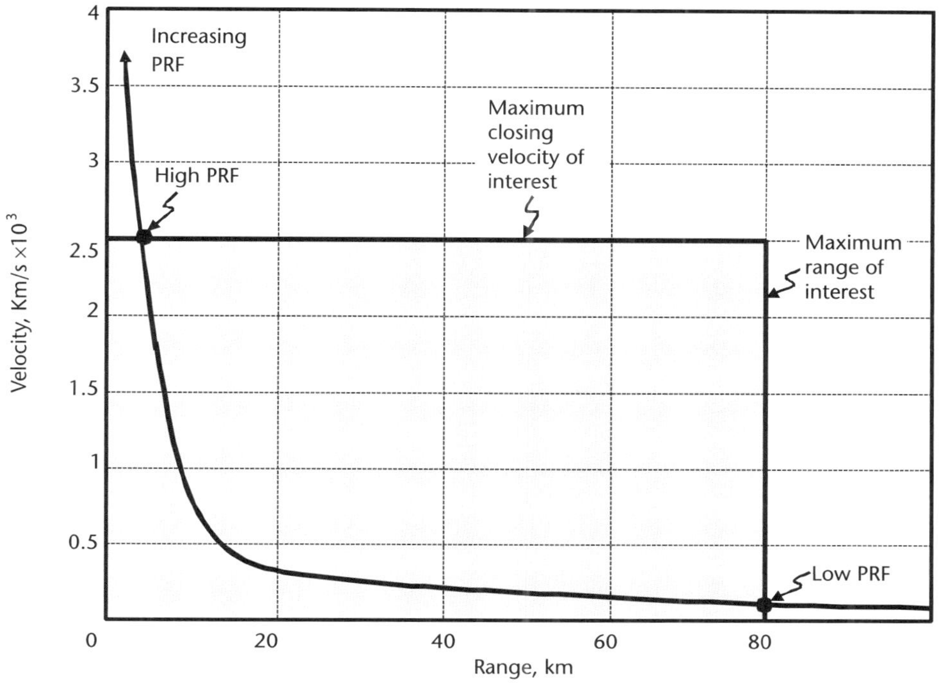

Figure 2.40 Unambiguous range vs. unambiguous velocity for air-to-air radars. (Reproduced with permission from G. Morris [14].)

require detection on three PRFs to ensure proper range ambiguity resolution in the case for which two targets at different ranges may produce some ghosts (this will be shown later). Targets detected on fewer than three PRFs are discarded.

In Figure 2.41 the blind zones for two PRFs are represented. The unambiguous range is inversely proportional to the PRF as $R_{un} = c/2\text{PRF}$. The blind region in the range dimension corresponds to the transmitted pulse width ($R_{\text{blind}} = cPW/2$) when the receiver is turned off. The unambiguous Doppler frequency f_{un} is equal to the PRF. The blind region shown is the area in which main-beam clutter is expected, because it is here assumed (as used in early designs) that the carrier frequency is locked to the main-beam clutter frequency and a notch filter of width equal to the maximum width[5] of the main-beam clutter is centered at the locked frequency. Currently MPRF signal processors translate the spectrum to center main-beam clutter at zero frequency. Target returns falling within the first shaded area are detected unambiguously. Target detections falling within all of the shaded areas of the range-Doppler domain are ambiguous in range or Doppler or both.

Range ambiguities are solved with the following considerations on the Figure 2.42, which shows the target returns respectively at two PRFs. The true delay time of the target is T_t, and T_{a1} and T_{a2} are the apparent delays at PRF_1 and PRF_2, respectively. The true delay T_t can be computed as a function of $T_1 = 1/\text{PRF}_1$ and $T_2 = 1/\text{PRF}_2$ in that:

5. The spectral width of main-beam clutter varies with the speed of the aircraft and with the radar pointing angle.

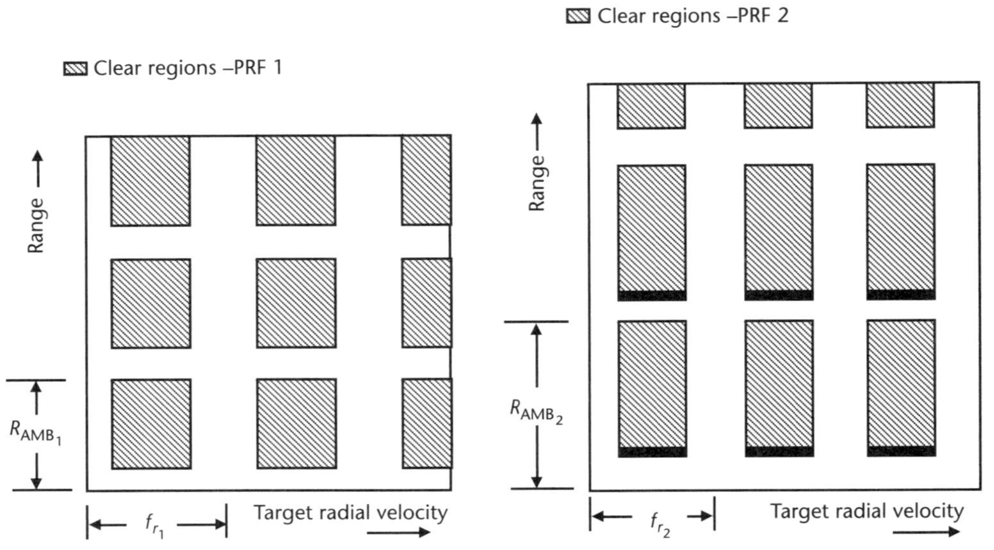

Figure 2.41 Range-Doppler domains at two PRFs. (Reproduced with permission from G. Morris [14].)

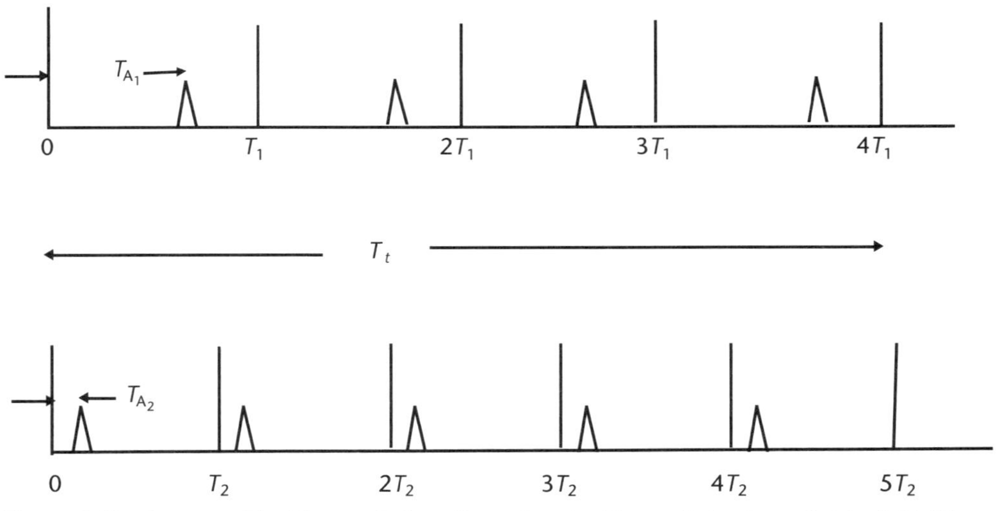

Figure 2.42 Range-ambiguities resolution. (Reproduced with permission from G. Morris [14].)

$$T_t = nT_1 + T_{a1} = mT_2 + T_{a2} \tag{2.65}$$

Ideally the equations can be solved by finding the proper couple of values m and n. In practice, the delay measurements are corrupted by interference, quantization errors, and noise, and no exact solution of m and n is possible. Rather a value of T_t that minimizes the difference is selected.

A simple algorithm for determining T_t is as follows:

1. Compute trial values of T_t using (2.65) for increasing m and n.

2. The true value of T_t is the one that produces the minimum error by both equations of the set.

This simple algorithm is implemented in the resolving digital circuit depicted in Figure 2.43.

Usually the PRFs are selected in accordance with the following criterion:

1. Select R_{us}, the desired system unambiguous range after range resolving.
2. Determine p PRFs, which are multiples of $PRF_0 = 1/2\ R_{us}$; that is, $PRF_1 = K\ PRF_0$, $PRF_2 = (K+1)\ PRF_0$, and so on until $PRF_p = (K + p)\ PRF_0$.

As mentioned, the resolution of range ambiguities requires the use of three PRFs, and the reason is the resolution of ghost targets, which are still possible from the detections of two targets in a two-PRF scheme.

Doppler ambiguities are resolved with the same method.

MPRF PD Search Mode

Because the target is generally ambiguous in both range and Doppler domains on any single PRF, several PRFs (usually seven to nine) are transmitted during a single antenna beam dwell on the target. Target detection on each PRF dwell is performed with use of a CFAR circuit. After resolution of range and Doppler ambiguities as shown previously, only targets for which an unambiguous range has been determined (upon detection by three or more PRFs) are declared detections and displayed. A typical display is shown in Figure 2.44 in which the vertical axis scale is the unambiguous range R_{us}. A typical MPRF radar search mode is hereafter reported [14]. A set of N PRFS to be transmitted, each comprising $>2^n$ valid pulses for Doppler analysis, is selected in order to reduce as far as possible the blind zones of the range-Doppler domains and to achieve an unambiguous system range R_{us}. The round-trip propagation time is $T_p = 2R_{us}/c$ (ms) and therefore $PRF_0 = 1/Tp$

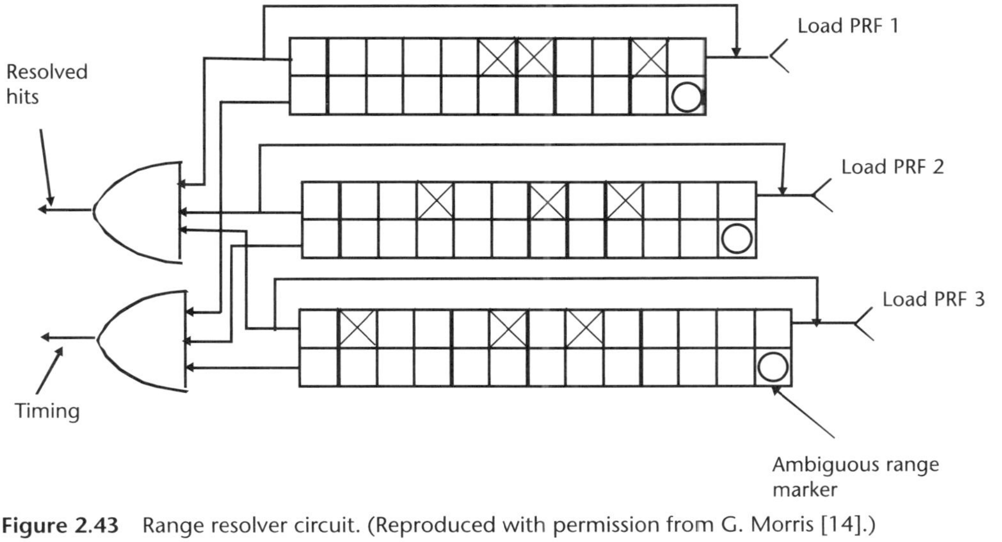

Figure 2.43 Range resolver circuit. (Reproduced with permission from G. Morris [14].)

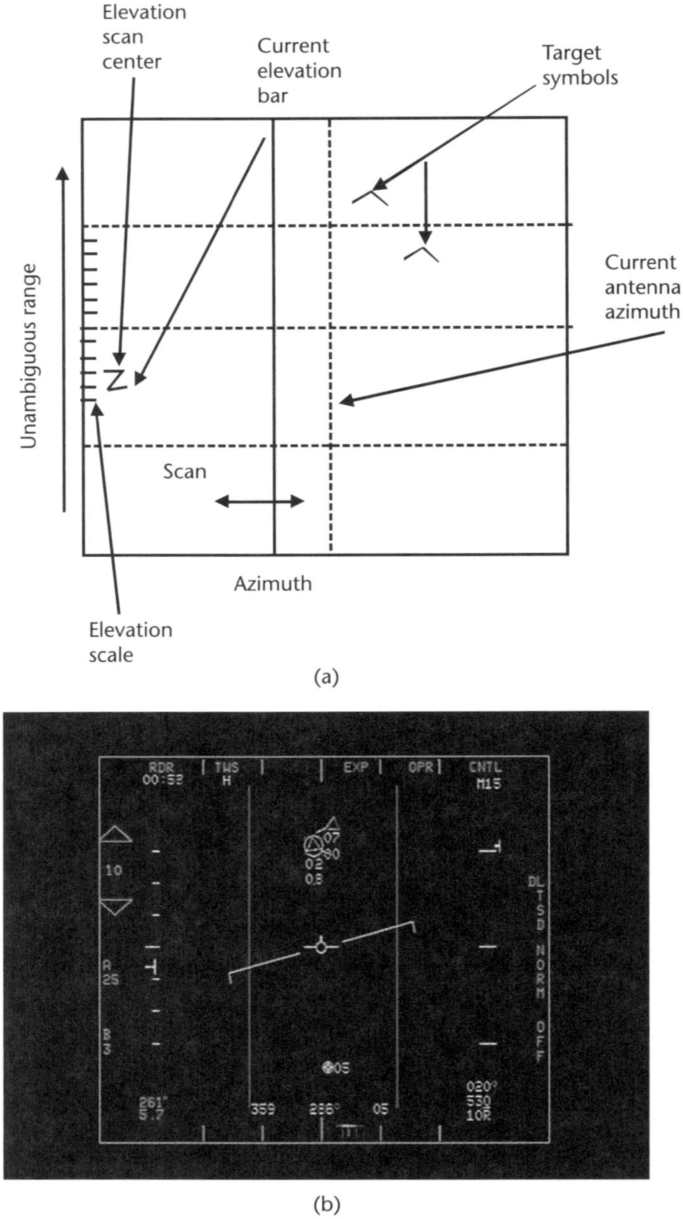

Figure 2.44 (a) Typical MPRF PD B-scan display. (b) GRIFO radar B-scan display. (Courtesy of Selex Galileo S.p.A.)

(kHz). So the N PRFs can be selected in order to span a bandwidth ranging from PRF_1 to PRF_N at step $1/Tp$ kHz. The overall waveform timing during a beam dwell in one direction is reported in Figure 2.45 for the case $N = 8$ and beam dwell time $T_D = 40$ ms.

The typical MPRF PD radar receiver processing is constituted by a bank of range bins in order to cover $T_{p,\max}$: Each range bin is provided with a bank of Doppler filters. The block diagram for MPRF PD radar receiver processing is shown in Figure 2.46. A picture of the GRIFO MPRF PD radar is shown in Figure 2.47.

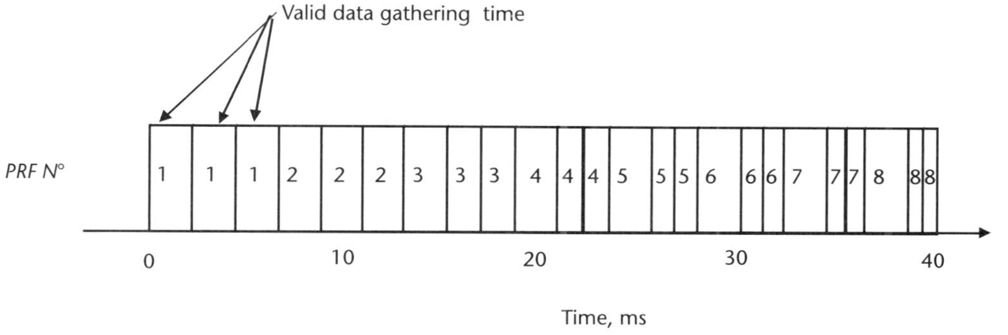

Figure 2.45 MPRF PD typical search mode PRF program for an antenna dwell. (Reproduced with permission from G. Morris [14].)

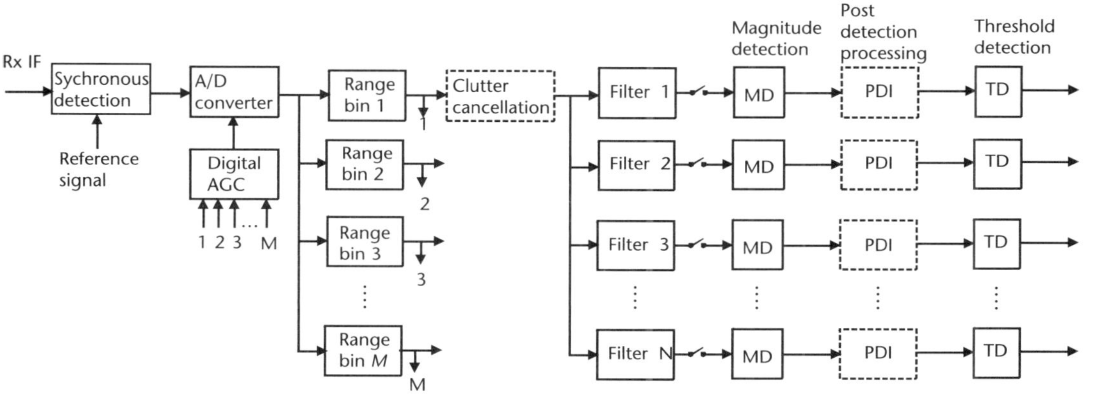

Figure 2.46 MPRF PD radar receiver processing block diagram. (Reproduced with permission from G. Morris [14].)

In conclusion, Table 2.8 lists the principal characteristics of the three PRF modes of airborne PD radars, which can be performed by a multimode radar.

2.3.9 Tracking Radars

The tracking radar is an essential component of weapons systems for its ability to provide accurate target location information in range, azimuth, and elevation. Its continuous observation of the target over time can provide the target's trajectory or *track* and predict where it will be next. This information is used by the weapons systems for precision aiming of guns and for missile guidance.

At least four types of radars can provide tracks of targets:

1. Single target tracker (STT),
2. Automatic detection and tracking (ADT),
3. Track-while-scan (TWS),
4. Phased array tracking.

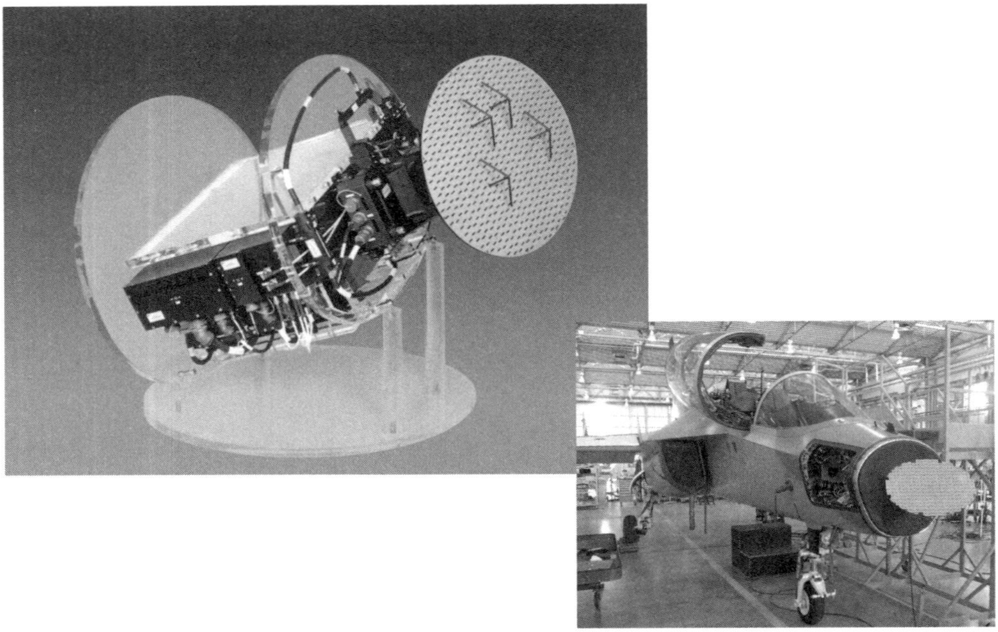

Figure 2.47 Pictures of two configurations of GRIFO MPRF PD radar. (Courtesy of Selex Galileo S.p.A.)

Table 2.8 Principal Characteristics of Airborne PD Radars

Type of Mode	*Typical System Function*	*Benefits*	*Limitations*
	High PRF Radar Modes		
Coherent	Velocity search (VS)	Low peak power	Highly ambiguous range
	Range while search (RWS)	Unambiguous Doppler	Eclipsing loss
	Range gated HPRF	Good look-down	Inaccurate ranging
		Illumination for semi-active missiles	Sidelobe clutter reduces tail-aspect detection
Type of Mode	*Medium PRF Radar Modes*	*Benefits*	*Limitations*
Coherent	Low-altitude tail-chase	Better than HPRF	Many PRFs and PWs
	A number of functions of HPRF and LPRF	Accurate ranging	Complex processing to solve range and Doppler ambiguities
Type of Mode	*Low PRF Radar Modes*	*Benefits*	*Limitations*
Coherent Doppler	Airborne/ground moving target detection	Precise range	Low P_{det} and high P_{fa} in look-down
Doppler beam sharpening	Improved resolution ground map for navigation	Ground maps	High peak power or PCR
Synthetic aperture	Stationary target detection	Long unambiguous range	Highly ambiguous Doppler
Noncoherent Ground map	Navigation	Range gating to reject sidelobe clutter	—
Terrain avoidance	Covert navigation	Simple data processing	—
Air-to-air ranging	Short-range gun and missile attack	—	—
Air-to-ground ranging	Bomb delivery	—	—
Terrain following	Covert navigation	—	—

STT radars are designated by a search radar that has detected the presence of a threat and designated it with approximate coordinates (at least range and azimuth). Upon target designation the STT performs a rapid search around the designated target's location to acquire the target within its narrow beamwidth. If the designation is 3-D, the STT can almost immediately acquire the target; otherwise, it performs a search in elevation and over a narrow azimuth sector. Figure 2.48 shows a typical STT acquisition scan.

As soon as the target has been acquired, the STT passes into tracking mode and provides the associated weapon system with the target's precise location coordinates. The required precision of an STT is on the order of 0.2 to 0.5 mr (rms) in angles and 1 to 5m in range. The smaller figures apply to gun's fire control systems.

STT types differ in angular precision, complexity, and cost and are listed here in increasing order:

1. Conical scan,
2. Lobe switching,
3. COSRO (conical scan on receive only),
4. LORO (lobe switching on receive only),
5. Monopulse.

STT types 1 through 4 use sequentially a single time-shared beam in at least four positions (more for types 1 and 3) to obtain an angle measurement in two coordinates. Type 5 (monopulse) uses the equivalent of four simultaneous beams to perform 2-D angle tracking.

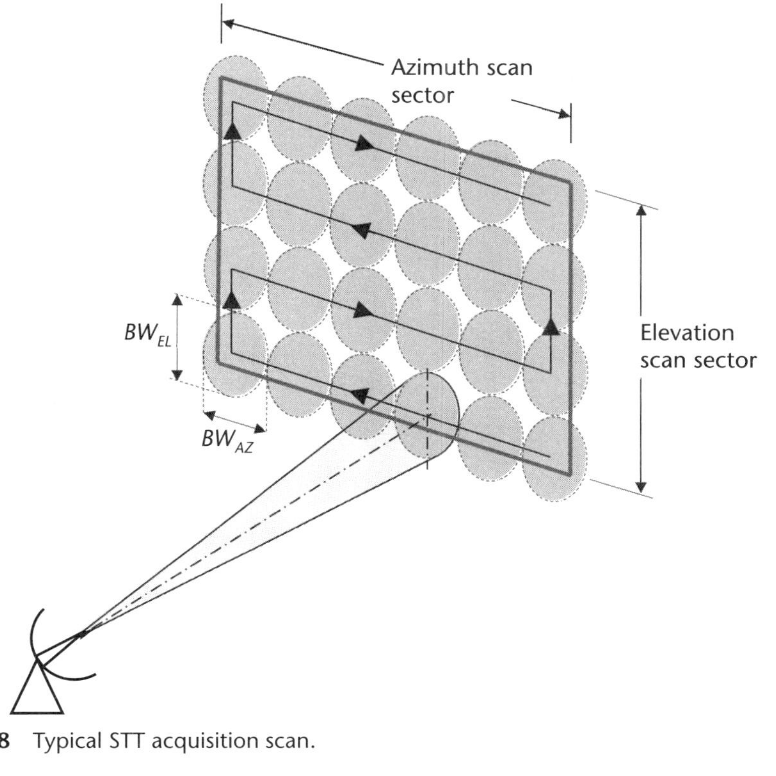

Figure 2.48 Typical STT acquisition scan.

2.3.9.1 Conical Scan

In this type of radar, the beam axis is not aligned with the antenna boresight, but nutates around it, describing a cone of width $2\theta_q$ (where θ_q is called the squint angle) as shown in Figure 2.49. A target along the axis of the cone will be illuminated by a constant gain, even if it is not the maximum, while targets located off-boresight will be illuminated by a gain that is modulated at the conical scan frequency $f_s = \Omega/2\pi$ as shown in Figure 2.50.

The depth of modulation and phase of the maximum with respect to two references (azimuth and elevation) provide the amplitude and the direction (in azimuth and elevation) of the angular error $\varepsilon(t)$, whose expression is:

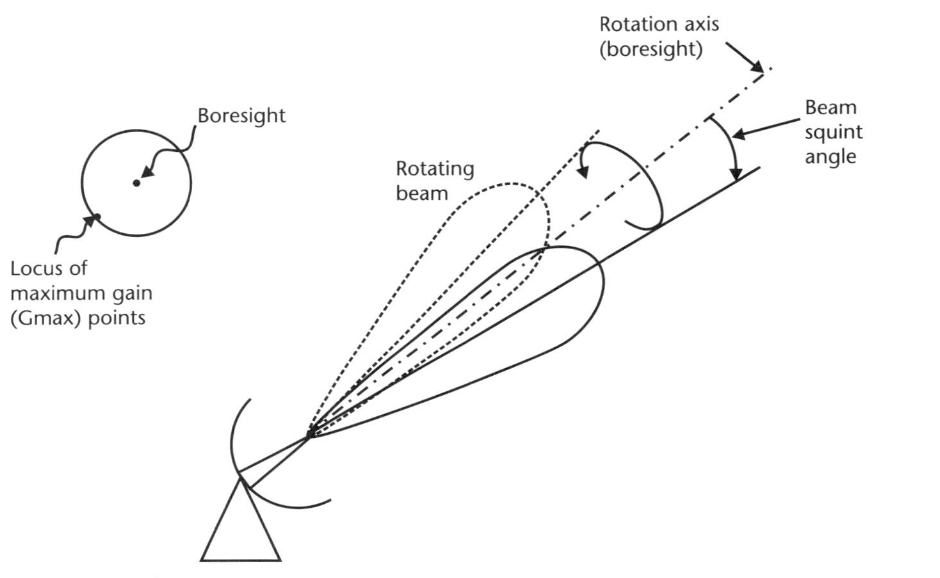

Figure 2.49 Conical scan antenna.

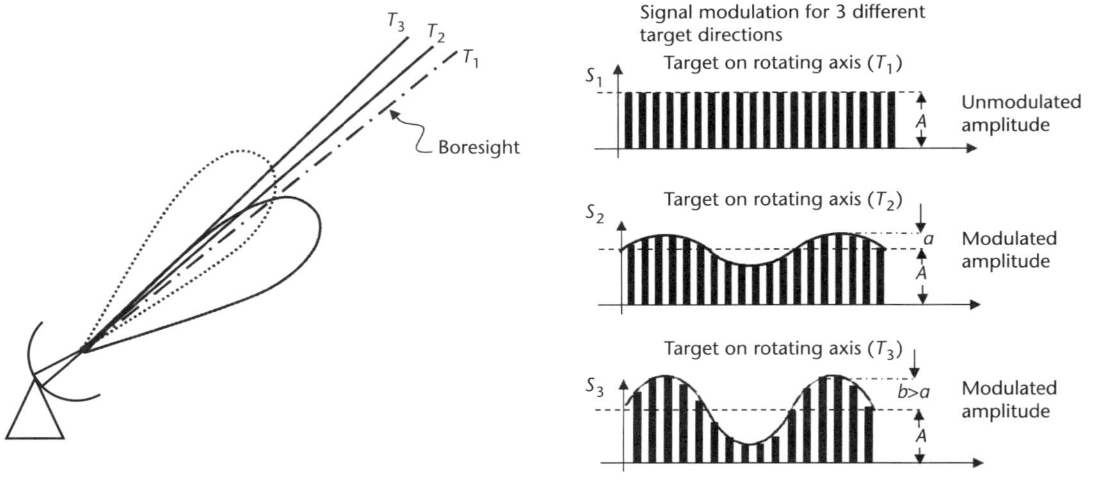

Figure 2.50 Conical scan angle error modulation.

$$\varepsilon(t) = k_s \Delta\theta_t \sin(2\pi f_s t + \phi) \quad (2.66)$$

The depth of modulation $k_s\Delta\theta_t$ depends on the ratio θ_q/θ_B, where θ_q is the squint angle and θ_B is the 3-dB antenna beamwidth as shown in Figure 2.51. The selection of the squint angle θ_q is a trade-off between the increase in the angular sensitivity k_s and the gain losses, which reduce the SNR and hence the detection capability. Usually θ_q is selected to be between 0.3 and 0.4 θ_B.

When in tracking mode the STT radar places a gate on the range axis of the receiver so as to allow output signals only within this gate of width approximately equal to the pulse width. The signal amplitude within the gate is controlled by an automatic gain control (AGC) loop. The gate is then automatically positioned at the target range by means of an automatic range-tracking closed loop, whose input is provided by the *split-gate* discriminator operating on the output video signal. This discriminator splits the video signal into two gates: an *early gate* and a *late gate,* where the two signals are integrated and then subtracted to produce the difference error signal as shown in Figure 2.52. The sign of the difference indicates the direction the two gates have to be moved in order to have the pair straddle the echo. The amplitude of the difference determines how far the pair of gates is from the center of the pulse (sometime called the *centroid*). When the error signal is zero, the two range gates are centered on the pulse and the position of the gates gives the target's range. The automatic range tracking loop is usually a second (or higher) order loop capable of range tracking in range maneuvering targets with accelerations in range on the order of some g (g = gravity acceleration = 9.8 m/s^2). A typical range tracking second-order loop is shown in Figure 2.53.

An ECCM method widely used by STT-type missile seekers, especially the ones on board sea-skimming missiles, is to use frequency agility pulse to pulse and different weights to the early-late gates of range tracking. By assigning a weight α to the early gate and β to the late gate, with $\alpha > \beta$, the range tracking is positioned in the front of the received pulse, thus reducing the effect of a possible range deception from a jammer (because the jammer pulse is always delayed with respect to the echo signal by the frequency setting time due to the frequency agility). This ECCM is called leading-edge tracking or antirange gate stealing and is very effective when applied with a logarithmic receiver. This ECCM will be further discussed in Chapter 6.

The video signal within the tracking gate at the output of the receiver is also sent to a coherent detector, which determines—after processing the amplitudes of many received pulses—the depth of modulation and the phase of the modulated signal with respect to two reference (azimuth and elevation) sinusoidal signals at scan frequency but 90° out of phase with each other. As mentioned earlier the maximum received amplitude indicates the amplitude of the angular error, while the phases of the maximum received amplitude with respect to the two reference signals indicate the direction (azimuth and elevation) of the error. This information is fed to the automatic servo loops that control azimuth and elevation of the antenna boresight in order to reduce to zero the angular error relevant to the target. The schematic block diagram of the conical scan STT is shown in Figure 2.54.

By using a second central fixed feed for transmission and using the notating feed only on reception, it is possible to avoid the amplitude modulation on trans-

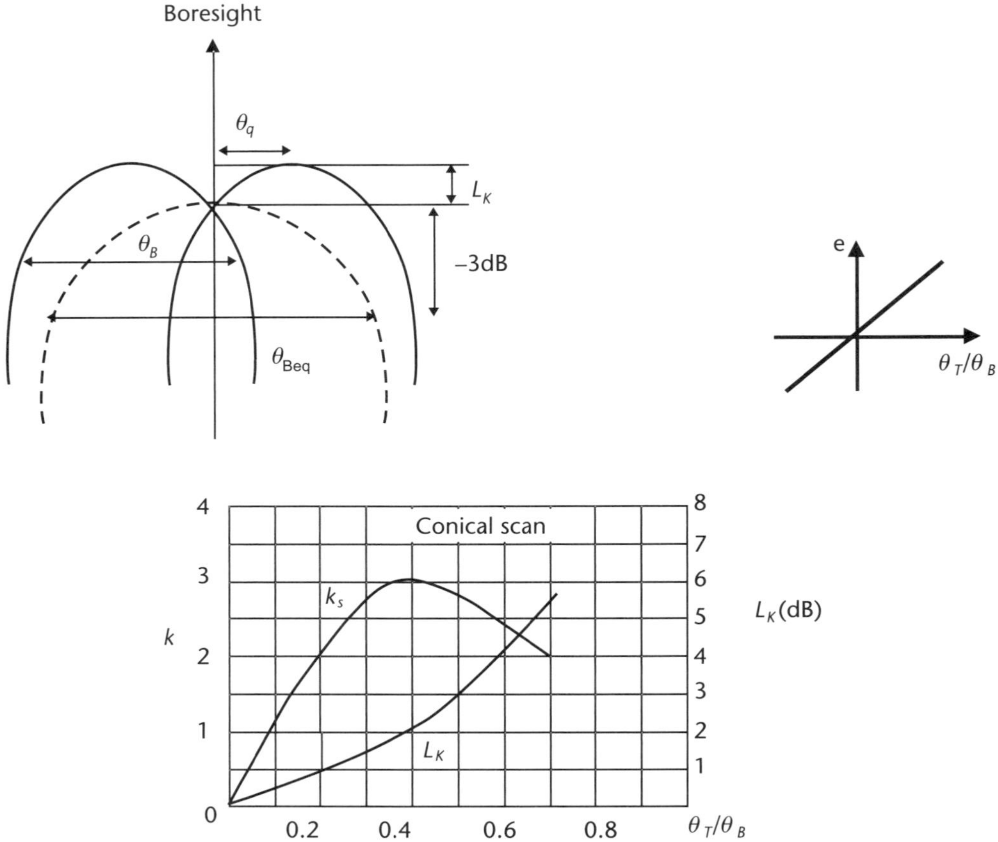

Figure 2.51 Conical scan angular slope and crossover loss. (From: [15].)

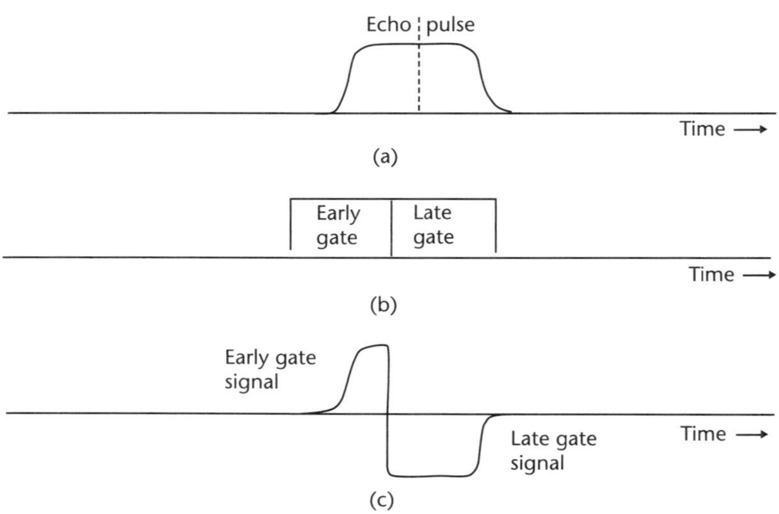

Figure 2.52 Split-gate discriminator.

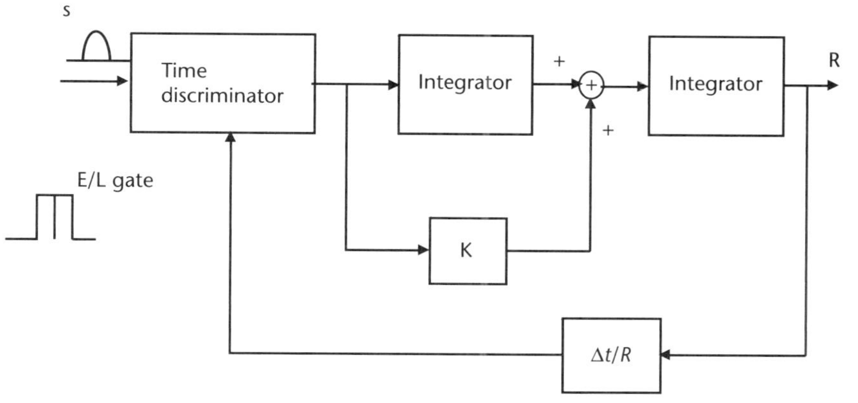

Figure 2.53 Range tracking loop.

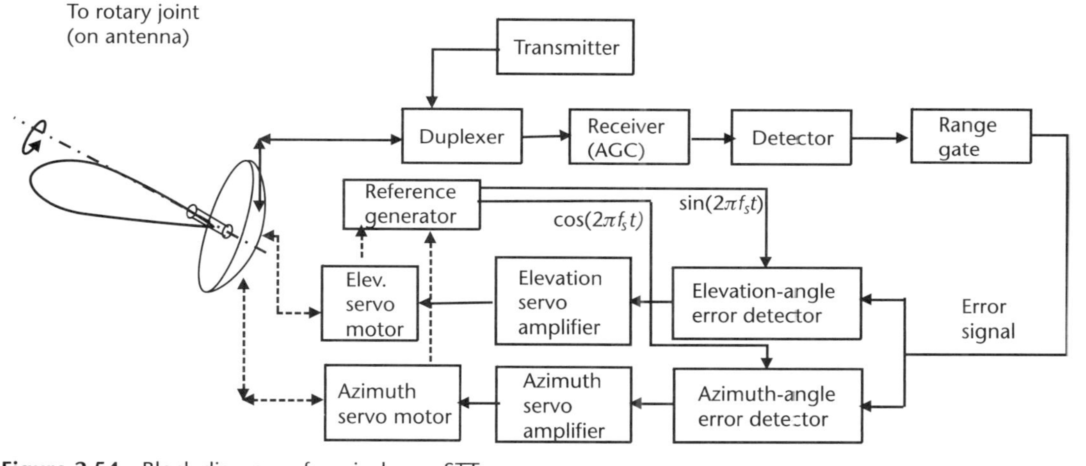

Figure 2.54 Block diagram of conical scan STT.

mission, which can be used by a jammer to produce angular errors to the STT. This arrangement characterizes the COSRO STT.

2.3.9.2 Lobe Switching

By switching a beam sequentially over four positions (two in azimuth and two in elevation), the error signal from a target not located on the switching axis (the boresight) can be determined by the difference between the received amplitudes at the two switched positions for each angular coordinate, as shown in Figure 2.55

The direction in which to move the beam in order to bring the target along the boresight is given by observing which beam position provides the larger amplitude. When the echo signals in the two beam positions are equal, the target is along the boresight direction. The lobe-switching radar thus is constituted by a square cluster of four feed horns illuminating a single reflector antenna. Each feed is illuminating the reflector in sequence, thus proving its squinted beam with respect to the antenna boresight.

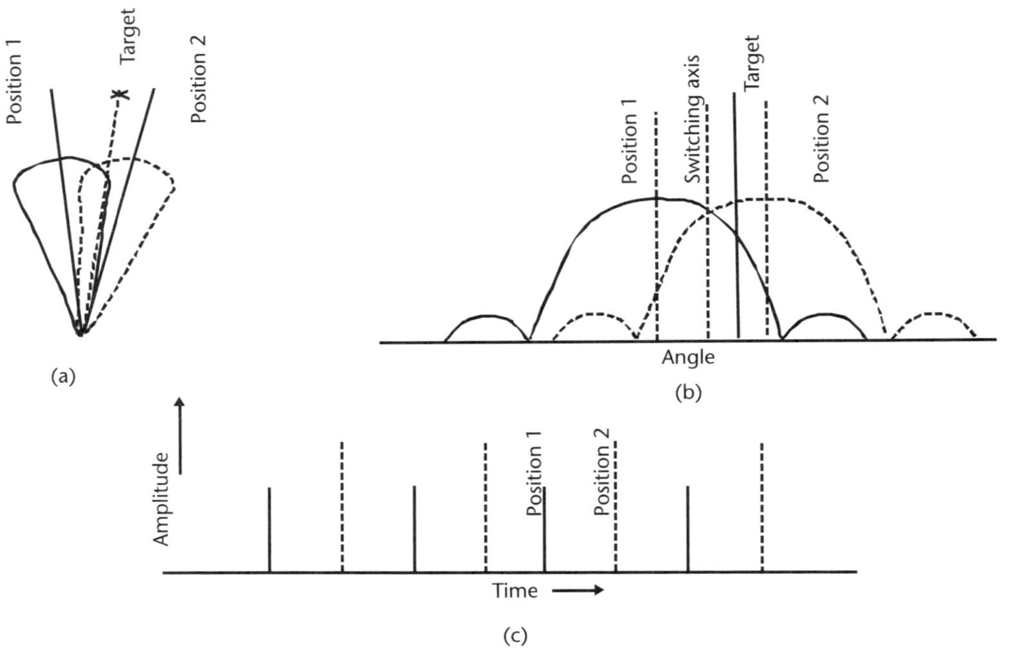

Figure 2.55 Lobe-switching antenna patterns and the error signal (for one angle coordinate).

A cluster of five feed horns can also be used, with a central feed used for transmission and the four outer feeds used for reception on a sequential basis. This arrangement, which characterizes a LORO STT, is used to avoid the amplitude modulation on transmission, which can be used by a jammer to produce angular errors to the STT.

Range tracking and angular tracking methods in these types of STT are the same as those for the conical scan STT.

2.3.9.3 Monopulse

A monopulse tracker obtains the information concerning the angular location of a target by comparison of signals received in two or four simultaneous beams. Angle measurement with simultaneous signals improves measurement accuracy because it avoids the degradation that results from signal amplitude changes (fluctuations) with time. The monopulse angle method is used in STT to provide an angle error signal in two orthogonal angle coordinates that are used to feed the closed-loop servo system to keep the boresight positioned in the direction of the moving target.

Two general methods are used to exploit monopulse angle measurement: *amplitude-comparison monopulse* and *phase-comparison monopulse*. They are discussed next.

Amplitude-Comparison Monopulse (ACM)

The concept of the ACM method is easily explained in only one angle coordinate by considering two overlapping antenna patterns with their main beams pointed in slightly different directions (squinted beams) as in Figure 2.56(a). The beams can be generated by two feeds slightly displaced in opposite directions from the focus of a parabolic reflector. The ACM method employs both the sum and the difference of the two squinted beams as shown in Figures 2.56(a) and (b). The sum pattern is used on transmission, while both the sum and the difference patterns are used on reception. The signal received with the difference pattern provides the magnitude of the angle error. The direction of the angle error is found by comparing the phase of the difference signal with the phase of the sum signal as can be derived from Figure 2.56(c).

By adding two more feeds in order to form a square cluster of four squinted feeds (two for each angle coordinate), the sum pattern is formed by the sum of the four squinted patterns, while two difference patterns are formed: one relevant to the azimuth and the other relevant to the elevation as shown in Figure 2.57. The comparator shown in Figure 2.57 is the circuitry, composed of hybrid-T waveguide components, that performs the addition and subtraction of the feed horn outputs to obtain the monopulse sum and difference signals

Almost all high-class monopulse STTs employ the Cassegrain reflector antenna, whose principle is shown in Figure 2.58. Its implementation as a Selex SI RTN 30X is shown in Figure 2.59. The introduction of this type of antenna with a polarization twist technique at the subreflector allowed the microwave circuitry to be placed behind the parabolic reflector and eliminated the antenna blockage problem.

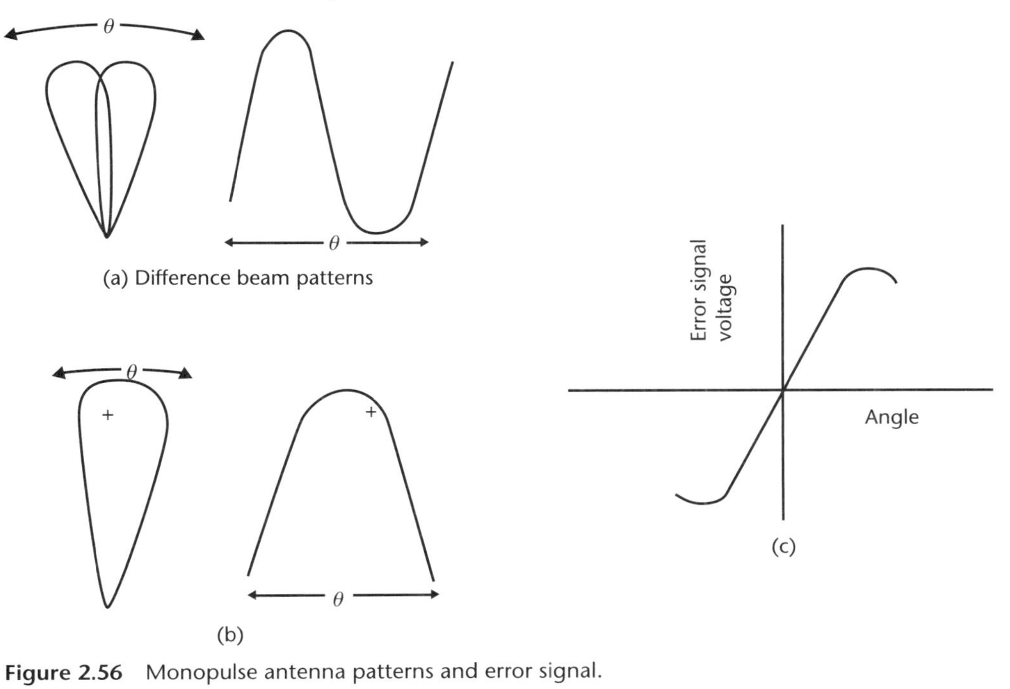

Figure 2.56 Monopulse antenna patterns and error signal.

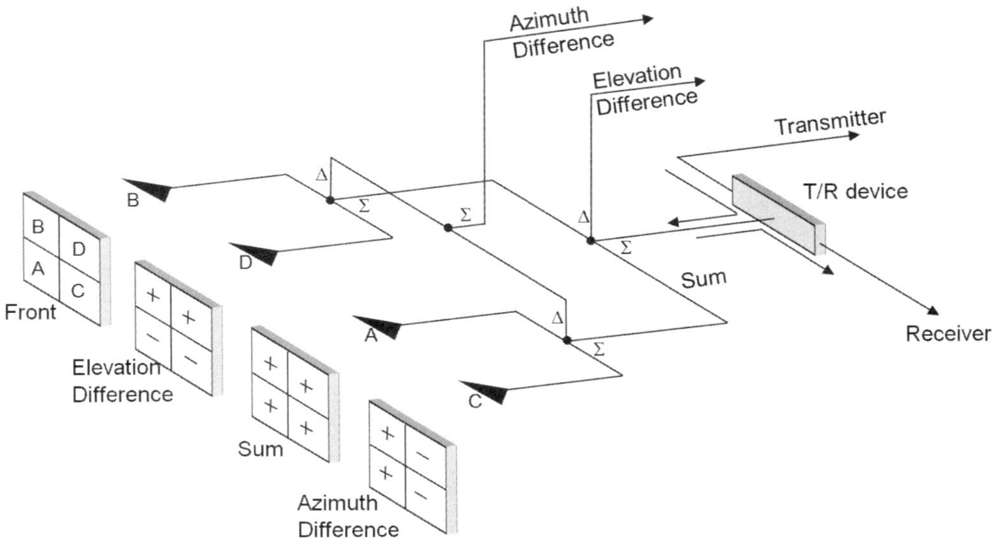

Figure 2.57 Monopulse square cluster of four squinted feeds (two for each angle coordinate). (From: [16], Figure 9.2.)

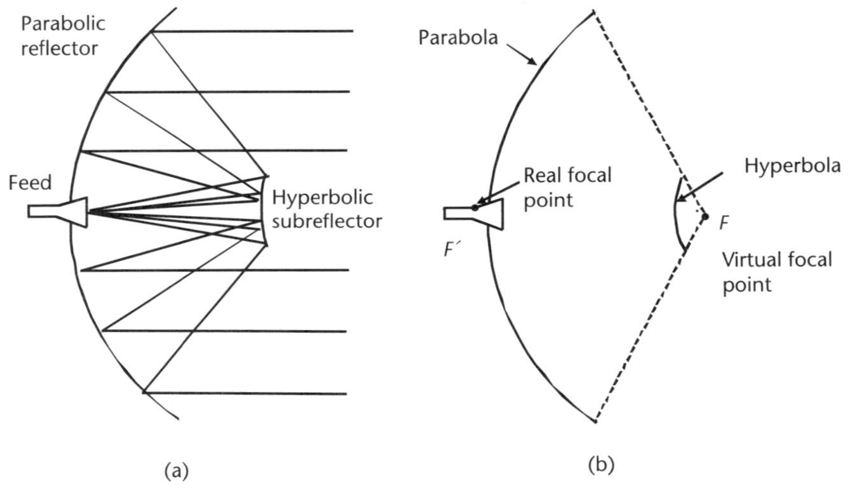

Figure 2.58 Cassegrain antenna: (a) actual and (b) concept.

Signals received from the sum and difference patterns are amplified separately and combined in a phase-sensitive detector to produce the angle-error signal. The sum signal also provides target detection and range measurement.

A simplified block diagram of a conventional monopulse STT is shown in Figure 2.60. As can be seen from the figure, the AGC loop and the range tracker are similar to those of a conical scan STT. The main difference here is that the AGC from the sum channel is fed back to control the gain of all three channels (Σ, Δ_{az}, Δ_{el}) so as to provide the STT with a constant angle sensitivity independent of changes in target cross-section fluctuations or changes in range and, most important, with immunity to amplitude jamming modulations in order to introduce angular errors.

Figure 2.59 RTN 30X (Courtesy of Selex SI.)

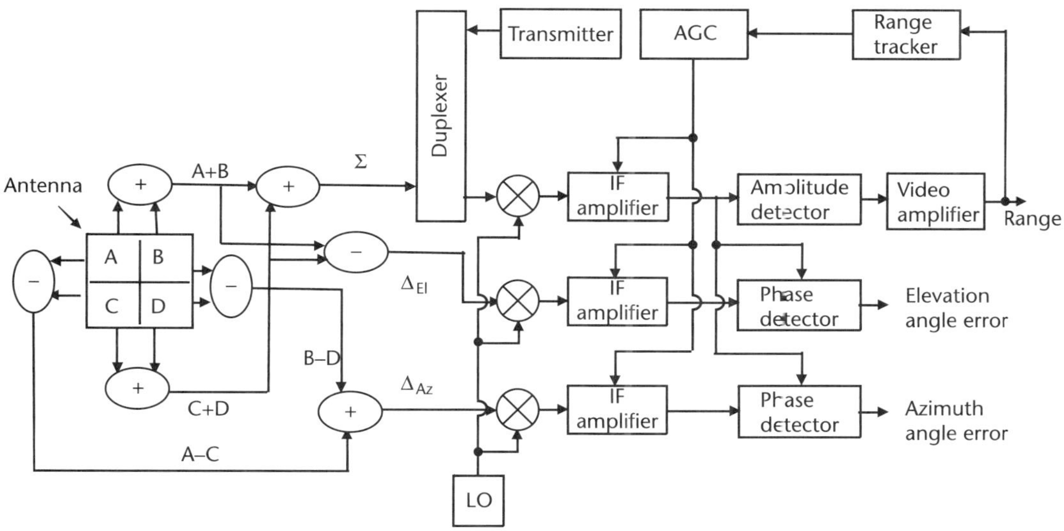

Figure 2.60 Block diagram of a conventional monopulse STT.

As can be noted from the monopulse STT block diagram, the sum signal Σ is used to demodulate coherently the two Δ signals and to provide the angle-error signal slopes signal in accordance with the following relationships:

$$\Delta_{az} = \underline{\Sigma} \cdot \underline{\Delta}_{az}/\Sigma^2 \text{ and } \Delta_{el} = \underline{\Sigma} \cdot \underline{\Delta}_{el}/\Sigma^2$$

where Σ and Δ denote the phasors of the respective signals, and the scalar product is used to determine both the normalized amplitude as well as the phase relationship between the two phasors and hence the sign of the Δ signal. The division by Σ^2 is achieved by means of the AGC as discussed above.

The ACM angular error sensitivity is dependent (similar to the conical scan method) on the ratio of the squint angle θ_q to the beamwidth θ_B of each squinted pattern, in that the greater the squint angle, the greater will be the slope of the error. However, as the squint angle increases, the on-axis gain of the sum pattern decreases and thus a trade-off has to be made between these two diverging requirements. The optimum value for squint angle θ_q was evaluated in Figure 4.7 of [3], assuming that the shape of the squinted beams can be modeled by a Gaussian function. The optimum squint angle was found to be $\theta_q = 0.31\theta_B$. This corresponds to a crossover of 1.2 dB down from the Σ beam peak.

The comparison of the slopes of the ACM and conical scan STT, as reported in Figure 4.7 of [3], shows the superior angular error sensitivity of the ACM, which can achieve angular accuracies on the order of 0.1 to 0.2 mrad (rms).

Phase-Comparison Monopulse (PCM)

In the PCM method two antenna beams are used to obtain an angle measurement in one coordinate, as for the ACM method. The two beams are parallel and cover the same region of space. The amplitude of the received signals are the same, but their phases are different by an amount $\Delta\phi = 2\pi d \sin\theta/\lambda$, where d is the spacing between the two antennas and θ is the direction of the arrival with respect to the normal to the baseline of the two antennas. The measurement of the phase difference of the two received signals provides the angle θ to the target. The overall antenna used to achieve the two angle coordinates is thus an array of four antennas arranged in two rows and two columns.

Angle information can be extracted by employing sum and difference patterns and processing the signals similarly to the ACM method. To get at the output of the phase detector, an error signal whose amplitude is a function of $\sin\theta$ a 90° phase shift has to be added to the difference signal as shown in Figure 2.61, which also shows the block diagram of a conventional PCM STT (in one angle coordinate).

The PCM method can be affected by the grating lobes (see Appendix C) of an array with two elements ($\theta_{grating} = \sin^{-1}(\lambda/d)$), where d is the separation (which is also the diameter of each antenna) between the two antennas. Thus, the separation d should be smaller than λ, but this in turn implies a low antenna gain. Due to this limitation the PCM STT has had very few applications in comparison to the ACM STT.

2.3.9.4 Limitations to Tracking Accuracy

Three types of error sources affect the angular and range accuracies of a STT:

1. Target-caused errors, such as amplitude fluctuations of the target (not affecting the monopulse STT); glint, which affects both range and angle accuracies; and Doppler scintillation and spectral lines;
2. External causes of error, such as multipath in low-angle tracking;
3. Internal sources or errors, such as receiver thermal noise.

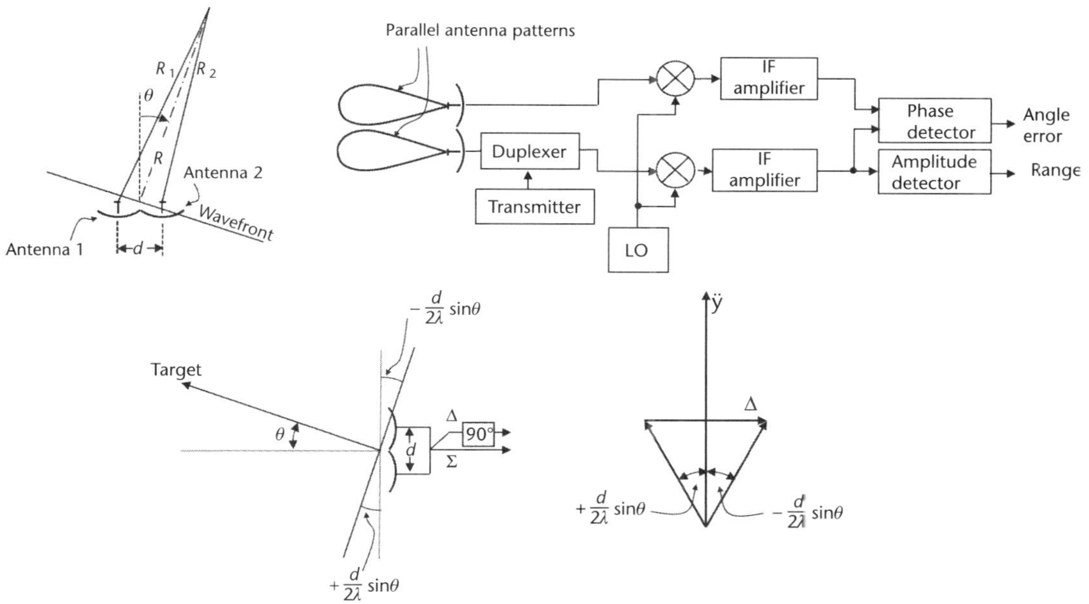

Figure 2.61 Block diagram of a conventional PCM STT (in one angle coordinate).

Of the above angular error sources, in this section we discuss only target glint and multipath in low-angle tracking because some ECM techniques (discussed in Chapter 6) tend to exploit the above limitations to increase the monopulse tracking errors to cause an angular break-lock to the STT.

2.3.9.5 Glint

Glint is the term commonly used to describe the fluctuation of a complex target's RCS, when it is composed of many scattering centers within the resolution cell of the radar. Each scatterer response arrives at the antenna with slightly different wave tilts, and these tilted wavefronts add vectorially across the aperture to give a composite disuniform and distorted wavefront with respect to the ideal planar wavefront coming from a single scatterer target. This produces an error on the angle-of-arrival measurement. Furthermore, the measured angle of arrival can sometimes cause the boresight of the tracking antenna to point outside the angular extent of the target, which can cause the radar to break track. The greater the extent of the target in angle as seen by the radar, the worse will be the angle measurement—which constitutes a major cause of angle errors at short ranges. Glint affects all types of STT radars.

Qualitative effects of angular glint can be derived by the *dumbbell target* model, constituted by two independent isotropic scatterers separated by an angular distance α as measured from the radar. In this discussion we apply the same symbolism of Section 6.3 of Chapter 6 in order to ease the reader into the evaluation of the cross-eye ECM technique

The two scatterers are symmetrically located (at $\pm\alpha/2$) with respect to the antenna boresight, as shown in Figure 2.62. The relative amplitude of the echo signals is assumed to be $a \leq 1$ and the relative phase $\phi \leq \pi$ rad. Differences in the phase of

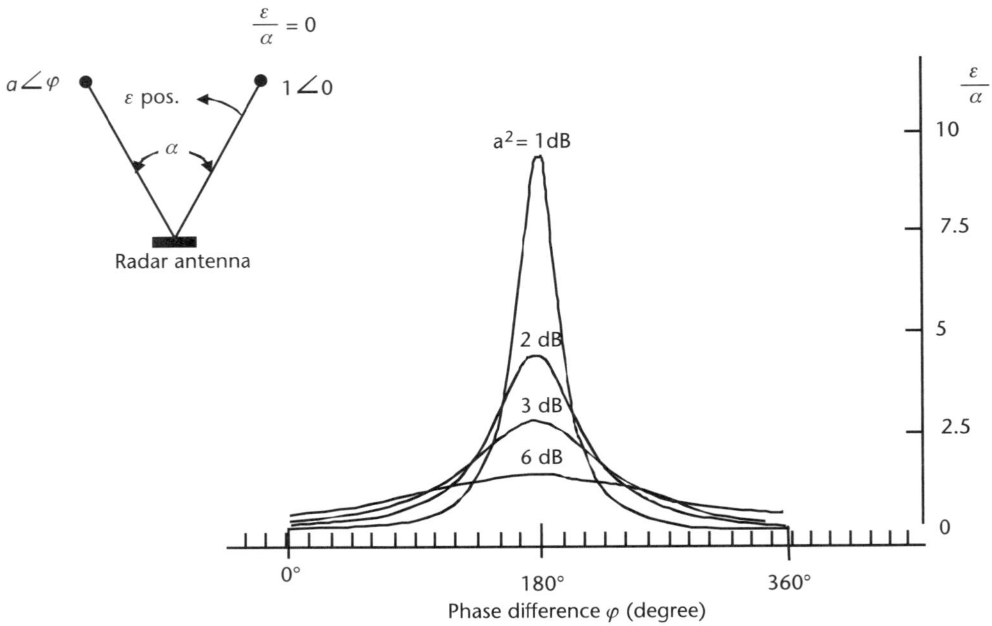

Figure 2.62 Glint produced by the dumbbell target model.

the two scatterers might be caused by slight differences in range between the two scatterers. The angular error $\Delta\theta$ as measured from the larger of the two isotropic scatterer is given by [17] (and will be also derived in Chapter 6) as:

$$\frac{\varepsilon}{\alpha} = \frac{a^2 + a}{1 + a^2 + 2a\cos\phi} \tag{2.67}$$

The equation is plotted in Figure 2.62. The larger of the two scatterers is located at $\varepsilon/\alpha = 0$ and the smaller at $\varepsilon/\alpha = +1$. Positive values of ε correspond to the angular region to the left of the larger scatterer, and negative values of ε correspond to the angular region to the right of it. When the echo signals from both scatterers are in phase ($\phi = 0$), we have $\varepsilon/\alpha = a\ (a + 1)$, sometimes called the center of gravity of the two scatterers. When the two echo signals are at the same amplitude ($a = 1$) and the phase difference is $\phi = \pi$ rad, (2.67) indicates $\varepsilon/\alpha \rightarrow -\infty$; that is, the antenna is driven well outside the bounds of the dumbbell target. Equation (2.67) has been derived for small angle errors; hence, it is not applicable for angular errors as large as $\varepsilon/\alpha \rightarrow -\infty$. Nevertheless the simple dumbbell target model describes the behavior to be expected by a STT radar subjected to glint from a complex target.

Equation (2.67) indicates that the tracking error ε is proportional to the angular extent of the target α as seen by the radar. This implies that the error becomes larger as the range becomes shorter.

A slightly more complex model than the two-scatterer target considers many individual equal RCS scatterers aligned uniformly along a line of length L perpendicular to the line of sight from the radar. The resultant RCS of such a target is assumed to behave according to the Rayleigh pdf and there is a probability of

13.4% that the apparent direction of the target measured by the radar is outside of the target region. A commonly used design equation is to consider the rms angular error σ_g due to glint to be $\sigma_g = 0.3L/R$, where L is the extent of the target and R the range to the target.

Glint also affects range tracking. By assuming a two-scatterer target model similar to the one used for angular glint with a separation in distance $c\Delta T/2$, the error ΔT_R due to range glint relative to the center of the two scatterers (where the early-late range-tracker should be centered) is:

$$\Delta T_R = \frac{\Delta T}{2}\frac{1-a^2}{1+a^2+2a\cos(2\pi f_0 \Delta T)} \tag{2.68}$$

Also in this case the rms range error σ_{gR} due to glint can be assumed to be $\sigma_{gR} = 0.15L_R$ where L_R is the span of the target in range.

The methods used most often to reduce glint are frequency agility and pulse compression. A change of frequency results in a change in the relative phase of the scatterers and hence in ε and ΔT_R. The frequency change must be sufficient to decorrelate the phase measurement (the phase must change for more than 2π rad) and thus must be on the order of $\Delta f > c/D$ (where c is the speed of light and D is the dimension of the target in the relevant axis). A number $n < 4$ of frequencies are sequentially transmitted. Then the relevant angle and range measurements weighted in accordance with their SNR and averaged, thus reducing the glint error by a factor $>n^{1/2}$.A better result is achieved by taking only the measurements at the frequency that results in maximum SNR, because this reduces the single-frequency glint by a factor n. The drawback of this remedy is that MTI or pulse Doppler processing cannot be used.

Range and angle accuracy are highly improved by pulse compression, which reduces the range cell resolution and hence the number of scatterers within it. High range resolution can be used in conjunction with MTI or pulse Doppler processing.

2.3.9.6 Receiver Noise

All theoretical expressions for the rms value of the error of a radar measurement are inversely proportional to the square-root of SNR. Because the maximum detection range of the radar is inversely proportional to $\text{SNR}^{1/4}$, the rms error values are therefore proportional to R^2. Receiver noise is a major factor limiting the accuracy of radar at long range where the SNR is small. The rms error in angle measurement has been given by [12] and is:

$$\sigma_n = \frac{k\theta_B}{k_s\sqrt{B\tau(S/N)(f_p/\beta_n)}} \tag{2.69}$$

where

k is the factor associated with beamwidth and dependent on the measurement method ($k = 1$ for monopulse and $k = 1.4$ for conical scan),

θ_B is the –3-dB beamwidth,

k_s is the slope of the angle-error signal at the boresight and is dependent on the measurement method (k_s = 1.57 for a good four-horn feed, and k_s = 1.5 for an optimized conical scan),

B is the signal bandwidth,

τ is the pulse width ($B\tau \approx 1$).

SNR is the signal-to-noise ratio per pulse,

f_p is the pulse repetition frequency,

β_n is the servo bandwidth ($f_p/2\beta_n$ is the number of pulses integrated).

The contributions of glint and receiver noise to the accuracy of both angular tracking methods (conical scan and monopulse) for STT radar as a function of range were evaluated by Barton [12] who showed that tracking accuracy deteriorates at both long and short ranges with the best accuracy occurring at the intermediate ranges.

At short ranges the prevailing error is due to the glint with an error that varies inversely with range; at large range the prevailing error is due to receiver noise, which causes the error to vary as the square of the range. The conical scan technique is affected by amplitude fluctuations, whereas the monopulse method is not.

2.3.9.7 Low-Angle Tracking

A radar that tracks at low elevation angles illuminates the target via two paths as shown in Figure 2.63. One is the direct path from radar to target and the other is the path with a reflection from the Earth's surface. It is as if the radar were illuminating two targets: the real one above the surface and its image below the surface. The situation is similar to the two-scatterer target model discussed with regard to glint, and in this case errors in elevation angle occur, which can be very large. Further, at low grazing angles, over a perfectly smooth reflecting surface (as can occur with a calm sea), the complex reflection coefficient $r = \rho\ exp\ (j\psi)$ is nearly $\rho = 1$ in amplitude and $\psi = \pi$ rad in phase, so that the signal reflected from the surface (and reaching the target image) is nearly equal in amplitude and out of phase with respect to the signal following the direct path to the target.

This situation is equivalent to the undesirable condition for the angle error due to glint. For this reason the tracking of targets at low elevation angles is subject

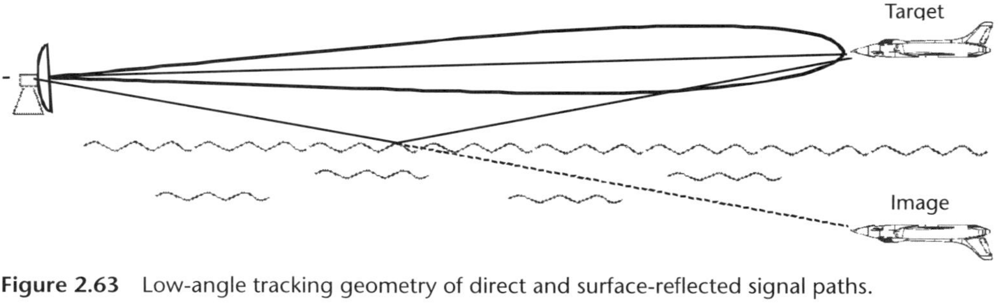

Figure 2.63 Low-angle tracking geometry of direct and surface-reflected signal paths.

to significant errors in the elevation angle; they can even cause loss of the target's track. The surface-reflected signal is also called a *multipath signal*, and the glint error due to low grazing angle tracking is called a *multipath error.* Multipath errors affect both radar guidance of missiles to targets low on the sea as well as the tracking of surface and shipboard STTs used for defense against low-altitude cruise missiles or sea-skimming missile attacks.

Howard [16], following the dumbbell target approach, expressed the target glint error as]similar to (2.67)]:

$$e = 2h\frac{\rho^2 + \rho\cos\phi}{1 + \rho^2 + 2\rho\cos\phi} \tag{2.70}$$

where e is the error (m), h is the height of the target (m), ρ is the amplitude of the reflection coefficient, and ϕ is the relative phase determined by the geometry of the direct and surface-reflected signal paths.

The effects of multipath signals depend on what part of the antenna pattern hits the Earth's surface. Howard [16] identified three regions in relation to elevation angle γ:

1. *Sidelobe region*, where only near sidelobes in elevation illuminate the surface. This occurs when the tracking is at close range. The angle accuracy begins to be degraded when the elevation angle $\gamma < 6\theta_B$ above the horizon. The expression of the rms multipath error in this region, which shows a cyclic behavior, is given by Howard [16]:

$$\sigma_\varepsilon = \frac{\rho\theta_B}{\sqrt{8G_{se}(peak)}} \tag{2.71}$$

   where
   σ_ε = rms elevation angle multipath error (mr),
   θ_B = one-way –3-dB beamwidth,
   ρ = amplitude of reflection coefficient,
   G_{se}(peak) = power ratio of the tracking-antenna sum pattern to the error pattern peak sidelobe level at the angle of arrival of the image signal.

   The cyclic behavior of the error is due to the maxima and minima determined by the combination of direct and surface-reflected signal paths in accordance with the geometry.
2. *Main beam region*, where the multipath effects begin to be severe because the elevation angle is about $\gamma \approx 0.8\theta_B$.
3. *Horizon region*, where elevation angle $\gamma \approx 0$, and the multipath error becomes very severe.

Many methods for reducing the large elevation angle error caused by the multipath signals at low angles have been demonstrated, but not a single method was shown suitable for all the applications where low-angle tracking is required, because each has its limitations. The methods discussed next, dealing mainly with the elimination of the large errors that occur when the main beam illuminates the surface, have had application.

Narrow Beamwidth

The reduction of the beamwidth θ_B reduces the situations in which the antenna beam illuminates the surface, but requires either larger antennas or higher frequencies. Both solutions are not always practicable. The method has been successfully adopted by employing two radars, one at the X-band (9 GHz) and the other at the K_a-band (35 GHz) using a single antenna system to provide operation at the two frequencies. Because the lower frequency radar has a longer detection range, it is used for target acquisition. Then precision low-altitude tracking is performed by the higher frequency radar.

Similar solutions have been applied with short-range STTs by adding on the same radar antenna pedestal some optical or IR sensors that have far better angle resolution. In these solutions the low-angle tracking is performed by the EO/IR sensor and range tracking by the radar.

Illogical Target Trajectory

Because aircraft or missile targets will not likely go below the surface of the Earth and are limited in their ability to accelerate in elevation (either upward or downward), radar tracking data indicative of unreasonable target behavior can be recognized and rejected. In some situations, the target might be flying so fast and the inertia of the antenna so large that the angle-error excursion caused by the multipath signal (usually at short ranges) is dampened.

Off-Boresight Monopulse Tracking

We can determine when the target is in the low-angle region by sensing large elevation-angle errors. Hence, the antenna can be locked at a small positive elevation angle (usually about 0.7 to $0.8\theta_B$) while continuing closed-loop azimuth tracking. With the beam fixed at a positive elevation angle, the rms tracking accuracy is worsened to typically $0.3\theta_B$, but wide swings of the antenna and loss of track are avoided.

2.3.9.8 Track While Scan in a Limited Sector Scan

This type of radar rapidly scans a relatively narrow angular sector, usually both in azimuth and elevation, performing search and track functions. Scanning may be performed with a narrow-beamwidth pencil beam that covers a rectangular sector in raster fashion or with two orthogonal fan beams, which scan the azimuth and elevation sector, respectively. TWS radars have been used in airport landing radars, airborne interceptors, and air defense systems.

With respect to the STT, there is no closed-loop positioning of the antenna, and the angle output is directly provided to the data processor as described for the surveillance radars. Hence, the TWS radar can provide simultaneous tracking of many targets within its sector of coverage, albeit with less accuracy than a STT.

In airborne-interceptor applications the TWS might be preferred when multiple targets are to be tracked with accuracy sufficient to launch fire-and-forget missiles, which contain their own guidance systems for homing in on the target.

TWS radars have been used in precision approach radars (PARs) or ground-controlled approach (GCA) systems to guide aircraft to a landing. In the past control of aircraft landing involved the use of TWS radars that had electromechanical scanners that scanned fan beams over a narrow sector of 20° azimuth and 7° elevation at a rate of twice per second. Current TWS radars have phased array antennas that use monopulse angle measurements.

In the air defense systems TWS radars have been used, especially by the former Soviet Union, for the control of weapons in SAM systems for both land- and ship-based systems. TWS radars that perform continuous regular scanning of a sector are vulnerable to angle jamming.

2.3.9.9 Tracking with a Phased Array Radar

Tracking with a phased array (PA) radar is more like that of a TWS radar or automatic tracking with a surveillance radar than the continuous track of a STT. The advantage of a PA radar is that it can have a much higher data rate than radars with mechanically scanned antennas and it can simultaneously track multiple targets within a large sector by time sharing a single antenna beam. This is possible because of the rapid inertialess (electronic scanning) beam positioning, which allows for switching of the direction of the beam in a couple of microseconds. Angle measurement is monopulse.

2.3.10 Synthetic Aperture Radars

Airborne and spaceborne SARs have become a highly used application to provide high-resolution ground mapping that is capable of allowing recognition of topographic features and objects (almost fixed) of interest on the ground or sea (such as buildings, tanks, armored vehicles, and ships).

The quality of the ground maps produced by radar is determined by the resolution cell size, which in the case of SAR, which is an airborne radar, has two components: *range* and *cross-range*. Because the features of the terrain may be oriented in any direction, ideally the two resolution dimensions should be equal, making the resolution cell a square pixel.

Although it is possible through pulse compression techniques (discussed in Section 2.3.4) to obtain very high range resolution, the achievement of a fine cross-range resolution requires a very large antenna on the order of tens to a hundred meters, which is not feasible on an aircraft. In lieu of an actual antenna, the required resolution is achieved through a synthetic aperture (or an array of equivalent required width) obtained by the forward motion of the airborne radar.

The SAR concept is illustrated in Figure 2.64, in which an airborne radar, with an antenna of diameter D and thus with a beamwidth $\theta_B = k\lambda/D$ (rad), transmits pulses with a period T_p. Each time a pulse is transmitted, the radar occupies a position a little further ($\Delta x = v\ T_p$) along the flight path (x-axis). By summing coherently the return from successive pulses, it is possible to synthesize a very long linear array.

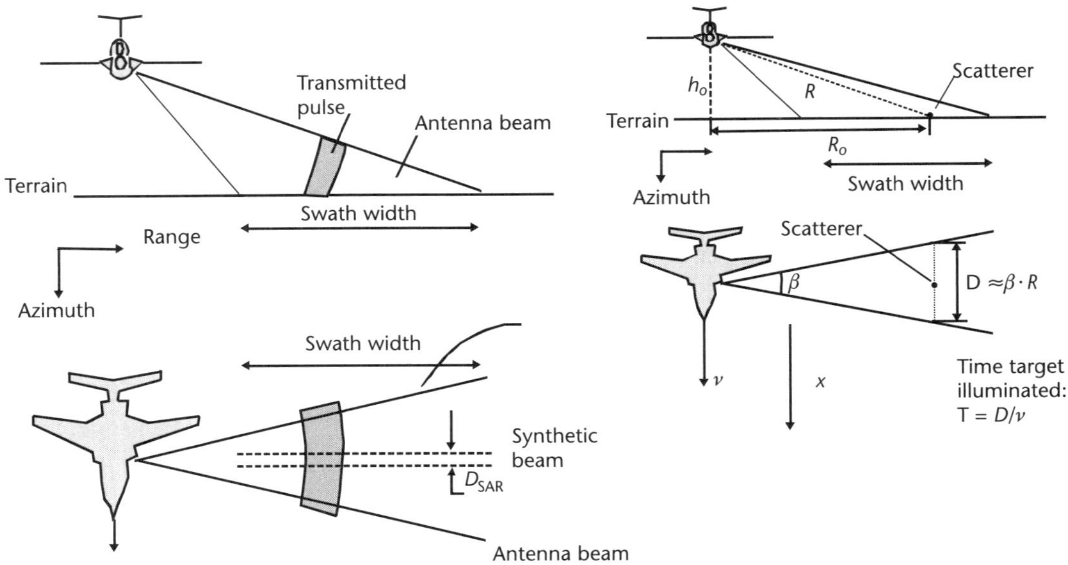

Figure 2.64 SAR principle and scanning configuration.

The fundamental difference between a real array and the synthetic one is that in the former the return from each scattering point is received simultaneously by all elements of the array, whereas in the latter the return is collected by the real airborne antenna sequentially over the period of time the radar takes to traverse the array.

The returns from a scattering point at distance R from the radar can be collected during the period $T = R\theta_B/v$ in which the point is illuminated by the antenna beam. Thus, the dimension of the synthetic array is $L_{SAR} = vT = R\theta_B$. With such a dimension for the array, the equivalent beamwidth is $\theta = k\lambda/(2L_{SAR})$, where the factor 2 takes into account that the phase contributions to the coherent sum (integration) are relevant to the two-way path radar to scatterer. Therefore, the cross-range resolution of the SAR is:

$$\delta_{Rcross} = R\theta = R\frac{k\lambda}{2L_{SAR}} = R\frac{k\lambda}{2R\theta_B} = \frac{D}{2} \tag{2.72}$$

and the SAR cross-range resolution is independent of range and wavelength and only proportional to the size D of the real antenna of the radar.

However, to achieve the preceding result, which is called *focused resolution*, some corrections have to be added to the phase of the returns due to the geometrical variation of the range of the point scatterer along the integration period T. The variation is shown in Figure 2.65, from which we derive the following corrections. These corrections have to be applied to the returns integration process in order to obtain a focused resolution:

$$\text{Range:} \quad R = \sqrt{R_0^2 + x^2} \approx R_0 + \frac{x^2}{2R_0} \tag{2.73}$$

where R is expressed with a Taylor series expansion. The term $x^2/\lambda R_0$ is called the *range curvature*.

$$\begin{aligned} \text{Phase:} \quad & \phi = \frac{4\pi}{\lambda} R = \phi_0 + \frac{2\pi x^2}{\lambda R_0} \\ \text{Doppler:} \quad & f_d = \frac{1}{2\pi}\frac{d\phi}{dt} = \frac{2vx}{\lambda R_0} = \frac{2v^2 t}{\lambda R_0} \end{aligned} \tag{2.74}$$

Thus, we can observe that the cross-range (azimuth) Doppler history has the same behavior as a chirped pulse (but on different time and frequency scale) and that SAR azimuth compression is similar to the pulse compression technique. As can be seen from Figure 2.65 the focusing algorithm applies to the linear region of the Doppler frequency history.

SARs have two main modes of operation (see Figure 2.66): either *swath* or *strip-map mode* or *spotlight mode*. In the former mode the antenna is fixed on one side of the aircraft and provides the image of a swath along the flight path. In the latter mode the antenna is always pointed to a specific area during the flight path. Spotlight mode provides more accurate images of the illuminated area than does swath mode.

SAR processing is a very intensive computational task. Real-time imagery requires very fast computing hardware that is not always available on board aircraft; hence, the alternative is to downlink the raw data to a processing center on the ground. In practice, real-time processing is always "*near*-real-time" because there is a small delay (seconds) between the collection of data and the display of the image.

SAR systems nowadays are used more often on small aircraft or UAVs that present an unstable platform motion; therefore, it is necessary to restore SAR image focusing and geometric accuracy. Corrections can be approached either on motion measurements using both inertial navigation systems (INSs; i.e., gyroscopes

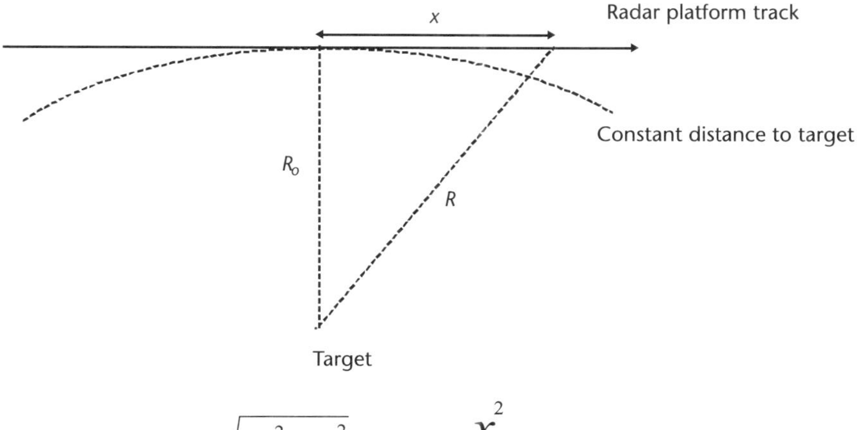

$$R = \sqrt{R_o^2 + x^2} \approx R_o + \frac{x^2}{2R_o}$$

Figure 2.65 Synthetic aperture focusing corrections.

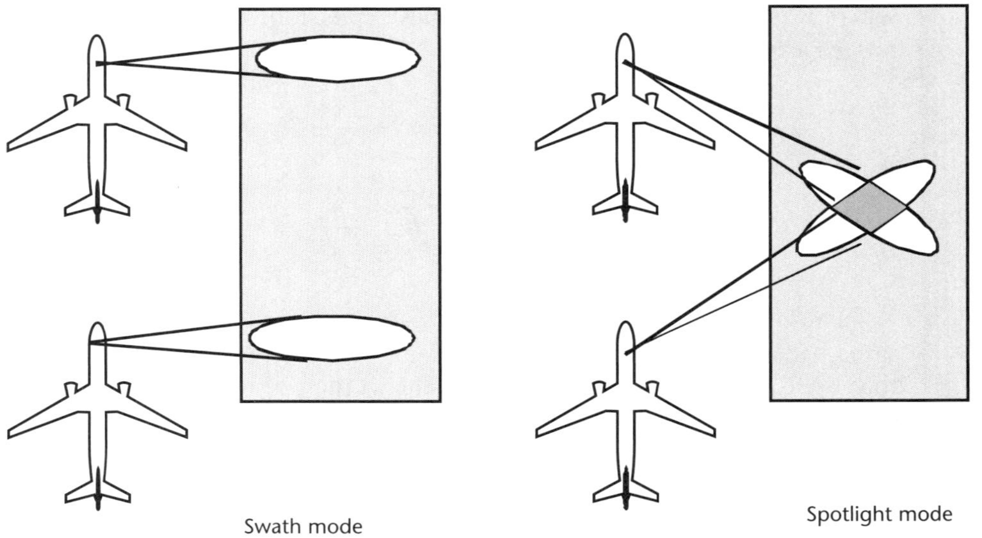

Figure 2.66 SAR modes.

and accelerometers) and GPS or with autofocus algorithms (i.e., error estimation from SAR data).

Additional SAR systems are the polarimetric SAR and the interferometric SAR. The polarimetric SAR exploits alternate transmission in H and V polarization. The returns are collected by two receiving channels in H and V polarization and processed (after being recorded) in both amplitude and phase difference in four channels: HH, HV, VH, and VV. The data from the scattering polarization matrix provide a more complete description of the targets: better contrast, discrimination between even and odd bounces of the terrain, and thus enhanced classification of them.

The interferometric SAR uses a couple of antennas at a distance Δx from each other placed on the belly of the aircraft. Through the differential phase measurement of the scatterer returns at the two antennas, it is possible to measure the relative height of the target (with respect to the terrain) as shown in Figure 2.67. Very accurate measurements of aircraft positions and movements are required to obtain a height map.

A further use of the SAR technique is exploited by fixed radars as a noncooperative target identification technique. Indeed, using the SAR algorithm to exploit the various returns of the target along a pass-course trajectory (in each return the target presents a different aspect to the radar, that is, a rotational movement around the vertical axis; see Figure 2.68) can provide a good image of the target. This technique is called inverse SAR (ISAR).

2.3.11 Bistatic Radars

A bistatic radar is, by definition, one in which the transmitter and receiver are deployed at two separate locations, which can change with time. The base length between the two sites is usually comparable to the ranges to the targets of interest. With this consideration over-the-horizon (OTH) radars, notwithstanding Tx and

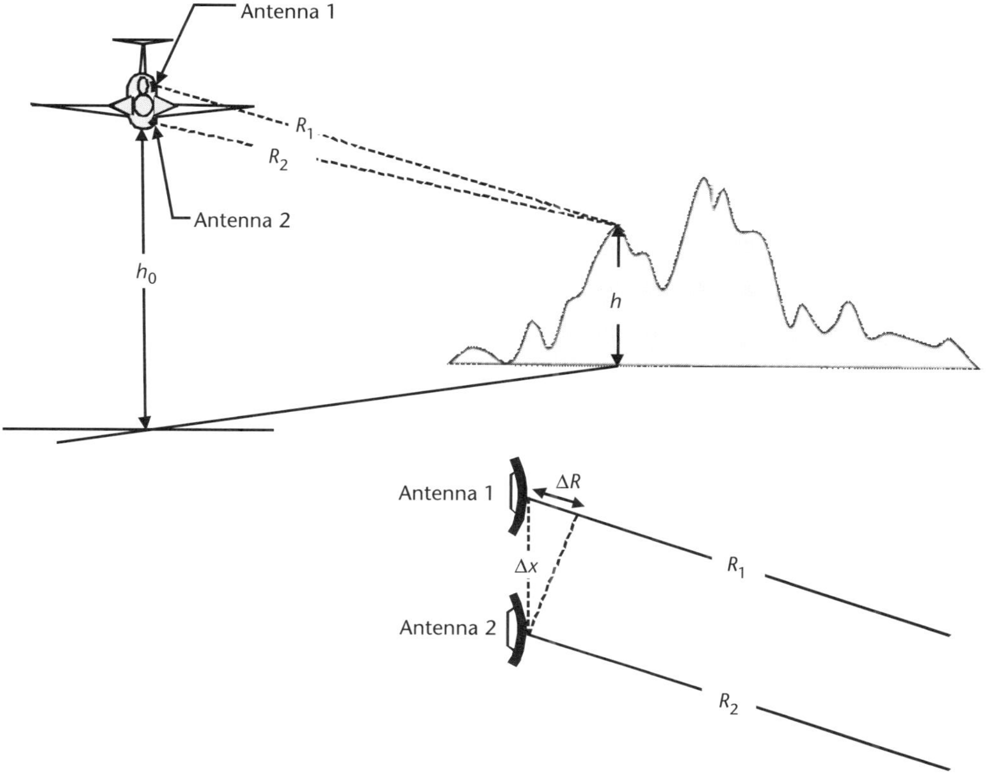

Figure 2.67 Interferometric SAR principle.

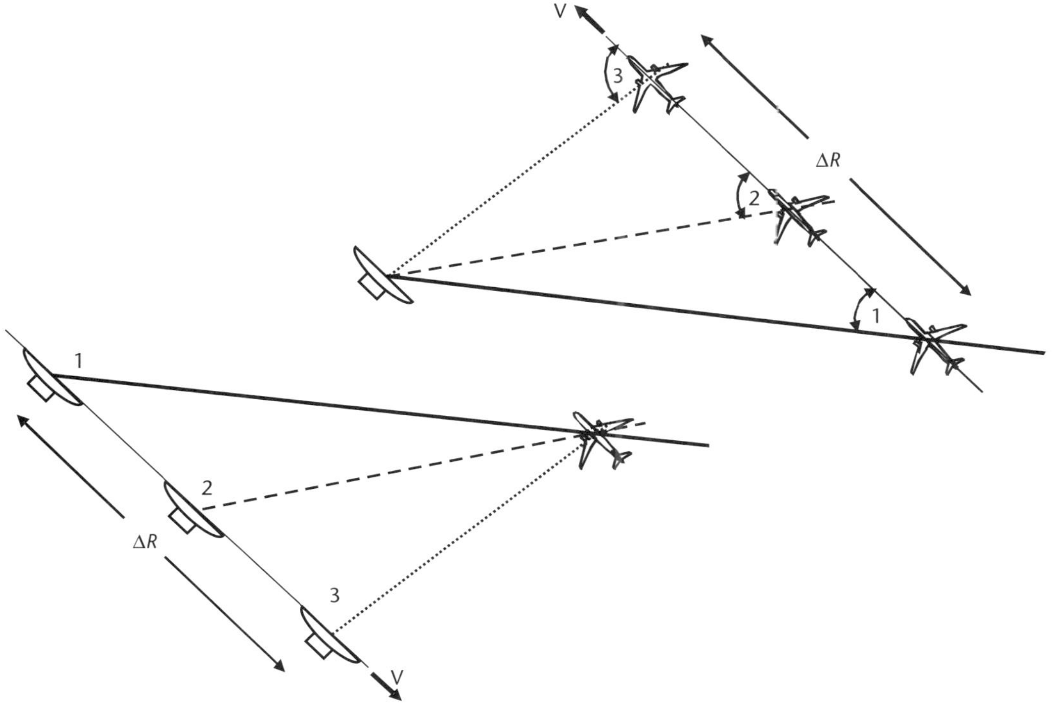

Figure 2.68 ISAR concept.

Rx that are separated by hundreds of kilometers, are considered monostatic as their range is thousands of kilometers. Bistatic radars have been designed, developed, and deployed for military, commercial, and scientific applications. Typical military applications include air and space surveillance and traffic surveillance. They are, however, niche applications.

Bistatic radars can operate with *dedicated transmitters,* which are designed for bistatic operation and controlled by the bistatic radar, or with transmitters of opportunity, that is, transmitters that are designed for other purposes, but found suitable for bistatic operation even when not controlled by the bistatic radar.

When the Tx of opportunity is from a monostatic radar, the bistatic radar is called a *hitchhiker*.

When the Tx of opportunity is from a broadcast station or a communication link, the bistatic radar has been called by many names such as *passive radar, passive bistatic radar* (*PBR*), *passive coherent location* (*PCL*), and *parasitic radar.* In military scenarios transmitters of opportunity are designated either *cooperative*, if it is an allied or friendly Tx, or *noncooperative*, if the Tx is hostile or neutral.

Multistatic radars are composed of a number of bistatic pairs or they can use triangulation of the scattered energy from a Tx illumination by taking direction-of-arrival (DOA) measurements from multiple receiver sites, as used by a SPASUR (space surveillance of noncooperative satellites) system.

The big advantages of the bistatic radar concept come from its immunity to countermeasures.

In conflicts that involve the intrusion of hostile aircraft, surveillance and tracking monostatic radars will be attacked with a wide range of countermeasures such as barrage and deception jamming, chaff, ARMs, and stealth (i.e., reduced RCS) aircraft.

The Tx is, of course, still vulnerable to direct physical attack, but a bistatic receiver can operate with target illumination by any of several Txs of opportunity and it is immune to ARMs.

In fact, when operating with dedicated Txs, the bistatic receivers are not vulnerable to jamming signals that are not incident upon them but directed toward their Txs. The use of adaptive nulling of the antenna pattern further reduces the incident jamming. Also, dedicated Txs may use a higher PRF than the usual surveillance radar, thus allowing Doppler processing to achieve discrimination of targets concealed within chaff clouds and corridors.

Bistatic and multistatic radars achieve good detection performance against stealth A/C because they can be deployed to benefit from the energy reflected away from its cone of silence. Also active stealth is ineffective because the correct nulling signal cannot be achieved at an unknown receiver location.

The location of the target in bistatic radars is determined by the signal delay measurement along the Tx-target-Rx path $R_T + R_R$ with respect to the signal TOA along the path Tx-Rx L and the Rx antenna bearing and elevation, respectively, relative to the baseline to the Tx and to the vertical axis.

The 3-D locus of the time delay is a prolate spheroid (with a horizontal cut that is an ellipse) with Tx and Rx as foci (Figure 2.69). The equation of the ellipses of constant time delay (or *isodelay contours*) is evaluated by assuming L to be the length of the base and $T = (R_T + R_R)/c$ to be the delay along the Tx-target-Rx, and is:

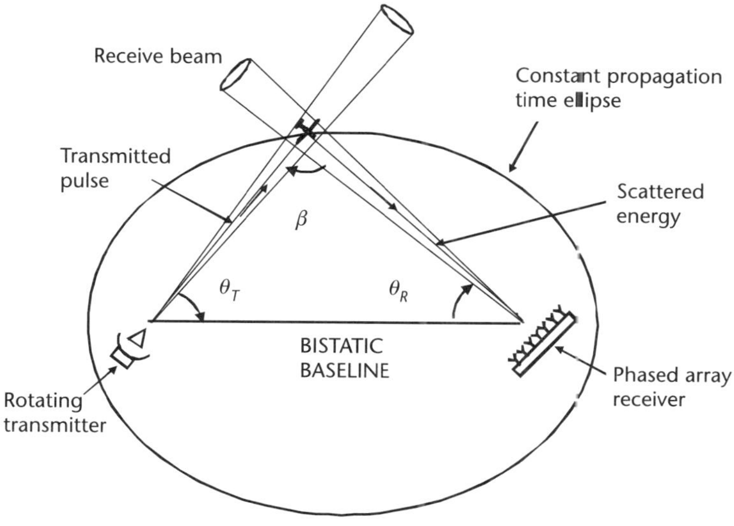

Figure 2.69 Geometry and principle of bistatic radar.

$$\frac{x^2}{(cT/2)^2} + \frac{y^2}{((cT/2)^2 - (L/2)^2} = 1 \tag{2.75}$$

The angles θ_T and θ_R are, respectively, the *look angles* or *line of sight* (LOS) to the target, which are taken as positive when measured clockwise with respect to the x-axis (the baseline). The *bistatic angle* $\beta = \pi - (\theta_R + \theta_T)$ is the angle between the two LOS at the target. A useful relationship is that the bisector of the bistatic angle is orthogonal to the tangent of the ellipse at the target. The tangent is often a good approximation to an isodelay contour.

The equation for the free-space bistatic radar detection range is derived straightforwardly from the monostatic radar's one:

$$P_r = \frac{P_t G_t}{4\pi R_T^2} \frac{\sigma}{4\pi R_r^2} \frac{G_r \lambda^2}{4\pi} \frac{1}{L_s} = \frac{C\sigma}{R_M^2} \tag{2.76}$$

where

Pt and P_r are, respectively, the transmitted and received power

G_t and G_r are, respectively, the transmitted and received antenna gain,

λ is the transmitted wavelength,

σ is the target RCS,

L_s are the overall system and propagation path losses,

$R_M = (R_T R_R)^{1/2}$ = benchmark range or equivalent monostatic range,

C is a constant that includes all the other parameters.

The free-space bistatic radar's benchmark range is a circle of radius R_M just as in the monostatic case (i.e., when the base length $L = 0$). For the general bistatic case, where $L > 0$, the free-space maximum detection contour becomes the familiar oval of Cassini. This oval provides a useful view of bistatic range coverage and it can be of four types in accordance with the ratio of base length to benchmark range L/R_M, as shown in Figure 2.70:

Benchmark [Figure 2.70(a)]: $L = 0$, monostatic case. Area $= \pi R^2_M$.

One oval [Figure 2.70(b)]: $L<2R_M$. Area $= \pi\,(R^2_M - (L^2/8\,R_M)^2)$. Max (min) range $= (R^2_M + L^2/4)^{1/2} \pm L/2$.

Lemniscate [Figure 2.70(c)]: $L = 2R_M$. Area $= 3\pi L^2/16$. Max (min) range $= (\sqrt{2}\pm 1)\,L/2$.

Two ovals [Figure 2.70(d)]: $L>2R_M$. Area $= \pi R^2_M (R^2_M/L^2)$. Max range $= R^2_M/L$, on R_X oval. For $L>3\,R_M$ the ovals are approximated as circles with radius $R_{Rav} = R^2_M/L$.

Expressions for receiver (and transmitter) ranges on the opposite oval are readily calculated through mirror-image symmetry.

Also note that the area of every bistatic oval is always less than the monostatic circle. Further the volume within which a bistatic radar can have both LOS

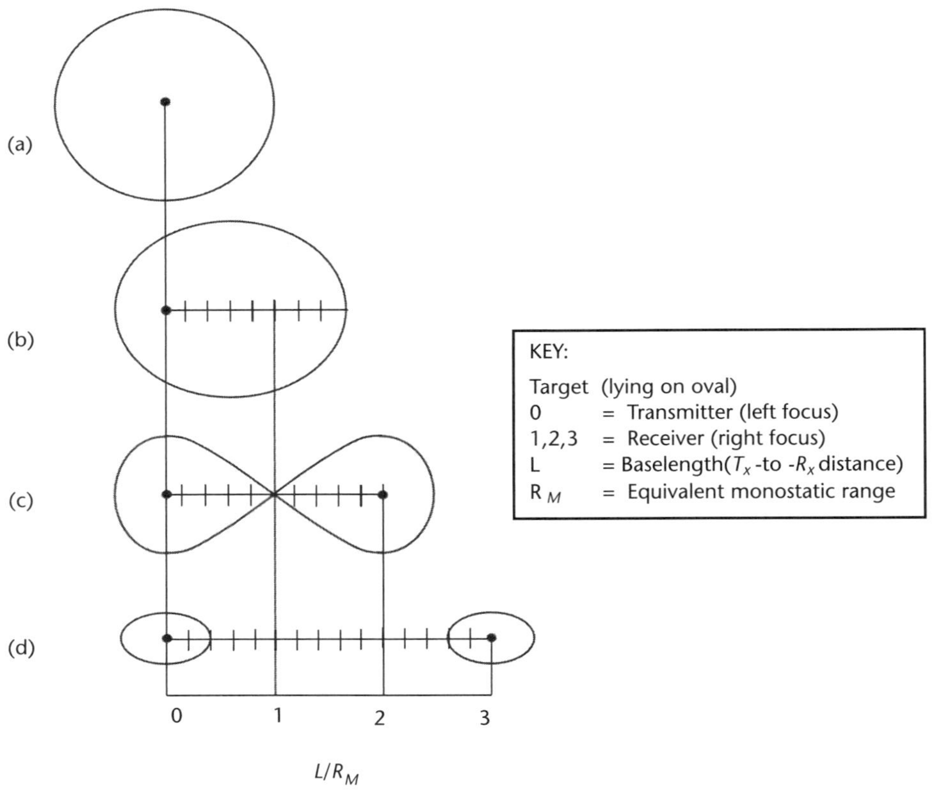

Figure 2.70 Normalized (L/R_M) ovals of Cassini, lying in the bistatic plane (i.e., the plane containing the Tx, Rx, and target. (From: [18].)

from the Tx and Rx to the target is severely affected by the wave propagation over Earth's curvature. In the case of a monostatic radar, the maximum horizon distance is evaluated (assuming the simple 4/3 Earth radius propagation model) as:

$$R_M = 130\left(\sqrt{H_t} + \sqrt{H_R}\right) \tag{2.77}$$

where all dimensions are in kilometers, H_t is the height of the target, H_R is the height of the Tx/Rx antenna, and the covered area is $A_M = \pi\, R^2{}_M$.

In the case of a bistatic area, the corresponding surface area for the same target at height H_t is provided by the overlap of two circles (as shown in Figure 2.71), one centered on the Tx with radius R_T given by

$$R_T = 130\left(\sqrt{H_t} + \sqrt{H_T}\right)$$

and the other centered on the Rx with radius R_R given by

$$R_R = 130\left(\sqrt{H_t} + \sqrt{H_R}\right)$$

The area of this overlap region is given in [19]: and has the form shown in Figure 2.71. The ratio of bistatic to monostatic target visibility area A_B/A_M as a function of baseline length L and target height (assuming $H_T = H_R$) is reported in [19].

2.3.11.1 Bistatic Radar Applications

The ovals of Cassini can be used to define three operating regions for bistatic radar: *co-site* ($L < 2R_M$), *Rx centered*, and *Tx centered* (respectively, in one of the two ovals for $L > 3R_M$).

By adding to the above classification the type of transmitter (dedicated, cooperative, and noncooperative), we arrive at Table 2.9.

A number of problems are encountered by radar hitchhiker operation (i.e., with the cooperative Tx of a monostatic radar) such as beam-on beam scan and the synchronization aspects. Indeed the hitchhiker receiver has to be able to reproduce the timings and frequencies of the Tx waveforms before it can measure accurately the time delays and Doppler shifts of target echoes. This is done by re-creating as

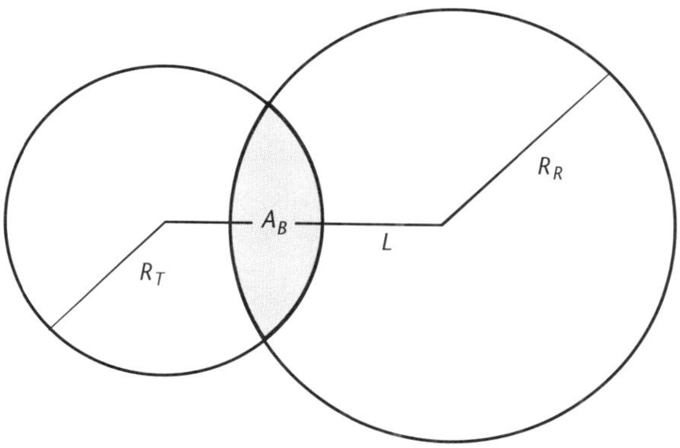

Figure 2.71 Area of overlapping region of bistatic radars.

Table 2.9 Bistatic Radar Applications and Examples (Adapted from: [18].)

Rx Operating Region	*Range to Base Length Relationship*	*Dedicated Tx*	*Cooperative Tx*	*Noncooperative Tx*
Co-site examples	$L < 2R_M$	Air surveillance [Sanctuary, Struna-1 (URSS)] Range instrumentation Satellite tracking [SPASUR, Graves (Fr)] Intrusion detection [bistatic radar for weapon location (BWRL)]	Air surveillance (Manastash ridge radar, Silent Sentry, HD-TV PBR) Range instrumentation (multistatic measurement system) Ionospheric measurements Wind measurements	Air surveillance [Klein Heidelberg (Ger, WWII)]
Rx–centered examples	$L > 2R_M$ $R_T >> R_R$		Short-range air surveillance (BAC) Silent air-to-ground attack (TBIRD) Planetary exploration (Covin Rest)	Short-range air surveillance
Tx–centered examples	$L > 2R_M$ $R_T << R_R$		Planetary exploration	Air threat monitoring Missile launch alert

a reference the Tx signal frequency and its modulations at the Rx site. Obtaining re-created frequency accuracy and stability of the reference signal for Doppler coherent processing requires frequency synthesizers locked to the rubidium atomic standard (stable to one part in 10^{11} per month). The block diagram of a typical hitchhiker radar is shown in Figure 2.72.

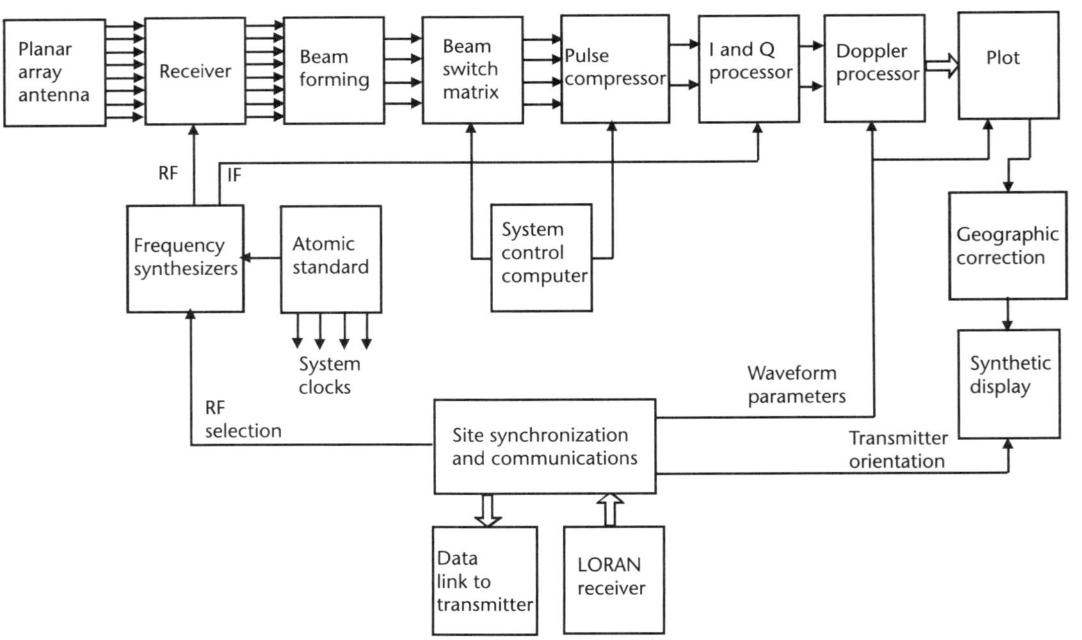

Figure 2.72 Block diagram of a typical bistatic radar receiver.

In beam scan-on-scan, when high-gain, narrow-beam scanning antennas are used by both Tx and Rx sites, inefficient use is made of the radar energy because only the volume common to both beams can be observed by the receiver at any given time (see Figure 2.73). The remedies to mitigate this problem are as follows:

1. *Step scanning*. This consists of fixing the Rx beam and waiting for the Tx beam to scan through the surveillance sector. The Rx beam is then stepped one adjacent beam for the next Tx beam scan and so on, until the Rx beam has stepped across the full surveillance sector. This remedy increases the surveillance frame time and is usually not acceptable for large-area surveillance. The remedy can be applied in situations when the base length is small and the Tx and Rx form a pseudo-monostatic geometry, as is the case for BRWL.
2. *Floodlight beams*. A floodlight beam can be used with either the Tx or Rx. The floodlight transmitter remedy obviously requires a dedicated Tx antenna to flood the surveillance sector continuously, while the Rx scans the sector with a high-gain antenna. This remedy achieves fair surveillance time frames. The floodlight receiver remedy can be used by a hitchhiker to flood a sector scanned by a high-gain Tx beam and to achieve as well as above a fair surveillance time frame.
3. *Multiple beams*. A bistatic receiver can use multiple simultaneous fixed Rx beams to cover the surveillance sector, which again restores the surveillance time frame. If the beams of the antennas all have the same gain, range performance is also restored. This remedy increases the cost and complexity

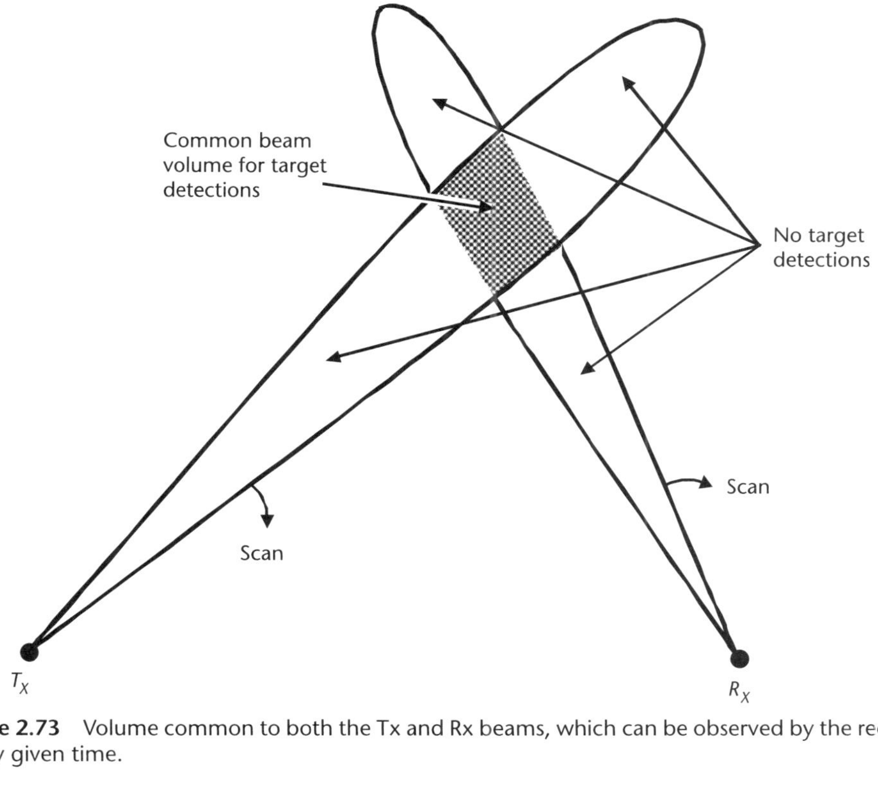

Figure 2.73 Volume common to both the Tx and Rx beams, which can be observed by the receiver at any given time.

of the receiver since a beamforming network and one signal processor per beam are required. The multibeam Rx can be used with any type of transmitter, including the floodlight Tx, where the reduced Tx gain can be compensated to increase the detection range by an increase of the dwell time on target from the multibeam Rx antenna.

2.3.11.2 Noncooperative Air Surveillance

Most passive bistatic radar (PBR) developments exploit broadcast transmitters as their source for radar illumination. Frequency modulation (FM) and high-definition television (HD-TV) terrestrial broadcast transmitters are the most attractive due to their high-power, noise-like, and relatively wide bandwidth signals for supporting air surveillance, otherwise assigned to VHF/UHF monostatic radars.

The 1985 trials of the Crystal Palace TV transmitter have dictated the requirements for the type of signal modulation to be used by subsequent broadcast transmitters: Doppler exploitation of stable, narrowband carrier lines in TV transmission and range/Doppler exploitation of the wider band, noise-like spectrum of FM transmission

The surveillance is covert, because even the transmitter is unaware it is being exploited, and provides counterstealth capabilities due to unavoidable A/C resonances in the VHF/UHF region. In this case the PBR has no control over their transmission and waveform properties, such as effective radiated power, transmission schedule, modulation type and content, and so forth.

The PBR's receiver provides target detection through a cross-correlation of the returned echo signal with a segment of the direct signal coming from the Tx. The cross-correlation is performed over the range of expected target echo time delays and Doppler shifts. Because the modulation content of many broadcast transmitters changes as a function of time, this complicates the matched filtering by the PBR's receiver. Indeed, a related modulation content inconvenience is provided by *dead air*, where no information is broadcast and the broadcast transmitter modulation is missing, thus increasing range measurement errors without limit. Dead air is a pause in modulation that can occur quite frequently (on the order of 1 Hz) when broadcasting a talk show or classical music. Consequently, once a track has been established, a nonlinear tracking filter may be needed to remove large error spikes. Further, especially in urban and suburban areas, the PBR can be affected by interference from the host emitter and other emitters, which can seriously degrade PBR performance.

In situ measurements of VHF and UHF bands in dense urban environments have shown an increase of up to 45 dB over the thermal noise. Normally it is about 25 dB (after spectral cancellation). The noise figures of the receivers are significantly higher than those obtained by the monostatic VHF/UHF radars and constitute a detection penalty for exploiting broadcast transmitters of opportunity.

The interference from the exploited Tx is called *direct-path breakthrough* and becomes particularly severe when the receiver is located in direct LOS of the transmitter, which must occur when surveillance of low-altitude air targets is required. If this direct path signal is not sufficiently attenuated, the received target signal becomes masked in range and often also in Doppler.

Direct-path breakthrough effects are similar to those of a spot noise jammer and can be characterized by an increase in the system noise temperature $T_s = F\ T_0$, where F is the receiver noise figure and $T_0 = 290°K$ is the standard temperature. To avoid the increase of T_s, it is necessary to provide an adequate decoupling (or attenuation) to the direct path signal. Typical required decoupling is on the order of 90 to 100 dB. This amount can be achieved by a combination of Earth masking, antenna shielding, and spatial and spectral cancellation. A brute-force remedy is to physically block the Tx signal from the receiver with a shroud or structure. Notwithstanding these masking and shielding techniques, spatial and spectral filtering always have to be added.

A typical modern air surveillance PBR is the CORA (COvert RAdar) developed by the Fraunhofer Institute (formerly FGAN), which operates with DAB and DVB-T transmitters of opportunity.

2.4 Communications

RF communications enable the exchange of information between two distant entities. As such, they play a vital part in the military field, which is based on real-time information reception at the headquarters from the battle area and distribution of commands from the latter to the deployed troops/assets in the battle area. This section provides an introduction to modern available communications technologies and techniques only to the depth necessary to understand the requirements for the EW equipment described in the following chapters that is relevant to the interception, classification, and jamming of hostile signals and receivers.

Military communication systems are either *broadcast* or *netted*. In broadcast communications information is generated at a single place (*source node*) and sent to many receivers (*sink nodes*). In netted communications, information is exchanged among the nodes of the net: Each node may act as a source or a sink (see Figure 2.74).

In tactical mobile networks there is often one node in the network that is referred to as the *net control station* (NCS). Usually this node is associated with the person responsible for the network and, hence, there is often more communication

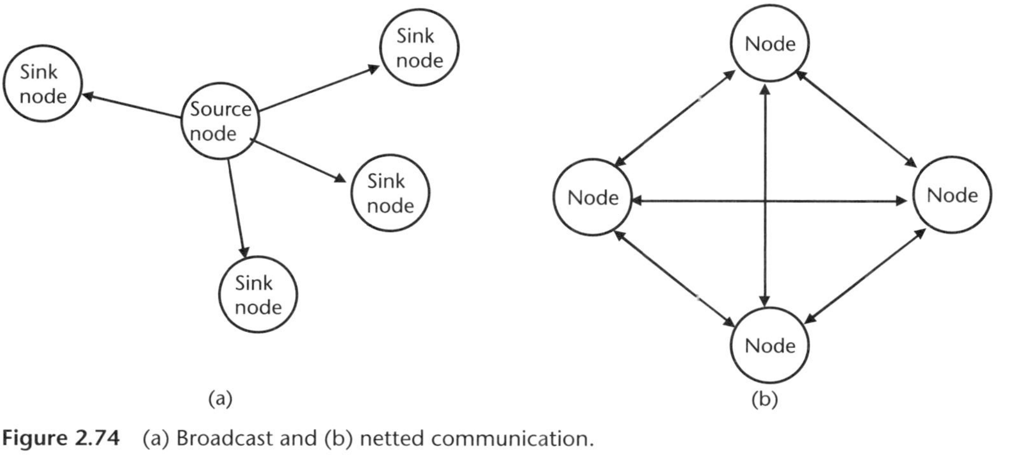

Figure 2.74 (a) Broadcast and (b) netted communication.

traffic from this node with respect to the other nodes. This is a feature that can be exploited by communication intercept equipment to identify the NCS.

The information is transported between radio equipment by modulating (i.e., modifying) a carrier signal. Modulation methods are of two types: analog and digital. The objective of the various available modulation techniques is to achieve *spectrum efficiency*; that is, to achieve a large amount of information within a given amount of bandwidth so as to make available the maximum number of channels within the bandwidth assigned to a service.

Three fundamental analog techniques are used to modulate a carrier signal, each of which is applicable to one of the basic parameters of a sinusoidal signal:

$$s(t) = A(t)\cos(2Gpf_c t + \varphi(t)) = A(t)\cos(\phi(t))$$

- Amplitude $A(t)$ (amplitude modulation, AM)
- Frequency f_c (frequency modulation, FM)
- Instantaneous phase $\Phi(t)$ (phase modulation, PM)

Because both frequency and phase are the argument of the cosine, the relevant modulations are also referred as *angle modulations*. The signal that is modified by modulation is called the *carrier signal*, whereas the information-bearing signal, which performs the modification, is called the *modulating signal* and is denoted by $m(t)$.

The analog AM and FM communication techniques are not very spectrum efficient; the PM techniques, especially in digital form, are more spectrum efficient and are therefore used in modern communication systems.

Digital modulation techniques are rapidly replacing, both in the military and civilian domains, the older analog communications. Digital techniques are more spectrum efficient and hence the precious VHF/UHF spectrum can be used to support more communication channels. Digital communications further satisfy the above requirement because they are operated at the higher frequency ranges of the spectrum and can meet two currently increasing demands.

Increased data volumes at higher data rates and larger occupation bandwidths.

The use of secure communications, which require modern encryption technologies.

Modern military communication systems employ digital modulations. However, in asymmetric conflicts it is still possible to come across communication systems that employ traditional analog modulation methods and therefore it is beneficial to the scope of this book to provide some brief notes about them.

At the receiver site the modulated carrier signal has to be demodulated to retrieve the information content, usually expressed with symbols. The two types of demodulation are *noncoherent* (in which no carrier phasing is used) and *coherent* (in which the carrier phase information is used). Coherent demodulation is more efficient than one noncoherent in recognizing the transmitted symbols of digital signalling, although at the expense of requiring more complex hardware implementations.

All demodulation techniques need to recover the symbol clock. The coherent PM demodulation techniques further need to know where the reference carrier phase is located in order to synchronize the receiver with the incoming signal.

Communications services are spread virtually all over the RF spectrum, the bands for which are shown in Table 2.10.

In the following subsections (and in Appendix D, in which the analog signal modulation techniques are described), the principal methods of digital signaling and coding will be tackled, followed by a discussion of secure communications systems entailing spread-spectrum and frequency-hopping techniques and a short description of the Military Information Distribution System (MIDS).

2.4.1 Access Methods

The use of a communication channel is ruled by the *access methods* of the communication systems. There are three principal access methods and each takes its name from one characteristic of the communication waveform:

1. *Frequency-division multiple access* (FDMA), in which the available frequency channels are divided among the users. Each user normally keeps the assigned channel until it needs to send messages and then it leaves it to a subsequent user. The FDMA scheme is shown in Figure 2.75(a).
2. *Time-division multiple access* (TDMA), in which the same channel is shared in time among many users. Switching among users happens so fast that the users do not perceive that the channel is shared. The TDMA scheme is shown in Figure 2.75(b). TDMA has a number of implementation schemes such as ALOHA, slotted ALOHA, CSMA, DAMA [20].
3. *Code-division multiple access* (CDMA),in which direct synthesis spread-spectrum (DSSS) or frequency-hopping spread-spectrum (FHSS) techniques are used to share among the users all of the frequencies of the communication system band. Each user is assigned a different spreading code.

Table 2.10 Frequency Band Designation and Services

Wavelength	*Frequency Band*	*Name*	*Designation*	*Prevalent Type of Service*
100– 10 km	3–30 kHz	Very low frequency	VLF	Radionavigation
10–1 km	30–300 kHz	Low frequency	LF	Maritime mobile
1–0.1 km	300–3000 kHz	Medium frequency	MF	Broadcast AM
100–10m	3–30 MHz	High frequency	HF	Maritime mobile
10–1m	30–300 MHz	Very high frequency		
	VHF	Broadcast FM, TV channels, aeronautical mobile		
1–0.1m	300–3000 MHz	Ultra-high frequency	UHF	TV channels, aero radionavigation, satellite navigation
4.2–3.75 cm	7–8 GHz	Super high frequency	SHF	Satellite communication
1.5–0.68 cm	20–44 GHz	Extremely high frequency	EHF	Satellite communication

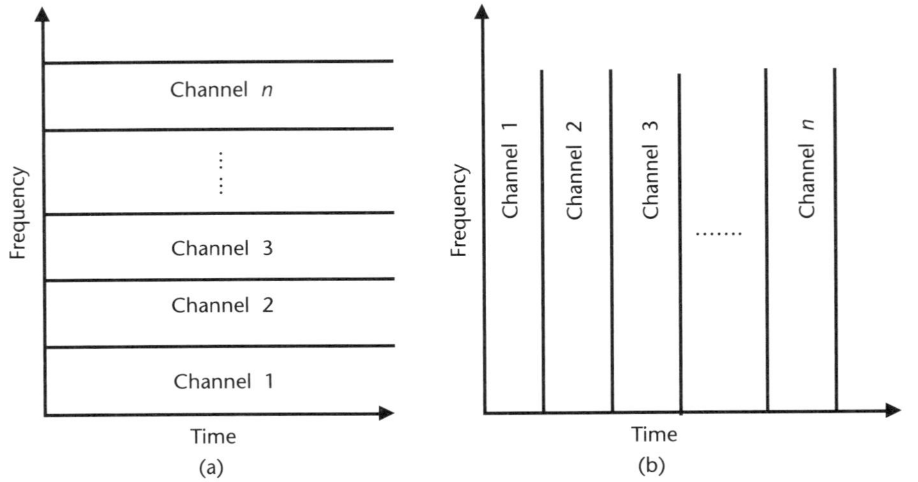

Figure 2.75 Principles of (a) FDMA and (b) TDMA access methods.

Examples of CDMA will be given when discussing the typical military communications system at the end of Section 2.4.

2.4.2 Digital Signaling

In recent years digital signaling techniques have become the preferred signal modulation communication techniques because they are more spectrum efficient than the analog ones. The analog signal containing the information to be transported by the communication system is sampled at some rate, and each sample is amplitude converted in digital form by ADCs. The digital representation of the signal is called the *pulse code* or *pulse word*. Pulse code modulation (PCM) is obtained by modulating each pulse code onto the carrier. PCM can be accomplished in many forms, among which the most used ones are (1) differential PCM (DPCM) and (2) adaptive differential PCM (ADPCM).

PCM assigns a number of digital bits to the amplitude of each digital sample of the analog signal. If the signal is not changing too rapidly from one sample to the next, there will be a redundancy of bits in the digitized amplitudes of the samples. To avoid such redundancy, DPCM encodes only the difference in amplitude between the two samples. Frequently there is only 1 bit difference between the samples: +1 bit if the next sample amplitude is larger than the current one; –1 bit if the next sample amplitude is smaller.

In a case where the signal amplitude may experience abrupt changes from one sample to the next, ADPCM senses such a difference over a limited number of successive samples and reacts by providing a larger number of bits to compensate or exceed the signal amplitude until the amplitude difference between samples is restored (i.e., amplitude tracking is accomplished and usual DPCM applies).

The most used digital modulations are:

1. *Amplitude shift key* (ASK), in which the amplitude of the signal is changed between two (or more) levels [see Figure 2.76(a)].

2. *Frequency shift key* (FSK), in which the carrier signal is changed between two (or more) frequencies [see Figure 2.76(b)].
3. *Phase shift key* (PSK), where a number of methods are used:
   - *Binary PSK* (BPSK), which uses two phase states (0 and π radians) corresponding to symbols 1 and –1 (or 0). The implementation of the modulation is as simple as multiplying the carrier by +1 (carrier phase unchanged) and –1 (which is the same as shifting the phase by π radians) [see Figure 2.76(c)].
   - *Quadrature PSK* (QPSK), in which two data bits are combined to form a single symbol, as shown in the relevant constellation of four states [see Figure 2.76(d)].
   - *Offset QPSK* (OQPSK), which also uses a four states constellation but they are offset by $\pi/4$ radians with respect to QPSK.

Note from the Figures 2.76(c) and)d) that there are amplitude variations in the BPSK and QPSK modulations where there are phase discontinuities, with maximum amplitude variations at phase shifts of π radians. This feature is not appropriate in implementations where the amplifiers are operating in saturation (such as in satellite and cellular phone communications) because the input amplitude variations produce intermodulated signals. The inconvenience has been rectified

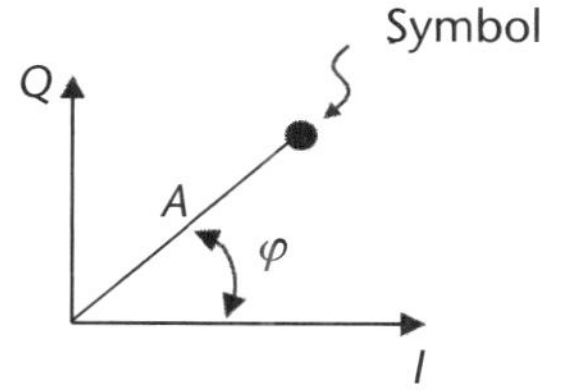

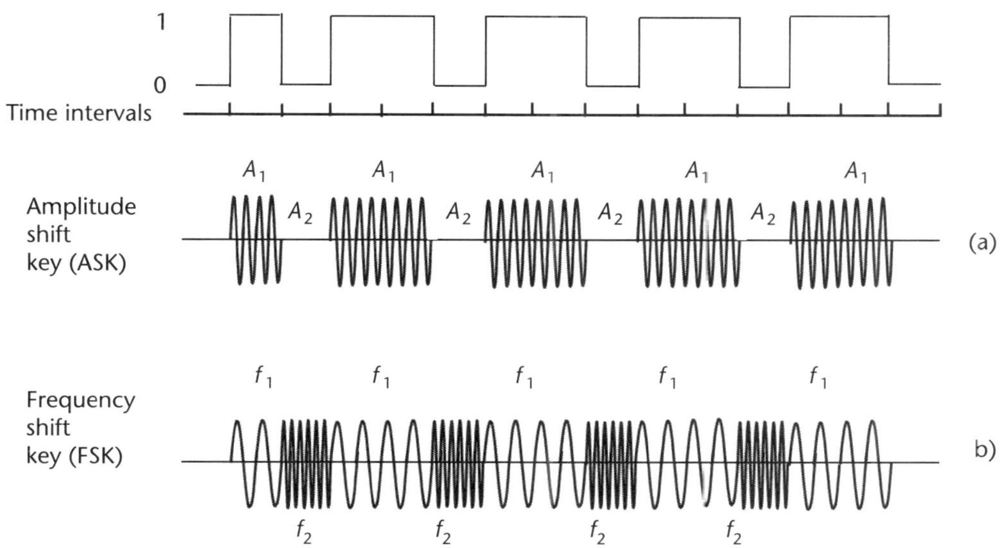

Figure 2.76 (a) ASK digital amplitude modulation. (b) FSK digital frequency modulation. (c) BPSK digital phase modulation. (d) QPSK digital phase modulation.

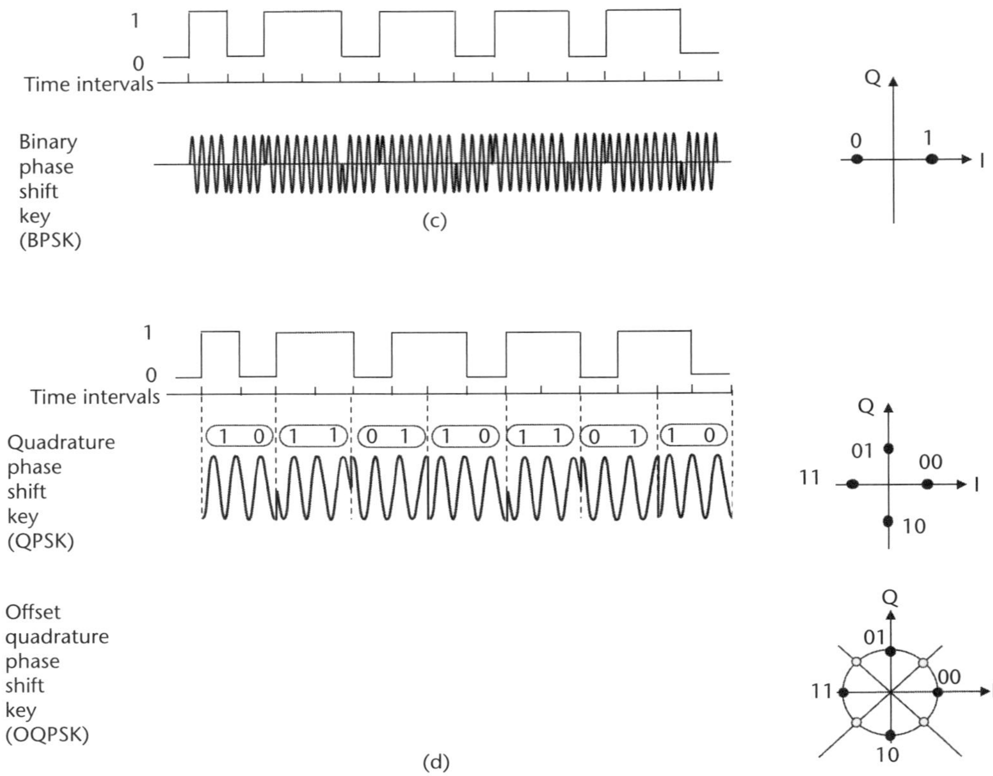

Figure 2.76 (continued)

by using the $\pi/4$ OQPSK where only $\pi/4$ and $3\pi/4$ phase transitions are allowed, preventing the signal states on the I-axis, where the amplitude crosses the zero.

Upon each symbol transition the constellation shifts counterclockwise by $\pi/4$ radians (which gives the prefix "offset" to this modulation). So the symbol which at clock period iT_c is at $\pi/4$ radians, and might correspond to data bit 00, at clock period $(i+1)T_c$, is at $\pi/2$ radians which still correspond to data bit 00. OQPSK modulation is used in some second-generation cellular phone systems such as the European Tetra.

A FSK modulation, called minimum shift key (MSK), is implemented with a frequency shift $(f_2 - f_1) = 1/2T_c$ (i.e., one-half of the modulating signal bit rate), which is the smallest shift that can maintain phase continuity. The MSK technique, with the application of a filter with a Gaussian transfer function of the type $H(F) = \exp(-\mu f^2)$ to the modulating signal (therefore the technique is called GMSK), is applied in the European second-generation cellular phone systems called Global System Mobile (GSM).

Further PSK techniques, the so-called 2^nPSK, combine n bits to form higher types of constellations for transmission. Current available techniques can be implemented up to $n = 8$.

One modulation technique that permits transmission of n bits per symbol is quadrature amplitude modulation (QAM) . This technique modulates both I and Q amplitudes in addition to the phase modulation. The I and Q values are transmitted as two simultaneous data streamed using the quadrature modulation scheme shown in the previous paragraph. In Figure 2.77 the constellation for QAM with 3 bits is shown ($2^3 = 8$ symbols). QAM modulations are more bandwidth efficient (i.e., they have higher data rates per unit bandwidth) and are used extensively for terrestrial radio links. However, because QAM techniques use more amplitude levels, they require linear amplifications on transmission, which are not suitable for satellite communications.

One high-speed (several megabits per second) modulation technique used for mitigating the typical negative effects experienced by mobile communication channels, such as multipath, interference, and impulsive noise, is orthogonal frequency-division multiplexing (OFDM), also called discrete multitone (DMT). In this FDM technique the adjacent channels are orthogonal to each other, but the overlapping sidebands on the adjacent channels do not affect the effective signaling of each channel. Thus, although each channel may carry a relatively low bit rate, the bunch of channels together may achieve a high bit rate.

As for any communication, signaling noise, either externally or internally induced, can cause a received symbol to be in error. The bit error rate (BER) of digital signaling is a function of the postdetection or demodulated SNR. The SNR required of most common digital modulations to achieve a BER of $<10^{-2}$ ranges from 8 dB (for BPSK) to 12 dB (for FSK) [20].

2.4.3 Secure Communications

Secure communications must resist unintentional and intentional (jamming) interference, detection by an adversary, and unauthorized extraction of information from a transmitted waveform. Spread-spectrum techniques are used to provide these secure communications requirements.

There are two fundamental spread-spectrum techniques: *frequency hopping* (FH) and *direct sequence* (DS). In the former, the transmitted carrier frequency of the signal is changed periodically, or "hopped," at a high rate, and covertness is achieved because only the associated receiver knows to which frequency the signal

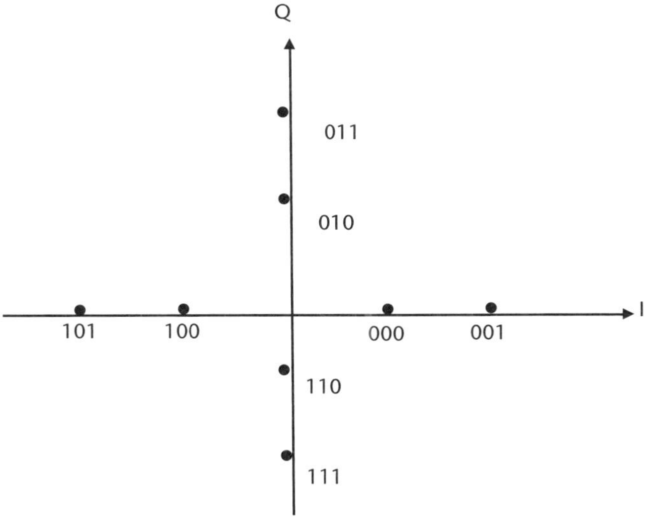

Figure 2.77 8QAM constellation.

has hopped. In the latter the energy in a relatively narrowband information signal is spread over a much larger bandwidth, so that in any narrowband portion of the spectrum occupied by the signal there is very little energy (frequently below the level of thermal noise present) in order to make the signal undetectable. Again, only the receiver knows the despreading key required to recover the narrowband information signal.

The principles and applied means of spread-spectrum (SS) techniques are used also in modern LPI radars and are discussed in the relevant previous sections, so that discussion is not repeated here. In the following sections, examples of SS communication systems are discussed. Namely, an example of a DSSS communication system (the Global Navigation Satellite System) is illustrated in Section 2.5 and an example of a FHSS communication system (the Joint Tactical Information System) is illustrated in Section 2.4.6.

2.4.4 Coding of Communication Signals

Coding is implemented either to remove redundancy in the source information (e.g., speech or images), in which case the coding is referred to as a *data compression* technique, or to improve the reliability of the communication (i.e., the probability that the correct message is received), in which case the coding is referred to as an *error control* technique.

Data compression techniques are used to reduce the content of any information to the minimum total number of bits to be transmitted. For example, in text or speech applications, the Huffman encoding technique is used in the allocation of the number of bits to represent a symbol because it assigns fewer bits to the symbols that are more frequently used. Further, phrases that are often repeated are addressed using pointers to previously sent phrases. These techniques are called *lossless* because no information in the transmitted message is lost. They can typically achieve a compression ratio on the order 10:1 [20].

Higher compression ratios are achieved in image transmission and storage techniques, which are widely used in both civilian and military domains. Redundancy in video signals comes both from the spatial redundancy within a single image as well as from the temporal redundancy from one image to the next.

Spatial redundancy in still pictures is addressed by the Joint Pictures Expert Group (JPEG) standard, which applies both the discrete cosine transform (DCT) method to transform a segment of 64 pixels of the picture (each pixel should be normally represented by a 10-bit word) and Huffman encoding to all the various segments to reduce the number of bits to represent the picture. The result is extremely efficient coding (called *intraframe coding*) of single images.

Temporal redundancy from one picture to the next comes from the fact that, at the speed of motion pictures, the objects in the image do not move much. Temporal redundancy is addressed by *interframe coding* such as the standards developed by the Motion Pictures Expert Group (MPEG) and called MPEG1,MPEG2, and MPEG4. More details on this subject can be found in [20].

The data rates of image transmission codes (except for the MPEG4 standard) require larger bandwidths (usually available at quite high carrier frequencies) than those used by military tactical radios in the battle area. Higher frequency equipment, however, is available at division and above levels, which make use of image communications.

2.4.4.1 Channel Coding

Channel coding is normally performed in communication and data storage systems for error control, because no electronic system is immune from noise generated either externally or internally to the system and noise bit errors inevitably occur. Some types of communications, such as digital speech, can tolerate a large number of bit errors before the speech message becomes unintelligible; other types, however, such as computer-to-computer exchanges, must be performed almost error free. In some cases it is sufficient to detect errors in order to discard the data containing the bit error and to request the data to be resent; in other cases, the data must be encoded so that, up to a certain extent, it is possible to correct them at the receiver. Error detection methods are discussed next.

Parity Bit Addition

One bit is added to the data word so that the number of 1's in a word is even (or odd). A parity check at the receiver immediately checks if one data word contains a bit error. The bit addition can also be applied when the data words are organized into a matrix, in which each element is constituted by a single bit of a word. In this case the parity bit is added to each row and to each column of the matrix, so that if a single error occurs, the bit in error is not only immediately detected (because a row and column in error identify a single bit at their crossing) but also corrected.

Cyclic Redundancy Check

The cyclic redundancy check (CRC) coding technique for error detection is widely used for automatic repeat request (ARQ) schemes, because it is easy to implement in hardware (and in fact there are integrated circuits available that compute the CRC). It is based on the notion that binary data streams can be treated as binary polynomials where the ordered bits c_i of the data word are the coefficients of the polynomial $d(x)$, which is called a *frame:*

$$d(x) = \sum_{0}^{n} c_i x^i$$

To each frame $d(x)$, a frame check sequence (FCS) is added on transmission. The FCS is constituted by the product of the remainder polynomial $r(x)$ obtained from the division of the frame polynomial $d(x)$ with a particularly selected polynomial $g(x)$:

$$\frac{d(x)}{g(x)} = q(x) + r(x)$$

and of the same polynomial $g(x)$, so that the transmitted data word is:

$$T(x) = d(x) - r(x)g(x)$$

On reception $T(x)$ is divided by $g(x)$ and if there are no errors in the transmission, then the remainder of this division is zero.

This code effectively detects bit errors and it has been estimated that when using a 17-bit polynomial the probability of 1 bit error undetected is 10^{-14} [20].

Checksum

With the checksum method, the set of data words (each word corresponds to a value $<2^n$, where n is the number of bits in the word) is summed together, without worrying about overflow, thus obtaining the checksum (modulus 2^n). The checksum is attached to the data, becoming part of the data set. When the set of transmitted data is summed on reception, the result is zero (modulus 2^n) unless there is an error.

This method provides a good probability of error detection. In fact, it has been evaluated that when summing a set of data words of 16 bits the probability that the checksum is not declaring an error in the presence of an inaccurate data set is $(1/64) \times 10^{-3} \leq 2^{-16}$.

Error correction is performed by two techniques: ARQ and forward error correction (FEC). The ARQ method, as seen previously for the CRC error detection method, adds extra bits to the transmitted data stream so that the check at the receiver can detect an error in the data stream. If that happens, a request to resend the data (NACK = not acknowledged) is made. This method of error correction is very ineffective for cases of low SNR in the received data; once a NACK response is sent, the probability of receiving a further data stream with an error is high and thus the channel can quickly get clogged resending messages received in error.

FEC methods add some bits to each digital word prior to conversion into the transmitted symbols in order to correct errors at the receiver. The bit addition reduces the bandwidth efficiency, so to send at the desired data rate, an increased bandwidth is required. This in turn implies that the noise power density is increased and the SNR lowered, thus increasing the BER. The goal of FEC design, therefore, is to reach an optimum trade-off between the number of bits to be added versus the bandwidth required to transmit the code words.

There are three major types of FEC:

1. *Block codes*, which encode large blocks of data at a time; the encoding depends only on the current block of data.
2. *Convolutional codes,* which encode a block of data, but the encoding depends not only on the current block of data but also on prior blocks of data.
3. *Cyclic codes*, where all code words are permutations of one original code word. Bose-Chaudhuri-Hocquenghem (BCH) codes and Reed-Solomon codes are examples of them.

A discussion about the performance of the preceding codes is beyond the scope of this book. We recommend the interested reader refer to [20, 21] for deeper insight into the matter.

The advantages of coding in terms of BER vs. SNR for several coding methods are described in [20], in which it is reported that very good performance is possible for low SNRs over uncoded data streams.

2.4.5 Typical Military Communication Systems

Military communication systems operate in allocated bands and are of three types [22]:

1. Combat net radios, which provide short- and medium-range voice (digitally modulated) communication and operate in the VHF band (28 to 88 MHz);
2. Tactical data links, which are used to transfer data within a C2 system from one node to another. They operate in two different bands, either UHF (225 to 400 MHz) or D-band (formerly the radar L-band, 960 to 1215 MHz (with some holes);
3. Satellite communications, which provide very-long-range data exchange between users.

These systems operate in different bands, such as UHF (225 to 400 MHz), SHF (7 to 8 GHz), and EHF (20 to 44 GHz), with different frequencies relevant to uplinks and downlinks.

Most combat net radios (CNRs) employ the FH technique, which can be of two types: *slow FH* (SFH) and *fast FH* (FFH), in accordance with the number of bits transmitted per frequency hop. The SFH technique is characterized by one or more bits transmitted within the same frequency hop, whereas the FFH technique is characterized by one or more frequency hops per transmitted bit. Both techniques cover the entire allocated bandwidth with a number of channels of bandwidth B_{ch}, which is 25 kHz in the VHF band and 50 kHz in the UHF band.

FH systems use a set of stable frequency synthesizers or direct digital synthesizers (DDSs) in both transmitting and receiving stations. The sequence of transmitted frequencies is generated on the basis of a PN code, which is known to both users. Demodulation of the transmitted message is accomplished after the receiver has accomplished the acquisition and synchronization phases of the transmitted code. Current CNR systems such as SINCGARS-V, Jaguar-V, and its successor Panther-2000-V operate in SFH mode with hopping rates ranging from 100 to >1,000 hop/s.

The principal tactical data link (TDL) in the NATO Alliance is the Military Information Distribution System, also known as Joint Tactical IDS (JTIDS) in the United States or Link 16. This system was developed at the end of 1980s to provide an interoperable, secure, digital communication system among the whole allied military network. MIDS operates in the 960- to 1215-MHz bandwidth, with some bandwidth holes allocated to other services (e.g., IFF and GPS). The available number of hopping frequencies within this bandwidth is 51 with a spacing of 3 MHz between them. The hopping frequencies are selected in accordance with a PN code with the rule that two adjacent transmissions must be separated by at least 30 MHz.

The TDL operates using the TDMA method, so that each user is assigned one or more adjacent timeslots to transmit its message to the network. The TDMA timing is periodic. Each period is called an *epoch* and has a duration of 12.8 minutes The epoch is composed of 64 *frames*, each with a duration of 12 sec. The frame is in turn composed of 1536 *timeslots,* each with a duration of 7.8125 ms. Thus the number of timeslots within an epoch is 98,304.

Each time slot is composed of:

- A jitter, which is a variable dead time used to make interception more difficult.
- A synchronization preamble followed by an identification band in which the receiver determines whether it is interested in the content of the message. The duration of the synchronization preamble is 0.936 ms.
- A message, which is spread over 109 information packets. Each packet is composed of 5 bits, which are mapped on 32 chips in accordance with 2-CPFSK digital modulation (thus applying internally to the message the DSS method), each having a duration of 200 ns, for a total duration of 6.4 μs. After each transmitted packet a dead time of 6.6 μs is applied. Thus, each message has a duration of 13 μs and the instantaneous bandwidth occupied by the message is 10 MHz. The total time of transmission is $109 \times 13 = 1.417$ ms.
- A propagation guard-time (on the order of 2 or 4 ms in accordance with the selected normal (300-nautical-mile [nm]) or extended (700-nm) network range.

A representation of the MIDS timing is shown in Figure 2.78.

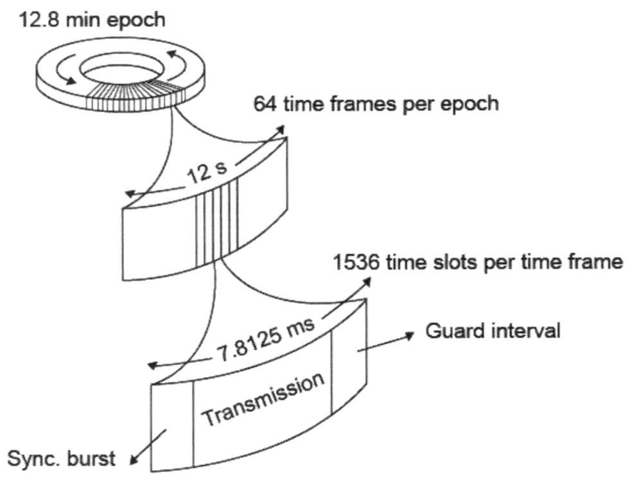

Mode	Guard Interval (ms)	Guard Interval Range Limit (nm)	Throughput After Error Correction (Kbps)	Hops/second
Standard Full Shot	4.4585	700	30	33,000
Packed-2 Full Shot	4.4585	700	59	33,000
Packed-4 Full Shot	2.0405	300	119	57,000

Figure 2.78 Representation of the MIDS timing.

Potentially, by employing the total guard-time to transmit on another "stacked net," the MIDS could operate at a frequency hop rate of $109 \times 1/(1.417 \times 10^{-3}) \approx$ 77,000 hop/s. In Figure 2.79 the actual hop rates are reported. Military satellite communications (SATCOM) operate in various bands: UHF (250 to 400 MHz), SHF (7 to 8 GHz), and EHF (20 to 44 GHz). In the past SATCOM required the use of a constellation of communication satellites, but currently the use of commercial low-Earth-orbiting (LEO) satellites is prevailing with special slots or services dedicated to the military (e.g., the use of INMARSAT commercial satellites for voice traffic).

The recent military satellite system MILSTAR makes use of geostationary-Earth-orbiting (GEO) satellites provided with uplinks at EHF (44 GHz) and UHF (300 MHz) and downlinks, respectively, at SHF (20 GHz) and UHF (250 MHz). Cross-links between satellites are provided at 60 GHz. These satellites practically operate in the LPI mode because of their narrow beamwidth at EHF and the use of FFH SS techniques.

2.5 Satellite Navigation Systems

Navigation refers to being able to determine the position and velocity of a moving platform or body with respect to a known reference. In this section, after briefly recalling the principles of operation of some radionavigation systems of interest in the military domain, we will focus on satellite navigation systems, which with their very accurate 3-D position measurements have become a powerful guidance system for unmanned weapons systems and munitions.

There are five basic radionavigation techniques:

1. Marker beacons: the reception of a beacon's signal indicates that the platform is in the vicinity of the transmitter.
2. Bearing and elevation technique, such as VOR (VHF omnidirectional radio range) or ILS (instrumented landing system), in which the reference station broadcasts a signal that varies with the direction of transmission, thus enabling the receiving platform to obtain a measurement of bearing and elevation to the reference station without the need for a directional antenna.
3. Passive ranging, or TOA system, such as the various GNSS (global navigation satellite systems), in which the reference station broadcasts a timing signal from which the platform can deduce its range from the transmitter. This requires that both transmitter and receiver clocks be synchronized.
4. Two-way ranging systems, such as DME (distance measuring equipment) in which the platform transmits a call to the reference station, which then transmits its signal back at a predetermined interval. This avoids the need for clock synchronization but introduces, limitations to the number of users and repetition rate.
5. Hyperbolic ranging system, such as LORAN (long-range navigation), in which the platform receiver measures the time difference (TD) between the received signals from two reference stations, removing the need to synchronize the receiver clock. The locus of platform position from the TD measurement is a hyperbola.

GNSS is the collective term for those navigation systems that provide the user with a 3-D positioning solution by passive ranging via radio signals transmitted by orbiting satellites. A number of those systems are currently available, but only three of them provide complete Earth coverage: GPS, GLONASS, and Galileo. The most well-known of them is the Navigation by Satellite Timing and Ranging (NAVSTAR) GPS, owned and operated by the U.S. government.

A description of the architecture, services, and determination of the user position and velocity from ranging measurements is beyond the objective of this book; we recommend interested reader refer to the book by P. Groves [23] for deeper insight in the subject. Here we are interested only in the structure of the GNSS signals, because they constitute a good example of a modern spread-spectrum communication system. Spread-spectrum modulations are used to obtain accurate range measurements (on the order of 2 to 6m rms) at receivers near the ground from weak received signals (as they are transmitted from satellites at a couple of tens of thousands of kilometers); thus, they are prone to severe interference.

GNSS signals are broadcast in a number of frequency bands within 1.1 to 1.6 GHz and are usually transmitted with quadrature modulation (see Appendix D). Right-handed circular polarization (RHCP) has been used for all GNSS signals. GNSS signals combine a carrier with a navigation data message $D(t)$ and a spreading or ranging code $C(t)$, both using BPSK modulation. The overall transmitted signal is expressed as:

$$s(t) = \sqrt{2P}C(t)D(t)\cos(2\pi f_c t + \phi_c) \tag{2.78}$$

where P = signal power; f_c = carrier frequency, ϕ_c = phase offset, and both C and D have an amplitude of ±1, varying with time.

The data message rate f_d typically varies between 50 and 500 chip/s (the duration of the message bit is $T_d = 1f_d$), while the spreading code rate ν varies between 0.511 and 10.23 Mchip/s (and its duration is $T_c = 1/\nu$). The energy per bit is thus $E = PT_d$.

The spread-spectrum communication scheme is represented in Figure 2.79 where $j(t)$ represents an interfering signal (unintentional or intentional, i.e., a jammer).

The spreading code $C(t)$ consists of a PRN sequence, which is known to the receiver. The effect of the spreading code is to increase the double-sided bandwidth of the navigation message to twice the spreading code chipping rate, while proportionately decreasing the power spectral density. The representations of the involved

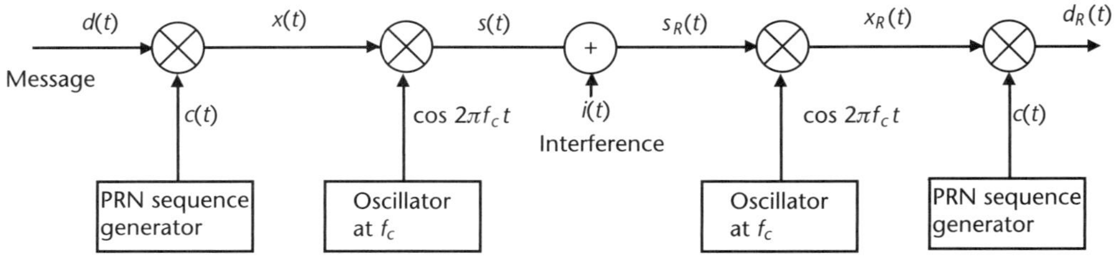

Figure 2.79 Scheme of coded signal communication.

signals $D(t)$, $C(t)$, and $x(t)$ are shown in Figure 2.80, where it is assumed that the interference is at the same carrier frequency of the signal.

In the receiver, the incoming spread-spectrum signal

$$r(t) = s(t) + j(t) = \sqrt{2P}C(t)D(t)\cos(2\pi f_c t + \phi_c) + j(t)\text{ccs}(2\pi f_c t)$$

is multiplied first by a replica of the signal from the transmitter oscillator ($cos(2\pi f_c t+\phi_c)$,that is, a synchronized replica, and then to a replica of the spreading code $c(t)$, thus obtaining $D_R(t)$ (for simplicity we assume $\phi_c = 0$):

$$\begin{aligned} D_R(t) &= \sqrt{2P}C^2(t)D(t)\cdot\cos^2(2\pi f_c t) + j(t)C(t)\cos^2(2\pi f_c t) \\ &= \frac{1}{2}\sqrt{2P}C^2(t)D(t)\left[1+\cos(4\pi f_c t)\right] + \frac{1}{2}j(t)C(t)\left[1+\cos(4\pi f_c t)\right] \end{aligned}$$

and by filtering out the high-frequency terms we have:

$$D_R(t) = \frac{1}{2}\sqrt{2P}C^2(t)D(t) + j(t)C(t)$$

The first term is just the correlation of the spreading signal and, as the two PRN generators are synchronized $C^2(t) = 1$ for all t, it is the message signal multiplied by a constant, while the second term is the interference signal $j(t)$ spread over the much larger bandwidth of the spreading signal. Thus, its energy in the

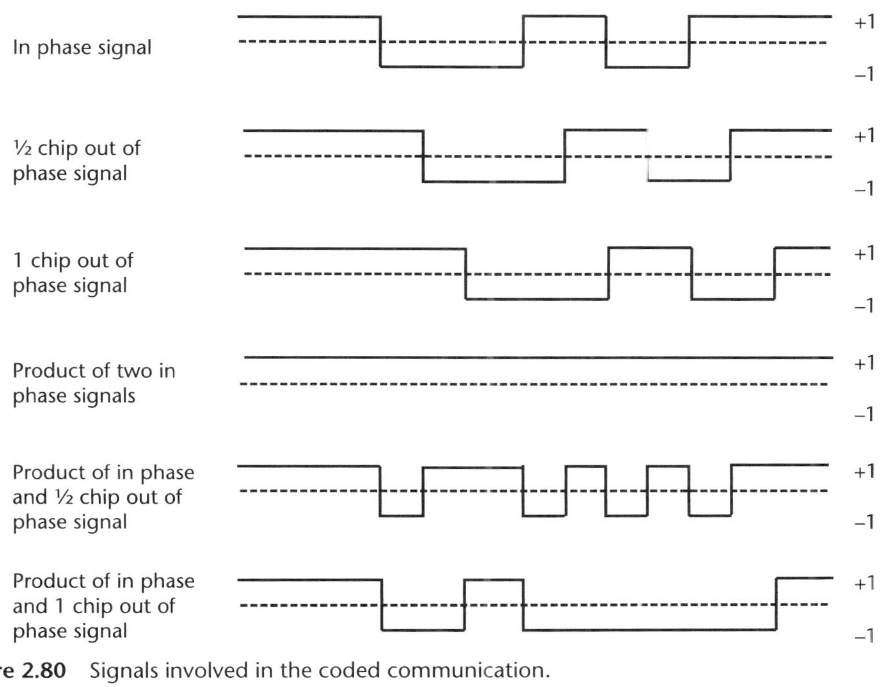

Figure 2.80 Signals involved in the coded communication.

message bandwidth $B_d = 1/T_d$ is quite low (if P_j is the power of the interference in a bandwidth B_j, then at the output of the receiver it is P_j/ν). The changes in the signal spectra along the complete modulation and demodulation process are shown in Figure 2.81.

If the signal and receiver-generated spreading codes are different, the correlation between them is much lower than if they are the same and aligned. Therefore, a number of different signals with different spreading codes may be broadcast simultaneously on the same carrier frequency, thus exploiting the channel CDMA technique.

The receiver selects the PRN spreading code relevant to the desired satellite signal. If the phase of the receiver-generated PRN code is not aligned with the transmitted signal, the receiver adjusts the phase so that all phases of its generated PRN code are searched until the correlation peak is found. This process is called *acquisition* of the transmitted signal. Once acquisition is complete, the receiver can operate at a lower SNR than acquisition because it is possible to vary the receiver-generated code phase slightly to reacquire the correlation peak. This process is called *tracking* of the transmitted signal. Also the carrier component of the GNSS signal is tracked to enhance the two processes of acquisition and tracking of the PRN code.

Many of the newer GNSS signals use FEC techniques for the data message $D(t)$. This introduces redundancy into the data and, although a higher symbol rate must be used to keep the message transmission unchanged for the duration, the data message can be successfully decoded in a poorer SNR environment.

Most newer GNSS signals use binary offset carrier (BOC) modulation instead of BPSK. This adds an extra component to the signal, the subcarrier $G(t)$, a square-wave signal at frequency f_g, which spreads the signal into two sub-bands, centered at $f_c \pm f_g$. The expression of the signal is:

$$s(t) = \sqrt{2P} \cdot G(t)C(t)D(t)\cos(2\pi f_c t + \phi_c)$$

BOC modulations can be used to minimize interference with BPSK signals sharing the same carrier frequency.

To separate the mainlobes of these sidebands, f_g must be at least equal to the spreading code chipping rate; that is, $f_g \geq \nu$. Usually the selection is $f_g = 2\nu$. BOC modulations can be of two types: *sine phased*, in which case the subcarrier transitions are in phase with the spreading code transitions, or *cosine phased*, in which

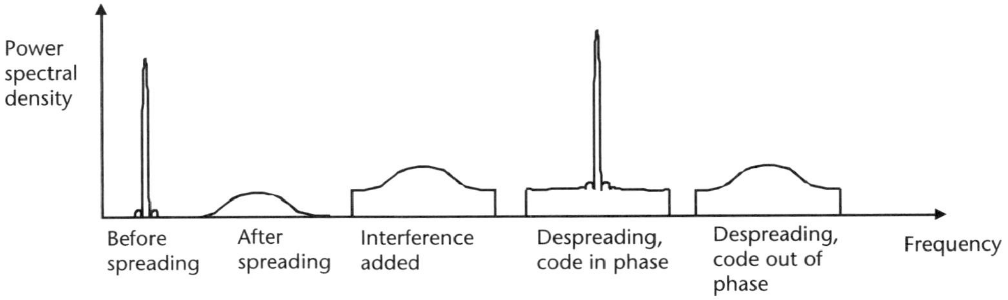

Figure 2.81 Changes of signal spectra along the modulation and demodulation process.

case the subcarrier transitions are a quarter of the subcarrier period out of phase with the spreading code transitions. BOC modulations are denoted as $BOC_s(f_g,\nu)$ if sine phased or $BOC_c(f_g,\nu)$ if cosine phased subcarrier functions, where f_g and ν are usually multiples of 1.023 MHz. Representations of $BOC_s(f_g,\nu)$ and $BOC_c(f_g,\nu)$ are shown in Figure 2.82.

2.6 Electro-Optical Thermal Imagers

EO thermal imagers are passive systems operating at wavelengths of 3 to 5 μm and 8 to 12 μm that exploit the temperature variation or gradient of the object scene to produce TV-like images at night as well as during the day. For this reason they are also called night vision devices.

The two types of EO thermal imagers are the forward-looking infrared (FLIR) systems and the infrared search and track (IRST) systems. FLIR (or EO imagers) are used for surveillance and target detection and classification in the presence of ground reflections. IRSTs are employed mainly in ground-to-air or air-to-air

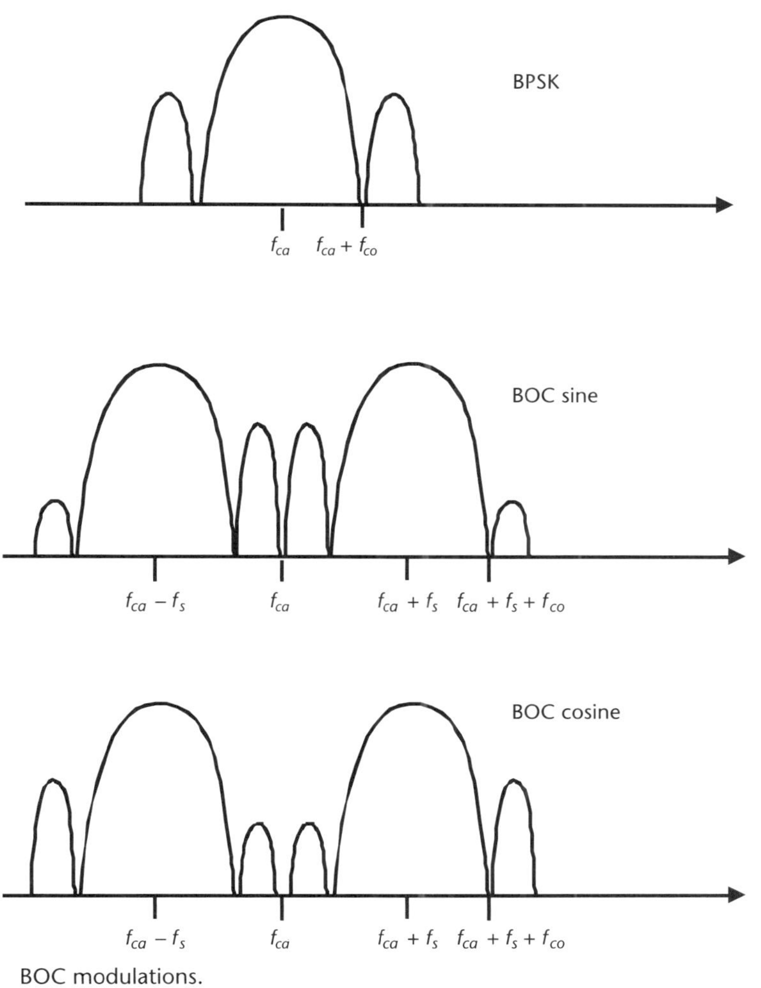

Figure 2.82 BOC modulations.

operations to detect "hot spots" in large volume searches with high angular accuracy and subsequent tracking in angle (either in lock-on or TWS mode) of designated targets such as airborne and spaceborne targets against sky backgrounds. IR missile and IR munition seekers can be considered special types of IRSTs.

The basic FLIR and night vision systems are composed of the following major parts, as shown in Figure 2.83: optical lens, scanners/detectors, signal processing, and display. Electromagnetic radiation from the source is collected by the optical lens and conveyed to the detector, where it is converted into an electrical signal. The conversion performed in the detector can be either by change of the resistivity (photoconductive detector) or by generation of an electric current (photovoltaic detector). The electrical signal is then processed to enhance the SNR and to define the target position within a reference frame.

The detector part can be constituted by a single detector, a linear array, or a 2-D array of detectors. With a single detector the object space has to be scanned by a couple of scanners, one rotating horizontally and the other rotating vertically. They scan the scene in two dimensions and reproduce sequentially the image of the object plane on the display. Typical scanning modes are raster or rosette as shown in Figure 2.84.

With a linear array of detectors the object space can be scanned in pushbroom or circular fashion.

For a 2-D detector array, more resolution cells are mapped in parallel into a mosaic of detectors, called a focal plane array (FPA),which is used to stare at the entire scene. Staring FPAs are used in surveillance and tracking applications that require higher resolution, such as air-to-ground or air-to-air combat.

The IR system detection performance (and relevant SNR) is evaluated through the following equation, in which—momentarily and for simplicity—the atmospheric and equipment losses are neglected and will be reintroduced later.

Let J_t (W/sterad) be the radiation intensity of a point source at distance R from the optical lens of the IR system. The irradiance H_t (W/cm^2) at the entrance aperture of the optics is:

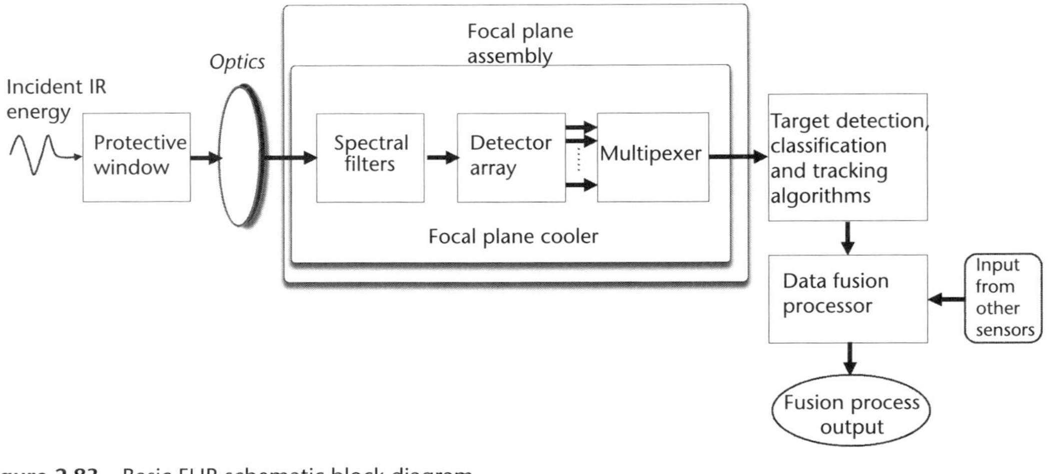

Figure 2.83 Basic FLIR schematic block diagram.

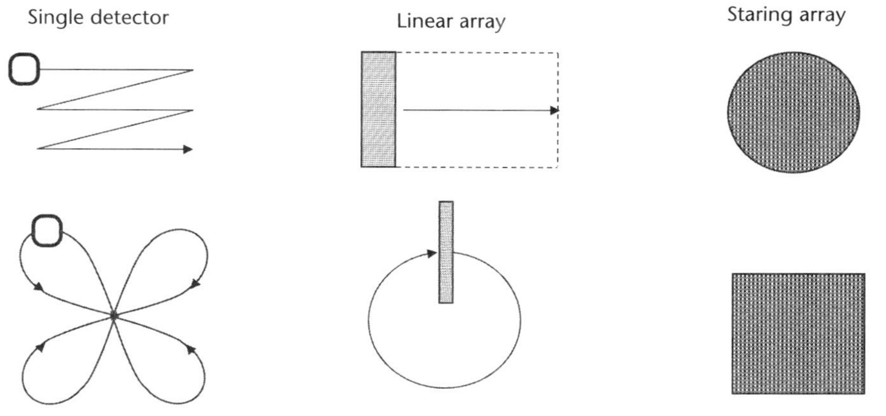

Figure 2.84 FLIR typical scanning modes.

$$H_t = J_r/R^2 \quad (\text{W/cm}^2) \tag{2.79}$$

Therefore, the input signal power S at the optical aperture A_e is:

$$S = A_e \frac{J_t}{R^2} = \frac{\pi d^2}{4} \frac{J_t}{R^2} \quad (\text{W}) \tag{2.80}$$

where d (cm) is the diameter of the lens.

The detector noise is given [2] by the relation

$$N = \frac{(A_d \Delta f)^{1/2}}{D*} \tag{2.81}$$

where

N = noise power level (W),
A_d = detector area (cm^2),
D^* = detector detectivity (cm Hz$^{1/2}$/W),
Δf = frequency bandwidth of the received signal (Hz).

The detector area A_d is related, for a squared detector, to the instantaneous field of view (IFOV) α_d and the focal length f_1 as follows:

$$A_d = (\alpha_d^{1/2} f_1)(\alpha_d^{1/2} f_1) = \alpha_d f_1^2 \tag{2.82}$$

The focal length, in turn, is related to the focal number $f/\#$ and the lens aperture diameter d by

$$f_1 = (f/\#)d \tag{2.83}$$

and thus

$$A_d = \alpha_d \left((f/\#)d\right)^2 \tag{2.84}$$

The detectivity is one of the primary parameters that characterizes the detector material. It is defined as the reciprocal of the noise equivalent power (NEP) of the detector, where NEP is the incident power at the detector that produce a $SNR_{rms} = 1$ at the detector output for a specified frequency bandwidth.

For a given material, the detectivity varies with the temperature of the focal plane (generally increasing at lower temperatures) and is selective as to the wavelength bandwidth at which energy is detected. Detector detectivity is proportional to the square root of the detector area A_d and the square root of the bandwidth Δf of the filter in the detector output. To remove the dependence from the detector area and filter bandwidth and to derive a performance parameter that is dependent only on the properties of the detector material, the detector detectivity (also called the specific detector detectivity) D^* is normalized with respect to A_d and Δf, both of which are assumed to have unit dimensions (1 cm^2 and 1 Hz). Thus,

$$D^*(\lambda, T) = (A_d \Delta f)^{1/2}/NEP \ \ (\text{cmHz}^{1/2}/\text{W})$$

Table 2.11 lists the operating parameters of selected IR detectors.

The bandwidth of the signal Δf is approximately the inverse of the time duration τ_d during which the detector receives the target energy. This time duration is equal to the frame time T_F when the total solid angle coverage Ω (sterad) of the sensor is divided by the IFOV α_d of each detector. Hence, $\Omega = n\alpha_d$ or, if a single detector of IFOV α_d is used, $\tau_d = T_F/n = T_F\,\alpha_d/\Omega$. Therefore as $\Delta f = 1/\tau_d$ the SNR can be evaluated from (2.80) and (2.82) and becomes, for a single detector,

$$\frac{S}{N} = \frac{\left(\dfrac{\pi d^2}{4}\right)\dfrac{J_t}{R^2}D^*}{(f/\#d)\sqrt{1/\tau_d}\sqrt{\alpha_d}} = \frac{\pi \cdot d \cdot J_t \cdot D^*}{4R^2(f/\#)\sqrt{\Omega/T_F}}\mathfrak{M} \tag{2.85}$$

To the above equation, excluding for the moment the atmospheric attenuation L_a (which is range dependent), a loss factor L_s has to be added to take into account

Table 2.11 Operating Parameters of Selected IR Detectors (Derived from: [24].)

Detector Material	*Operating Mode*	*Usual Focal Plane Operating Temperature (K)*	*Peak Wavelength (μm)*	*Bandwidth (μm)*	*Detectivity D* at 500 K (cm Hz$^{1/2}$/W)*
PbS	PC	300	2.4	1.1–3.5	10^{11}
PbSe	PC	300	3.9	1–5.0	10^{10}
InSb	PV	77	4.9	2–5.5	7×10^{10}
HgCdTe	PV	195	4.5	1–5.5	4×10^{10}
HgCdTe	PC	77	10.5	8–14	2.5×10^{10}

PV, photovoltaic; PC, photoconductive.

the various system losses that are dependent to the system (i.e., they are due to optics, electronics, detector, monitor, scanning, and bandwidth) and is called the modulation transfer function (MTF). Each of the above-mentioned causes provides a loss factor $0.5 < k_i < 1$ and therefore typically $L_s = \prod k_i = 0.25$.

The atmospheric attenuation L_a is quite severe in this part of the spectrum and strongly reduces the detection range. Typical attenuation losses vs. range in the 8- to 12-μm bandwidth [2] are on the order, respectively, of 0.2 to 0.9 dB/km for clear weather conditions (visibility range > 4 km) and of 3 to 4 dB/km for foggy conditions (visibility range ≈ 1 km).

Equation (2.85) calculates the point target detection performance of an IRST relevant to "hot spots" against a clear background and is used to determine the minimum resolvable temperature (MRT) in the thermal imaging systems; that is, from the SNR value above and the detection curves reported in Appendix B, the two detection conditions—P_d (detection probability) and P_{fa} (false alarm probability)—can be evaluated once one of the two conditions has been selected.

The emissive signature of objects is ruled by Planck's radiation law $W_b(\lambda,T)$, which describes the energy radiated at all wavelengths by blackbody objects (ideal emitters with emissivity $\varepsilon = 1$) as a function of the temperature of the object:

$$W_b(\lambda,T) = \frac{2\pi c^2}{\lambda^5\left(\exp\left(\frac{hc}{\lambda kT}\right) - 1\right)} \quad \left[\mathrm{W/(cm^2 sr\ \mu m)}\right] \tag{2.86}$$

where

λ = wavelength (μm),

T = object temperature (K),

h = Planck's constant = 6.625×10^{-34} [W·s^2 or J·s],

c = speed of light = 3×10^8 (m/s),

k = Boltzmann's constant = 1.38×10^{-23} (J/K).

For a given blackbody temperature, the wavelength, at which the maximum power is emitted is found by equating to zero the partial derivative of (2.86) with respect to the wavelength λ, which provides the Wein's displacement law for the wavelength λ_M at which the maximum power density occurs:

$$\lambda_M T = 2897.8\ (\mu\mathrm{m K}) \tag{2.87}$$

The blackbody emittance and λ_M as a function of temperature are shown in Figure 2.85(a).

The radiation intensity J_t of a target can be evaluated using Stefan-Boltzmann's law, which states that the radiated power density W (W/cm^2) from a unit surface is proportional to the fourth power of the surface temperature T:

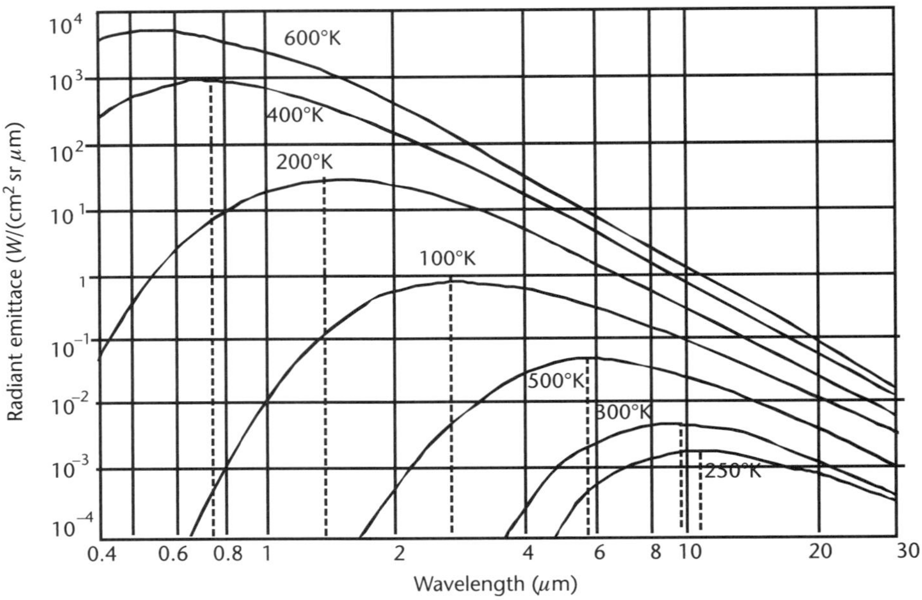

(a)

(b)

Figure 2.85 (a) Blackbody emittance and λ_M as a function of temperature. Blackbody emittance is given by Planck's radiation law. Vertical dashed lines mark the wavelength of maximum energy emission for the indicated blackbody temperature (b) PIRATE IRST (Courtesy of Selex Galileo.)

$$W = \varepsilon \sigma T^4 \quad (\text{W/cm}^2) \tag{2.88}$$

where

ε = target emissivity,

σ = Stefan-Boltzmann's constant = 5.67×10^{-12} (W/cm^2 K^4),

T = temperature (K).

Assuming that the target of area A (cm^2) is a Lambertian radiator, we have

$$J_t = W\,A/\pi = \varepsilon\sigma T^3 A/\pi \quad \text{(W/sterad)} \tag{2.89}$$

Detection of a warm ground target is always performed in the presence of a colder background, which power C adds to the system noise. The evaluation of background power is conducted by considering the spectral emittance of a thermal target, expressed as follows:

$$W(\lambda, T) = \varepsilon_c(\lambda) W_b(\lambda, T) \tag{2.90}$$

where

$W(\lambda,T)$ = background spectral emittance [W/(cm^2 sr μm)],

$\varepsilon_c(\lambda)$ = background emissivity,

$W_b(\lambda,T)$ = relative blackbody spectral emittance (Planck's law).

The spectral emittance $W(\lambda,T)$ of distributed thermal targets (background), such as a grass-covered field at two temperatures, is obtained by measurements as shown in Table 2.12 (reproduced from [16])

The background power [where the limits of the integral represent the response bandwidth $(\lambda_2 - \lambda_1)$ of the detector] is:

$$C = \frac{\pi d^2}{4}\alpha_d \int_{\lambda_1}^{\lambda_2} \varepsilon_c(\lambda) W_b(\lambda, T) D^*(\lambda) \cdot d\lambda$$

and thus, taking into account (2.85), the clutter to noise power ratio C/N is:

Table 2.12 Differential Thermal Emittance $\frac{\partial W_b}{\partial T} = \int_{\lambda_1}^{\lambda_2} \frac{\partial W_b(\lambda,T)}{\partial T} d\lambda$ [W/(cm^2 K)] of Grass-Covered Field at Several Background Temperatures (Reproduced from: [24].)

λ_1 (μm)	λ_2 (μm)	$T_1 = 290$ K	$T_2 = 310$ K
3	5.5	2.73×10^{-5}	4.72×10^{-5}
4	5	1.26×10^{-5}	2.23×10^{-5}
8	12	1.77×10^{-4}	2.17×10^{-4}
10	14	1.42×10^{-4}	1.65×10^{-4}

$$\frac{C}{N} = \frac{\pi \cdot d \cdot \alpha_d L_s L_a}{4(f/\#)} \sqrt{\frac{nT_F}{\Omega}} \int_{\lambda_1}^{\lambda_2} \varepsilon_c(\lambda) W_b(\lambda, T) D^*(\lambda) \cdot d\lambda \tag{2.91}$$

where L_s represents the system losses and L_a represents the optical transmission efficiency .

An important detection performance parameter of optical systems operating against a background is the noise equivalent temperature (NET), which is defined as the difference in temperature of the background that produces a variation of the background-to-noise ratio of unity, that is, $\Delta(C/N) = 1$.

When reproducing the object scene on the display of the equipment, the local variation of C/N from the surrounding background is important to detect the presence of a warmer target.

The parameter NET can be obtained from the above considerations by using the data of Table 2.12 to evaluate the differential background spectral emittance $\Delta W(\lambda,T)$, which is, from (2.90):

$$\Delta W(\lambda, T) = \frac{\partial W(\lambda, T)}{\partial \varepsilon} d\varepsilon + \frac{\partial W(\lambda, T)}{\partial T} dT \cong \frac{\partial W(\lambda, T)}{\partial T} dT \tag{2.92}$$

having taken into account that both background emissivity $\varepsilon_c(\lambda)$ and detectivity $D^*(\lambda)$ are constant with the temperature. So the change in the background-to-noise ratio due to an increase in temperature is:

$$\Delta(C/N) = \frac{\pi d \alpha_d L_s L_a}{4(f/\#)} \sqrt{\frac{nT_F}{\Omega}} \Delta T \cdot \int_{\lambda_1}^{\lambda_2} \frac{\partial W(\lambda, T)}{\partial T} D^*(\lambda) \varepsilon(\lambda) \cdot d\lambda \frac{1}{2}$$

and the required NET is

$$NET = \Delta T\big|_{\Delta(C/N)=1} = \frac{4(f/\#)}{\pi d \alpha_d L_s L_a} \sqrt{\frac{\Omega}{nT_F}} \frac{1}{\int_{\lambda_1}^{\lambda_2} \frac{\partial W(\lambda, T)}{\partial T} D^*(\lambda) \varepsilon_c(\lambda) \cdot d\lambda} \tag{2.93}$$

The NET parameter is very useful in the design and evaluation of the performance of IRST systems. For FLIR systems, which are reproducing an object scene, a better parameter for the evaluation of performance is the minimum resolvable temperature, which is discussed in the following section.

A typical airborne IRST, the PIRATE, installed on board the European Typhoon is shown in Figure 2.85(b).

2.6.1 Minimum Resolvable Temperature

If we want a target resolution distance T_x from an observation point at distance R we need an IFOV expressed in mrad (see Figure 2.86):

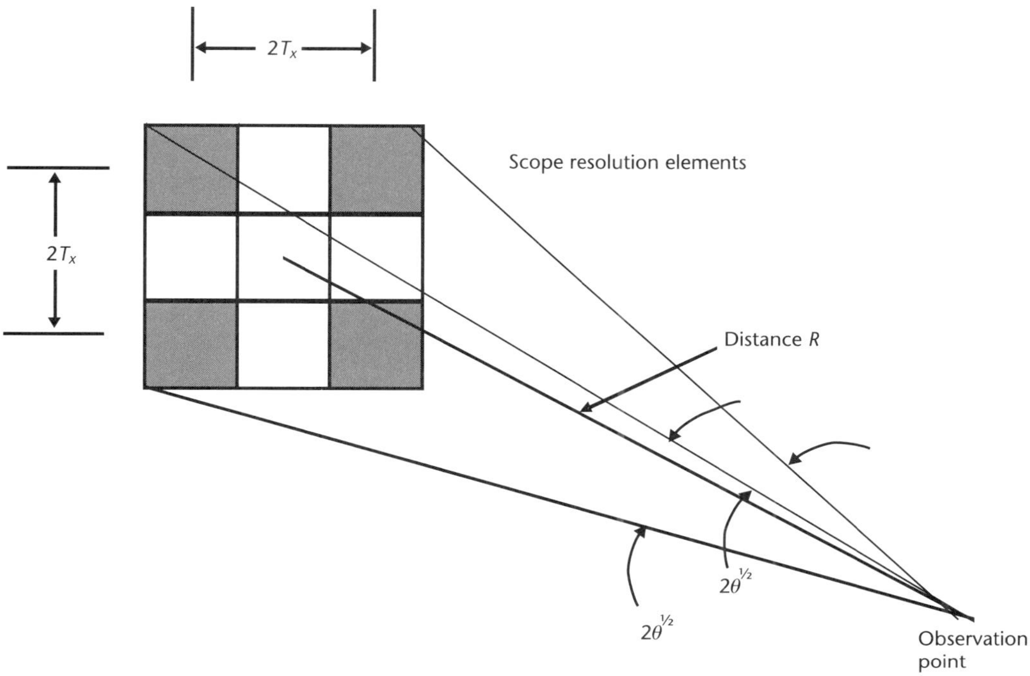

Figure 2.86 Minimum resolvable temperature.

$$\sqrt{\theta_d} = 10^3 \, T_x/R \ \text{(mr)} \tag{2.94}$$

Using the angular resolution of (2.94) we can define the spatial frequency f_s as:

$$f_s = \frac{1}{2\sqrt{\theta_d}} \ \text{[cycle/mr]} = \frac{1}{2}\frac{R}{T_x} \cdot 10^{-3} \tag{2.95}$$

By inspection of Figure 2.86 the spatial frequency f_s can be considered to be the rate of change of light-to-dark features with respect to a distance R.

Since the NET of the target is calculated per the detector resolution cell $\sqrt{\alpha_d}$, the equivalent display NET, called MRT, has to be modified by the ratio $\sqrt{\alpha_d/\theta_d}$; that is:

$$MRT = NET\sqrt{\alpha_d/\theta_d} \tag{2.96}$$

If the observer integrates a number of successive frames n_i at the frame rate f_F in the observation time T_{obs}, we have $n_i = f_F T_{obs}$ and thus the minimum integrated resolvable temperature becomes:

$$MRT_i = MRT/\sqrt{n_i} = NET\sqrt{\frac{\alpha_d}{f_F T_{obs} \theta_d}}$$

From this equation, by substituting both NET with its expression (2.91) and $\theta_d = 1/f_s^2$, and taking into account that to achieve at least 50% probability of detection of a target in the display scope a multiplying factor of 3 to the NET is required, we obtain the MRT expressed as a function of the spatial frequency f_s as follows:

$$MRT = \frac{6NET \cdot f_s \sqrt{\alpha_d}}{f_F T_{obs}} = \frac{24(f/\#)}{\pi L_a d}\left(\frac{f_s}{L_s}\right)\sqrt{\frac{\Omega}{\alpha_d n T_F T_{obs} f_F}} \cdot \frac{1}{\int_{\lambda_1}^{\lambda_2} \frac{\partial W(\lambda, T)}{\partial T} D^*(\lambda) \cdot d\lambda} \quad (2.97)$$

This expression for MRT, which includes all system losses in the term L_S, is called the modulation transfer function (MTF) of the system (which also can be expressed as a function of the spatial frequency f_S [2]). The MRT is an important parameter in the design and evaluation of the performance of FLIR systems. The MRT is, in fact, generally defined as the required threshold temperature difference between a square target and the background in order for an observer to perceive the target through the EO sensor.

The MRT, therefore, includes the effects of sharpness, contrast rendition, and SNR as a function of the spatial frequency f_S.

Johnson [25] determined through experiments with observers the number of line pairs of pixels required by an observer to detect, determine symmetry and orientation, recognize or classify, and identify tactical targets. These discrimination levels are reported in Tables 2.13 and 2.14. From these tables it can be observed that only one line pair across the minimum target dimension is required for detection, while for the target recognition, which is the most frequently used observation, up to four line pairs are required, as shown in Figure 2.87.

Table 2.13 Object Discrimination Level

Discrimination Level	*Interpretation*
Detection	Target presence
Orientation	Target symmetry or asymmetry and its orientation are discerned
Classification or recognition	Target class (e.g., truck, tank, building) is discerned
Identification	Target is described to the limit of observer's knowledge (e.g., type of truck, tank, or building)

Table 2.14 Johnson's Object Discrimination Criteria for a One-Dimensional Array

	Resolution per Minimum Dimension in Line Pairs			
Target Broadside View	*Detection*	*Orientation*	*Classification*	*Identification*
Truck	0.9	1.25	4.5	8.0
Stalin tank	0.75	1.2	3.3	6.0
Centurion tank	0.75	1.2	3.5	6.0
Jeep	1.2	1.5	4.5	5.5
Soldier (standing)	1.5	1.8	3.8	8.0
Average	1.0 ± 0.25	1.4 ± 0.35	4.0 ± 0.8	6.4 ± 1.5

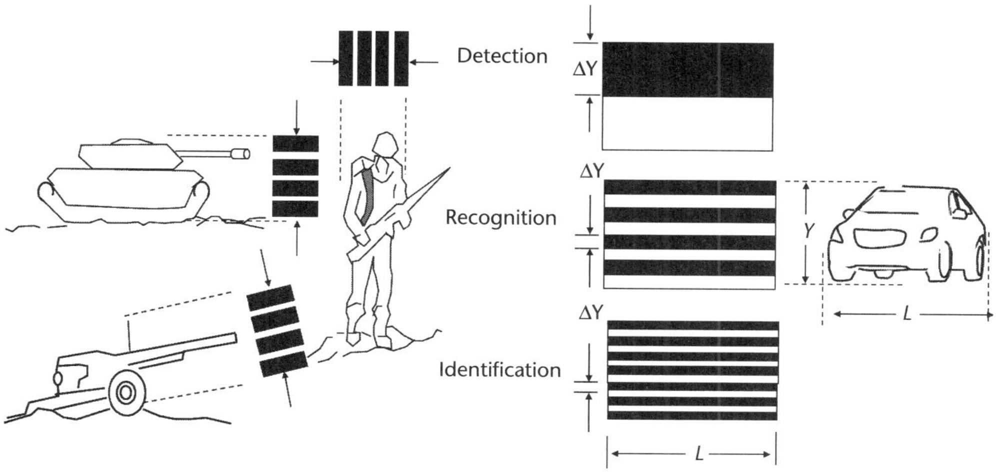

Figure 2.87 Line pair across the minimum target dimension required for detection.

2.6.2 IR Missile Seekers

IR missile seekers are passive, and use heat energy emitted by the target in the IR spectrum to track the target. As already shown in the previous section when discussing Planck's law for the blackbody spectral emittance, the hotter an object, the larger the energy it emits, and the lower the wavelength.

The hot engine parts of an A/C are typically at temperatures ranging from 1300 to 2000 K with peak energy emission wavelengths ranging from 2 to 4 μm. First-generation IR missiles used uncooled lead-sulfide (PbS) detectors that have a peak detectivity at 2 μm. This type of missile is therefore limited to stern engagements, since the missile seeker must track the hot turbine in the engine tailpipe of the target A/C.

More sensitive cooled detectors, based either on an indium-antimonide (InSb) material, which has a peak detectivity in the range of 3 to 5 μm, or on a mercury-cadmium (HgCd) material, which has a peak detectivity in the range of 8 to 12 μm, were later introduced, which allowed for successive IR missiles to achieve all-aspect track capability of a target A/C (from engine exhaust to the warm skin of the A/C).

Along with the detector spectral diversity, the successive generations of IR seekers are using a variety of techniques to find and track targets in the presence of flare countermeasures as shown in Table 2.15 for the four generations of IR seeker listed there.

Early seekers (first and second generation) used a spinning reticle of various shapes in front of the detector to reject the background radiation and encode the target position. In the successive generations the reticle is gradually being replaced by techniques that either sequentially scan the target with small IFOV detectors or stare at the target with a mosaic of detectors in a FPA configuration, which acts as an imager.

Further, bicolor discrimination techniques, using two sets of detectors, each responding in a different spectral band, allow through the ratio of the two responses at the seeker for discrimination between a decoy (flare) or a real target aircraft (see Figure 2.88). Table 2.16 lists the characteristics of some IR missiles.

Table 2.15 IR Seeker Missile Evolution

Generation	*Years*	*Type of Reticle*	*Presence of CCM*	*Missile*
First	1960s–1970s	Chopping	None	SA-7, AA-2
Second	1970s–1980s	Chopping	Yes	SA-13, AA-7
Third	1990s	Pseudo-imagers and multicolor	Yes	SA-18, AA-11
Fourth	2000s	Full imagers and multimode	Yes	Keiko SAM

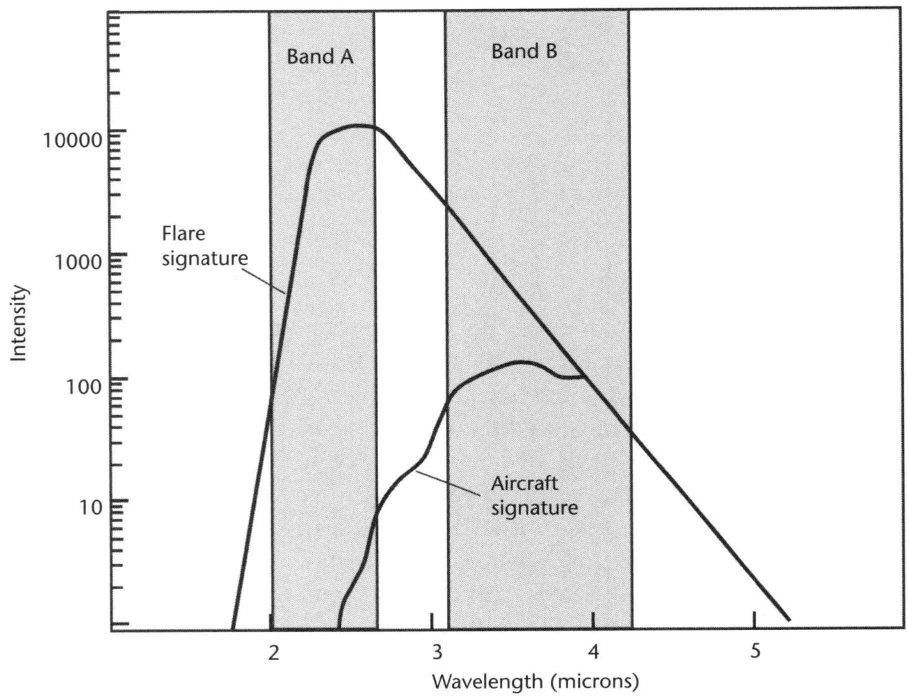

Figure 2.88 Principle of bicolor discrimination technique.

Table 2.16 IR Missile Characteristics

Name	*Type*	*Target Acquisition*	*Seeker Type*	*Firing Range (m)*	*Target Altitude (m)*	*Speed (m/s)*	*Missile CCM*
SA-7	SAM	Visual	Uncooled IR	4,200	2,300	385	None
SA-18	SAM	Visual	—	5,200	3,500	—	Bicolor
SA-16	SAM	Visual	Cooled IR	5,200	3,500	570	Bicolor
Redeye	SAM	Visual	Uncooled IR	5,500	2,700	522	None
Stinger	SAM	Visual	—	4,500	3,800	718	ConScan
Mistral	S/AAM	FLIR, A/C radar	Cooled IR	6,000	4,500	852	Bicolor
Sidewinder	AAM	FLIR, A/C radar	—	8,000	NA	—	Flare rejection
Alamo	AAM	FLIR, A/C radar	—	40,000	NA	—	—

2.6.2.1 Types of Seekers and Reticles

There are many types of reticles for seeker and in the following subsections only the principle of operation of five of them will be illustrated.

Seeker with Rotating Reticle

Figure 2.89 illustrates a seeker with a wagon wheel type rotating reticle. The spinning reticle in front of the detector chops the detected long continuous signal into a series of pulses. Phase information in the chopped signal indicates the angular direction of the point target from the seeker axis, while its amplitude indicates the magnitude of the error

A spinning reticle with only half the disk transparent (Figure 2.90) tends to produce a nearly steady signal on extended radiation sources such as clouds, while retaining the pulse signal from the target point source. The target signal can be extracted using a simple filter tuned to the reticle scan rate.

Seeker with Stationary Reticle and Rotating Lens

Another types of seeker is one with a stationary reticle and rotating lens displaced with respect to the seeker, as shown in Figure 2.91).

Conical Scan Seeker

Figure 2.92 shows a conical scan seeker. In this seeker, the reticle is fixed and a secondary mirror is tilted and spun. The target image produces a circular path around the outer edge of the reticle. If the target is centered in the seeker scan, the detector gives a constant frequency square wave. If the target is not centered, the circular path is somewhat offset and the detector gives a FM square wave, which has a depth of modulation that is directly proportional to the angular error from the center of the seeker. The error signal is demodulated to provide the input to the target-tracking servo.

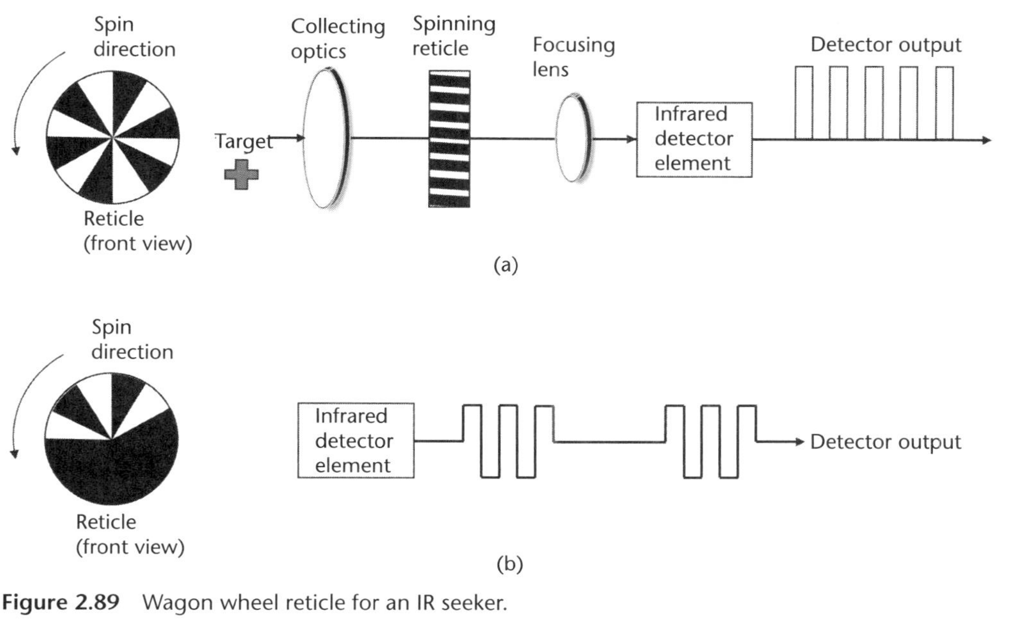

Figure 2.89 Wagon wheel reticle for an IR seeker.

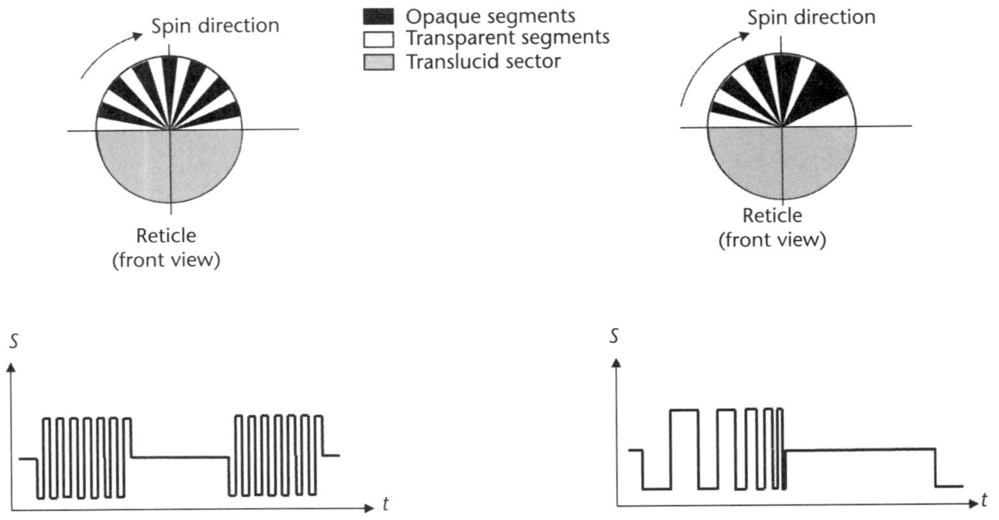

Figure 2.90 Spinning reticle for an IR seeker with only half the disk transparent.

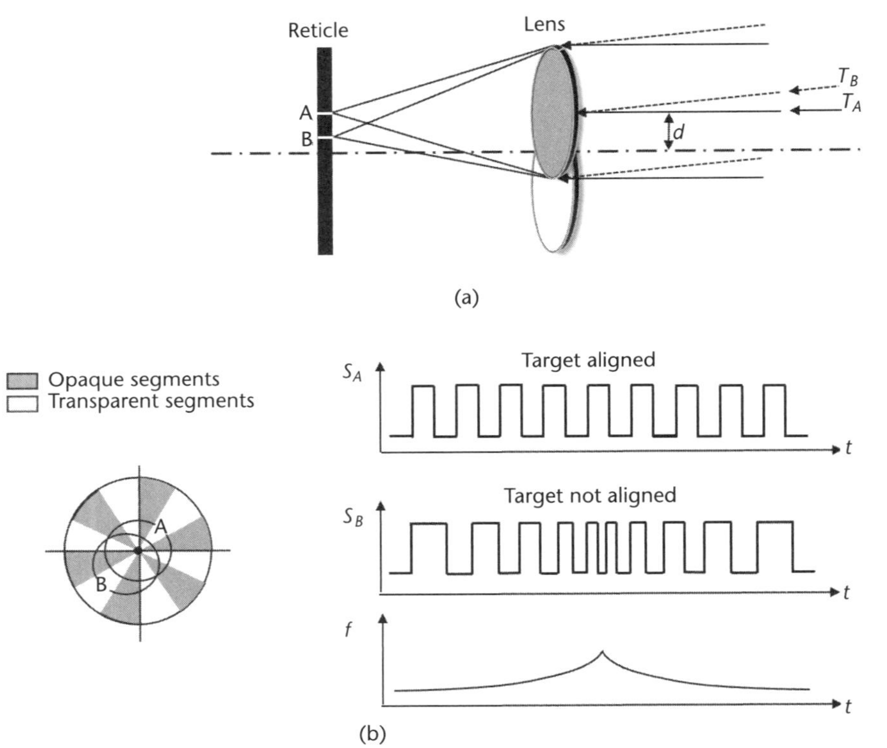

Figure 2.91 Seeker with stationary reticle and rotating lens displaced with respect to the seeker.

The conical scan optics are usually designed to spin the target very close to the edge of the reticle. In this way the largest FM depth for a given angular error is provided and thus a more accurate tracking loop is achieved. Resistance to

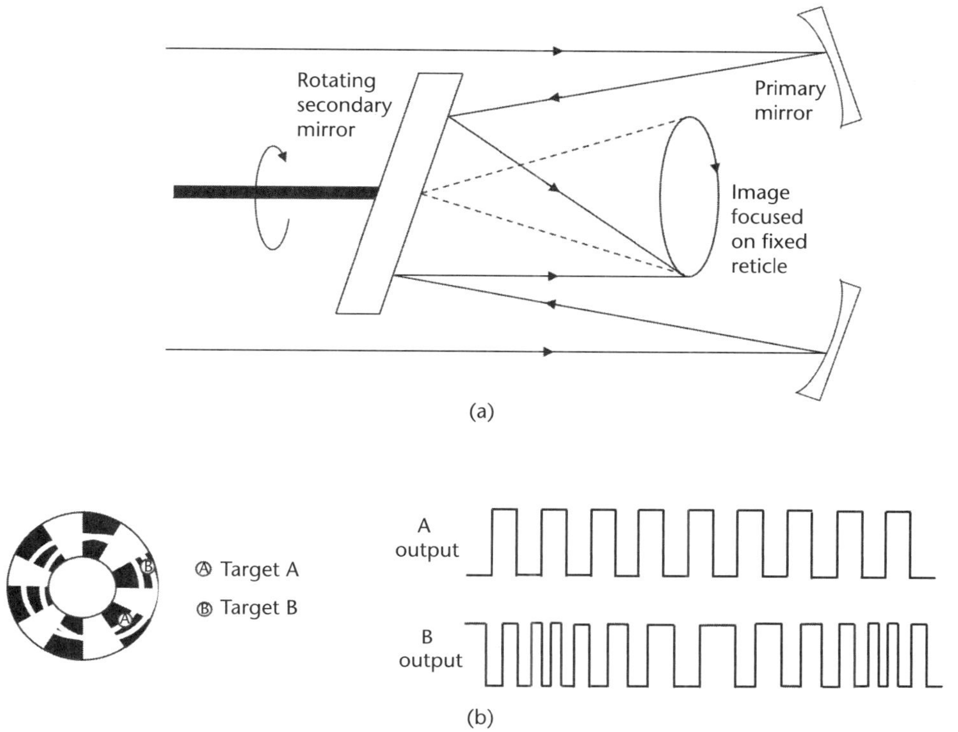

Figure 2.92 Conical scan seeker. (Adapted from [22].)

flare is achieved because the flares tend to drop off the reticle much faster than the spin-scan period.

Rosette Scan Seeker

This seeker belongs to the pseudo-imaging type that does not build an image but instead uses the spatial information from the scan to isolate the target from the background or the flare.

In this seeker the detector is fixed and aligned with the optics boresight axis, while the seeker FOV is scanned by a smaller IFOV that is rotated by means of a couple of prisms at two slightly different rotations speeds Ω_1 and Ω_2.The shape of the small IFOV path within the seeker FOV is similar to the petal of a flower, hence the name of this seeker, which is illustrated in Figure 2.93.

When the seeker boresight is aligned with the target, the position of the two rotating prisms at the detection time t_0 indicates the two target angular errors θ and ϕ (azimuth and elevation), evaluated as follows:

$$\begin{aligned} \theta &= \frac{FOV}{2}(\cos\omega_1 t_0 + \cos\omega_2 t_0) \\ \phi &= \frac{FOV}{2}(\sin\omega_1 t_0 + \sin\omega_2 t_0) \end{aligned} \tag{2.98}$$

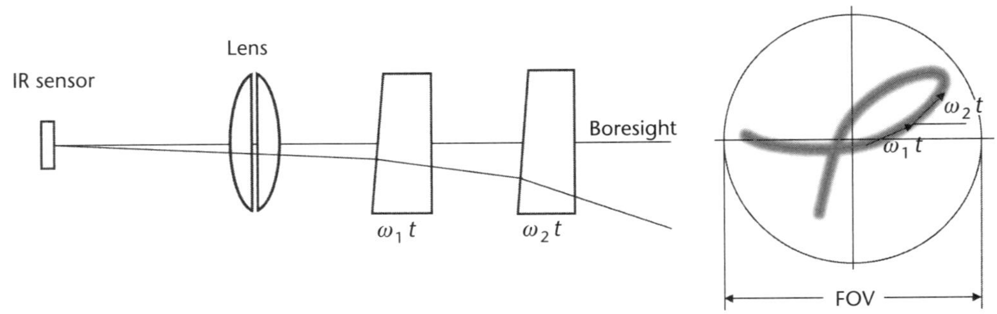

Figure 2.93 Rosette scan seeker. (Adapted from [15].)

This type of seeker is practically immune to jamming produced by AM IR lamps or lasers and further it allows for determination of the position of different targets within the FOV, thus discriminating, by means of kinematics filtering, the target from flare decoys.

Imaging Seekers

This type of seeker uses a FPA of detectors that stare at the scene and provide a TV-like spatial map. Imaging seekers are inherently resistant to flares, because flares simply do not look like an extended aircraft with exhaust plume.

The most difficult aspect of imaging seekers is their image processing, because a typical FPA of 128 × 128 pixels with 10 bits each and a sampling rate of 100 Hz provides about 16 Mbit/s data throughput.

2.6.2.2 IR Missile Detection Range

The IR missile seekers have the same behavior as an IRST relevant to detection performance, in that the latter depends on several factors such as the target's temperature, the atmosphere transmittance, the system optics, and the detector's sensitivity.

The main targets of the IR SAMs or AAMs are the exhaust nozzle and plume of the jet engine of the aircraft. Currently, IR seekers are also used in sea-skimming antiship missiles to support the microwave seeker in providing better tracking accuracy and resistance to countermeasures. In this case the main target is the ship's engine nozzle. The temperature of the A/C tailpipe is on the order of 1,000 to 2,000 K, while the plume temperature can range from 700 to 1000 K. When the A/C's afterburner is operational, the plume temperature can approach that of the hot tailpipe.

A secondary target is the aerodynamic heating of the aircraft skin. The skin's temperature depends on the aircraft speed according to the following expression:

$$T = T_0(1 + 0.164M^2) \quad \text{(K)}$$

where M equals the Mach number, which is the aircraft speed/sound propagation speed in the air, and T_0 is the standard temperature, which is assumed to be 250 K. Typical supersonic A/C skin temperatures range from 250 K ($M \approx 1$) to 500 K($M = 2.5$).

The temperature of the target establishes the wavelength of the emissions, which peaks in accordance with (2.86) in the 2- to 5-μm range for tailpipe detection and in the 8- to 12-μm range for A/C skin detection.

From (2.86) and (2.87), by assuming a tailpipe temperature T_{tp} of 700 K, a nozzle area of 1,600 cm^2, and emissivity ε of 0.9, we obtain a radiation intensity $J_t = 693$ W/sterad and by assuming a skin temperature $T_{sk} = 414$ K (A/C speed $M = 2$), an A/C area $A = 10$ m$^2 = 10^5$ cm^2, and emissivity $\varepsilon = 0.05$ for aluminum, we obtain $J_t = 255$ W/sterad.

The atmospheric attenuation (dB/km) for 23-km visibility at the midband for the 3.5- and 10-μm-wavelength bandwidths is reported in Table 2.17.

With reference to (2.85), here rewritten to express the range R, we have:

$$R = \left[J_t \cdot L_a(R) \cdot \left(\frac{\pi d}{4(f/\#)} \right) \cdot D^* \cdot \left(\frac{L_s}{S/N} \sqrt{\frac{T_F}{\Omega}} \right) \right]^{1/2} \tag{2.99}$$

where all the parameters have already been discussed and need only to be selected in order to get the IR missile detection range for a high-speed fighter aircraft. This equation has been written as the product of five terms:

- The first is relevant to the target and has been evaluated above for the two bandwidths where detection is maximized.
- The second term represents the atmospheric loss, which is dependent on the range and therefore the above equation has to be solved iteratively. As shown in Table 2.17, the atmospheric attenuation decreases with altitude, therefore the detection ranges are larger in an air-to-air combat than in a ground-to-air combat.
- The third term is relevant to the optics of the seeker and its parameters define the IFOV of the seeker.
- The fourth term is D^* (cm Hz$^{1/2}$/W), the detectivity, measured with the bandwidth of 1 Hz, of a detector with a square aperture of 1 cm^2. Its value depends on the detector material and the selected cooling method. As an example, the current preferred IR missile detector material InSb requires cooling to liquid-nitrogen temperatures (77 K) to be sensitive in the 3- to 5-μm bandwidth with $D^* \approx 9 \times 10^{10}$ (cm Hz$^{1/2}$/W).

Table 2.17 IR Atmospheric Attenuation (dB/km) for 23 km Visibility

	Band	
Altitude (km)	*3.5 μm*	*10 μm*
0.2	0.88	0.57
2	0.37	0.17
5	0.16	0.05
10	0.05	0.04

- The fifth term is relevant to the parameters of the system, such as the system losses L_s, the overall FOV Ω, the frame period T_F, and the required SNR for the detection conditions (P_d,P_{fa}).

Typical modern IR missile detection ranges are 6 to 12 km for an A/C plume and 4 to 6 km for an A/C skin, depending on the altitude.

2.6.2.3 IR Missile Seeker Counter-Countermeasures

Except for the first-generation IR seekers, successive generations have employed CCMs against the most used countermeasure: flares, the operation of which will be described in Chapter 6.

In general, IR CCMs operate in two phases: The first phase is the detection of the flare within the seeker's FOV and the second phase is the actuation of some method to reject the flare.

A variety of techniques can be used to detect flares:

- *Rise-time measurements:* An abrupt, sharp rise-time increase in received energy indicates the presence of a flare in the seeker's FOV.
- *Two-color detection:* A sudden change in the ratio of the received energies in the two bandwidths indicates the presence of a flare in the seeker's FOV.
- *Kinematics filters:* In a beam-aspect engagement, these filters can sense the sudden change in the LOS rate when trying to track the flare due to the rapid deceleration that flares incur (as a result of their high aerodynamic drag) as soon as they are dispensed from an aircraft.

Once the presence of the flare has been detected within the seeker's FOV, the rejection of the flare consists of exploiting the effect that the flares separate to the rear of the aircraft and two hot spots are distinguishable on opposite sides of the FOV. The rejection methods that can be used are:

- *Use of a memory technique:* The seeker rejects its freshly detected track and keeps its motion relative to the target, waiting for the flare to leave its FOV.
- *Biasing gimbals:* Biasing of the seeker gimbals will drive the seeker in the direction of the target motion. This action causes the flare to leave the FOV earlier.
- *Use of an attenuation filter:* An attenuation filter is placed in the quadrant of the FOV below and to the rear of the target in order to reduce any energy received from the flare. If the attenuation amount is sufficient, the seeker will continue to track the target.
- *Use of FOV gating:* In a rosette scan seeker, some time after the flare is dispensed, the target and the flare will be no longer in the same lobe of the scan. By computing the relative motion of the target, the missile is able to determine on which lobes the target should appear. Thus, information from all other lobes can be ignored, allowing the missile to keep its track on the target.

2.6.3 Missile Approach Warner

MAWs supplement radar warning receivers (RWRs) in tactical aircraft EW systems, because both have to warn the aircraft of an approaching missile and provide its direction of arrival and possibly its time to impact. RWRs alert aircraft to the approach of radar guided missiles, whereas MAWs warn aircraft of an approaching short-range IR/EO guided missile attack. Tactical aircraft must have access to angular and raw distance information in order to counter the missile attack by either dispensing flares or employing active countermeasures (see Chapter 6).

MAWs may be either passive or active. The former can operate either in the IR or ultraviolet (UV) band to detect the exhaust plume signature from the missile's rocket engine and to provide the missile's direction of arrival. The latter is in general a pulse Doppler radar that detects and tracks missiles by means of their skin return.

The radar has the advantage of providing all-weather range and range-rate data that enable the evaluation of the time to impact. Disadvantages involve the difficulty of discriminating a small RCS target from strong clutter returns and the possibility that, when operating in continuous mode, the radar will act as a beacon for an ARM. The latter disadvantage can be dealt with by having a passive MAW to cue any radar operation. Therefore, in some cases the two types of sensors are combined in an integrated MAW system, which provides the benefits of both sensors.

Passive MAW sensors, either IR or UV, detect the missile's exhaust plume at some point prior to burn-out[6]. They are in general staring systems, which use a mosaic of detectors to generate an image of the target scene and provide accurate angular data, on the order of 1°, and a rough range estimate on the basis of the received signal energy.

IR MAWs are susceptible to background (clutter) false alarms; therefore, in recent systems two-color FPA detectors have been used in order to exploit spectral-spatial-temporal clutter rejection techniques.

UV detectors are designed to detect the burning of solid rocket propellants. Detection can be accomplished with low false alarms in the "solar blind" region between 0.22 and 0.28 μm, because the Earth's ozone layer (usually from 30 to 50 km from the Earth's surface) absorbs most of the solar radiation in this band. Thus, at altitudes below the ozone layer (and below, say, 20 km) the signature of the burning of solid rocket propellants is detectable above an extremely low background.

Discussion of active MAW sensors, which are pulse Doppler radars, is included in Section 2.3.7.

2.7 Laser Radar Systems

Laser radar systems operate like microwave radars but at a much higher frequency—on the order of 20 to 200 times higher. The higher frequency has the beneficial effect of smaller components and of an extremely high angular accuracy.

6. In fact, the missile, after the launch boosting phase and midcourse phase, where it produces an energetic, highly detectable signature, may have exhausted its engine such that in the approaching phase the aircraft may have no more plume.

The atmospheric attenuation, however, is significant at these high frequencies and hence laser radars built to operate on the ground have a limited range (about 10 km). Spaceborne laser radars, exploiting the absence of atmospheric attenuation, may have ranges of thousands of kilometers.

Laser radars, due to their very narrow beamwidth (on the order of 1 mrad), are not apt for volume searches; that task is better accomplished by IR passive systems. The coarse angular position of the detected target provided by the latter is used to cue the laser radar, which provides the range and radial velocity of the target.

Laser radars, primarily because of atmospheric transmittance windows (see Figure 2.3) and the availability of detectors and laser sources, operate in general at two wavelength bands: around 1 μm (1 μm $= 10^{-6}$ m) with solid-state (semiconductor) lasers and around 9.2 to 10.8 μm with gas (CO_2) lasers.

The military applications of laser systems are:

- Designators, which illuminate targets with a laser beam so to guide munitions against them;
- Rangefinders, which measure with a high accuracy the distance of the target from an observer;
- Radar seekers, which are integrated within munitions (or missiles) to identify and autonomously guide the munitions to the desired target;
- Target speed measurements, which are taken when used in a heterodyne receiver configuration to determine the Doppler frequency shift of the target.
- Differential absorption measurements (differential absorption laser, DIAL) in nuclear, biological, and chemical (NBC) operations, which measure the concentration of a particular gas in the atmosphere through the differential attenuation of the transmitted signal at two wavelengths. The first wavelength is tuned to the absorption line of the selected gas, and the second wavelength is far from the first one. The amount of attenuation is proportional to the gas concentration in the atmosphere.

The two last applications are conducted with CO_2 gas lasers that can be tuned over the large bandwidth mentioned earlier.

2.7.1 Laser Target Designation and Ranging

Target designators are used in ground battlefield scenarios either on board aircraft, such as in the LANTIRN (Low Altitude Navigation and Targeting Infra-Red System for Night) system, or on ground-based equipment to provide range and bearing information to the munitions launched against a target. Laser target designators are relatively large and heavy systems (up to 20 kg) that have a range of about 10 km. Some air-launched special munitions such as the Maverick laser-guided bomb and the Hellfire missile contain laser seekers that locate and guide the weapons to the target.

Laser rangefinders are smaller (about 3 to 4L) and lighter in weight than the designators. They resemble a pair of binoculars and have an effective range of about 1 km. Most of the currently available laser target designators and rangefinders are operating at the 1.064-μm wavelength (Nd: YAG), which is not a completely eye-safe wavelength [20] because the human eye is sensible from 0.4 to 1.2 μm.

The recent advent of small, rugged, and uncooled mid-IR band semiconductor laser diodes has shifted the usable wavelength bandwidth beyond 1.4 μm, which is the limit wavelength for unsafe human eye operation.

Typical semiconductor laser materials are:

- Ho:YAG, emitting at 2.09 to 2.10 μm,
- Tm:YAG, emitting at 2.32 μm,
- Er:YAG, emitting at 2.94 μm,
- Dy:YLF, emitting at 4.34 μm,

where Ho = holmium, Tm = thulium, Er = erbium, Dy = dysprosium, YAG = yttrium-aluminum garnet, and YLF = yttrium-lithium fluoride.

In military applications the laser radars locate and classify targets and guide munitions to them. In general, they use three types of imagery to perform their task:

- Range images, by processing the backscattered signals from the target scatterers;
- Elevation images, which delineate the height profile of the scene within the FOV;
- Intensity images, which are produced by the differences in the reflectivity of the objects within the FOV.

2.7.2 Laser Radar Receivers

The two types of receivers are *direct* and *heterodyne*. Direct receivers are similar to the EO detectors, with the difference that they measure the backscattered energy from the target. The schematic block diagram of a direct receiver is shown in Figure 2.94, which illustrates both the imaging functions as well as the servo-mechanism for scanning the mirrors of the optics to acquire the target within its FOV.

In a heterodyne receiver a portion of the laser transmitted beam is diverted into the frequency shifter. The shifter provides a local oscillator frequency to mix with the received laser signal, which is then translated to a lower frequency where it can be amplified to enhance the receiver's sensitivity. The schematic block diagram of a laser heterodyne receiver is shown in Figure 2.95. This scheme allows for coherent laser radar operation, which not only provides intensity and range information but also the Doppler frequency shift proportional to the target radial velocity. Coherent laser operation is based on a CO_2 laser at 10.6 μm, which produces a very stable long-term frequency.

2.7.3 Laser Radar Range Equation

The signal power received by a laser radar from a target at range R can be derived in a very similar manner to that used for microwave radar (as described in Section

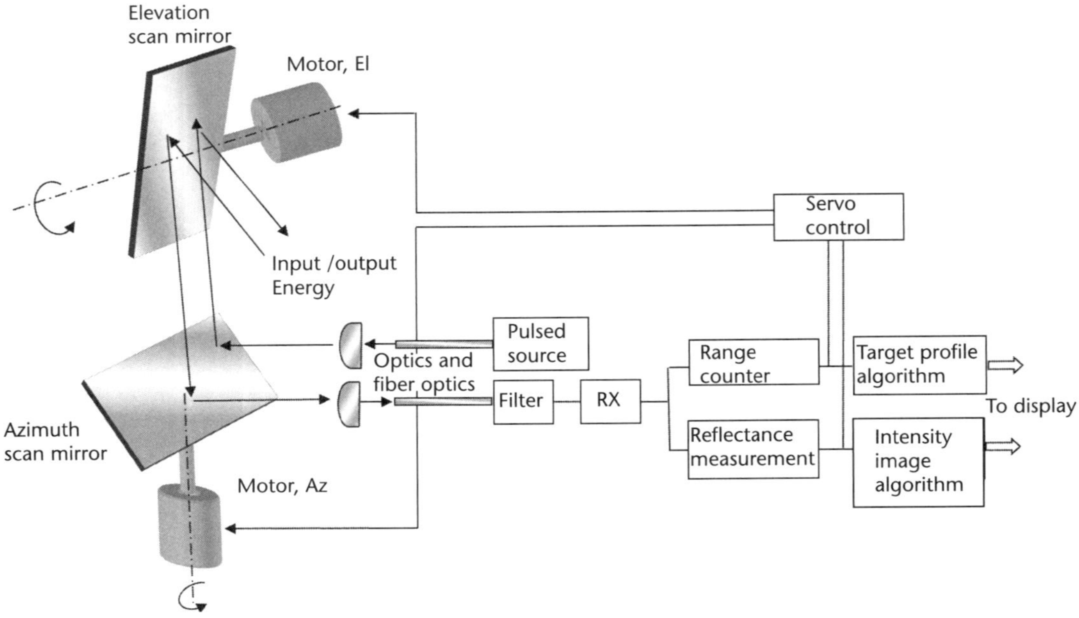

Figure 2.94 Direct laser radar receiver.

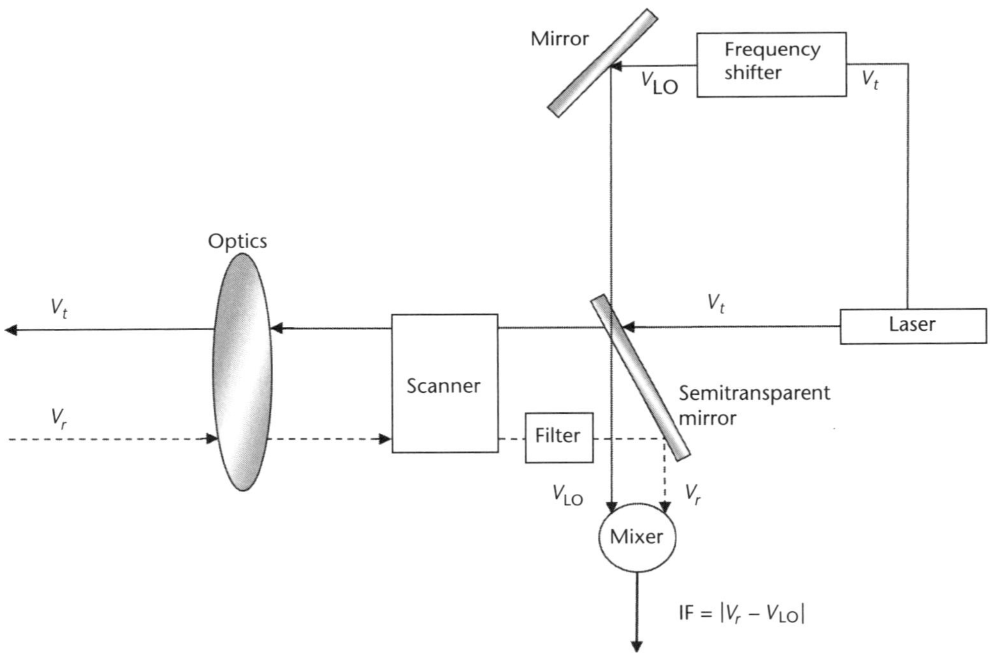

Figure 2.95 Heterodyne laser radar receiver.

2.2). We will use the same symbolism as used there but recalling that in the IR domain lenses and optics are used instead of antennas.

The received signal power P_r can be expressed as the product of five terms as follows:

$$P_r = \frac{P_t}{4\pi R^2} \cdot G_{\text{lens}} \cdot \frac{\sigma}{4\pi R^2} \cdot \frac{\pi D_0^2}{4} \cdot \frac{1}{L} \tag{2.100}$$

The first term expresses the distribution of the point source laser power P_t over a spherical area $4\pi R^2$. The second term expresses through the gain G_{lens} the focusing effect of the lens, which directs the laser power in the direction of the target and not isotropically over 4π sterad. The gain of the lens can be expressed also as $G_{\text{lens}} = 4\pi/(\pi\theta_B{}^2/4)$, where the beamwidth θ_B provided by the lens is multiplied by the factor $(\sqrt{\pi}/4)$ to take into account the uniform illumination of the aperture (which is narrowing the beam). The third term expresses the portion of the incident power intercepted by a point target and isotropically reirradiated (the so-called RCS). The fourth term describes the capture area of the receiving aperture (usually a circle of diameter D_0), and the fifth factor represents both the losses incurred in the two-way atmospheric path as well as the system losses.

The main difference with respect to the microwave radar equation is with the RCS of the target. A first consideration is that in the microwave domain the radar range equation is applicable in the far field of the antenna, which is defined as:

$$R_{FF} = 2D^2/\lambda \tag{2.101}$$

where R_{FF} (m) is the far-field distance, D (m) is the larger dimension of the antenna, and λ (m) is the transmitted wavelength. In the case of the laser domain for $\lambda = 1$ μm $= 10^{-6}$ m, even for a small dimension of the lens diameter such as $D_0 = 0.05$m, we have $R_{FF} = 5{,}161$m. Therefore, taking into account that the actual lens diameter can be two to three times larger, we have to understand that laser radar operation is always conducted in the near field of the lens.

A second consideration is that of the extremely narrow beam of the laser radar, which is proportional to the ratio of the wavelength to the lens diameter:

$$\theta_B = \left(\sqrt{\pi/4}\right)\lambda/D_0 \tag{2.102}$$

which means that for $D_0 = 0.05$m and $\lambda = 1$ μm, a beamwidth $\theta_B = 0.0176$ mrad.

With such a beamwidth, at a distance of 10 km the beam produces a spot of diameter $d = R\theta_B = 0.176$m, which is usually much less than the dimension of any useful target.

The last consideration implies that an actual target is never completely irradiated by the laser beam and therefore the backscattering cannot be isotropically radiated over a full sphere but only in the forward hemisphere of the target, that is, no more $\sigma/(4\pi R^2)$ but $\sigma/(2\pi R^2)$. Hence the RCS is defined as:

$$\sigma = \frac{\pi\rho}{4}(R\theta_B)^2 \tag{2.103}$$

where ρ is the target reflectivity [i.e., the complement to 1 of the emissivity $\varepsilon(\rho = 1 - \varepsilon)$]. As an example, unpainted metal is much more reflective at the mid–IR band than the reflected background.

Another often-used assumption relevant to the distribution of backscattered power from the target is that of Lambertian scattering, which states that the reflected flux per unit solid angle is proportional to the cosine of the angle between the direction of interest and the normal to the surface. So the total backscatter power is not distributed over 2π sterad but only over π sterad, because it can be deducted from Figure 2.96, which brings us to the following computation:

$$\int_0^{\pi/2} \cos\theta \cdot dA = \int_0^{\pi/2} \cos\theta \cdot 2\pi \sin\theta \cdot d\theta = \pi \tag{2.104}$$

Therefore, the backscatter term in the laser radar range equation becomes

$$\frac{(\pi/4)(R\theta_B)^2 \rho}{\pi R^2} = \frac{\theta_B^2}{4}\rho \tag{2.105}$$

and (2.100), taking into account (2.102) to (2.104), becomes

$$\text{Pr} = \frac{P_t D_0^2 \theta_B^2 \rho}{\pi R^2 L} \tag{2.106}$$

This equation states that the signal power from the actual target is inversely proportional to the square of the distance to the radar.

In the case of a target of very small size (dimension $d < R\theta_B$) (2.100) holds except for the consideration of Lambertian scattering (which produces a reduction from σ to $\sigma/2$); hence, the equation becomes

$$\text{Pr} = \frac{P_t \sigma D_0^2}{2\pi R^4 \theta_B^2 L} \tag{2.107}$$

In the case of the laser target designator, the laser beam at target distance R_1 is usually larger than the size of the target (in order to avoid the need for the search of the target) and the laser receiver is positioned at a distance $R_2 \neq R_1$. Therefore,

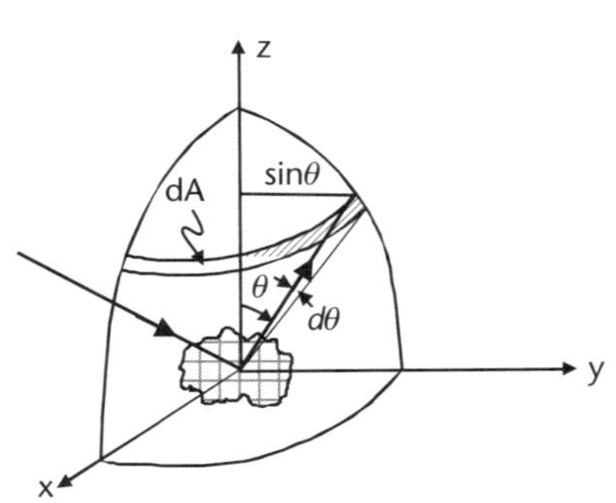

Figure 2.96 Lambertian backscatter.

the laser target designator equation is very similar to the microwave bistatic or semiactive one and is as follows:

$$P_r = \frac{P_t}{4\pi R_1^2}\frac{4\pi}{(\pi/4)\theta_B^2}\frac{\sigma'}{2\pi R_2^2}\frac{\pi D_0^2}{4}\frac{1}{L} \tag{2.108}$$

Note that in the above equation the RCS of the target is expressed as σ' to indicate that it is different from the one measured when the transmitter and the receiver are in the same direction.

2.7.4 Target Detection

As shown in the case of microwave radar, the detection of a target is impaired by the presence of noise in the receiver. In what follows we report on the noise level of the two types of laser receivers discussed earlier.

In the case of direct detection, the laser backscatter energy impinges directly on the detector, whose noise level is (as shown in Section 2.6.2)

$$N = (A_d \Delta f)^{1/2} D^* \tag{2.109}$$

where A_d is the detector area, Δf is the bandwidth of the signal, and D^* is the detectivity of the detector material.

The noise level of a heterodyne receiver can be computed from quantum mechanics, which indicates that the power spectral density of an ideal amplifier can be expressed as

$$\Phi(f) = \frac{hf}{\exp(hf/kT)-1} + hf \quad [\text{w/Hz}] \tag{2.110}$$

where

h = Planck's constant = 6.6256×10^{-34} (W/s^2),

f = frequency (Hz),

k = Boltzmann's constant = 1.38×10^{-23} (w/s K),

T = absolute temperature (K).

Equation (2.110) is plotted in Figure 2.97. From the figure's graph, the noise level of an heterodyne receiver can be expressed as:

$$N_h = \frac{hf}{\eta}B \tag{2.111}$$

where B is the receiver bandwidth and η is the quantum efficiency (0.3 to 0.5).

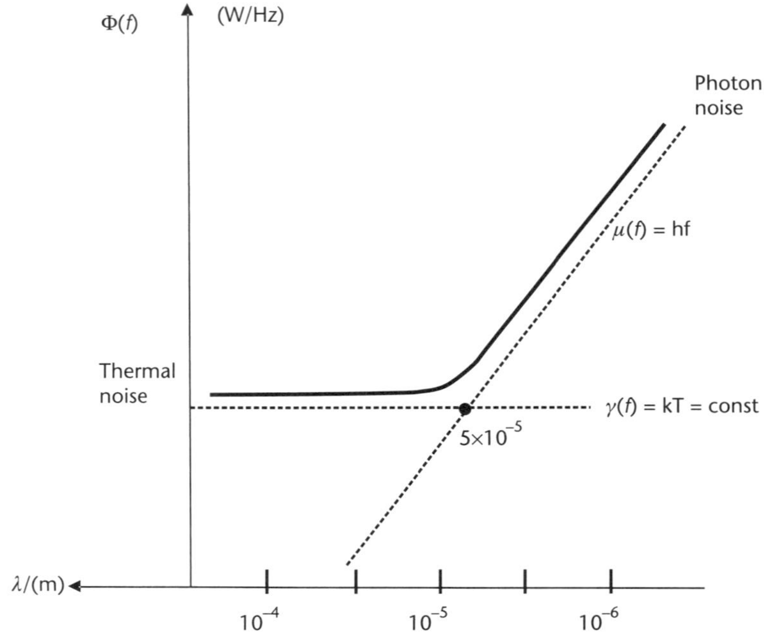

Figure 2.97 Noise power spectral density of an ideal amplifier (from Quantum Mechanics Equations).

Laser radar targets can be divided into two general categories: (1) speckled targets, in which case they are composed of many similar reflectors; and (2) glint targets, in which case they have a dominating reflector. The two types of targets correspond, respectively, to the scintillating targets and fluctuating targets studied in microwave radar detection theory (see Swerling models 1 and 3) and indeed they follow the same probability density distribution of their echo value: The speckled targets follow a Rayleigh distribution and the glint targets follow a Gaussian distribution.

Having determined the ratio P_r/N = SNR and the target category, the probability of detection is evaluated through the usual detection curves (Swerling models of microwave radars) after having fixed a threshold over the noise power level in accordance with achieving a selected probability of false alarms due to noise samples only.

Relevant to the measurement of target kinematics parameters the laser radar operates in a manner that is identical to that of microwave radars and all of the considerations developed in this chapter about radar waveforms and signal processing also apply to laser radars.

References

[1] Vaccaro, D., *EW Receiving Systems*, Norwood, MA: Artech House, 1993.

[2] Hovanessian, S., *Introduction to Sensor Systems*, Dedham, MA: Artech House, 1988.

[3] Skolnik, M., *Introduction to Radar Systems*, 3rd Ed., New York: McGraw Hill, 2001.

[4] Rihaczek, A., *Principles of High Resolution Radar*, Norwood, MA: Artech House, 1996.

[5] Galati, G., and Crescimbeni, R., "Basic Concepts on Detection, Estimation and Optimum Filtering," Chap. 1 in G. Galati (Ed.), *Advanced Radar Techniques and Systems*, London: Peter Peregrinus Ltd., 1993.

[6] Galati, G., and Lombardi, P., "Design and Evolution of an Adaptive MTI," *IEEE Trans. on Aerospace and Electronic Systems,* Vol. AES-14, 1978, pp. 899–905.

[7] Lewis, B., Kretschmer, F., and Shelton, W., *Aspects of Radar Signal Processing*, Dedham, MA: Artech House, 1986.

[8] Lewis, B., "Range-Time Sidelobe Reduction Technique for FM-Derived PC Codes," *IEEE Trans. on Aerospace and Electronic Systems*, Vol. AES-29, July 1993, pp. 834–840.

[9] Lynch, Jr., D., *Introduction to RF Stealth*, Raleigh, NC: Scitech Publishing, 2004, Chapters 1 and 5.

[10] Golay, M., "Complementary Series," *IRE Trans. on Information Theory*, Vol. IT-7, April 1961, pp. 82–87.

[11] Bar-Shalom, Y., and Li, X.-R., *Estimation and Tracking: Principles, Techniques and Software*, Norwood, MA: Artech House, 1993.

[12] Barton, D., *Modern Radar System Analysis*, Dedham, MA: Artech House, 1988, Chapter 11.

[13] Pace, P., *Detecting and Classifying Low Probability of Intercept Radars*, Norwood, MA: Artech House, 2004.

[14] Morris, G., *Airborne Pulsed Doppler Radar*, Dedham, MA: Artech House, 1988.

[15] Neri, F., *Introduction to Electronic Defense Systems, 2nd Ed.,* Raleigh, NC: Scitech Publishing, 2006.

[16] Howard, D., "Tracking Radars," Chapter 9 in M. Skolnik (Ed.), *Radar Handbook,* 3rd ed., New York: McGraw Hill, 2008.

[17] Nathanson, F., *Radar Design Principles*, 2nd Ed., New York: McGraw Hill, 1991, Sec. 8.8.

[18] Willis, N., "Bistatic Radar," Chapter 23 in M. Skolnik (Ed.), *Radar Handbook,* 3rd ed., New York: McGraw Hill, 2008.

[19] Dunsmore, M., "Bistatic Radars," Chapter 11 in G. Galati (Ed.), *Advanced Radar Techniques and Systems*, London: Peter Peregrinus Ltd., 1993.

[20] Poisel, R., *Introduction to Communication Electronic Warfare Systems,* Norwood, MA: Artech House, 2004.

[21] Torrieri, D., *Principles of Secure Communication Systems,* Second Edition, Norwood, MA: Artech House, 1992.

[22] Schleher, D., *Electronic Warfare in the Information Age,* Norwood, MA: Artech House, 1999.

[23] Groves, P., *Principles of GNSS, Inertial, and Multisensor Integrated Navigation Systems*, Norwood, MA: Artech House, 2008.

[24] Klein, L., *Millimiter-Wave and Infrared Multisensor Design and Signal Processing*, Norwood, MA: Artech House, 1997.

[25] Johnson, J., "Analysis of Image Forming Systems," paper presented at Image Intensifier Symposium, Fort Belvoir, VA, October 1958, Report AD-220160, pp. 249–273.

CHAPTER 3

Electronic Warfare RF Band Sensor Systems

3.1 Introduction

EW RF band sensors are those systems capable of detecting, classifying, identifying, and locating the RF band emitters of surveillance systems, weapons systems, and communication links of the military systems present in the environment.

Typically the radar-type emitters of surveillance and weapons systems operate at high frequencies (usually >1 GHz) to exploit directive antennas for target direction measurements. They also use phase- or frequency-modulated waveforms either of the periodic pulse type or CW type to exploit pulse compression techniques on reception to obtain accurate measurements of target distance. The carrier frequency of the signal waveform can be changed pulse to pulse or by a group of pulses (in accordance with the required pulse coherent integration performed by the radar receiver) within a large bandwidth, called the frequency agility bandwidth, to provide an ECCM technique. Currently there is some revival in the Western world of the use of lower frequency radar emitters (usually <1 GHz) to exploit antistealth and passive covert radar capabilities.

The communications-type emitters of the communication links typically operate at lower frequencies (usually <1 GHz), although currently they are expanding to higher frequencies to exploit larger signal bandwidths and the number of channel bandwidths. They usually transmit quite long (with respect to the radar emitter ones) amplitude/phase/frequency-modulated CW waveforms, seldom pulse bursts, and exploit spread-spectrum (SS) and FH techniques, which are the equivalent communications terms for the pulse compression and frequency agility techniques of radar, only both are exploited as anti-interference or ECCM techniques.

Until recently (and still currently in ground-based systems), sensors devoted to radar-type emitters and those devoted to communications-type emitters had a substantial difference based on the characteristics of the emitter waveforms (as just described and shown in Chapter 2) with regard to sensor architecture, the techniques employed, and their operation. The use of new technologies allows us to exploit, for airborne and shipboard applications, a common architecture for the EW sensors of both types of emitters, thus providing a so-called EW spectrum sensor. This chapter is therefore split into sections that deal with radar EW sensors, communications EW sensors, and EW spectrum sensors.

3.2 EW Radar Band Sensors

The architecture of the EW radar band sensor is different from that of the EW communications band sensor in accordance with the radar signal waveforms (usually periodic short-duration pulses) and the main functions performed by the sensor, which have to be executed in a very short time due to the extreme rapidity with which a weapon system becomes a threat. The functions and classifications of the relevant sensors are as follows:

- A *radar warning receiver* (RWR) warns of the presence of threat emitters and gives an indication of the direction of arrival (DOA) within the sensor spatial and frequency coverage.
- An *electronic support measurement* (ESM) receiver detects DOA and classification of all the radar emitters within the sensor spatial and frequency coverage.
- An *electronic intelligence* (ELINT) receiver provides prolonged and accurate measurements of all of the characteristics of a radar emitter (waveform and antenna patterns) in order to provide the data necessary for its analysis and modeling of the associated weapon system as well as for its identification to be logged in the emitter libraries of the two previous sensor types

The typical architecture for these past generation sensors is briefly discussed next in order to provide the reader with some information about the sensors' structure before introducing the main challenges to be faced by EW sensors, which include the following:

- *Receiver sensitivity*, that is, the minimum signal power level at which signal detection and its parameter measurements are achieved by the sensor;
- *Probability of interception* (POI) of all of the emitters (or only the threat emitters) present within the sensor spatial and frequency coverage within a predefined time period (time of interception, TOI).
- The *ability to discriminate the various emitters* within the spatial and frequency coverage and thus to classify the radar emitter mode (i.e., the characteristic parameters of its waveform and antenna scan type) in order to identify the emitter and its associated weapons system.

The architecture of the past generation EW radar band sensors is discussed in Section 3.5.1, before proceeding to a description of the architecture of the new generation of EW radar band sensors that are considered in detail in this chapter.

3.2.1 RWR Architecture

Basic EW passive surveillance equipment used in applications that do not require high sensitivity is represented by the RWR. Its task is the provision of an alarm to the aircraft pilot/platform commander relevant to the main-beam illumination of an emitting threat in the surrounding environment. The following functions are performed by the RWR:

- Detection of the threatening emitter,
- Coarse measurement of the DOA of the emission relative to the platform longitudinal axis (this measurement requires many directional antennas with associated receiver channels in a monopulse fashion),
- Measurement of the instantaneous threat signal parameters,
- Deinterleaving and sorting of the threat signal within a not highly dense signal environment,
- Identification of the threat signal by comparison with properly prepared data libraries.

The typical schematic block diagram of a wide-open (WO) RWR (i.e., complete RF coverage of the required threat emitter spectrum), which is the most commonly used RWR architecture, is shown in Figure 3.5 (refer to Section 3.4 for greater

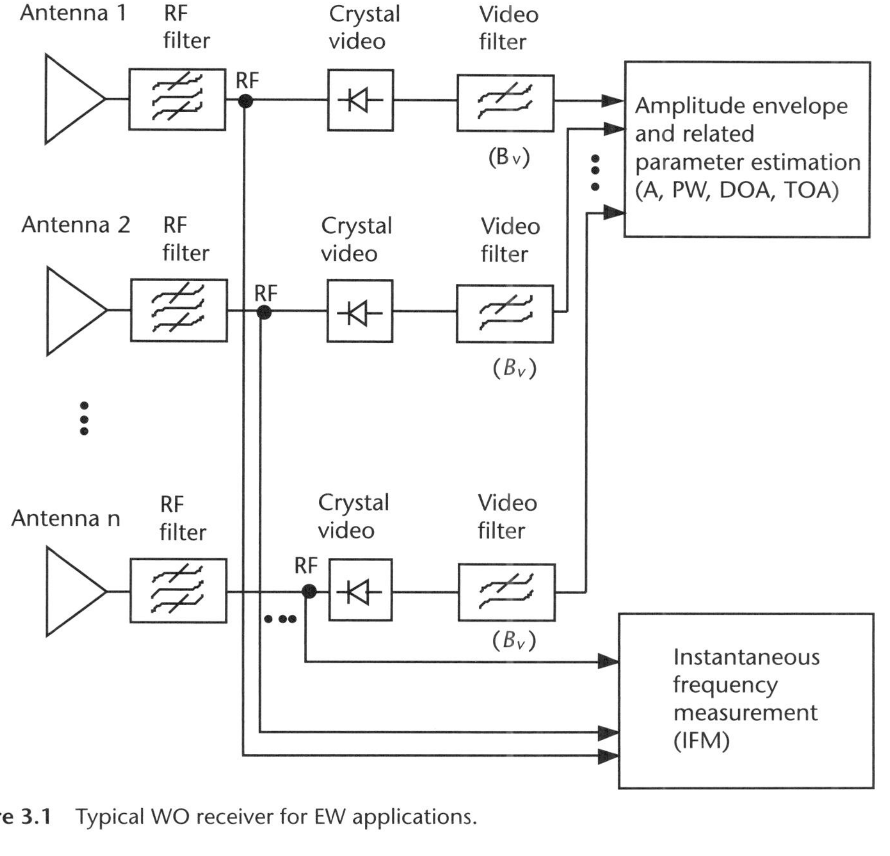

Figure 3.1 Typical WO receiver for EW applications.

detail). In this architecture the crystal detector (which usually acts as a square-law detector) channel allows detection over the entire RF bandwidth as limited by the input RF filter, but removes any information relevant to the frequency and phase of the emitter waveform. The video filter performs a smoothing in order to enable the detection of pulses having from a given (minimum) PW to larger PWs. A crystal detector channel is associated with each directive antenna of the equipment, and the relative measured pulse amplitudes are processed to provide the following instantaneous emitter pulse parameters:

- Pulse amplitude (*A*),
- Pulse width (PW),
- Direction of arrival (DOA),
- Time of arrival (TOA).

The RF signal, before detection, is also sent to an instantaneous frequency measurement (IFM) device (see Section 3.5), which provides the signal frequency estimation (frequency *F*). The IFM is usually a quite expensive device due to the very large bandwidth required and the microwave circuitry involved. Most early WO RWRs were without IFM.

All of the measurements of the instantaneous emitter pulse parameters are gathered within a pulse descriptor word (PDW), which is used in the pulse deinterleaving process (clustering of all the received PDWs belonging to each of the various emitters present in the environment) and emitter sorting process, which provides the evaluation of long time emitter signal parameters such as the pulse repetition interval (PRI), antenna scan type (AST), and antenna scan relevant period (ASP). These two processes are provided by the equipment processor (not shown in Figure 3.1), which also performs threat identification through comparison of the sorted emitter data with the threat emitter data available within the mission threat library.

3.2.2 ESM Architecture

In applications such as ESM equipment, in which in addition to the provision of threat warning a high sensitivity is required to detect both mainlobes and sidelobes of all type of emitters present in the environment (and accordingly a very dense signal environment is expected), a superheterodyne (SH) receiver or a mixed SH + WO receiver architecture can be used. The architecture of an ESM employing a SH stage is shown in Figure 3.2. In this architecture the overall RF bandwidth is divided in a number of sub-bands that are individually shifted into a fixed IF bandwidth by one of a number of local oscillator (LO) frequencies as generated by a synthesizer or a comb generator that can sweep the overall RF bandwidth. The required selectivity provided by this architecture has to be traded off against the requirement of a very high POI against all types of emitters present in the environment. The trade-off is usually performed by adopting a large IF bandwidth (larger than the available radar frequency agility bandwidth) and by rapid LO switching schemes. This type of receiver is usually quite expensive. So to reduce the cost of a complete SH architecture, also applied to the many monopulse DOA measurement channels, the latter are based on the WO architecture.

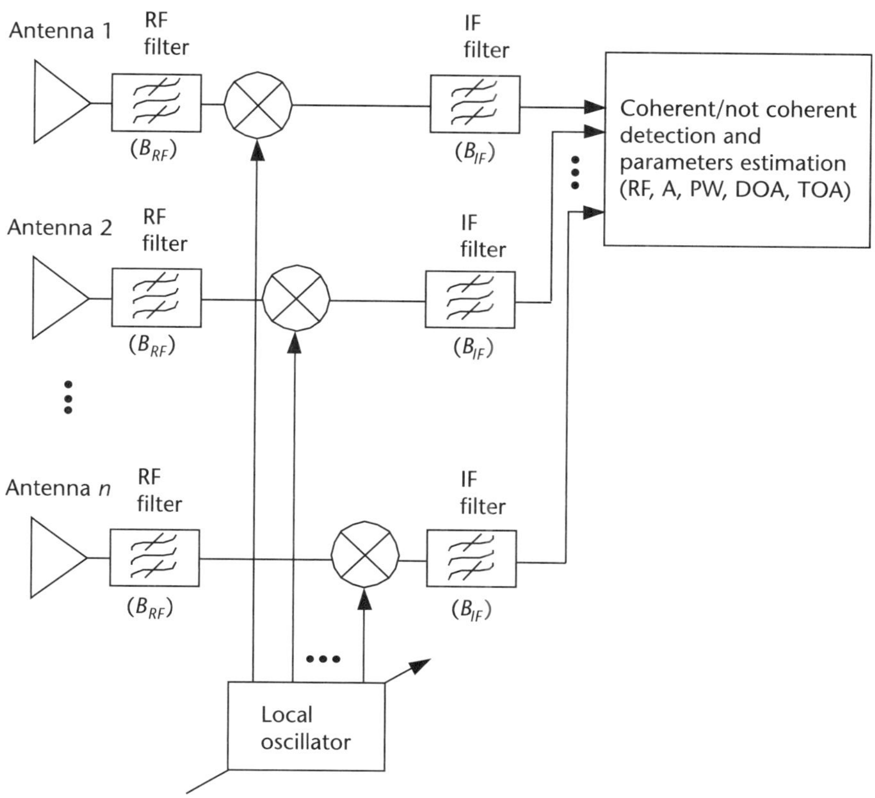

Figure 3.2 Typical SH receiver.

3.2.3 ELINT Architecture

The third type of EW passive surveillance equipment is the ELINT receiver. The task of this equipment is to selectively acquire and analyze, both instantaneously as well as long term, the signal characteristics of an emitter in order to gather sufficient data to provide to the data libraries of RWR and ESM equipment for its identification by those equipment.

The ELINT equipment, as shown in Figure 3.3, is based on the SH receiver architecture but with a number of selectable IF bandwidths. IF filters, coupled to a high-gain directive antenna, are able to achieve the very high sensitivity required to analyze the signals also transmitted from the far lobes of the radar antenna. The selectable IF bandwidth is also required to analyze peculiar intrapulse and interpulse signal modulations, which may provide individual emitter identification (emitter fingerprinting).

3.3 EW Sensor Sensitivity

The sensitivity of a sensor is defined as the minimum received signal power at which the signal is detected and its relevant instantaneous parameters are measured. A radar-type EW sensor is required to detect a radar emitter (airborne or surface) well before the latter can detect the EW sensor platform (aircraft or ship) or site

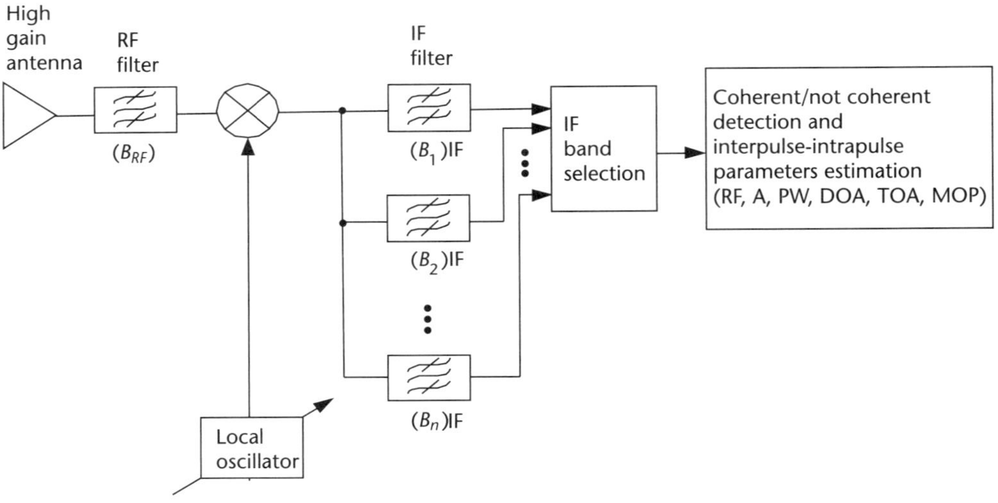

Figure 3.3 ELINT receiver.

(with its associated weapon system). This requirement is measured by the range advantage factor (RAF) and was quite easily met in the past until the advent of the so-called low probability of intercept radar [1], which has required a number of technological innovations of the radar-type EW sensor architecture.

The RAF is defined as the ratio of the range R_e at which the sensor detects the radar emitter to the range R_r at which the radar detects the EW sensor platform, characterized by its RCS:

$$RAF = R_e/R_r = 1 + a \quad a > 0 \tag{3.1}$$

To protect its platform, the required excess sensor range $aR_r = (R_e - R_r)$ is dependent on the product of the closing velocity v_c, which is the sum $v_c = v_e + v_r$ of the relative velocities of the two platforms (sensor and radar), and the warning time T_w, which is the time allotted to the EW sensor platform commander to be prepared to react against the radar threat; that is:

$$R_e = R_r + v_c T_w \tag{3.2}$$

With the above conditions in mind, the required radar-type EW sensor sensitivity can be evaluated through the range equations of both the radar and sensor as relevant to the engagement situation depicted in Figure 3.4 in which the EW sensor is placed on a ship (a platform with a large RCS) and the radar emitter is placed on an aircraft moving toward the ship with a velocity almost equal to v_c.

The radar detection range equation expresses [2] the received signal power S_r at the radar as the product of three terms: the first representing the radar transmitted power propagating to range R in the form of a spherical wave, the second representing the backscattered power from the target RCS propagating back to the radar at range R in the form of a spherical wave, and the third representing the radar receive aperture $A_r = G_r\lambda^2/4\pi$:

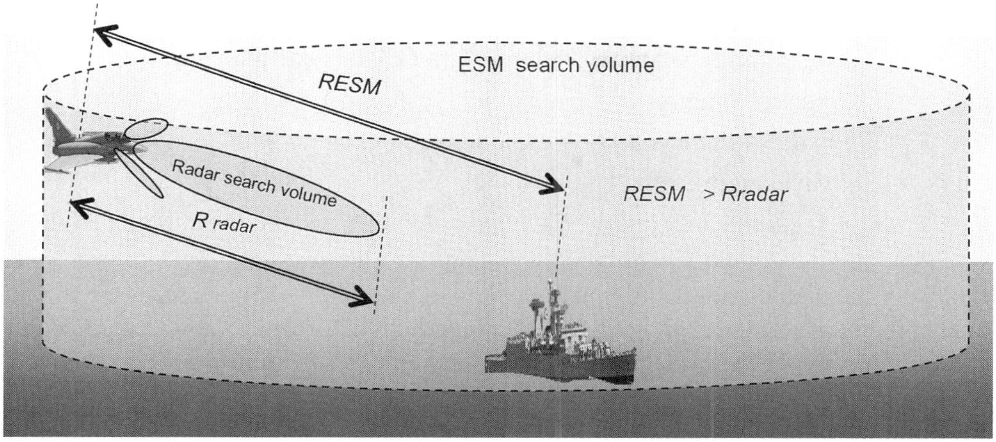

Figure 3.4 Typical geometry of an engagement between a radar and an EW sensor.

$$S_r = \frac{P_t G_t / L_t}{4\pi R^2} \frac{RCS}{4\pi R^2} \frac{G_r \lambda^2 / L_r}{4\pi} \tag{3.3}$$

where:

S_r = power at the radar receiver (W),
R = distance between the radar and the EW sensor platform (m),
G_t = gain of the radar Tx antenna,
G_r = gain of the radar Rx antenna,
λ = wavelength (m),
RCS = EW sensor platform radar cross section (m^2)
L_t = factor expressing the radar losses from the transmitter to the Tx antenna,
L_r = factor expressing the radar losses from the Rx antenna to the receiver.

From Chapter 2, we know that radar detection occurs when S_r exceeds, by an amount dependent on the required probability of detection P_D (usually ≥90%), a threshold set above the video detected radar noise power floor in order to get a false alarm (detection produced by a noise sample), probability P_{FA} (usually $<10^{-6}$). Let us define this power value as S_{r0}.

The EW sensor detection range equation expresses, in a manner similar to that above, the received signal power S_e at the EW sensor as the product of two terms, the first representing the radar transmitted power propagating to range R in the form of a spherical wave and the second representing the EW sensor receive aperture $A_e = G_e\lambda^2/4\pi$:

$$S_e = \left[(P_t G_t / L_t)/(4\pi R^2)\right]\left[(G_e \lambda^2 / L_e)/(4\pi)\right] \tag{3.4}$$

where

S_e = power at the EW sensor receiver (W),

R = distance between the radar and the EW sensor platform (m),

G_t = gain of the radar Tx antenna,

G_e =gain of the EW sensor Rx antenna,

λ = wavelength (m),

L_e = factor expressing the EW sensor losses from the Rx antenna to the receiver.

The detection of a signal in sensors is performed by setting at the output of the receiver detector a threshold above the receiver noise power level N in order to achieve a false alarm probability P_{FA} (usually $<10^{-6}$) and a probability of detection P_D (usually $\geq 90\%$) of a signal of sufficient power S. Let us define this power value as S_{e0}.

To evaluate the RAF, as well as the sensitivities of the two involved types of equipment, we have to remember from detection theory (see Appendix A) that for the assumed values of the two probabilities ($P_D = 90\%$, $P_{FA} = 10^{-6}$) the SNR relevant to a nonfluctuating signal power is

$$SNR_0(P_D = 90\%, P_{FA} = 10^{-6}) = 13 \text{ dB} = 20 \tag{3.5}$$

Actually for an EW sensor to get a full reliable signal with instantaneous parameter measurement a larger SNR is required (e.g., the SNR required for a fair DOA estimate is >18 dB). However, for the considerations that we will develop we can consider SNR_0 to be valid for both types of equipment.

For the radar we can now equate S_{r0} and hence (3.3) with the required detection value provided by detection theory as follows:

$$S_{r0} = (\text{Radar Receiver Noise Power}) \times SNR_0 = (kTB_rF_r)SNR_0 \tag{3.6}$$

where

F_r = noise figure of the radar RF receiver chain,

k = Boltzmann's constant $k = 1.38 \times 10^{-23}$ W/(Hz K),

T = 290 K = standard temperature,

B_r = equivalent noise bandwidth at the end of the radar processing chain, which is the transmitted signal bandwidth reduced by the multiple signal integrations (both predetection integration exploiting the signal coherency as well postdetection integration exploiting the signal duration in time, the so called time on target).

We can express this bandwidth as

$$B_r = B_t / G_p \tag{3.7}$$

where B_t is the radar signal transmitted bandwidth and G_p is the radar processing signal gain achieved as mentioned earlier. Current available processing gains range from 10 to 1,000 (10 to 30 dB), thus actual B_r are on order of 0.1 to 0.001B_t [3]. We can proceed in a similar way for the EW sensor:

$$S_{e0} = (\text{EW Sensor Receiver Noise Power}) \times SNR_0 = (kTB_eF_e)SNR_0/G_{pe} \tag{3.8}$$

where

F_e = noise figure of the EW sensor RF receiver chain;

B_e = equivalent noise bandwidth of the EW sensor RF receiver chain, which is usually not matched to any peculiar radar waveform and depends on the sensor architecture;

G_{pe} = EW sensor processing signal gain.

By equating (3.3) with (3.6), and similarly (3.4) with (3.8), and expressing the results in terms of the detection ranges of the radar R_r and the EW sensor R_e, we obtain:

$$R_r = \left[\frac{P_tG_tG_r\lambda^2 \cdot RCS \cdot G_p}{(4\pi)^3 L_tL_r(kTB_tF_r)\cdot SNR_0}\right]^{1/4} \tag{3.9}$$

$$R_e = \left[\frac{P_tG_tG_e\lambda^2 G_{pe}}{(4\pi)^2 L_tL_e(kTB_eF_e)\cdot SNR_0}\right]^{1/2} \tag{3.10}$$

Before considering the RAF, note that the radar antenna mainlobe is not always directed toward the EW sensor, so to illuminate the latter with its maximum gain G_t it can illuminate it through the antenna sidelobes with a gain as follows:

$$G_{te} = G_t/SLL < G_t \tag{3.11}$$

where the sidelobe level (SLL) expresses the reduction gain factor at radar antenna sidelobes, which is usually quite high in modern radars.

In the following we substitute G_{te} in lieu of G_t in (3.10) to enlarge the scope of our considerations. Thus, (3.10) becomes:

$$R_e = \left[\frac{P_tG_{te}G_e\lambda^2 G_{pe}}{(4\pi)^2 L_tL_e(kTB_eF_e)\cdot SNR_0}\right]^{1/2} \tag{3.10$'$}$$

By taking the ratio of (3.10′) to (3.9) we obtain the RAF:

$$\frac{R_e}{R_r} = \left[\frac{P_tG_{te}^2G_e^2G_{pe}^2\lambda^2 B_tF_rL_r}{4\pi \cdot RCS \cdot kTB_e^2F_e^2L_e^2 \cdot SNR_0 \cdot G_tG_rL_tG_p}\right]^{1/4} \tag{3.12}$$

To gain insight into this equation, let us make some assumptions that have been shown to be valid in many cases:

- $G_t = G_r$ (i.e., the radar Rx antenna gain is the same as the radar Tx antenna gain, which is the case for a single-beam Rx antenna),
- $L_t = L_r$ (i.e., the radar receiver losses are nearly the same as the radar transmitter losses),
- $F_e \sim 5F_r$ (i.e., the EW sensor receiver noise figure is much larger than the radar one, due to the wider receive bandwidth),
- $L_e \sim 1$ dB = 1.2,
- $SNR_0 = 13$ dB = 20.

With the above assumptions and rearranging (3.12), we have:

$$\frac{R_e}{R_r} = \left[\frac{K}{4\pi} \frac{P_t \lambda^2 G_{te}^2 G_e^2 G_{pe}^2 B_t}{G_t^2 G_p B_e^2 \cdot RCS} \right]^{1/4} \tag{3.13}$$

where $K = [(F_r / F_e^2\, L_e^2)(1/(kT\, SNR_0))] = 83\ dB_{m/MHz}$.

Let us now evaluate the RAF in some typical cases.

Case 1: *Past generation ground surveillance radar and past generation EW sensor (WO preamplified crystal detector of sensitivity class –60 dB$_m$) on an aircraft.* The radar parameters are:

$P_t = 100$ kW = 80 dB$_m$; $G_t = G_r = 35$ dB; $B_t = 1$ MHz = 0 dB$_{MHz}$;
$\lambda = 0.1$m = –10 dB$_{mt}$; $G_p = 13$ dB; $F_r = 3$ dB; $L_t = L_r = 2$ dB; RCS = 5 m^2 = 7 dB$_{m2}$

With the above parameters the radar detection range evaluated with (3.9) is:

$R_r = 22.75$ dB$_{km}$ = 188.4 km = 101.8 Nm

The EW sensor parameters are:

$G_e = G_{pe} = 0$ dB; $B_v = 20$ MHz; $B_{RF} = 16000$ MHz; $F_e = 10$ dB; $L_e = 2$ dB
$B_e = (2\, B_v\, B_{RF})^{1/2} = 800$ MHz = 29 dB$_{MHz}$

With these parameters the EW sensor detection range evaluated with (3.10) is:

$R_e = 35.5$ dB$_{km}$ = 3548 km = 1916 Nm

and from (3.10′) in the case of $G_{te} = G_t$/SLL = 0 dB, the EW sensor detection range is:

$R'_e = 18$ dB$_{km}$ = 63 km = 34 Nm

Thus, the RAF on the radar mainlobe and sidelobes is, respectively, RAF_{ML} = 18.8 and RAF_{SL} = 0.33, which means that in any case the EW sensor platform is well warned in advance only of the mainlobe of the radar emitter.

A similar evaluation shows that an even much simpler past generation EW sensor (non-preamplified WO crystal video detector of sensitivity class --45 dB_m) can provide RAF_{ML} = 2.64 and RAF_{SL} = 0.046 against the same radar emitter.

Case 2: New generation LPI radar and past generation EW sensor (WO preamplified crystal detector of sensitivity class –60 dB_m) on a frigate ship. The radar parameters are:

P_t = 100W = 50 dB_m; $G_t = G_r$ = 35 dB; B_t = 500 MHz = 27 dB_MHz; λ = 0.03m = –15 dB_mt;
G_p = 30 dB; F_r = 3 dB; $L_t = L_r$ = 2 dB; RCS = 1,000 m^2 = 30 dB_m2

With these parameters the radar detection range evaluated with (3.9) is:

$$R_r = 16.0\ \text{dB}_\text{km} = 40\ \text{km} = 21.6\ \text{Nm}$$

The EW sensor parameters are the same as for case 1 above:

$$G_e = G_{pe} = 0\ \text{dB};\ B_v = 20\ \text{MHz};\ B_{RF} = 16000\ \text{MHz};\ F_e = 10\ \text{dB};\ L_e = 2\ \text{dB}$$
$$B_e = (2\ B_v\ B_{RF})^{1/2} = 800\ \text{MHz} = 29\ \text{dB}_\text{MHz}$$

With these parameters the EW sensor detection range evaluated with (3.10) is:

$$R_e = 15.5\ \text{dB}_\text{km} = 35.5\ \text{km} = 19.2\ \text{Nm}$$

and from (3.10′) in the case of $G_{te} = G_t/\text{SLL}$ = 0 dB the EW sensor detection range is

$$R'_e = -2\ \text{dB}_\text{km} = 0.63\ \text{km} = 0.34\ \text{Nm}$$

Thus, the RAF for the radar mainlobe and sidelobes is, respectively, RAF_{ML} = 0.88 and $\text{RAF}_{SL} \sim 0$, which means that the EW sensor platform is not warned in advance of the presence of the radar emitter. The situation for the EW platform is even worse in the case of a very low probability of intercept radar (VPLIR) [1], which decreases its transmitted power and modifies its waveform to keep a constant received echo power while approaching the target (i.e., $R_e \sim 0$ km).

Case 3: New generation LPIR and new generation EW sensor (a number of large band digital receivers connected to a wideband high-gain phased array antenna of sensitivity class –90 dB_m) on a frigate ship. The radar parameters are the same as those of case 2 above:

P_t = 100W = 50 dB_m; $G_t = G_r$ = 35 dB; B_t = 500 MHz = 27 dB_MHz; λ = 0.03m = –15 dB_mt; G_p = 30 dB; F_r = 3 dB; $L_t = L_r$ = 2 dB; RCS = 1000 m^2 = 30dB_m2

With these parameters the radar detection range evaluated with (3.9) is :

$$R_r = 16.0\ \text{dB}_\text{km} = 40\ \text{km} = 21.6\ \text{Nm}$$

The EW sensor parameters are as follows:

$G_e = 15$ dB; $G_{pe} = 5$ dB; $B_e = 20$ MHz $= 13$ dB$_{\text{MHz}}$; $F_e = 10$ dB; $L_e = 2$ dB

With these parameters the EW sensor detection range evaluated with (3.10) is:

$$R_e = 33.5 \text{ dB}_{\text{km}} = 2239 \text{ km} = 1209 \text{ Nm}$$

and from (3.10′) in the case of $G_{te} = G_t/\text{SLL} = 0$ dB, the EW sensor detection range is

$$R'_e = 16 \text{ dB}_{\text{km}} = 40 \text{ km} = 21.6 \text{ Nm}$$

Thus, the RAF for the radar mainlobe and sidelobes is, respectively, $\text{RAF}_{ML} = 30.2$ and $\text{RAF}_{SL} = 1$, which means that the EW sensor platform is warned well in advance about an attack from the radar emitter platform and insufficiently warned of the presence of the radar emitter in the environment.

3.3.1 Conclusions

From (3.13) we note that good RAF can easily be achieved by a small RCS EW platform against past generation radars (high P_t and low G_p) even with very wide noise equivalent bandwidth WO receivers (see case 1 above) while the RAF is difficult to be achieved as shown in case 2 if:

- Detection has to occur on the sidelobes of the radar $(G_{te}/G_t) < -30$ dB.
- The radar transmitter power P_t is very much decreased (on the order of 1W or less for the new generation of stealth or VLPIRs) and consequently also the radar processing gain G_p is highly increased.
- The radar frequency is increased (λ is decreasing).
- The EW sensor platform has a large RCS.

The directions along which the new generation of EW sensors are moving in order to regain the RAF are as shown in case 3:

- Increasing the antenna gain G_e through adoption of phased array antennas,
- Reducing the equivalent noise bandwidth B_e by using new architectures based on large-bandwidth channelized digital receivers,
- Adoption of new algorithms that provide fair detection processing gain G_{pe} on the LPI radar waveforms.

3.4 Probability of Interception

The first task to be accomplished by EW equipment on board a platform when accessing a new territory is the interception of the threat emitters present in the environment. Interception is fundamental to the warning function of EW equipment and it comprises signal detection and measurement of the signal parameters and DOA of the threat emitter signal, followed by the identification of the emitter (and hence of its associated weapons system) and possibly a rough estimate of its distance from the defended platform, in order to assess its dangerousness. Presence and DOA of a peculiar or unexpected threat emitter can induce the platform commander either to divert the platform route, so as to avoid engagement with the threat, or to keep the original route if the platform is provided with sufficient ECMs to tackle the threat.

Let us consider three simple EW equipment architectures to introduce the evaluation of a sensor's POI:

1. An omni wideband (WB) antenna, covering the complete azimuth plane (with some extent in the elevation plane) and an associated WB receiver covering the complete required RF bandwidth. A simple modified architecture envisages, in lieu of the omni antenna, for or more WB directional antennas that cover the complete azimuth plane, but provide through the differential signal amplitude, as received from the antennas, a rough DOA estimate.
2. A WB antenna of the above type associated with a SH narrowband (NB) receiver that is swiftly swept across the whole WB.
3. A rotating high-gain WB antenna associated with either a WB receiver or a SH NB receiver. This solution offers high sensitivity and is in general used in ELINT equipment to provide emitter detection and continuous parameter measurements at the emitter antenna sidelobes.

At first glance interception may seem an easy task to be accomplished because it would be sufficient to provide the EW platform with a WO receiver, capable of instantaneously surveying the whole azimuth and threat frequency band. Indeed, this was the solution for the RWR in the early days of EW (say, the years 1950 to 1970) when there were few radar threats in the environment and most of them used low-duty-cycle and high effective radiated power (ERP). (Duty cycle d is defined as the ratio of pulse duration or PW to the PRI, i.e., d = PW/PRI and ERP is the product of transmit power P_t and Tx antenna gain G_t, i.e., ERP = $P_t\ G_t$). New threats are characterized by a higher duty cycle and, hence, lower peak power and ERP for a given level of detection performance (which depends on average power). Further, the current environment is characterized by a largely increased number of radar threats. Increased numbers of more modern radars have led to environments with high pulse densities, measured as M pulses per unit time:

$$M = N_r PRF_{av} \tag{3.14}$$

where N_r is the number of radars present and PRF_{av} is the average PRF, which can approach 100 kHz when HPRF radars are present. Thus, M can reach values of millions of pulses per second. In addition, the use of long pulses leads to overlapping of threat pulses such that low-level pulses can be masked, and reception of long-duration waveforms can be interrupted by high-level pulses from other radars. Pulse masking in turn presents a large difficulty for the WB receiver in the evaluation of the PRI and in the determination of *PRI law* (i.e., the time behavior of the PRI), which characterizes the radar emitter and allows its identification.

A way to solve this problem is the adoption of a SH NB receiver for which band B_s is swiftly shifted in steps of duration T_d (the dwell time in milliseconds) within a scanning period T_{scan} (fraction of a second) across the complete required RF coverage. This solution offers the advantage of strongly reducing the instantaneously received pulse density and pulse overlapping as well as providing a higher sensitivity with respect to the WB open receiver. The drawback of this solution (beyond some detection of quite large out-of-band spurious signals to be solved with the adoption of a WB large signal detector) is the detection of scanning emitters, which do not illuminate continuously with their main beam the EW sensor but only periodically (ASP) and in accordance with their scanning law (AST). Most of the surveillance and target acquisition radars are of this type and thus both ASP and AST are important parameters for the identification of the radar emitter.

To evaluate the POI of any of the two types of periodic EW sensor scan techniques (either stepped center frequency sweeping of a bandpass filter or the continuous rotation of the antenna 3-dB beamwidth) with respect to radar scans, we can schematize the problem, following the approach of [2], as the evaluation of the time coincidences (windows overlaps) between two periodic pulse trains, as depicted in Figure 3.5:

1. The radar signal window width, sometimes called the time on target, is given by $\tau_R = T_R\theta_{R3dB}/2\pi$, where T_R is the radar antenna scan period and θ_{R3dB} is the radar beamwidth;
2. The EW sensor window width is either the dwell time of the sensor bandpass, given by $\tau_E = T_E B_E/B_S$, or $\tau_E = T_E\theta_{E3dB}$, where T_E is the sensor scan period, B_E is the bandwidth of each frequency step within the scanned bandwidth B_S, and θ_{E3dB} is the sensor beamwidth.

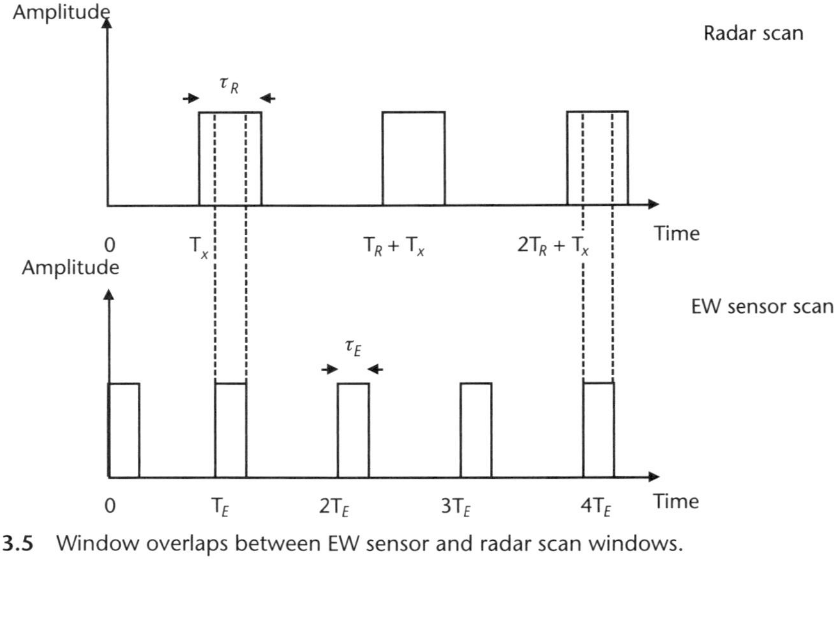

Figure 3.5 Window overlaps between EW sensor and radar scan windows.

Before solving the general case of different window durations $\tau_E \neq \tau_R$, let us assume that $\tau_E = \tau_R = T$ and assume this duration as the unit time, so that $T_E = kT$ *and* $T_R = hT$ are the periods of the EW sensor and radar scan, respectively, and that $T_x = pT$ is the relative phase between the two waveforms with respect to an arbitrary time reference

From Figure 3.5 note that a coincidence occurs if two integers m and n can be found such that:

$$mT_E + T_x = nT_R \tag{3.15}$$

that is,

$$mk + p = nh \tag{3.15'}$$

Equation (3.15′) can be solved using the theory of linear congruencies, which states that a solution exists if, given three integers a, b, and c, the greatest common divisor (gcd) of two integers (a,b) [i.e., $gcd(a,b)$] is a factor of the third integer; that is, it divides the third integer c. Furthermore, if there is a solution, many additional solutions can be obtained from the first one by adding to it multiples of the least common multiple (lcm) of (a,b) [i.e., $lcm(a,b)$]. Let's clarify this statement by considering the following three integers:

$$k = 12,\ h = 15, \text{ and } p = 6$$

Then $gcd(12,15) = 3$, which is a factor of $p = 6$, and $lcm(12,15) = 60$. The first solution for (3.15′) gives $m = n = 2$ and the following ones are $m_i = n_i = 2 + 60i$. Going back to (3.15′) we can therefore state that a coincidence occurs only if $gcd(k,h)$ is a factor of p, and if there is a coincidence, then multiple coincidences will occur periodically with a period equal to $lcm(k,h)$. The fraction of time C, called the *coincidence fraction*, for which the two pulse trains coincide is:

$$C = \tau / T_p \tag{3.16}$$

where $T_p = lcm(T_E, T_R)$is the coincidence period of the two windows.

As for the evaluation of the POI, we are interested in examining the coincidence fraction C over long periods of time. So we can consider without loss of exactness the phase T_x between two pulse trains as $T_x = 0$ (e.g., by selecting the reference time as the one of the first coincidence).

The case of different window durations $\tau_E \neq \tau_R$ can be deduced from the previous simple case by defining a unit window duration such that:

$$\tau_R = \alpha T \text{ and } \tau_E = \beta T \tag{3.17}$$

and considering the two original windows as decoupled respectively into α and β equal adjacent simple windows of the previous case, as shown in Figure 3.6. Therefore, we now have to consider the coincidence fraction c by exhaustively examining $\alpha\beta$ pairs of elementary pulse trains and counting the number of such pairs that have coincidences within the period T_p.

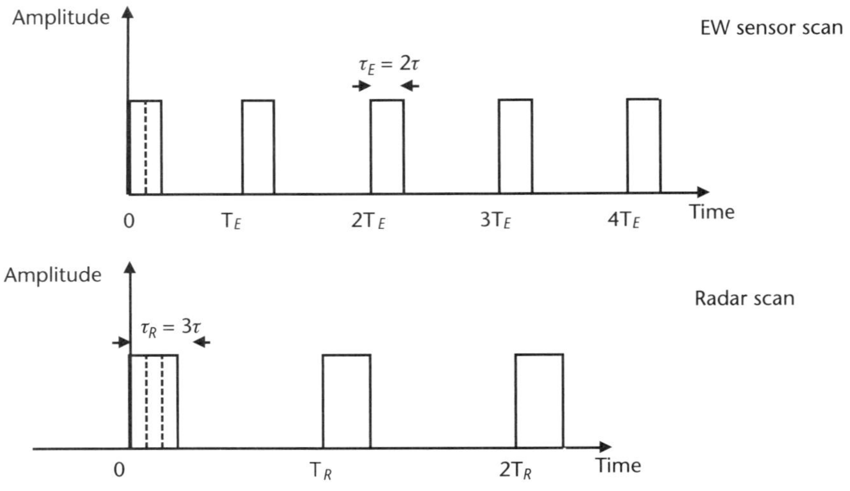

Figure 3.6 Generic radar and EW sensor windows ($T_E \neq T_R$; $\tau_R \neq \tau_E$; $\alpha = 2$; $\beta = 3$).

It can be shown [2] that the coincidence fraction is

$$C(T_E, T_R) = \alpha\beta/(\text{hk}) = \tau_E \tau_E/(T_E T_R) \tag{3.18}$$

the mean period between coincidences is

$$T_p = T_E T_R/(\tau_R + \tau_E) \tag{3.19}$$

and the average duration of the coincidences is

$$\tau_P = \tau_R \tau_E/(\tau_R + \tau_E) \tag{3.20}$$

Reference [2] also treats the case of triple pulse train coincidences as if it is the case of an EW sensor (ELINT) searching for a scanning radar by means of a high gain antenna, rotating at high speed, connected to a stepped frequency scan receiver.

To achieve a reliable POI with pulse trains providing small coincidence fractions, it is intuitive that it is necessary to use an observation period that lasts for many T_p. A good approximation for the evaluation of POI versus observation time (although not strictly valid for periodic pulse trains as reported above) can be deduced according to [2] with the following assumptions:

- The probability of a coincidence $p(t)$ is independent from one time to the next.
- The probability of a coincidence at time $t + \Delta t$ [i.e., $p(t + \Delta t)$] is increased with respect to $p(t)$ by the ratio of the increment in time Δt to the mean period between coincidences T_p, that is:

$$p(t + \Delta t) - p(t) = (1 - p(t))\Delta t/T_p \tag{3.21}$$

By considering the limit for $\Delta t \rightarrow 0$, (3.21) becomes a differential equation:

$$d\,p(t)/dt = \lim_{\Delta t \to 0}[p(t+\Delta t) - p(t)]/\Delta t = [(1 - p(t)\Delta t/T_p] \tag{3.21'}$$

which has a solution of the type $p(t) = A + Be^{-\alpha t}$, where the constants A, B, and α can be evaluated from the boundary conditions as follows:

- $p(0) = A + B = C$ as at any instant of time and thus also at $t = 0$ the probability of coincidence is equal to C.
- $p(\infty) = 1 = A$; that is, after a long time the probability of coincidence should approach unity and consequently $B = C - 1$.
- Further by substituting (3.21′) into (3.21), we have $\alpha = 1/T_p$.

Therefore, (3.21) becomes

$$p(t) = 1 - (1 - C)e^{1t/Tp} \tag{3.22}$$

As mentioned earlier, this equation is only approximately correct for the case of perfectly periodic pulse trains as the coincidences occur either periodically at multiples of T_p or not at all and in particular for them $p(T_p) = 1$, while (3.22) gives ~0.63 for small C.

Equation (3.22) allows us to determine in many practical cases (where perfect periodic pulse trains are not present) the time required to achieve a specified value of POI as a function of the coincidence fraction c. This time, when a POI = 90% is specified, is also called the *time to intercept* (TOI). Indeed by expressing (3.22) in logarithmic form and solving for the time t/T_P, we have

$$t/T_P = \ln(1 - C) - \ln(1 - POI) \tag{3.23}$$

Plots of (3.23) as a function of c are reported in [2].

To provide insight into the theory discussed, let us work out two numerical examples of POI evaluation.

Example 1: An EW sensor with a WB antenna (so that it is not necessary to search in azimuth) and a very fast stepped scan frequency receiver able to cover a 1,000 MHz bandwidth in search steps of 100 MHz at a 10-ms duration is used to intercept a surveillance radar with an antenna of 2° azimuth beamwidth rotating at 10 revolutions per minute (rpm). The EW sensor pulse train is characterized by

$$\tau_E = 10 \text{ ms and } T_E = (1000 \text{ MHz}/100 \text{ MHz})\ 10 \text{ ms} = 0.1 \text{ sec}$$

while the surveillance radar is characterized by

$$T_R = 6 \text{ sec and } \tau_R = (2°/360°)\ 6 \text{ sec} = 0.033 \text{ sec} = 33.3 \text{ ms}$$

In accordance with (3.18) to (3.20) we have the coincidence fraction

$$C(T_E, T_R) = \tau_R \tau_E/(T_E T_R) = (0.0333 \times 0.01)/(0.1 \times 6) = 0.55 \times 10^{-3}$$

the mean period between coincidences

$$T_p = T_E T_R / (\tau_R + \tau_E) = (0.1 \times 6)/(0.0333 + 0.01) = 13.86 \text{ sec}$$

and the average duration of the coincidences is

$$\tau_P = \tau_R \tau_E / (\tau_R + \tau_E) = (33.3 \times 10 \times 10^{-3})/(33.3 + 10) = 7.7 \text{ ms}$$

and from (3.23) we get that the TOI = $2.3T_p = 31.9$ sec.

If the EW sensor receiver above had a lower frequency step, such as 10 MHz, its pulse train would be characterized by τ_E = 10 ms and T'_E = (1000 MHz/10 MHz) 10 ms = 1 sec, so that $C' = 0.55 \times 10^{-4}$; $T'_p = 138.6$ sdc; $\tau_P = 7.7$ ms; and TOI′ = $2.3T'_p = 319$ sec.

Example 2: An EW sensor with a WB receiver (so that it is not necessary to search in frequency) and with a narrow azimuth beamwidth of 1°, rotating at 60 rpm, is used to intercept a surveillance radar with an antenna of 2° azimuth beamwidth rotating at 10 rpms. The EW sensor pulse train is characterized by

$$T_E = 1 \text{ sec and } \tau_E = (1°/360°)\ T_E = 2.8 \text{ ms}$$

while the surveillance radar is characterized by

$$T_R = 6 \text{ sec and } \tau_R = (2°/360°)\ 6 \text{ sec} = 0.033 \text{ sec} = 33.3 \text{ ms}$$

In accordance with (3.18) to (3.20), we have the coincidence fraction

$$C(T_E, T_R) = \tau_R \tau_E / (T_E T_R) = (0.0333 \times 0.0028)/(1 \times 6) = 9.2 \times 10^{-5}$$

the mean period between coincidences

$$T_p = T_E T_R / (\tau_R + \tau_E) = (1 \times 6)/(0.0333 + 0.0028) = 167.6 \text{ sec}$$

and the average duration of the coincidences is

$$\tau_P = \tau_R \tau_E / (\tau_R + \tau_E) = (33.3 \times 2.8 \times 10^{-3})/(33.3 + 2.8) = 2.6 \text{ ms}$$

and from (3.23) we get that the TOI = $2.3T_p = 385.5$ sec.

From example 1 we can deduce that the parameters of a stepped scan receiver have a large impact on the EW sensor reaction time (that is, the time employed by the sensor to declare that it has detected a scanning threat). To reduce the employed time to the reaction times usually required by military services (TOI on the order of 2 to 5 ARP), receivers are designed and developed with a larger overall receiver bandwidth and are capable of searching with large frequency steps at very fast stepped scan speeds.

In practical equipment a number of frequency scan strategies, which have an impact on the sensor architecture, can be exploited to reduce the TOI to the required values. A strategy used in receivers that exploit fixed frequency steps within the entire RF bandwidth consists of performing a couple of fast stepped frequency scans over the entire required RF bandwidth and, after having intercepted all of the steady (i.e., nonscanning) emitters, then performing slower stepped frequency scans, but allocating within each dwell time at a frequency step a small fraction of the dwell time to visit the bandwidths where, in accordance with pre-mission information, a scanning radar emitter may be present. The fraction of dwell time required is limited to a couple of PRI of the possible emitter and is sufficient to assess the presence of the main beam of the emitter.

Upon this detection the receiver dynamically reprograms its stepped frequency scan strategy in order to allocate a longer dwell time at the emitter frequency at about a time T_R apart so as to perform complete and reliable acquisition of all of the emitter waveform parameters. Once this result has been achieved, the stepped frequency scan strategy is unchanged until a new detected scanning emitter requires further reprogramming.

Of course, any stepped frequency scan strategy is effective if the assumptions at its base are valid and verified through many runs of simulated scenarios mimicking the real expected ones. It is recommended that a sensor architecture be adopted that can support an adaptive frequency scan strategy that is capable of coping with the dynamic changes of the scenario and providing high POI on all the emitters present in the environment.

3.5 EW Radar Band Sensor Architectures

3.5.1 Architecture of Past Generation Intercept Receivers

In the EW domain five basic configurations for intercept receivers were used in the past to cover the required wide radar threat frequency band. Those configurations are listed here in decreasing order of diffusion:

1. Crystal video receiver (CVR), either WO or WB channelized;
2. Superheterodyne receiver (SHR), either swept NB or swept WB;
3. Channelized receiver (CHR), where the required WB frequency coverage is divided into a large number of high dynamic range NB receiver channels;
4. Transform receivers (TR), such as microscan, Bragg cells, or compressive receivers;
5. Cued receivers, which are hybrids of the previous configurations.

In all of the WB receiver configurations, an IFM receiver (see next section) is employed to provide the signal carrier frequency data to the EW sensor.

In this section the above listed frequency receiver configurations are assessed in terms of:

- Instantaneous bandwidth coverage,
- Sensitivity,
- Handling of simultaneous signals,
- Dynamic range,
- Instantaneous signal parameter measurements and relevant accuracies,

- Configuration complexity.

The assessment summary of the above configurations is reported on in Table 3.1.

3.5.1.1 The Instantaneous Frequency Measurement Receiver

The IFM receiver had an important role in the EW equipment of the past generation and is still applied in current generation equipment, although the tendency is to replace it with a digital receiver (DRx). The IFM receiver concept can be understood from looking at Figure 3.7.

The RF input $s_i(t) = A\cos(\omega t)$, where $\omega = 2\pi f$ and f is the instantaneous frequency carrier, is hard-limit amplified and then split in two equal amplitude signals, one of which is delayed with respect to the other through a delay line of length ℓ, which provides a time delay $\tau = \ell/c_p$ (c_p is the EM wave propagation speed in the delay line medium). The two signals are then applied to a phase detector network (composed of a phase correlator, four diode detectors and two differential amplifiers [4]).

The signal amplitudes at the output of the network can be shown to be:

Table 3.1 EW Frequency Receiver Comparison Table

	Receiver Type				
Parameters	*CVR*	*SHR*	*CHR*	*TR*	*Cued*
Instantaneous B_{RF}	Excellent	Poor to fair	Good to excellent	Good	Excellent
Sensitivity	Poor to fair	Excellent to good	Excellent	Good	Excellent
Simultaneous signal handling	Poor	Poor to fair	Good	Good	Good
Dynamic range	Poor to good	Excellent	Good	Good	Excellent
Signal parameter measurement accuracy	Fair	Good	Good	Good	Excellent
Configuration complexity and cost	Low to medium	Medium to high	High to very high	Medium to high	High

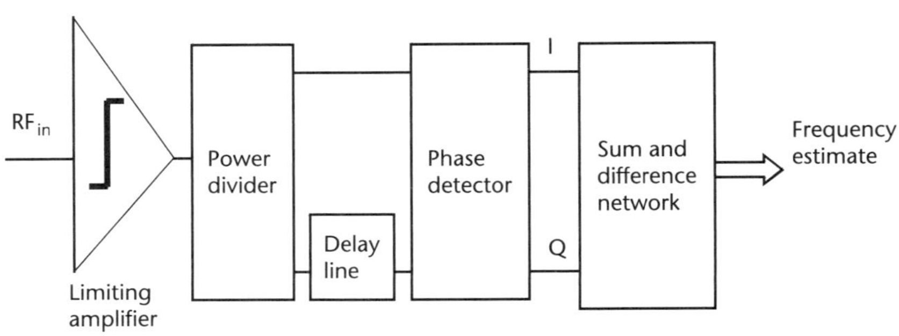

Figure 3.7 IFM receiver functional block diagram.

$$S_{o1} = B \cos(\omega\tau)$$
$$S_{o2} = B \sin(\omega\tau)$$

which can be represented in a xy plane as the two components of a vector of amplitude B and angle $\varphi = \omega\tau$ with respect to the x-axis. Because τ is a constant, the angle φ is a measure of the signal carrier frequency f

$$\varphi = 2\pi f\tau \tag{3.24}$$

The bandwidth BW of the IFM device is limited by the length ℓ of the delay line; in fact, the bandwidth is determined by

$$(\omega_{max} - \omega_{min})\,\tau = 2\pi \text{ or } BW = f_{max} - f_{min} = 1/\tau = c_p/\ell$$

The frequency measurement accuracy is dependent on the time delay τ as can be shown by taking the derivative of (3.24):

$$\delta f = \delta\varphi/(2\pi\tau) \tag{3.25}$$

in which $\Delta\varphi$ is the phase measurement accuracy, which is dependent on the phase correlator design and usually about 10 degrees. Because, however, τ is selected small to achieve a large BW, the frequency accuracy Δf becomes very broad.

Good frequency measurement accuracy is obtained by using in parallel several phase correlators, each with time delay τ_n larger than the multiple of the time delay τ_1, which is selected to achieve a large BW. The most used configuration is the harmonic configuration where the time delays τ_n are 2^{n-1} multiples of the time delay τ_1. Usually up to four time delays are selected, whereas the shorter time delays are used to resolve the frequency measurement ambiguities and the largest time delay $\tau_4 = 8\tau_1$ provides the fine frequency measurement accuracy $\Delta f_4 = \Delta f_1/8$.

Of course, the largest usable time delay must be shorter than the minimum PW that the equipment is required to measure, otherwise the delayed and undelayed portions of the minimum PW signal will never overlap within the phase correlator with the largest time delay.

The functional block diagram of the harmonic IFM configuration and its operational characteristics are given in Figures 3.8 and 3.9, respectively.

In the above-mentioned configuration, the higher the number of bits a correlator generates in the frequency measurement, the higher the required SNR, as the increased quantization of the signals $\cos(\omega\tau)$ and $\sin(\omega\tau)$ encompasses small quantized amplitude steps, which can be affected by noise.

The usual frequency measurement accuracy of octave bandwidth multiple correlator IFM receivers is on the order of 2 to 10 MHz (in accordance with the frequency band).

The IFM receiver provides good signal sorting in low-density signal environments and can handle wideband modulation if high-speed sampling of the output signal is applied. Its major disadvantage is its limited dynamic range, which produces a confused frequency measurement in the case of overlapping pulses of different amplitudes and frequencies, as shown in Figure 3.10.

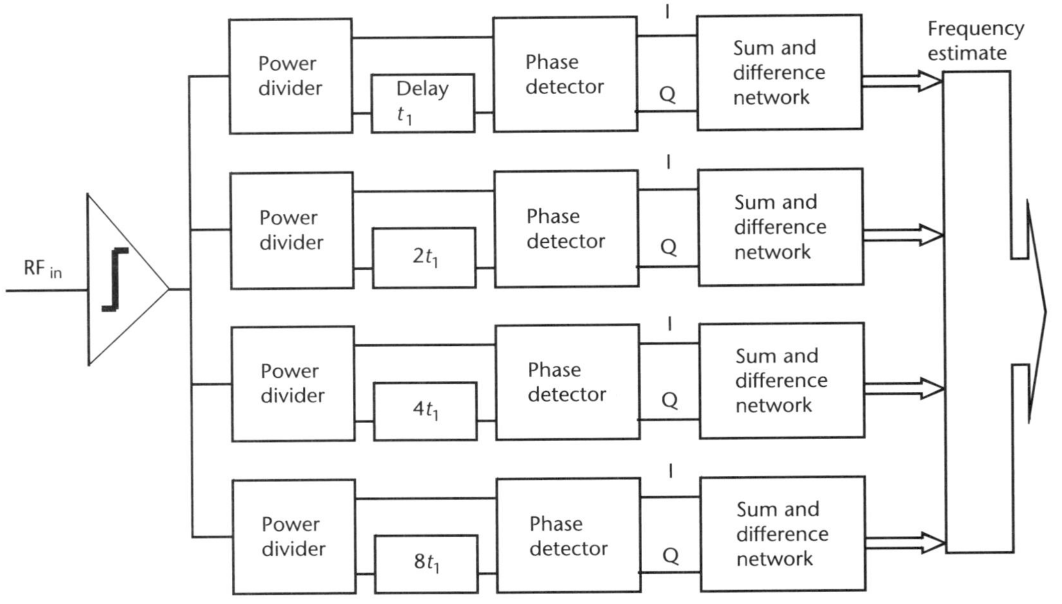

Figure 3.8 Functional block diagram of a four-base harmonic IFM receiver.

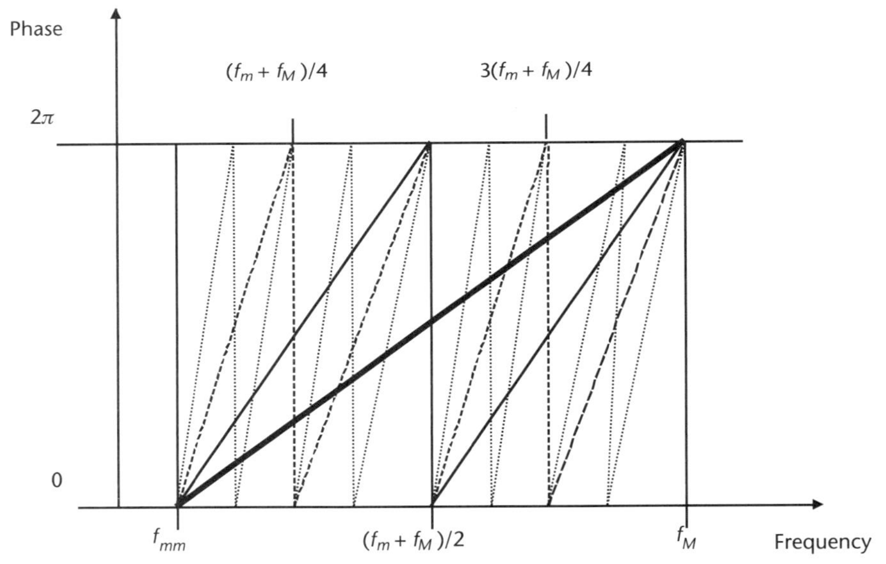

Figure 3.9 Operational characteristics of a four-base harmonic IFM receiver.

The evaluation of the IFM frequency measurement error in the case of overlapping signals is derived in a straightforward manner from (3.24) by considering that amplitude limiting does not affect the phases of the two input signal phasors, which are respectively:

$$\alpha_1 = 2\pi f_1 \tau$$

and

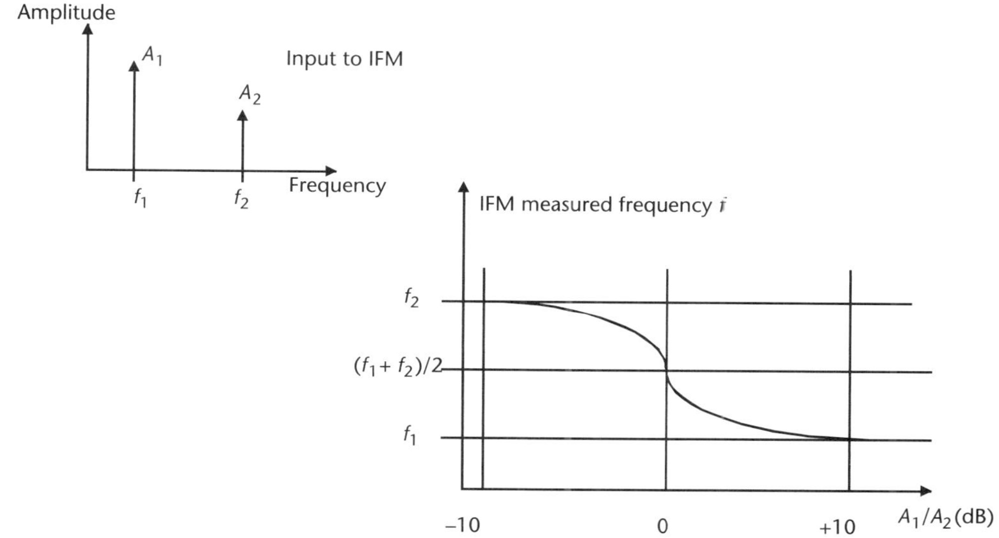

Figure 3.10 IFM measured frequency vs. amplitude ratio A_1/A_2 (dB) for the case of overlapping pulse signals.

$$\alpha_2 = 2\pi f_2 \tau$$

The resultant phasor, the sum of the two input phasors, has a phase $\alpha = 2\pi f\tau$, which can be derived by inspection of Figure 3.10. The resultant phasor has the following coordinates in the xy plane:

$$\cos\alpha = (A_1/A_2)\cos\alpha_1 + \cos\alpha_2$$
$$\sin\alpha = (A_1/A_2)\sin\alpha_1 + \sin\alpha_2$$

Therefore,

$$\alpha = \tan^{-1}(((A_1/A_2)\sin\alpha_1 + \sin\alpha_2)/((A_1/A_2)\cos\alpha_1 + \cos\alpha_2))$$

and

$$f = (\tan^{-1}\alpha)/(2\pi\tau)$$

3.5.1.2 Crystal Video Receivers

CVRs are the most common past generation EW receiver. They are used primarily in RWR equipment because they cover either a WB, such as the I- or J-band or both of them, or the complete RF radar threat band ranging from the D- to J- or K_a-band. In this latter case they are defined as wide-open receivers.

Old CVRs were characterized by low sensitivity, adequate to provide RAF for a high ERP threat emitter mainlobe. In the early time of electronic warfare, the environment was characterized by few low-duty threat emitters transmitting pulses at audio band PRF, so the emitter was recognized or identified by its pulse train PRF.

Threat emitter PRF was reported as an audio alarm to the pilot's earphones. With the increasing number and technological evolution of threats in more recent times, threat carrier frequency became essential to provide threat identification. Current generation CVRs show both a higher sensitivity due to the addition of an amplifier in front of the detector and a frequency measurement ability due to the addition of an IFM receiver.

The block diagrams of a modern WO CVR used as RWR equipment and of a WB CVR used as medium-class ESM equipment are shown in Figure 3.11.

The CVRs can measure the following signal (pulse) instantaneous parameters:

- Pulse amplitude (*A*),
- Pulse width (PW),
- Direction of arrival (DOA),
- Time of arrival (TOA),
- Frequency (*F*).

Because the CVR is small, usually it is installed very close to the antenna in the DOA channel.

Because the video amplifiers following the detector are usually ac coupled (to avoid bias levels) a detected CW signal would produce a dc voltage that will be not passed by those amplifiers. A common solution to detect CW signals is to add an amplitude on/off RF modulator in front of the detector that, when activated, produces pulsed video signals of long PWs that pass through the ac-coupled amplifiers.

The associated processor provides emitter deinterleaving and sorting and computes through the collection of emitter streams of pulses the following further parameters of the emitter:

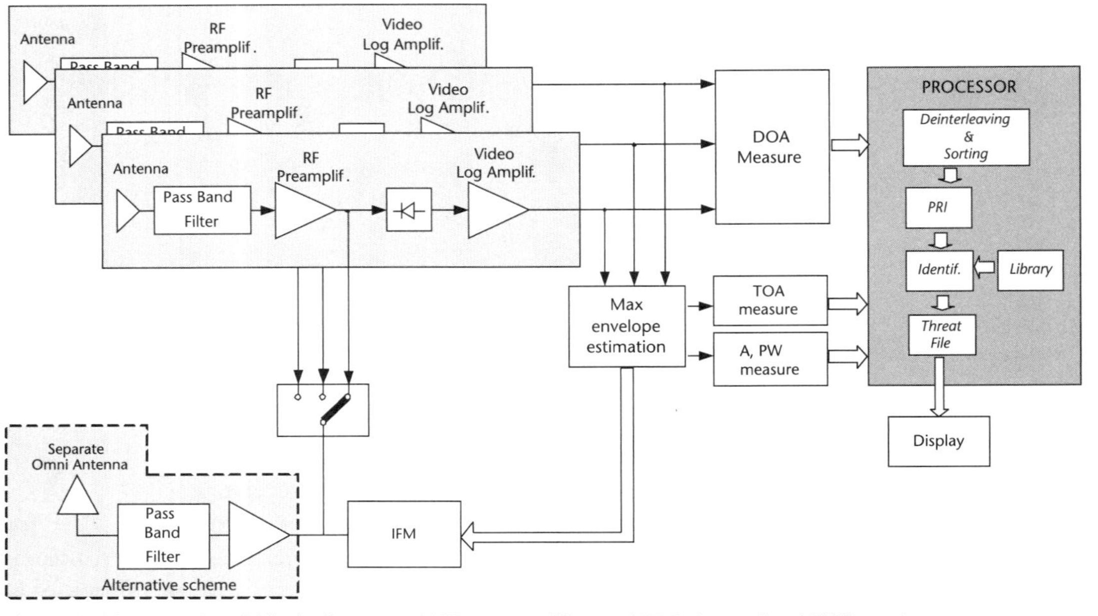

Figure 3.11 Functional block diagram of WB preamplifier and IFM channelized RWR equipment.

- PRI and its behavior along time (PRI law),
- ASP and its behavior along time (AST),
- Frequency behavior along time (frequency type).

Typical performance for a WO CVR and a WB CVR plus IFM receiver is reported in Table 3.2.

3.5.1.3 Scanning Superheterodyne Receiver

The SH receiver is the most commonly used receiver in communications and radar equipment for its high sensitivity and frequency selectivity. In EW applications, NB SHs are used primarily in ELINT equipment to isolate an emitter signal from the environment and measure its fine-grain information without any interfering signal, whereas large band (LB) scanning SHs are used to search for and detect threat emitters in highly dense pulse scenarios because their selectivity strongly reduces the received pulse density.

Current SH receivers use a digitally controlled preprogrammed (in accordance with mission data) scanning strategy over a number of bandwidths across the RF coverage (the bandwidths may also be adjacent so as to cover the required RF coverage). Usually the dwell time in a bandwidth is longer in dense threat bandwidths and shorter in bandwidths where there are few threats.

SHs are characterized by a high sensitivity (due to its reduced instantaneous RF bandwidth with respect to the CVR one) but, because they are time multiplexed, they have intercept limitations on the scanning emitters, unless appropriate scan dynamic programming is applied.

The schematic block diagram of a NB and of a LB SH is depicted respectively in Figures 3.12 and Figure 3.13.

3.5.1.4 Channelized Receivers

This receiver architecture covers the required RF bandwidth through a number of parallel contiguous channels, which have an elementary bandwidth that roughly matches the expected threat emitter spectrum (order of magnitude from 50 up to 200 MHz). Each channel is provided with all of the instantaneous pulse signal parameter measurement devices.

Table 3.2 Typical Performance of a WO CVR and a WB CVR plus IFM Receiver

Performance	
RF Coverage	2–18 GHz (with IFM), 28–40 GHz (without frequency measurements)
Video bandwidth	20 MHz
Sensitivity	–45 dBm typ (WO), –60 dBm typ. (with preamplifier)
Instantaneous dynamic range	40 dB
Total dynamic range	70 dB (with preamplifier)
Frequency measurement accuracy	2–10 MHz with harmonic IFM

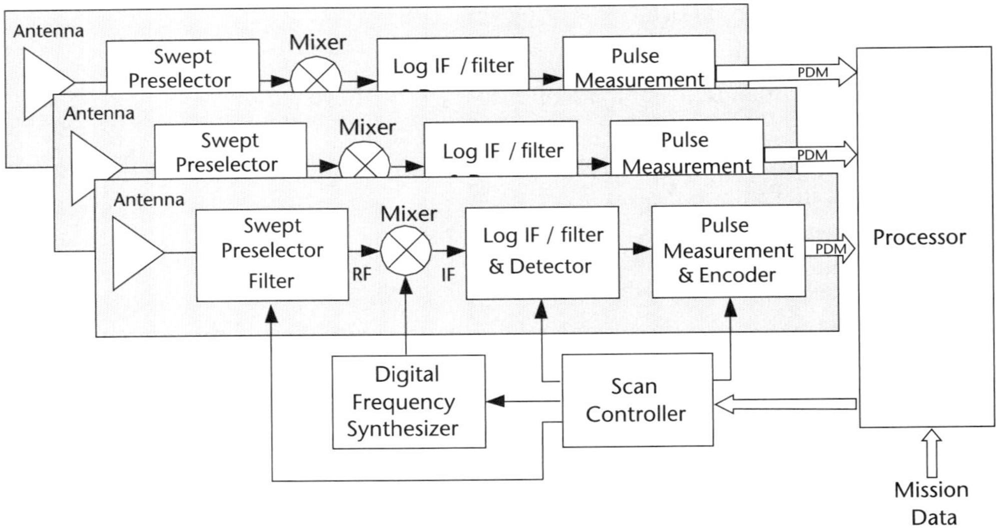

Figure 3.12 Schematic block diagram of a NB SH.

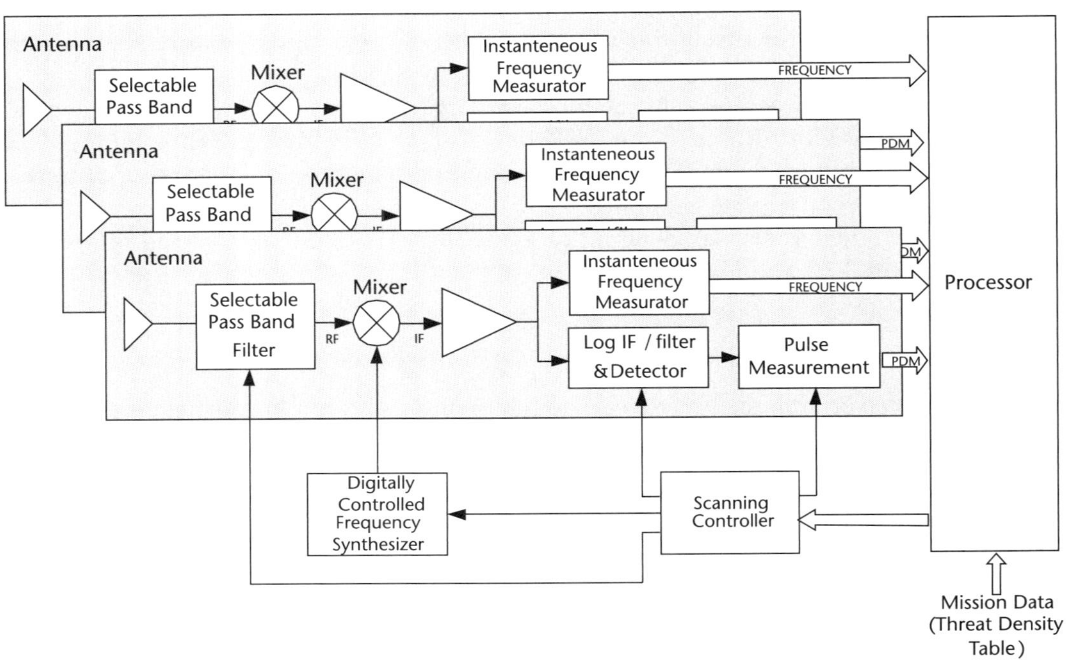

Figure 3.13 Schematic block diagram of a LB SH.

This architecture is very complex and bulky, thus it has been used only on ground-based and large transport aircraft ELINT equipment [1]. The important advantage of this architecture is its ability to handle multiple simultaneous threat signals that overlap in time (as is the case in highly dense RF band segments) and produce the poor performance of IFM devices. This situation is depicted in Figure 3.14, in which two simultaneous signals fall in separate high dynamic channels and

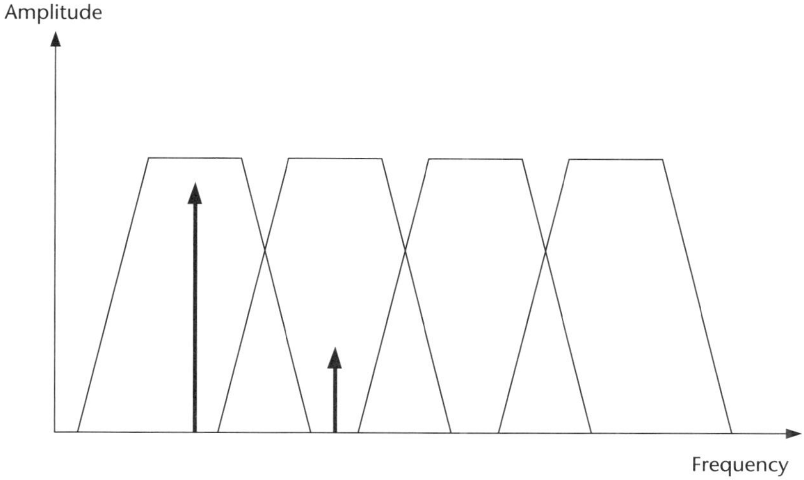

Figure 3.14 Discrimination of two simultaneous signals, overlapping in time, provided by the channelized receiver.

are accurately processed. Accurate processing occurs even if the signals fall in two adjacent channels, thanks to the very sharp bandwidth roll-off of each channel.

Practical configurations of the channelized architecture (in terms of reduced complexity and volume) were implemented by exploiting SAW filter channelization at the IF output of a WB SHR, to be used in the dwells at very crowded RF band segments along the frequency sweep to cover the complete required frequency coverage.

3.5.1.5 Transform Intercept Receivers

This type of receiver, which was implemented with various technologies (such as Bragg cells or microscan compressive receivers), approximates the operation of a Fourier transform. The Bragg cell receiver, which is depicted in Figure 3.15, exploits the acousto-optical characteristics of some materials, such as a crystal of lithium niobate (LiNb). In these crystals an incident beam of coherent light is generated by a laser source. Upon interaction with a sound wave, while traveling transverse to the light beam generated by a piezoelectric transducer fed by the input IF signal, it is deflected at an angle proportional to the input IF.

The underlying physical principle is as follows: The sound wave, which propagates in the LiNb crystal at a very short wavelength $\lambda_B = v_B/f_{IF}$ (as the propagation speed v_B of acoustic waves in the medium is much smaller than the propagation speed in the air), produces a compression and rarefaction of the medium (and hence of its index of refraction of the light) at step λ_B. With the laser beam inclined with respect to the normal to the crystal according to the so-called "Bragg angle," the diffracted light produced by the modulated (at wavelength λ_B) index of refraction consists of only two component beams: an undeflected (and hence useless) beam (called of order 0) and a deflected beam (called of order 1) by an angle β (proportional to f_{IF}) with respect to the incident laser beam.

Through a proper optical lens arrangement, the deflected beam is focused in its focal plane, where a number of adjacent CCD photodetectors are placed and for

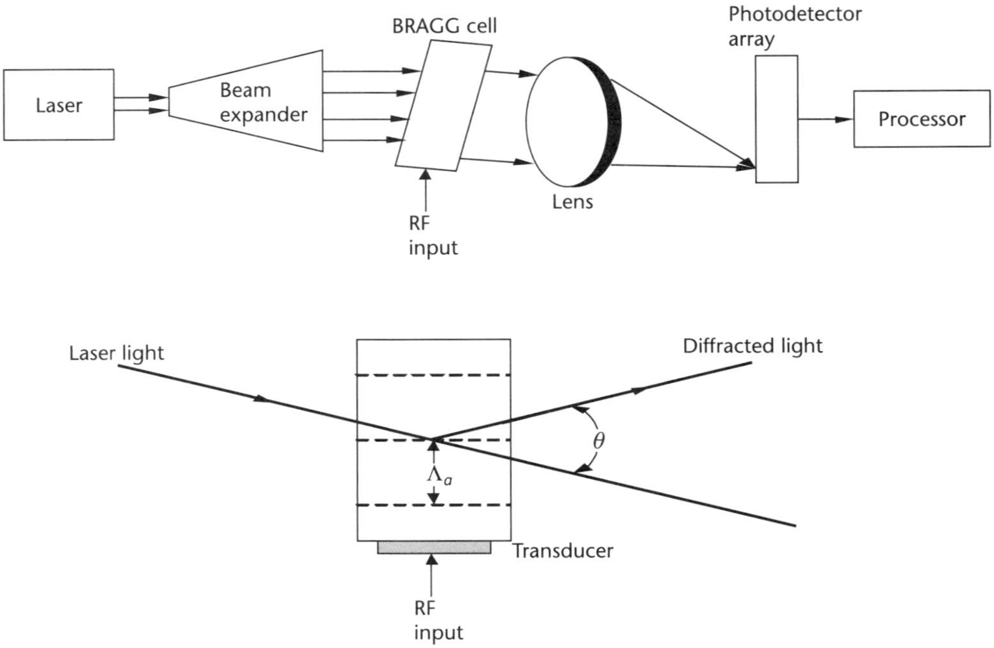

Figure 3.15 Pictorial block diagram and principle of operation of a Bragg cell receiver.

which the output amplitude is proportional to the power spectrum of the incoming IF signal. In this respect the Bragg cell operates like an instantaneous spectrum analyzer in that it reproduces (according to a scale factor) the incoming IF signal spectrum, thus discriminating the various simultaneous emitter signals present at the input of the receiver.

The Bragg cell receiver had a period of great use in the 1980s, especially in ground applications, but its use was quickly abandoned as its performance became impaired by the vibration environment of mobile platforms. Typical Bragg cell receiver performance is covered in Table 3.3.

3.5.2 Architecture of New EW radar Band Sensors

The tremendous evolution of electronic high-speed digital technologies in terms of sample and hold (S/H) circuitry, ADC, and DSP devices has produced in all sensors and actuators, but particularly in EW ones, serious changes in their architecture.

Table 3.3 Bragg Cell Receiver Performance

Center IF (GHz)	1.5–2.5
Bandwidth (GHz) B_{IF}	0.5–1
Transit time (μs)	0.1–0.3
Channel width (MHz) B_{ch} ($B_{ch} \sim 1/PW_{min}$)	20
Number of channels ($m = B_{IF}/B_{ch}$)	25–50
Instantaneous dynamic range (dB)	40
CCD read-out time (μs)	0.05–0.5

These changes have provided better equipment performance and efficiency and have reduced volume and power consumption [5].

In the case of EW sensors, the novel device at the base of the new architecture is the digital receiver (DRx). A typical DRx block diagram is shown in Figure 3.16. The "basic" structure of a DRx is composed of the following:

- An ADC, which provides discrete amplitude levels at discrete "time instants."
- A "processing engine" implemented on an application specific integrated circuit (ASIC) or field programmable gate array (FPGA). The FPGA has the advantage of being reprogrammable according not only to application needs but also to "mission" needs.
- A "storage memory" for signal samples recording. This option allows offline software processing to performed by a processor not hosted on the DRx.

To allow multichannel operation, the following functions are required to perform differential measures of phase, amplitude, and TOA:

- A synchronizer that aligns the samples and the processing between different DRx channels,
- Some I/O devices for exchanging data.

A DRx board is shown in Figure 3.17.

DSP operations implemented within a DRx depend on the requested application; however, some basic techniques can be individuated. First, consider the concept of IF sampling, which means that the signal is sampled not at the baseband but at a proper IF. As an example, given a sampling frequency RSAMP of 1 GHz, the frequency range 500 MHz ÷ 1000 MHz can be selected instead of the 0 ÷ 500 MHz one. This technique is an extension of the Nyquist theorem. In fact, the actual sampled bandwidth is always 500 MHz, even if it is allocated above one-half of the sampling frequency. A proper anti-aliasing (AA) filter should be provided to select the interested frequency range.

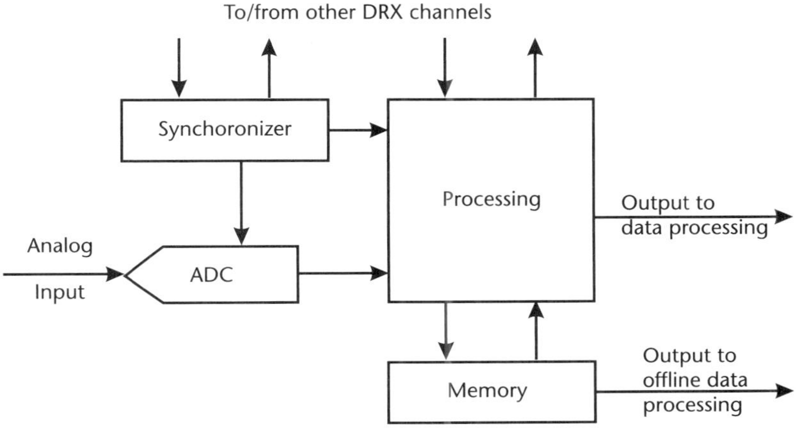

Figure 3.16 Concept of IF sampling and digital I/Q extraction.

Figure 3.17 Double multibit DRx board with 1-GHz instantaneous bandwidth. (Courtesy of Elettronica.)

The concept of operation of IF sampling is shown in Figure 3.18. The incoming signal is filtered in the IF bandwidth, which ranges from RSAMP/2 to RSAMP. The sampling creates replicas in the 0÷RSAMP/2 bandwidth, and finally the digital processing provides the complex baseband signal, for example, composed of in-phase (I) and in-quadrature (Q) components.

The main advantages of IF sampling are as follows:

- IF sampling is performed in IF, so it saves an analog conversion stage.
- Downconversion, carried out with digital techniques, avoids the "traditional" analog I/Q mixing, which poses heavy constraints of phase/amplitude matching.
- Operation in the bandwidth RSAMP/2 to RSAMP avoids the presence of second order and higher harmonics within the useful bandwidth.

The ADCs employed in IF sampling are required to not only have the ability to work with high sampling frequencies but also with a wide analog bandwidth (which leads to a small time aperture jitter) in order to simplify the IF stages.

One of the most important techniques that is often used in modern digital receiver architectures is the uniform filter bank based on the discrete Fourier transform (DFT), which is usually implemented by using fast Fourier transform (FFT) algorithms. The use of a uniform filter bank is very important in a number of EW applications. The implementation cost, which is quite high with analog devices, is lowered when digital processing is used. The actual implementation of a uniform

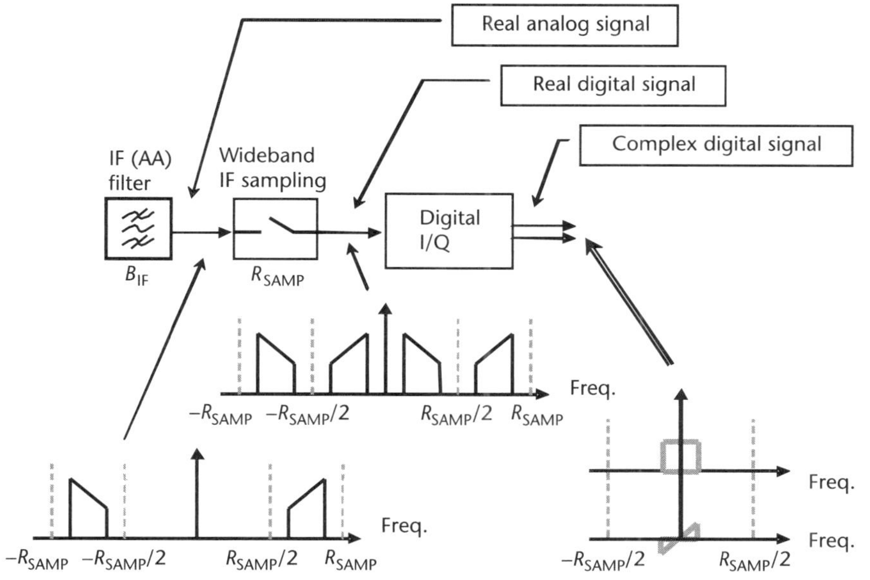

Figure 3.18 Concept of IF sampling and digital I/Q extraction.

filter bank, as shown in Figure 3.19, is achieved by downsampling (by a factor *M*) and multiplexing in *M* parallel flows of the incoming wideband digital signal. Then, after proper weighting, the samples are sent to an *M*-point FFT. In this way *M* complex signals are obtained but just *M*/2 are retained since, being the real input signal, the first *M*/2 FFT outputs are the complex conjugates of the second *M*/2.

The generic *k*th complex signal is the complex envelope of the signal having a bandwidth equal to RSAMP/M and centered on the frequency *k* × RSAMP/M.

Details on uniform filter bank implementation in digital circuits can be found in [3]. Advantages of this FFT-based technique are modularity and simplicity. Further, if a higher value of *M* is required, the rate at which the FFT has to be computed

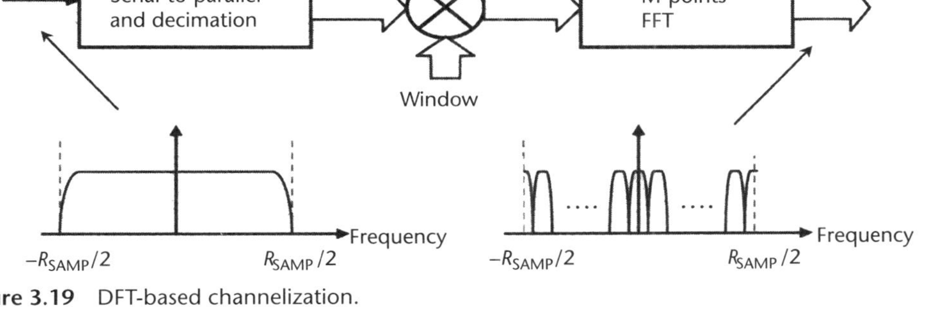

Figure 3.19 DFT-based channelization.

decreases (is equal to RSAMP/M), which means that the clock speed can remain the same, while the overall complexity increases a little due to the additional stages needed to evaluate a longer DFT. This type of processing is most useful when a large bandwidth has to be monitored (ESM and RWR systems).

When the carrier frequency of a digital signal is known (together with a rough indication of its bandwidth), it can be demodulated with a digital downconverter (DDC), which acts as an analog downconverter but with the obvious advantage of programmability in both frequency and bandwidth. A numerically controlled LO and a variable-bandwidth filter are needed to do that.

The disadvantage of digital filters is their cost because they often require a large computational effort to actually be made by a number of multipliers, even if they are designed with finite impulse response (FIR) or infinite impulse response (IIR).

In contrast, the use of the so-called cascaded integrated comb (CIC) filters avoids all multiplications; they actually do only differences and sums combined with downsampling. The reader can find an interesting treatment in [3]. CIC filters are very useful in wideband digital processing techniques where the use of computational resources has to be minimized. They can be used together with FIR filters to reduce the sampling frequency as far as the filter bandwidth decreases. A CIC-based DDC is shown in Figure 3.20.

The output of a DDC is always a complex baseband signal sampled at a given rate. Until now, the described processing was needed for signal filtering, either in the case of "surveillance" (e.g., when the signal frequency is not known and the full bandwidth has to be exploited) or in the case of "cueing" (e.g., when the signal has been detected and proper downconversion and filtering are needed in order to do a "fine" demodulation).

The algorithms employed in the treatment of signals can be divided into detection algorithms and parameter estimation algorithms. Both types of algorithms can be applied either to the signal amplitude or to its phase. Amplitude-based processing has been traditionally adopted by EW receivers and phase processing has

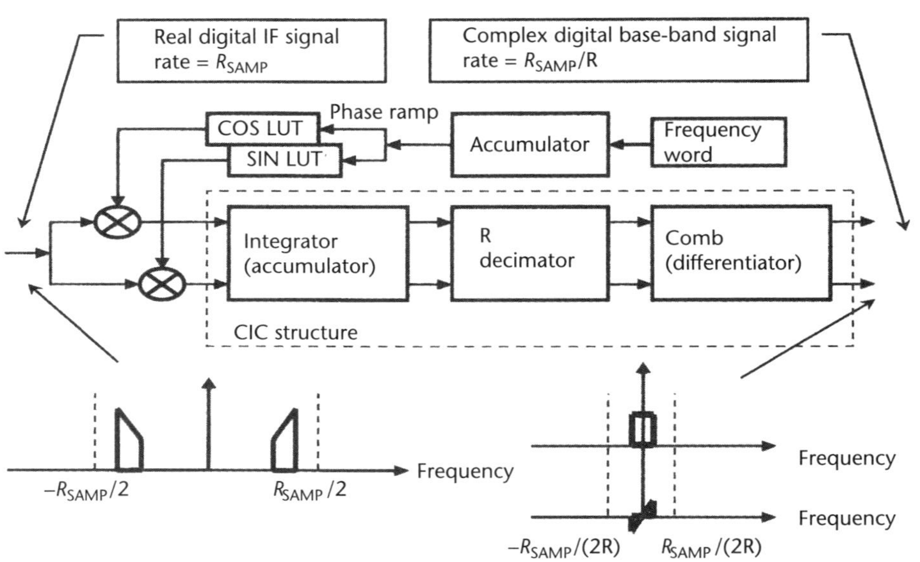

Figure 3.20 DDC based on CIC filtering (LUT = look-up table).

became the selected solution of coherent DRx. Detection/estimation problems are quite easy to treat due to the availability, within the DRx, of the complex baseband signal. One of the most used algorithms in treating complex digital signals is the CORDIC (coordinate rotator digital computer). The CORDIC algorithm (see [5]) is a very tricky operation to do "rotations" of a complex number. The CORDIC is mainly used for rectangular-to-polar conversions, but it can also be used to perform other nonlinear operations.

The implementation of detection algorithms has been made easy by DSP. Optimum detection strategies are quite easy to implement; as an example, the detection scheme illustrated in Figure 3.21, is based on the likelihood function (see [3]) and can be summarized in a nonlinear operation (the evaluation of the likelihood function) followed by a threshold device. This approach can be easily implemented in a digital system when nonlinear transformation is implemented using LUTs.

Estimation algorithms are as easy to implement as are the detection algorithms: averages, filtering, and nonlinear operations can be easily performed via digital techniques, so estimation problems are well suited for digital implementations. In addiction, the theoretical Cramer-Rao bounds on signal parameters are achievable starting from discrete-time observation, as reported in [3]. DRx devices find application in all kind of EW receivers.

The architecture of a WO Rx, based on the so-called "monobit" DRx, is shown in Figure 3.22. A monobit DRx employs a 1-bit ADC, which allows ultra-wideband sampling (on the order of tens of gigahertz). One-bit analog-to-digital conversion performed on an IF signal is equivalent to bandpass hard-limiting; this operation, well known from the literature (see [4]), does not affect the signal phase. So, after digital channelization (performed with the techniques described in Section 3.5.3 above), we can extract the carrier frequency and the phase information. A monobit DRx has the following two main advantages with respect to traditional IFM:

- The device cost is much lower with respect to an analog IFM, which involves state-of-the-art RF technology and usually has to cover the 0.5- to 18-GHz bandwidth.

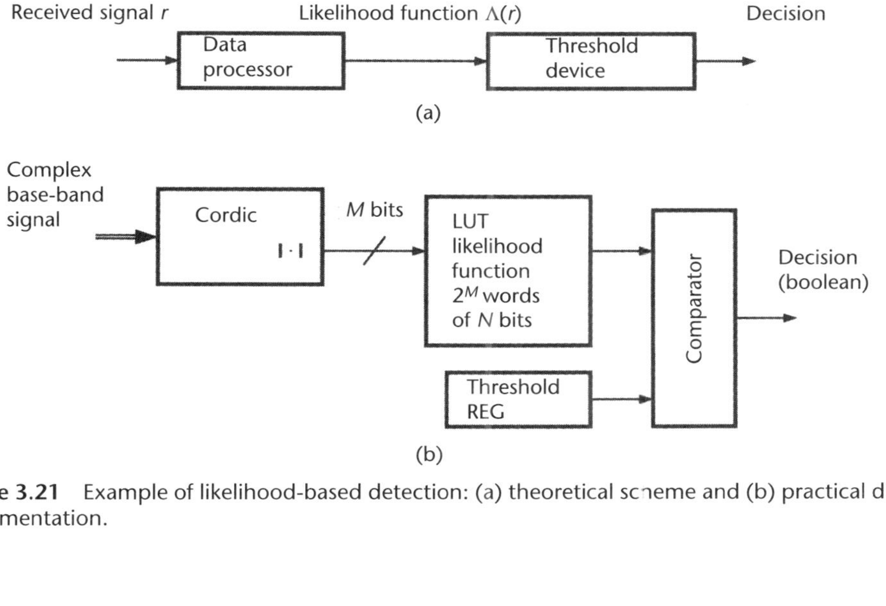

Figure 3.21 Example of likelihood-based detection: (a) theoretical scheme and (b) practical digital implementation.

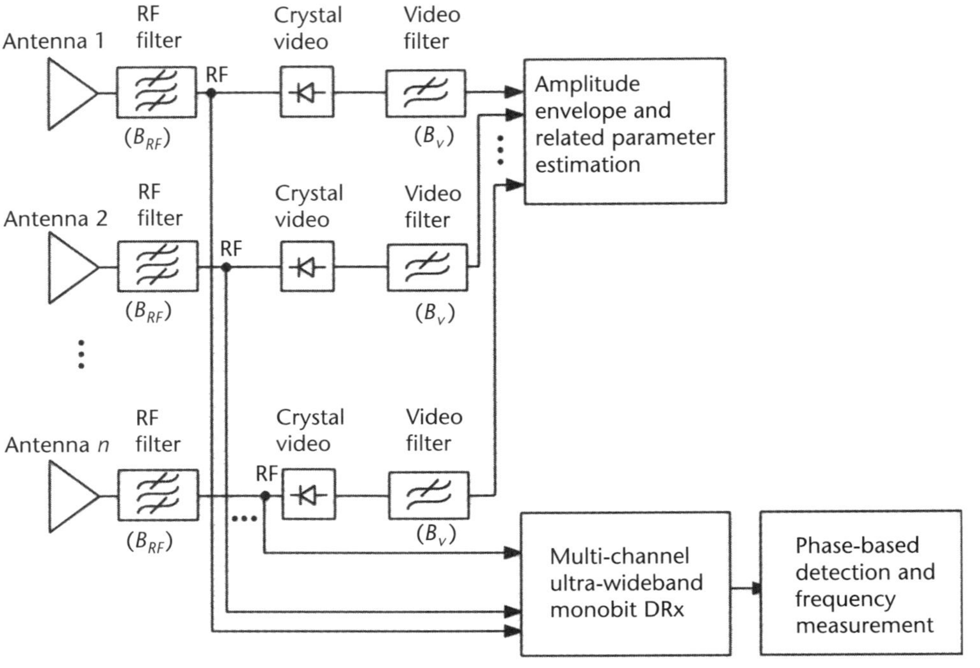

Figure 3.22 WO Rx based on bonobit DRx.

- Phase resolution may be better than with analog IFM, so phase processing can easily detect modulation-on-pulse (MOP) as Barker or Frank codes and, in general, PSK modulations.

The architecture of an ESM SH receiver that uses digital technology is depicted in Figure 3.23. A variable LO sweeps the overall RF bandwidth. The IF bandwidth is channelized with digital techniques so as to allow multiple emitter processing. To show the capability of a DRx-based ESM, a scenario with three simultaneous pulsed emitters is shown in Figure 3.24. An oscilloscope or spectrum analyzer provides, respectively, only time- or frequency-domain representation.

A SH DRx provides, thanks to digital channelization, a time-frequency representation, as depicted in Figure 3.25 since:

- The complex of digital channels, all taken at the same time lag, provides a frequency description.
- Each channel output, along the sequential time lags, gives the time history of signals present in its spectrum portion.

A fully digital implementation of an ELINT receiver is shown in Figure 3.26.

The aim of a detection function is to start the fine analysis on the selected signal. To this purpose, signal recognition must be performed in the detection function. The detection can be performed either after channelization or before, depending on the traffic level and/or the required sensitivity (taking into account that the ELINT receiver can have a very high sensitivity).

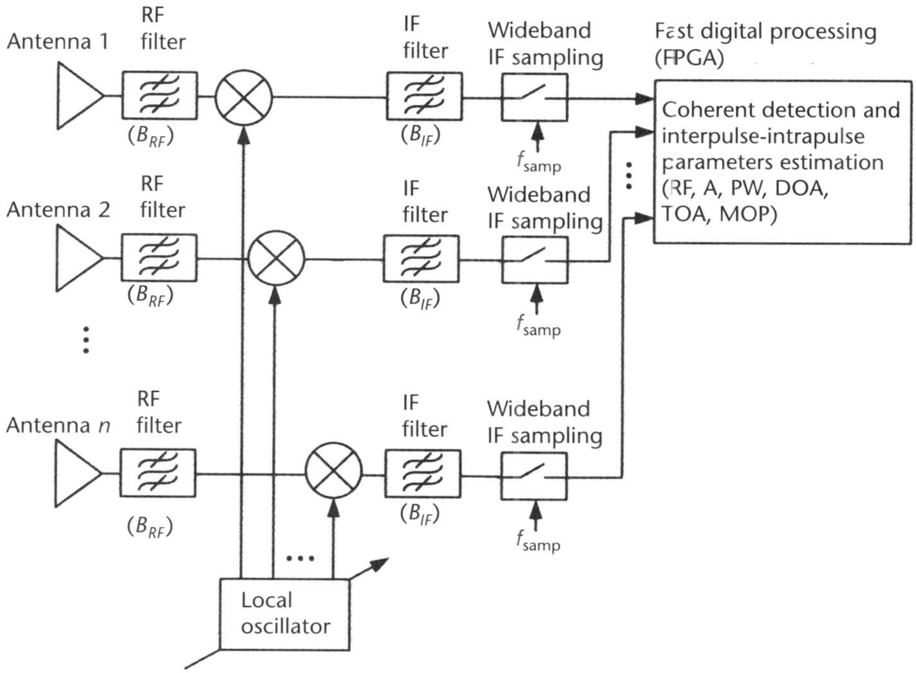

Figure 3.23 SH Receiver.

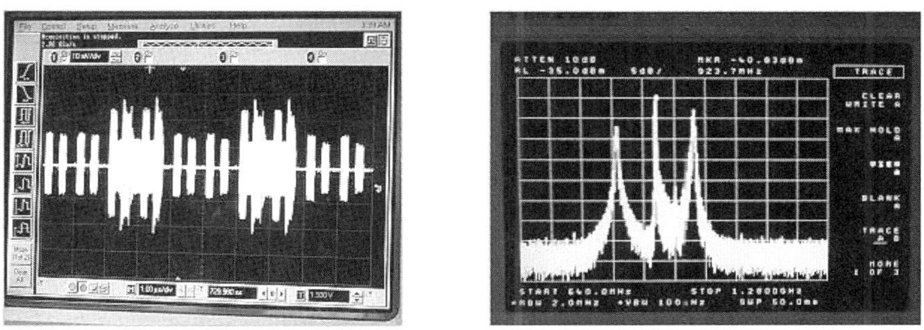

Figure 3.24 Three-emitter scenario: (left) oscilloscope view and (right) spectrum analyzer view. (Courtesy of Elettronica.)

Once the detection has occurred a coarse frequency should be provided to LO2 (numerically controlled) in order to extract the selected signal. The analysis performed in the last block of Figure 3.27 typically provides the following data:

- Distribution of received pulses within a time frame,
- PRI and PRF agility data,
- Modulation on pulse (MOP) analysis,
- Unwanted modulation on pulse (UMOP) analysis.

All of these data are stored in order to recognize the class of the radar or even the specific radar (UMOP analysis). The block diagram of a double-channel SH

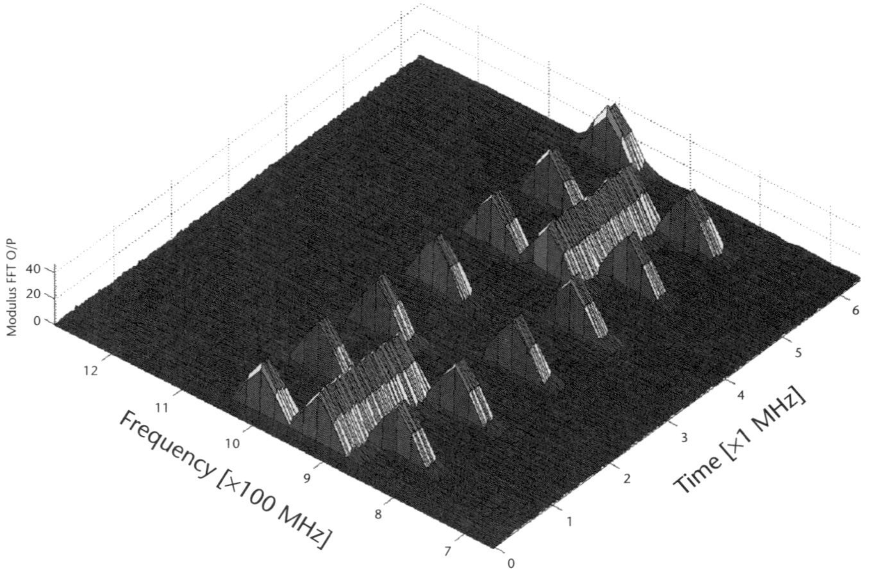

Figure 3.25 Three-emitter scenario: Time-frequency representation performed with a DRx (only the envelope is reported).

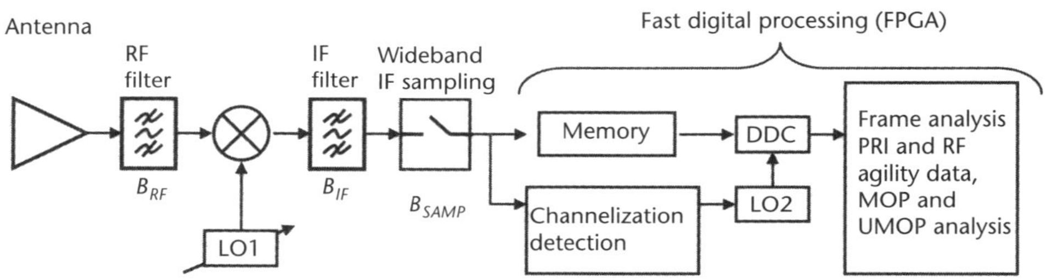

Figure 3.26 Digital ELINT receiver.

receiver is shown in Figure 3.27. The double-channel architecture is fundamental for differential measurements: they can be carried out in amplitude, phase, or time, depending on the application, as will be discussed in Chapter 4, relevant to direction-finding techniques.

3.5.3 DSP Technologies

The development of digital receivers is strictly related to semiconductor technology. Actually the choice between a digital receiver and a Bragg cell receivers (see Appendix A) was driven by the improvement of semiconductor technology. Technology made available some characteristics of the key component; mainly ADC and FPGA, which allows high-performance DRxs to be built. These characteristics are:

- High-speed sampling,
- Large bandwidth and spectral purity of the ADC,
- High-speed systolic processing with large computation resources (FPGA),

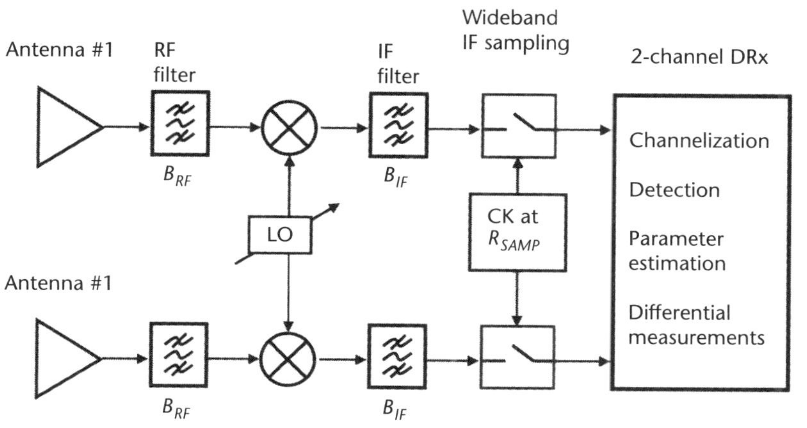

Figure 3.27 Double-channel SH receiver.

- High-speed connectivity.

In the following discussion we refer mainly to the channelized SH receiver because it is the most demanding.

The maximum sampling frequency is the main characteristic of the ADC. It is directly related to the DRx theoretical bandwidth that is one-half of the sampling frequency. As shown in the previous section, the bandwidth of a DRx with a sampling frequency RSAMP is RSAMP/2. This band can be allocated not only from 0 to RSAMP /2 (baseband), but from RSAMP to 1.5 RSAMP, and so on.

To select the band, a filter is required. This filter (the antialiasing filter) reduces the DRx bandwidth; in fact, its transition bandwidth should be deducted from RSAMP/2. Usually the transition bandwidth is 10% of the cutting frequency, which means that if a DRx works with a maximum input frequency of f_s, its bandwidth will be 40% of f_s; if, however, the RDx works up to 2 RSAMP, its bandwidth will be 30%.

Other limitations to the high-frequency operations come from the input stage of the ADC. This stage amplifies the input signal and holds it during the conversion process. The requirements of this section are:

- Large analog bandwidth,
- Flatness,
- Spectral purity.

The first two items influence the maximum input frequency of the DRx, whereas the third one restricts the dynamic range.

A high input frequency is preferable for two reasons: (1) to increase the bandwidth of the DRx and (2) to sample the signal directly in IF or even in RF, thus simplifying the microwave section. At present the performances of the ADC allow a DRx bandwidth of almost 1 GHz to be reached; with a dynamic range of 40 to 50 dB, the input bandwidth could be allocated from RSAMP/2 to RSAMP.

In the following an example of ADC will be given. The device is the e2v AT84AS004 shown in Figure 3.28. It contains not only a 10-bit flash converter

Figure 3.28 Typical ADC device. (Courtesy of E2V.)

with maximum sampling frequency of 2 GHz, but also a demultiplexer (DEMUX) that reduces the speed of the samples in order to allow a connection with the FPGAs that are usually employed to perform the signal processing.

Figure 3.29 plots the performance of a typical ADC. The first plot of Figure 3.29(a) shows the flatness of the device. Note that the analog bandwidth is 3 GHz and the flatness from 1 to 2 GHz is almost 1 dB. The effective number of bits (ENOB) vs. the input frequency is shown in Figure 3.29(b): the ENOB in the band from RSAMP /2 to RSAMP is better than 7 bits. Figure 3.29(c) shows the spurious free dynamic range (SFDR) vs. the input frequency. This parameter evaluates the spectral purity of the device; it is the ratio between the signal power and the highest spurii at the output of the ADC. The SFDR gives an upper bound to the DRx dynamic range; in fact, the minimum detectable signal should be above the spurii level in order to avoid false alarms. The plot shows that the SFDR is always around 55 dB, and, taking into account a safety margin of 5 or 10 dB, a dynamic range of 45 to 50 dB can be reached.

Also a monobit ADC could be characterized by its analog bandwidth and its sampling frequency. At present, these devices can sample at a maximum rate of 10 to 12 GHz. Thus, they have a maximum bandwidth of <6 GHz. A number of monobit DRxs are used in parallel with proper clock shifts in order to cover the required band from 2 to 18 GHz.

The vast success of the DRx is due to the improvement of the processing devices used in its architecture. Indeed, the data rate produced by the ADC is not manageable by any general-purpose DSP, so ad hoc processors have been developed. Two possible HW platforms are available for these processors: the ASIC and the FPGA. The main differences between them are the speed and the programmability. The ASIC can reach higher processing speeds than the FPGA, but the FPGA is reprogrammable. With respect to costs, the ASIC has a big nonrecurring cost, whereas the FPGA shows a higher recurring cost (cost per unit). The choice between the two HW platforms depends on a number of factors:

- The number of receivers to be built,
- The need to reprogram the device,
- The required processing speed.

In the following discussion, we refer mainly to the evolution of the FPGA but we expect that the performance trends of both the FPGA and ASIC will be similar.

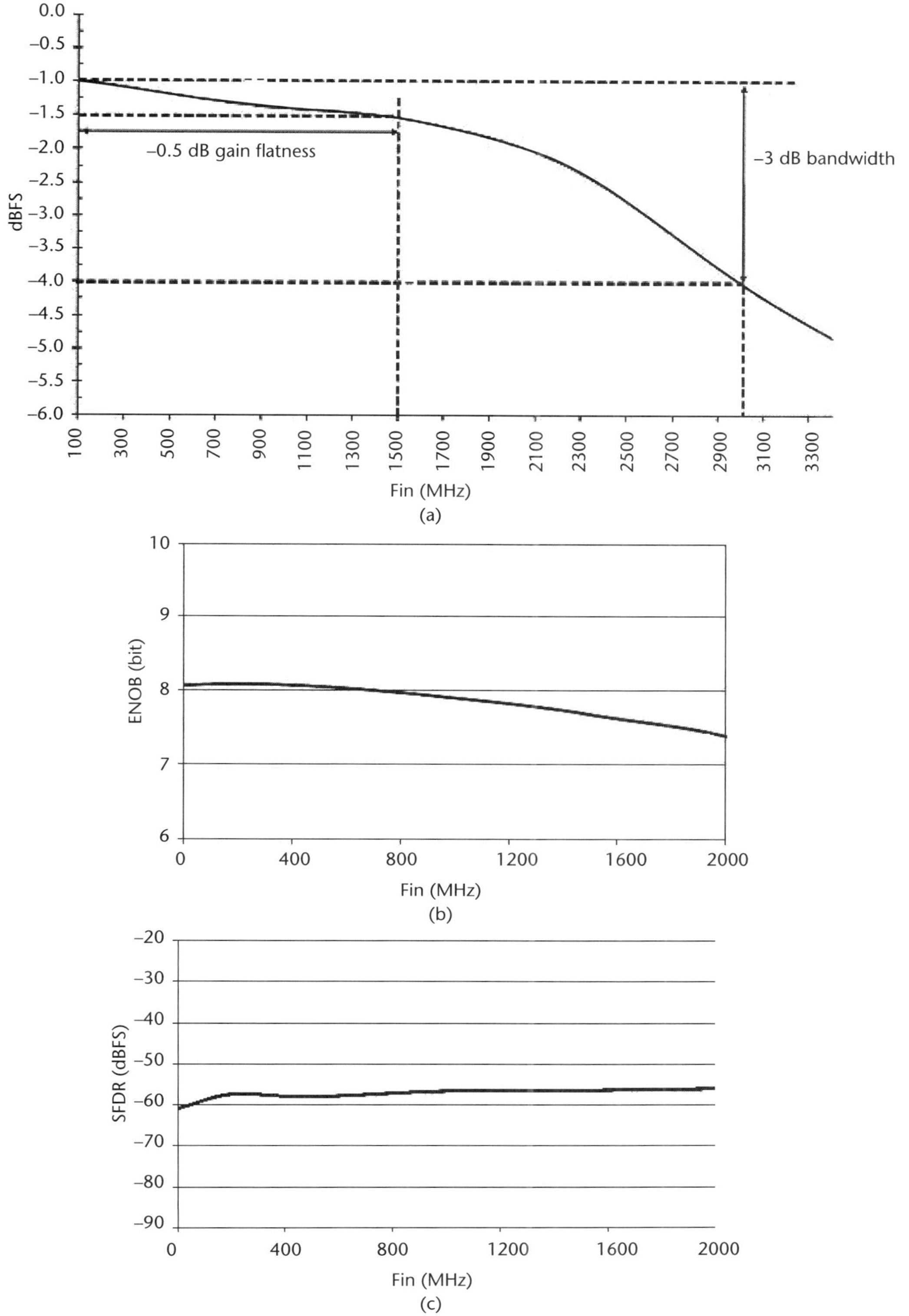

Figure 3.29 Performances of a wideband ADC: (a) gain flatness, (b) ENOB, and (c) SFDR over the bandwidth.

The three items that distinguish a processing device are:

- Computational power,
- Memory,
- Connectivity.

The FPGAs are substantially systolic devices, which means that they are composed of a number of simple elements (cells) that can be configured and connected to obtain a complex net. The characteristics of these devices are:

- Number of cells,
- Complexity and processing speed,
- Connections between them.

Moreover the latest families of FPGAs contain embedded special functions (called DSPs or MACs) that are mainly composed of multipliers and adders. These functions allow FIR filters, correlators, and so on, to be built. The introduction of these functions within the FPGAs have specialized these devices in signal processing. Both cells and embedded special functions contribute to the computational power of the FPGA.

The DRx memory can be used as computational element (LUT) or as memory to store data and parameters. The connectivity between different devices (even if on different boards) can be performed via high-speed serial transceivers. The new silicon technologies increase the density of the FPGA and consequently increase both the memory and the computational power. The connectivity improves the preceding characteristics by increasing the speed of the serial transceiver (up to 6.5 Gbps).

At present, the Xilinx Virtex5 FPGA can contain one complete channelized DRx with 1 GHz of bandwidth. The new Xilinx Virtex 6 FPGA family improves on that performance by a factor of 3 (based on a comparison between the V5 SX95 and the V6 SX315; see Figure 3.30).

At present, the Xilinx Virtex5 FPGA can contain one complete channelized DRx with 1 GHz of bandwidth. The new Xilinx Virtex 6 FPGA family improves on that performance by a factor of 3 (based on a comparison between the V5 SX95 and the V6 SX315; see Figure 3.30). The increased computational power, memory, and connectivity may be used in several ways:

	Virtex 5			Virtex 6	
Part Number	XC5VSX35T	XC5VSX50T	XC5VSX95T	SX315T	SX475T
CLB Flip-Flops	21,760	32,640	58,880	394K	595K
Block RAM/FIFO w/ECC (36Kbits each)	84	132	244	704	1,064
DSP48E Slices	192	288	640	1,344	2,016
Low-Power Transceivers	8	12	16	24	36

Figure 3.30 Performance comparison of the Xilinx Virtex 5 and Virtex 6 FPGAs.

- To improve the signal processing algorithms in order to enhance DRx performance in terms of sensitivity, number, and accuracy of measurements (e.g., frequency resolution).
- To increase the storage capacity in order to record signals to be subsequently analyzed.
- To use the connectivity improvements to communicate among several DRx channels in order to enhance direction-finding (angular estimation) techniques.

In conclusion, the silicon devices are constantly improving and this in turn increases the DRx bandwidth, sensitivity, and signal parameter measurement accuracies, providing a large improvement in the performance of next-generation EW equipment.

3.6 Detection and Classification of LPI Radars

LPI radars are characterized by [6]:

- Very low transmitted power,
- Low sidelobe antennas,
- Transmitted waveforms spread over a wide frequency BW and over long time intervals (long pulses or even CW-type transmissions),
- Coherent integration processing on reception of a high number of pulses or time intervals.

Many WB modulation techniques are currently available that provide LPI waveforms, such as:

- Linear (i.e., chirp) and nonlinear frequency modulation (LFM-NLFM);
- Discrete phase coding modulations (PSK) such as generalized Barker codes, Frank codes (which provide a good stepwise approximation of LFM), polyphase codes and polytime codes;
- FH modulations such as the Costas codes;
- PN waveforms.

We recommend the interested reader refer to [6] for more insight into the structure of these modulation techniques.

The classical filtering method widely adopted in current ESM DRxs is based on FFT processing (as shown in Section 3.5). This approach, which provides excellent detection performance in the case of nonmodulated or limited BW (i.e., BW narrower than the FFT frequency bin size B_n) modulated pulses, is no longer effective in case of large BW modulated pulses (BW > B_n) and thus more sophisticated filtering procedures are needed.

Two distinct signal processing approaches are relevant to either the case where we simply need to detect the presence of an LPI radar signal in the environment or the case where we also need to classify the emitter by measuring its parameters (for ELINT or analyses to select the best jamming program to counter that emitter). Obviously the two approaches can be used in sequence in highly sophisticated ESM equipment.

To detect only LPI radar signals, a cross-correlation technique can be used [7] that provides the processing gain PG = SNR_{out}/SNR_{in} at the output of the receiver scheme shown in Figure 3.31.

In the above scheme the output of the generic FFT bin M can be expressed as

$$r_M(k) = s(k) + n_c(k) + j\,n_s(k) \tag{3.26}$$

where $n_c(k)$ and $n_s(k)$ are discrete independent additive white Gaussian noise (AWGN) discrete-time processes for which samples the following expressions hold :

$$\begin{aligned} &E\{n_c[k]\cdot n_c[h+k]\} = (N_0/2)\cdot\delta[h] \\ &E\{n_s[k]\cdot n_s[h+k]\} = (N_0/2)\cdot\delta[h] \\ &E\{n_c[k]\cdot n_c[h]\} = 0 \ \ \forall k,h \end{aligned} \tag{3.27}$$

and the signal is expressed as

$$s[k] = \sqrt{E\cdot T_c/PW}\cdot\exp[j\cdot 2\pi\cdot f_R\cdot kT_c + j\cdot\Phi(kT_c)] \tag{3.28}$$

where the phase modulation with T_{chip}, equal to PW/N_{chip}, is the length of the single subpulse and is assumed as follows:

$$\Phi(t) = \sum_{k=0}^{Nchip-1} \varphi[k]\cdot rect_{Tchip}(t - k\cdot Tchip) \tag{3.29}$$

where $rect_{T_{chip}}(t - kT_{chip})$ is a pulse of unit amplitude and duration $Tchip$ centered at the time $(t - kT_{chip})$.

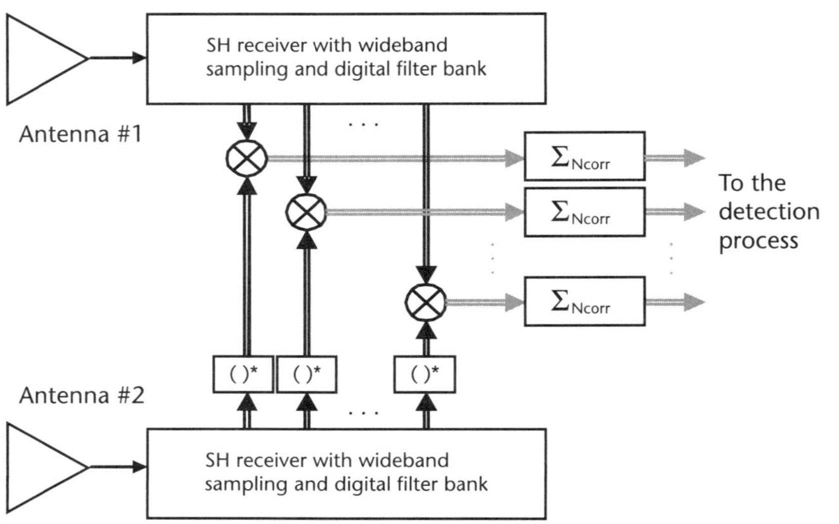

Figure 3.31 ESM processing block diagram.

The sequence is assumed to be a discrete sequence of random variables: are assumed to be independent, identically distributed (i.i.d.) and uniformly distributed over 2π. The power signal density is expressed as:

$$E\left\{|S(f)|^2\right\} \cong E \cdot Tchip \cdot \text{sinc}\left[\pi \cdot (f - f_0) \cdot Tchip\right]^2 \tag{3.30}$$

The SNR_{out} at the output of the cross-correlator, when a signal is present, is defined as:

$$SNR_{out} \equiv \frac{E\left\{\text{Re}\left[XC_{[M]}\right]\right\}^2 + E\left\{\text{Im}\left[XC_{[M]}\right]\right\}^2}{Var\left(\text{Re}\left[XC_{[M]}\right]\right) + Var\left(\text{Im}\left[XC_{[M]}\right]\right)} \tag{3.31}$$

where E{.} is the expectation or the average over signal plus noise correlation samples over the time interval T_{CORR}, and Re [.] and Im [.] are, respectively, the real and imaginary part of the cross-correlation.

The preceding definition allows for SNR_{out} evaluation not only from a theoretical point of view but also in the case of simulation and measuring; in fact, once several (complex) values of the cross-correlator output are determined a simple statistical analysis provides the SNR estimate.

The SNR_{out}, taking into account (3.30) and (3.31), becomes:

$$SNR_{out} = \begin{cases} \dfrac{B_N \cdot PW \cdot SNR_{IN}{}^2}{(T_{CORR}/PW) + 2 \cdot SNR_{IN}} & T_{CORR} > PW \\ \\ \dfrac{B_N \cdot T_{CORR} \cdot SNR_{IN}{}^2}{1 + 2 \cdot SNR_{IN}} & T_{CORR} \leq PW \end{cases} \tag{3.32}$$

In the case of LIP pulsed signals, when T_{CORR} exceeds PW, the corss-correlation receiver integrates only the product of noise samples in the period T_{CORR}-PW.

In Figure 3.32 the behavior of SNR_{out} vs. SNR_{in} (3.22) taken from [7] is reported for the following LPI PSK signals:

- CW unmodulated signal (e.g., a pure tone) has also been used as a benchmark.
- CW signal obtained by repeating a Barker 13 (implemented with B-PSK modulation) with T_{chip} = 100 ns and repetition period equal to 1.3 μs.
- CW signal obtained by repeating Frank 64 (8-PSK modulation) with T_{chip} = 100 ns and repetition period equal to 6.4 μs.

The following parameters are used for cross-correlator processing:

- B_n = 80 MHz → T_C = 12.5 ns.
- N_{CORR} = 800 → T_{CORR} = 10 μs.
- SNR_{in} (which represents the SNR at the output of the channelizer, e.g., over B_n = 80 MHz) ranging from –20 dB up to 30 dB).

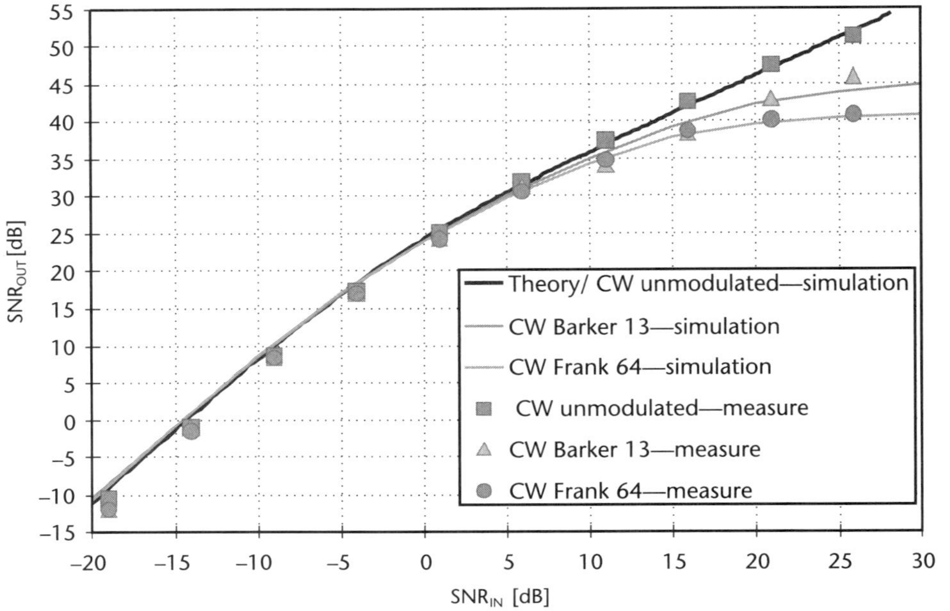

Figure 3.32 Cross-correlation processing gain on LPI PSK modulated CW signal.

For the classification of the LPI radar signals, several filtering methods are available in the literature and have been analyzed in [6] and [8], such as :

- Filtering methods based on the cyclostationary signal properties such as the FFT accumulation method (FAM), for which the processing block diagram is shown in Figure 3.33.
- Strip spectral correlation analysis (SSCA), for which the processing block diagram is shown in Figure 3.34.
- Time-frequency analysis based on the Wigner-Ville distribution (WVD) and the quadrature mirror filter (QMF), for which the processing block diagram is shown in Figure 3.35.

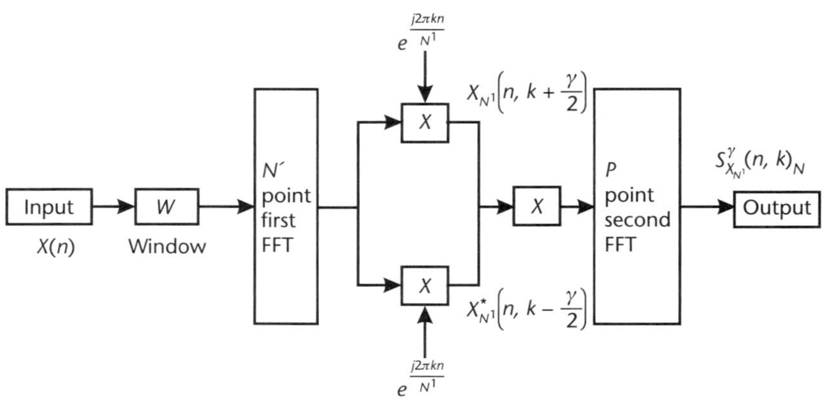

Figure 3.33 FAM processing block diagram.

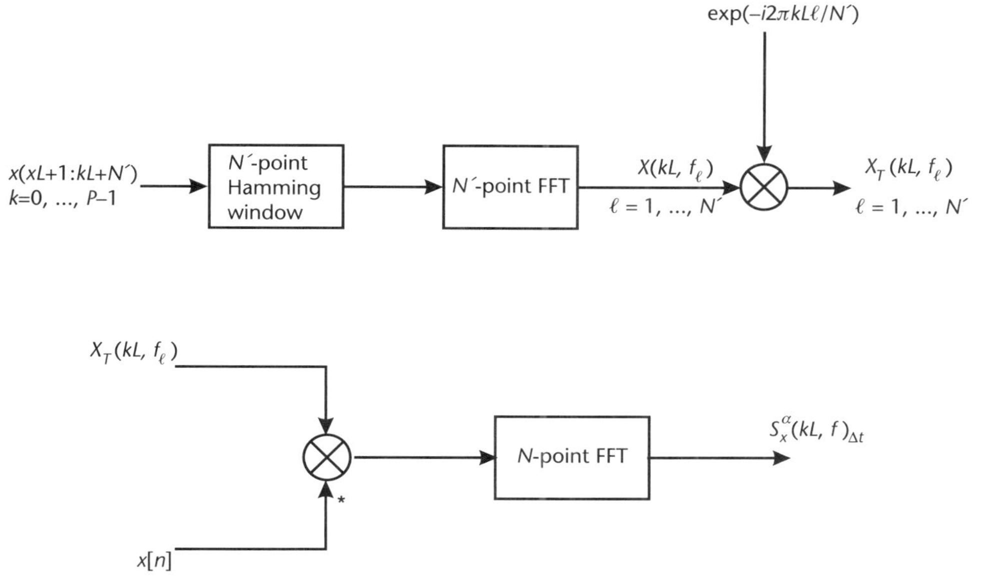

Figure 3.34 SSCA two-stage processing block diagram.

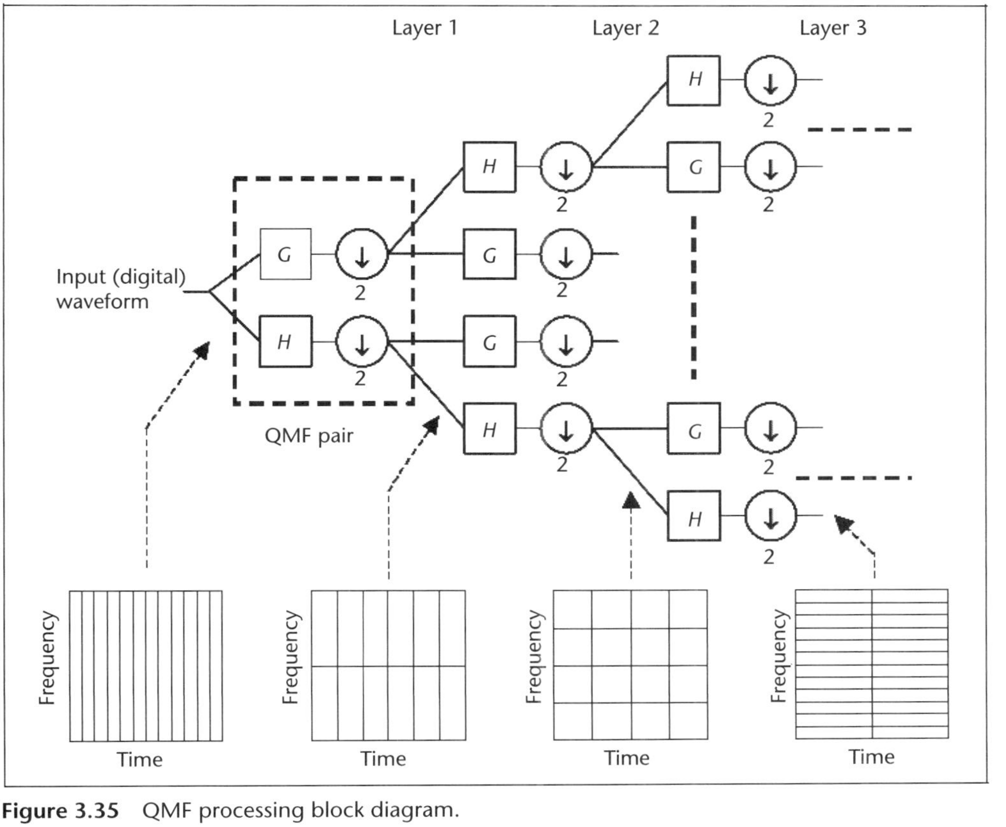

Figure 3.35 QMF processing block diagram.

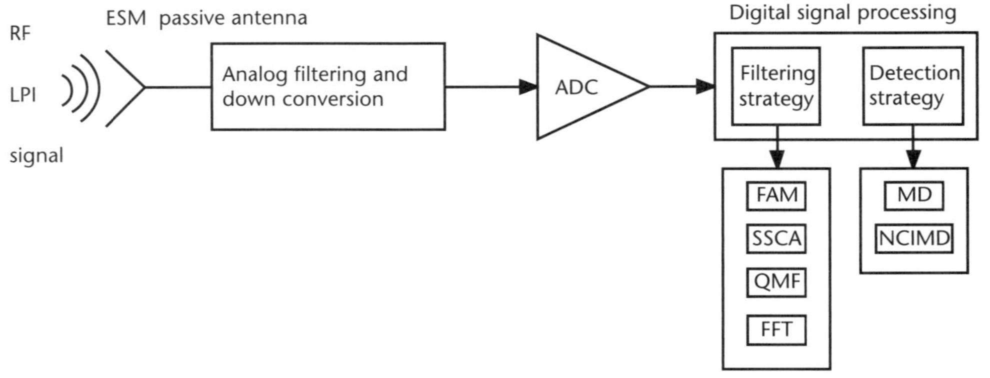

Figure 3.36 Block scheme of the simulated receiver chain with the different filtering methods and the NIMD.

The output of the methods above are matrices, and each method is peculiarly effective with respect to some LPI waveform. A comparison of the detection effectiveness of the above processing methods with respect to three different LPI waveforms (LFM, Frank code, and Costas code) was performed in [8], where the probability of detection was further maximized by applying a noncoherent integration of *N* maximum detection outputs (NIMD) for each filtering method. The relevant results are herewith reported (by permission of the authors) in Figures 3.36 through 3.39. The comparison shows that the QMF method (with a mini-

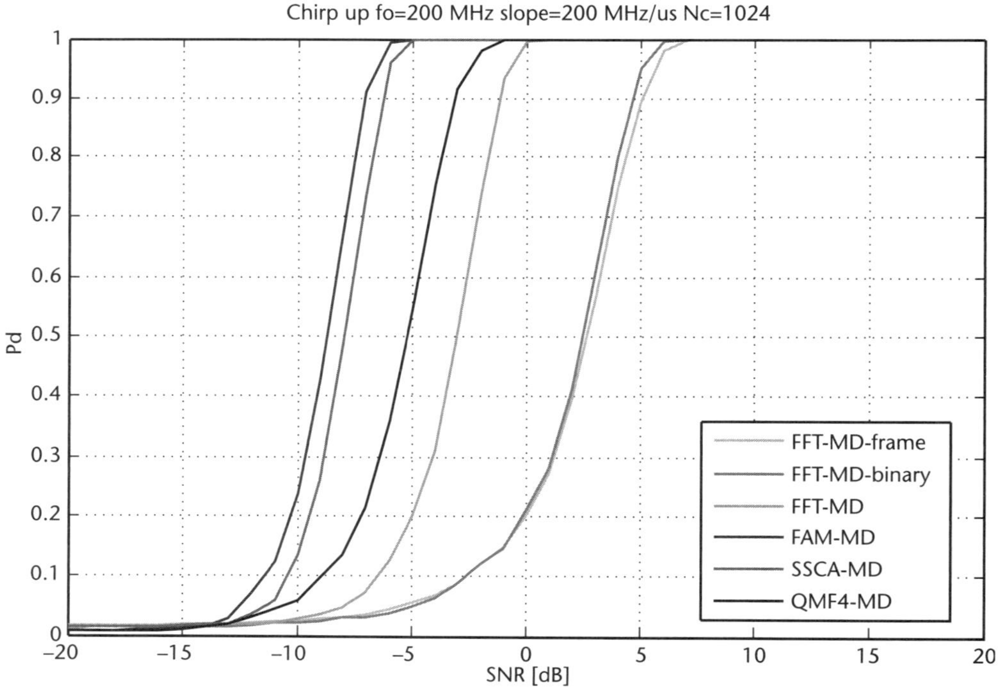

Figure 3.37 Detection performance of the different filtering methods relevant to the chirp (LFM) signal.

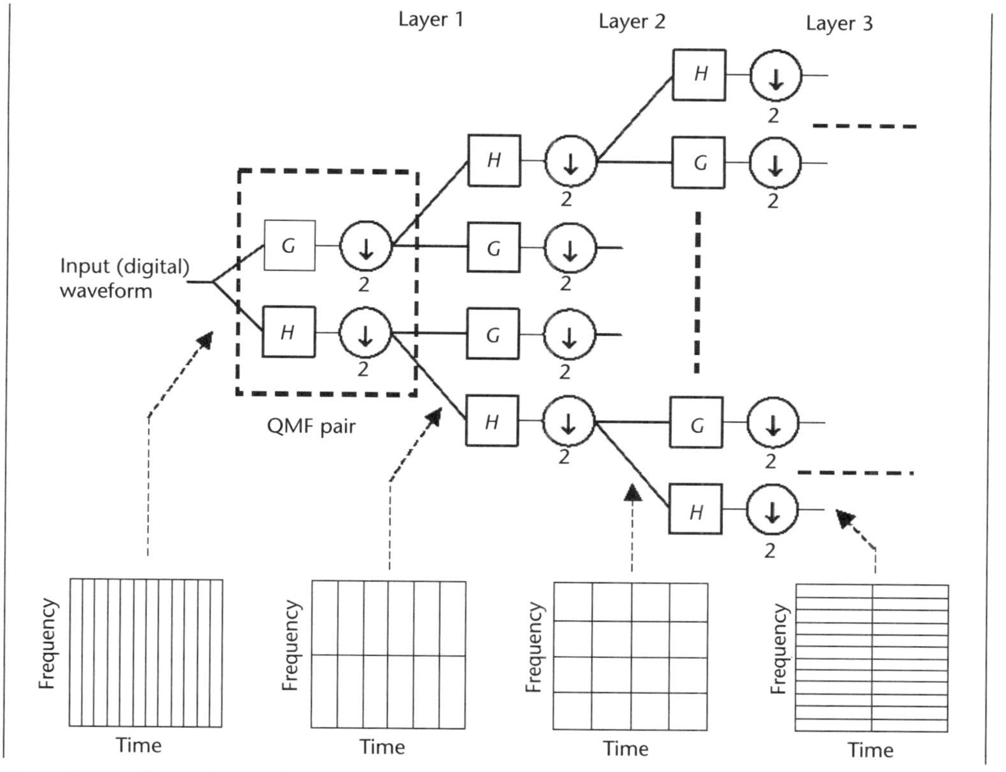

Figure 3.38 Detection performance of the different filtering methods relevant to the Frank coded signal.

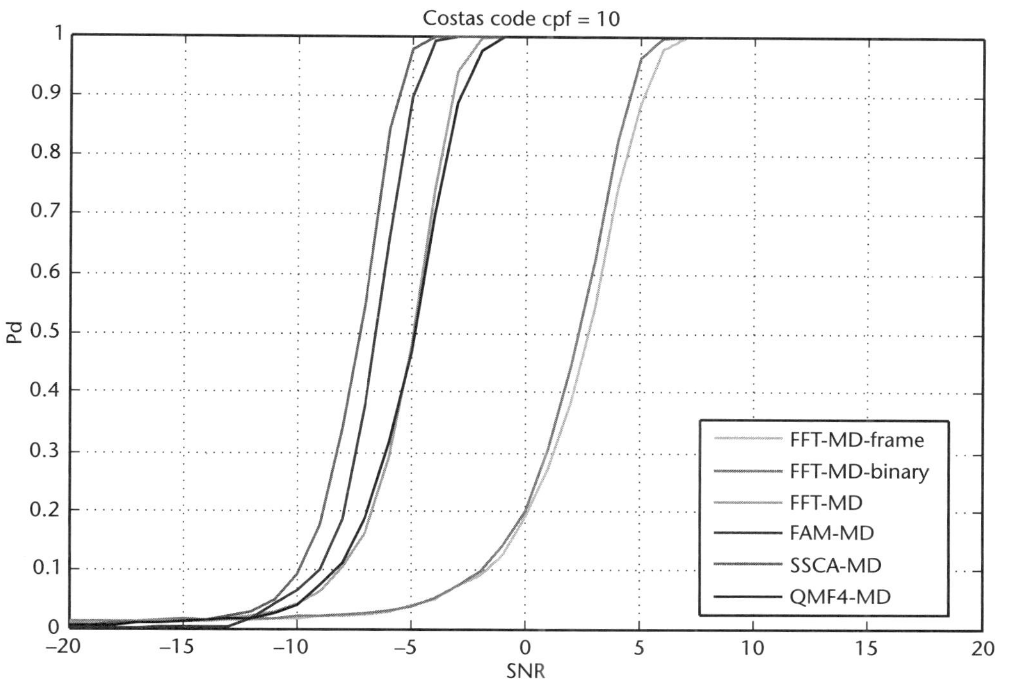

Figure 3.39 Detection performance of the different filtering methods relevant to the Costas coded signal.

mum of four layers) is the more effective filtering method for the three examined waveforms.

A word of caution must be mentioned in that the above-quoted results were evaluated in the absence of any interfering signals, a circumstance that never happens in real environments. Therefore, the methods must discussed have to be applied in conjunction with interference mitigating procedures (e.g., narrow beam spatial selectivity or adaptive beam pattern nulling).

Finally, a recommendation from Section 12.2 of [6] should be mentioned here: In terms of the classification of the LPI waveforms, that classification should be performed by employing in parallel a number of methods each followed by its own neural network (NN) that is trained to recognize the signal modulation among the numerous LPI modulations through the time-frequency or bifrequency images.

3.7 Emitter Deinterleaving and Sorting

Current EM scenarios are constituted by a very dense and complex mixture of different types of simultaneous signals, for which the amplitude and frequency are distributed over very wide ranges of values as shown in Figure 3.40. As soon as an emitter signal (usually a pulse) is detected, a number of its instantaneous parameters are measured and transformed into digital words in order to form a pulse descriptor message (PDM), which will be used within the sensor processor for the *deinterleaving* process, a process constituted by knowledge-based algorithms running on high-speed digital hardware, which discriminates and groups the PDMs of the pulses belonging to the same emitter, as received within a time frame T_F, thus

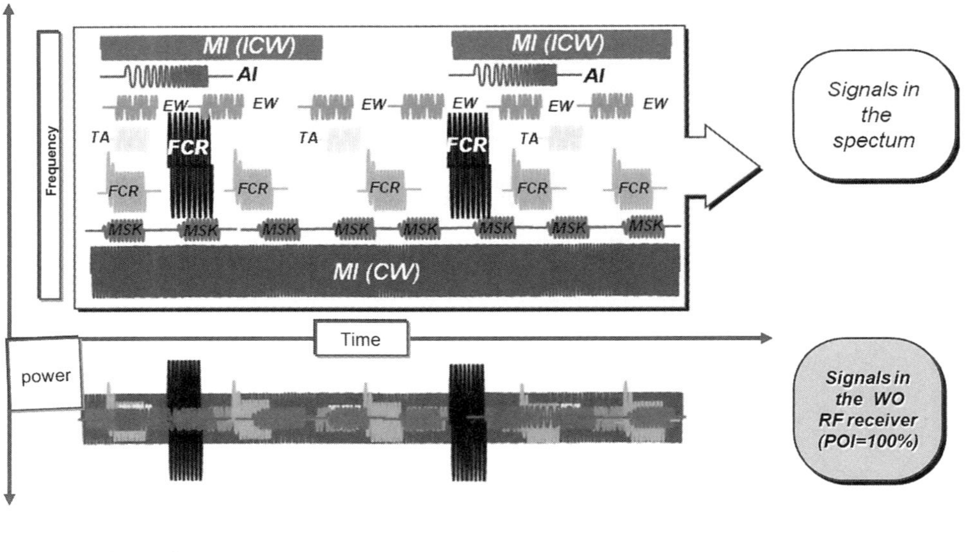

Legenda:

-MI:	Missile Illuminator	-MSK:	Active Missile Seeker
-ICW:	Interrupted Continuous Wave	-CW:	Continuous Wave
-EW:	Early Warning Radar	-AI:	Airborne Intercept Radar
-FCR:	Fire Control Radar	-TA:	Target Acquisition Radar

Figure 3.40 Typical symmetrical conflict EM scenario handled by a channelized Rx and a WO Rx.

extracting them from the PDMs received by the other emitters. In earlier sensor generations, this process performed by specialized digital processing boards capable of high processing speeds, but currently it is performed by common processor boards, due to the availability of high processing speeds and throughput with these boards.

The instantaneous pulse parameters usually employed for deinterleaving are as follows:

- Amplitude (A),
- Direction of arrival (DOA, usually only azimuth),
- Carrier frequency ($FREQ$),
- Time of arrival (TOA, usually the time at which a pulse detection occurs),

and for sophisticated deinterleaving also:

- Pulse width (PW),
- Modulation on pulse [MOP, an indication that can be either the simple presence of a modulation within the pulse or the very type of the modulation (phase or frequency)].

The deinterleaving process operates on all of the PDMs received within T_F by producing tri-dimensional histograms along the DOA, FREQ, and PW axes (see Figure 3.41). PDMs grouped within a *cell*, which has dimensions that are related to the pulse instantaneous parameter measurement accuracies of the sensor, are likely to belong to the same emitter. To take into account the possible frequency agility of the emitter, a rapid check is usually performed on equal DOA and PW and different TOA in order to add the relevant PDM to the cell.

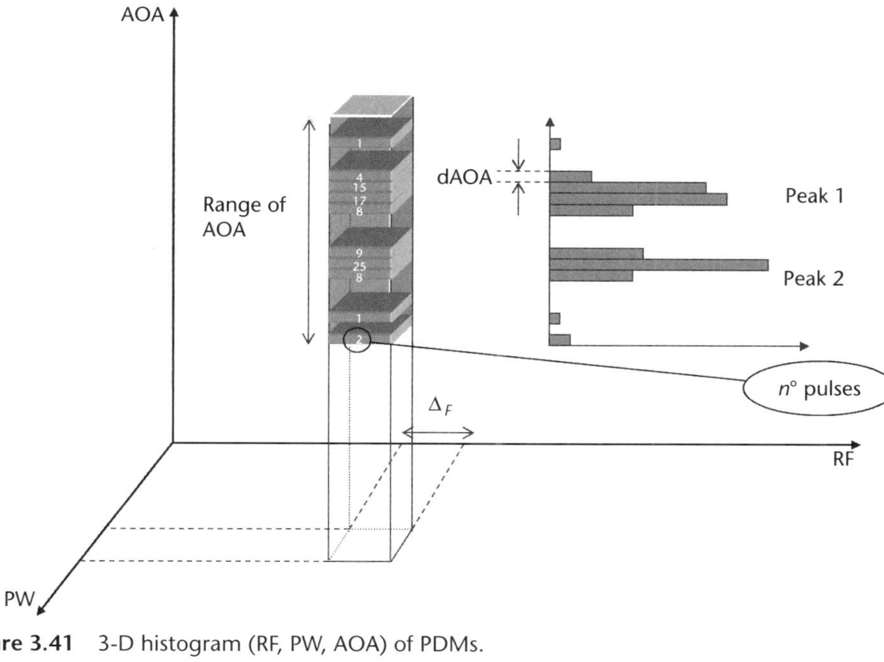

Figure 3.41 3-D histogram (RF, PW, AOA) of PDMs.

The PDMs associated with a cell are further checked according to MOP (otherwise they are discarded from the cell) and then ordered, according to increasing TOA, to determine the PRI and the PRI law of the emitter by using various algorithms appropriate to each known PRI law. Sometimes, if available from the EW platform mission data, templates of the PRI law of known emitters are compared with the TOA ordered PDM file of the cells in order to determine the matching to known emitters. This above-described process is called *emitter sorting,* because it provides at its completion a list of potential emitters detected within T_F.

Both of the processes just described are as much as possible *independent* of intelligence data and make intensive use of clustering algorithms on all received PDMs, using all of the signal parameters provided by the receiver architecture (DOA, RF, PW, PRI, MOP) in order to deinterleave and sort both *known as well as unknown emitters.*

At the completion of the emitter sorting process relevant to a time frame, a synthetic emitter file is provided to the EW processor, which updates its track file of previously sorted emitters with fresh emitter data and introduces new emitter entries in the track file. The emitter track file is then compared with the data available in the emitter mission library in order to perform the identification of the emitter mode and then associate it with the relevant weapons system.

A schematic block diagram of the pulse deinterleaving and emitter sorting processes is shown in Figure 3.42.

3.8 Emitter Identification

After an emitter has been sorted and reported in the active emitter file (AEF) of the ESM system, it has to be compared with the emitter data recorded in the library (relevant to known emitters) in order to determine its most likely identity. As mentioned in the previous section, a sorted emitter is reported in the AEF with a set of measured data, including the four measured primary parameters (RF, PRI, PW, and ASP) and relevant notations

- *RF notations:* FIX, HOP, BST, SLD, AGL, CPX, and so on;
- *PW notations:* FIX, HOP, BST, SLD, AGL, CPX, and so on;
- *PRI notations:* FIX, STG, SWT, JIT, SLD, COD, CPX;

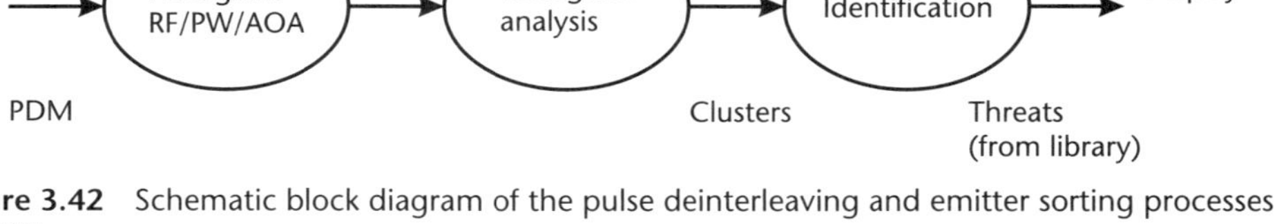

Figure 3.42 Schematic block diagram of the pulse deinterleaving and emitter sorting processes in the EW sensor.

and the emitter secondary parameters, relevant to the presence of MOP, and relevant notations, which are:

- *FMOP:* CHIRP, SAW, TRI, SIN, and so on;
- *PMOP:* BARKER, FRANK, UNK;
- *AMOP.*

The comparison of each of these parameters is performed with the corresponding parameter of each likely recorded emitter of the library. Likely recorded emitters are those that have passed through a frequency filter (which rejects all of the recorded emitters outside the frequency carrier interval of the measured emitter) and the notation filter (which rejects all of the recorded emitters that have notations different from the corresponding ones of the sorted emitter). This comparison produces for each primary and secondary emitter parameter, which we will generically designate as i, a score, called $\text{Score}_{i,j}$, relevant to the corresponding parameter of the j recorded emitter of the list of likely recorded emitters.

$\text{Score}_{i,j}$ is evaluated as the percentage overlap between the measured parameter interval and the corresponding parameter interval of the recorded emitter, incremented by the parameter measurement accuracy of the actual receiver. As an example, let us consider the parameter PRI and assume that:

- The sorted emitter measured PRI interval is I_{PRI} ($I_{\text{PRI}} = \text{PRI}_{\text{max}} - \text{PRI}_{\text{min}}$).
- The j recorded emitter PRI interval is $L_{\text{PRI},j}$.
- The measurement PRI accuracy of the receiver is Δ_{PRI}.

The PRI parameter score of the j recorded emitter is therefore:

$$\text{Score}_{PRI,j} = I_{PRI} / (L_{PRI,j} + 2\delta_{PRI}) < 1 \tag{3.33}$$

The Total Score of each selected j recorded emitter is evaluated as the weighted sum of the scores of each measured parameter, where the weights $w_{i,j}$ are the product of two factors:

$$w_{i,j} = Fn_{i,j} Fm_{i,j} \tag{3.34}$$

where $Fn_{i,j}$ is relevant to the parameter notation, and Fm_i is relevant to the measurement compatibility. Hence:

$$\text{Total Score } (j) = \sum i(w_{i,j} \text{Score}_{i,j}) \tag{3.35}$$

If Total Score exceeds a selected threshold, defined as the confidence level, the j recorded emitter can be considered a likely identity of the sorted emitter. Of course, the most likely identity of the sorted emitter is assigned to the recorded emitter, which achieves the maximum total score. This identification process is depicted in Figure 3.43.

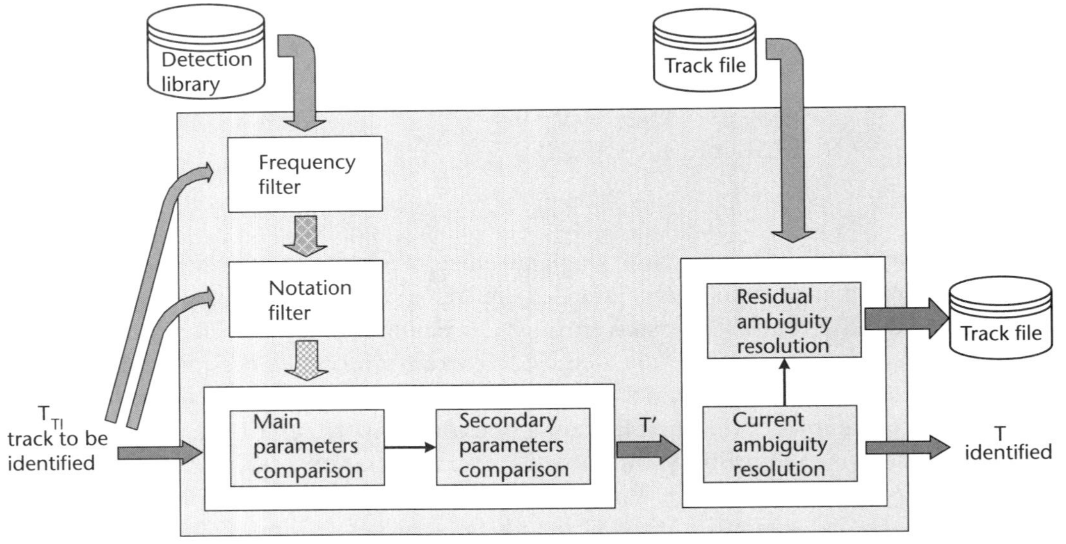

Figure 3.43 Flow diagram of the emitter identification process.

3.9 Communications ESM

Today's military operations are characterized by the need for robust and secure communications networks that can transfer multimedia information in real time to fixed and mobile stations, thus providing all the elements of the network with a capillary fast command chain and an updated complete situation awareness.

Communications intercept systems are termed CESM (communications ESM) if they have to provide quickly emitter signal information to combat troops or COMINT (communications intelligence) if they have to analyze the signals over an extended period of time in order to provide intelligence on the emission characteristics and contents.

The tasks of CESM equipment are the search for, interception of, classification of, identification of, and direction finding of communications signals in the surrounding environment in order to activate further EW operations (such as emitter location, signal analysis, or jamming of the relevant network). These tasks are made difficult by the presence of a multitude of friendly and civilian communications and by the adoption by modern military communications equipment of LPI techniques, such as SS direct synthesis of digital coded signals (mostly encrypted) and frequency hopping.

The tasks of COMINT equipment is to record the emitter signals for subsequent possible decryption of its contents and for the analysis and evaluation of its LPI characteristics and ECCM features in order to identify the most appropriate jamming techniques.

3.9.1 CESM

Communications services are virtually spread all over the RF spectrum. Table 3.4 lists the frequency band denominations.

Table 3.4 Frequency Band Designations

Wavelength	*Frequency Band*	*Name*	*Designation*
100–10 km	3–30 kHz	Very low frequency	VLF
10–1 km	30–300 kHz	Low frequency	LF
1–0.1 km	300–3000 kHz	Medium frequency	MF
100–10m	3–30 MHz	High frequency	HF
10–1m	30–300 MHz	Very high frequency	VHF
1–0.1m	300–3000 MHz	Ultra-high frequency	UHF
100–10 mm	3–30 GHz	Super-high frequency	SHF
10–1 mm	30–300 GHz	Extra high frequency	EHF
1–0.1 mm	300–3000 GHz	Optical	Optical

To cope with the extremely crowded communications band characterized by many simultaneous fixed frequencies, burst type, modern FH, and SS signals, the CESM equipment architecture can be implemented either with a number of wide–bandwidth channelized receivers covering the required frequency bands or with a number (equal to the number of receiving antennas) of wide-bandwidth SH channelized receivers, whose band is rapidly stepwise scanned over the interested RF bands. This latter architecture is the most used in practice and will be described in the following discussion.

HF band CESM equipment, due to the large wavelength dimension in this band, needs very large antennas (usually an array of antennas displaced over a large area) in order to obtain an accurate AOA. The employed DF algorithms are either the Watson-Watt method with use of Adcock-type antennas [i.e., four equal vertical elements (monopoles) displaced at the vertexes of a square and interconnected in pairs along the diagonals] or the correlative phase method (see Section 4.3.1) reinforced with some MUSIC algorithm to solve the multipath DF ambiguities [9].

Thanks to refraction by the ionosphere of the HF signal through the measurement of the elevation AOA, and assuming the height of the refraction layer (D, E, F_1, or F_2) is known, it is possible to evaluate the distance of the emitter on the ground, thus implementing a single-site emitter location.

The architecture of the HF band CESM equipment is much similar, with the exception of the type and number of DF antennas and of the size of the IBW to be swept across the HF band, to the one of the CESM equipment operating in the VHF/UHF bands (30 to 3000 MHz), which is illustrated in Figure 3.44.

The CESM equipment is composed of the following units:

- An antenna group, composed of a number of sub-band antenna arrays; usually three sub-bands are used such as 30 to 300 MHz, 300 to 1000 MHz, and 1000 to 3000 MHz;
- Antenna front-end receiver and sub-band antenna arrays switching matrix;
- RF receiver;
- Processor and interfaces;
- Auxiliary units (HMI, databases and emitter libraries, recording devices).

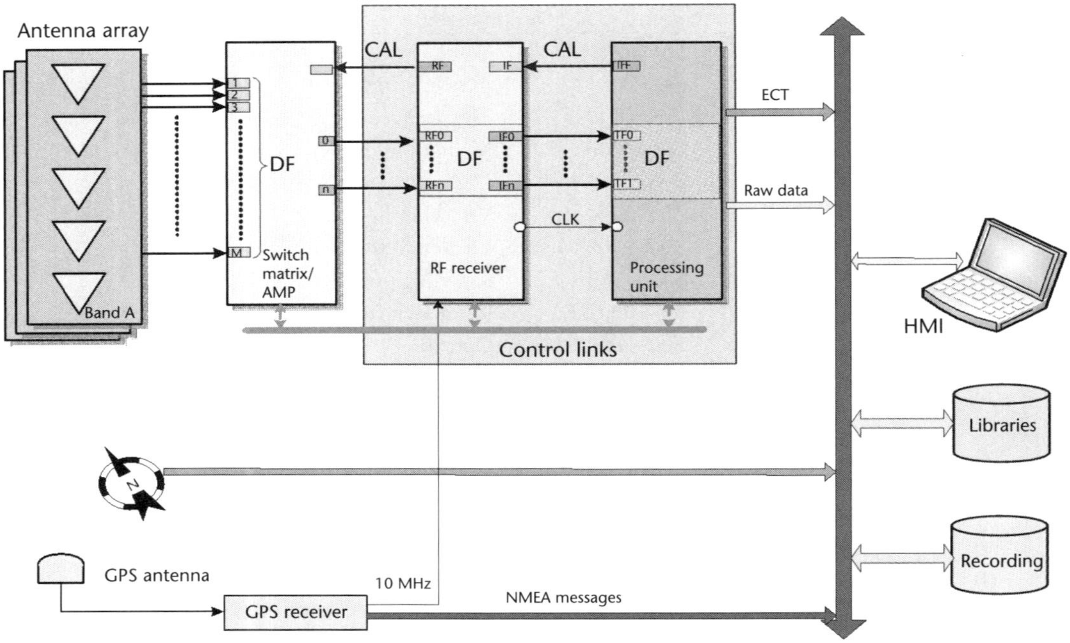

Figure 3.44 Schematic block diagram of typical V/UHF CESM equipment.

The CESM equipment when operating in ground sites is usually composed of a GPS receiver in order to provide its site geographical coordinates and to be synchronized with the other CESM sensors of the emitter location network (see also Section 4.5).

The employed DF method is usually the correlative DF method (as reported in Section 4.3.1) as applied to the sub-band uniform circular arrays (UCAs) with an odd number of vertical dipoles (to reduce the number of phase measurement ambiguities due to harmonic frequencies and thus increase the measurement bandwidth of each sub-band array). Typical AOA accuracy is on the order of 1° to 2° rms over each sub-band.

In Figures 3.45 and 3.46, respectively, the principle of the correlative DF method (as reported in Section 4.3.1) and the picture of a typical CESM antenna group are shown.

The antenna front-end (AFE) and sub-band arrays switching matrix provide the equipment with preselector filtering and first low-noise amplification of the received RF signals from each antenna of the UCA before they are translated to IF in the relevant RF receiver via mixing with a proper LO signal. In theory, it would be possible for the correlative phase DF measurement to use only two channels at a time and then switch sequentially the two RF receiver channels over the other pairs of antennas of the UCA. This DF scheme was used in the past to reduce the cost of the equipment, but the current increased FH rate of modern emitters made it ineffective, because monopulse DF is essential in classifying this type of emitters.

Of the various features of the analog part of the RF receiver, the important ones are as follows:

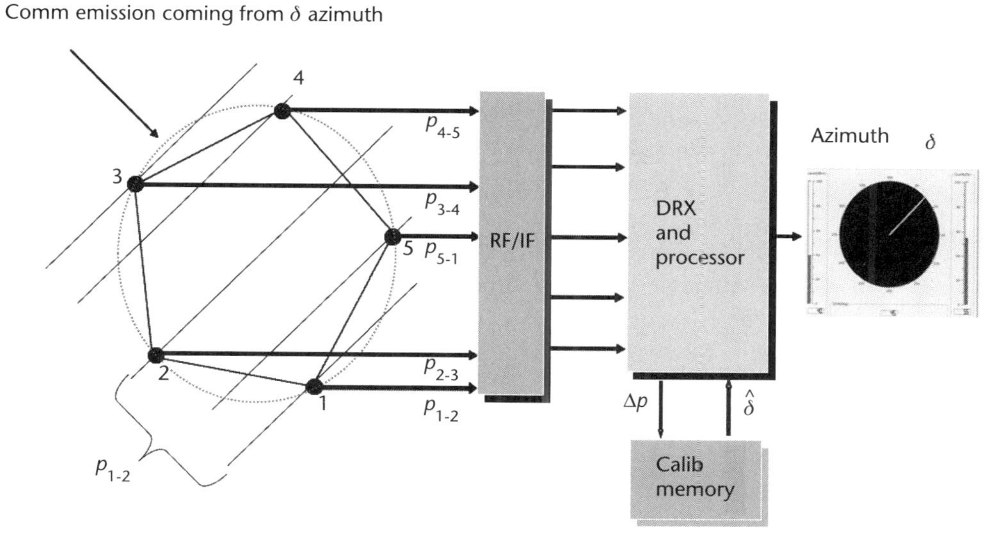

Figure 3.45 Principle of the correlative DF method.

Figure 3.46 A typical CESM antenna group with three sub-band UCAs of five vertical dipoles. (Courtesy of Elettronica S.p.A.)

- The tuning time of the synthesizer determines the stepwise sweeping speed of the receiver IBW (on the order of >40 MHz) over the whole frequency coverage bandwidth. The shorter the tuning time, the higher the probability of intercepting burst transmissions. Current values for this parameter are on the order of tens of microseconds.
- The instantaneous dynamic range (IDR) enables discrimination of small-amplitude signals in the presence of very-large-amplitude signals (mostly close to the equipment site). Usually IDR > 60 dB.
- The noise figure of the RF chain (from antenna to the IF input) is on the order of 8 to 10 dB across the frequency coverage bandwidth.

The IF signals of each receiving channel are digitally converted through a 12- to 14-bit ADC (in order to achieve an effective IDR > 60 dB) and processed by a number of FPGAs, which provide detection (uniform filter bank) and phase measurement algorithms. The schematic block diagram of two adjacent (in the UCA) channels is shown in Figure 3.47. The figure also shows the windowing applied to the digital data stream to reduce the sidelobes of the subsequent FFT detection. The IBW is swept stepwise at high tuning speed across the required frequency band in order to achieve a high POI and to determine the distribution of the emitters with frequency. The IBW tuning speed is controlled by the mission data, so that either the IBW dwell times can be increased at frequencies where FH emitters are envisaged (to correlate frequency detections at the same angle along the dwell time) or the number of frequency bins in the IBW can be increased/reduced in accordance with the envisaged signal bandwidth at some frequencies (to approximate the matched filter bandwidth of those signals). Figure 3.48 shows a typical steady IBW frequency scanning strategy.

After detection, with frequency carrier and DF measurements of the emitter signal, the CESM equipment has to perform signal classification to help determine the adversary's EOB by indicating which type of communication equipment is in use.

Signal classification encompasses many types of processing, such as:

- Signal type recognition (e.g., analog or digital);
- Modulation recognition if analog (e.g., AM, FM, PM);
- Type of digital signal (e.g., BPSK, QPSK);
- Type of multiplexing (e.g., FDM, PPM).

Two algorithmic approaches can be applied to perform signal classification [4]:

1. Signal feature extraction followed by pattern recognition processing based on neural networks, which are able to discriminate among signal types;

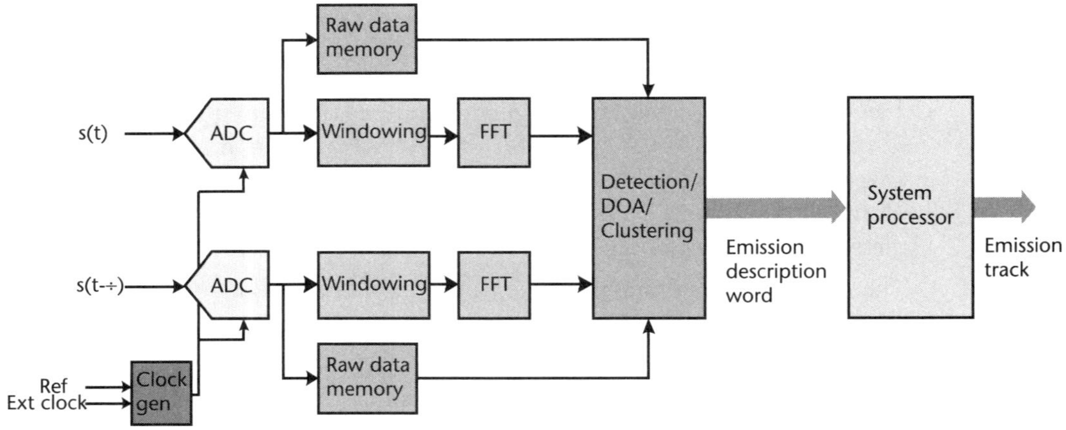

Figure 3.47 Two DRx channels performing multichannel signal detection and differential phase measurement from two antennas within the IBW equipment.

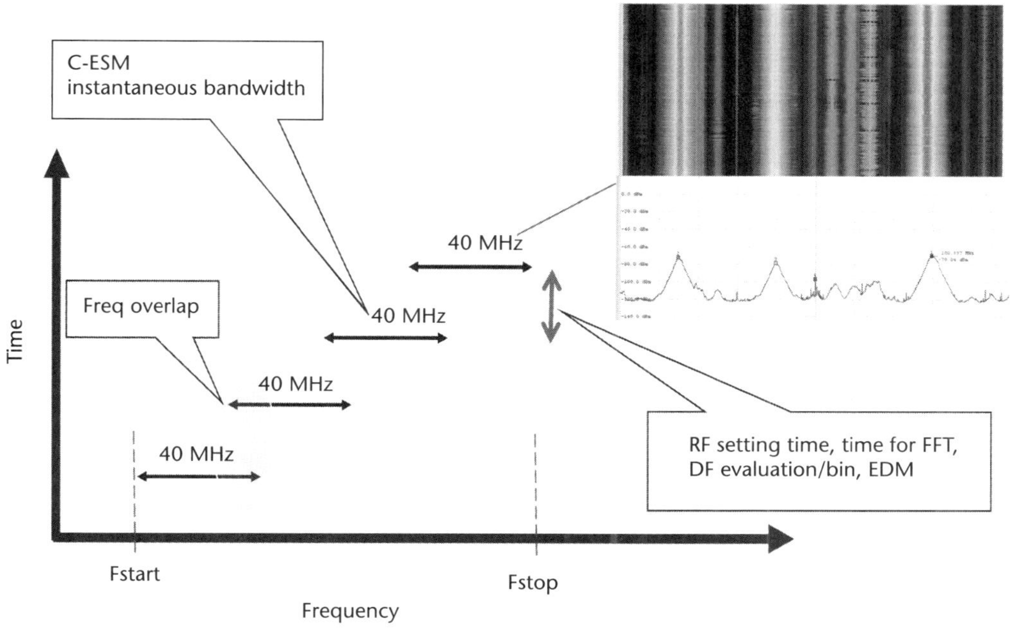

Figure 3.48 Typical CESM equipment IBW scanning strategy.

2. The decision theoretic approach, which determines the signal type on optimization of the established likelihood ratios based on statistical properties of the signals.

Among the signal classifiers of both approaches, here we simply cite a few and invite the interested reader to deepen his knowledge by reading Chapter 10 of [4] and the relevant literature reported in its references. For pattern recognition processing:

- Nandi-Azzouz classifier, which is applicable for discriminating among 13 analog and digital modulation types: AM, FM, MFSK, MASK DSB, LSB, USB, MPSK, VSB, and combined (amplitude and phase) modulation. Figure 3.49 shows the flow diagram of this classifier and Table 3.5 defines the parameters used in the figure..
- Assaleh-Farrell-Mammone classifier, which is applicable for discriminating among digital modulation types.

For the decision theoretic approach:

- Kim-Polydoros classifier, which is applicable for discriminating between BPSK and QPSK;
- Sills classifier, which is applicable for discriminating among three types of PSK signal (BPSK, QPSK and 2^3PSK) and among three types of QAM signal (2^4QAM, 2^5QAM, and 2^6QAM).

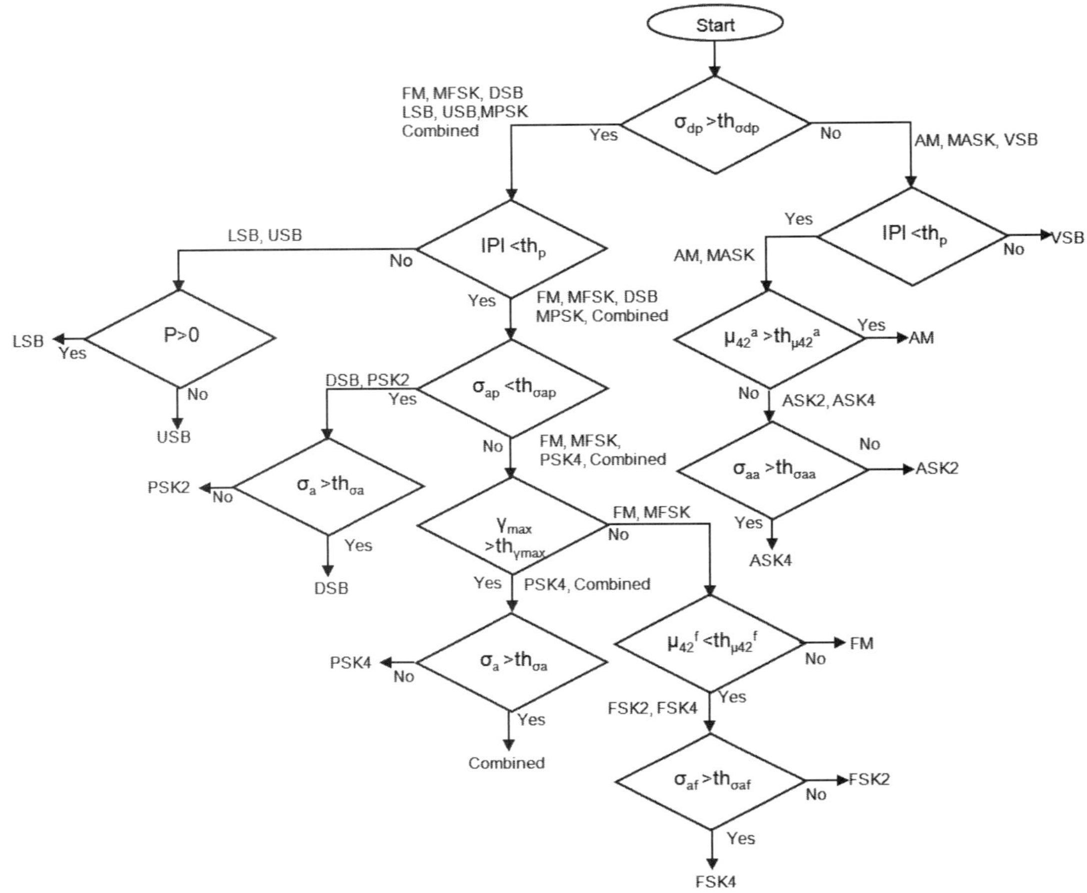

Figure 3.49 Nandi-Azzouz classifier processing flow diagram (see definition of parameters in Table 3.5).

Table 3.5 Definition of Parameters Used in Figure 3.50

Parameter	*Definition*
γ_{max}	Maximum of the PSD of the normalized centered instantaneous amplitude
σ_{ap}	Standard deviation of the absolute value of the instantaneous phase
σ_{dp}	Standard deviation of the instantaneous phase
P	Symmetry of the spectrum of the signal
σ_{aa}	Standard deviation of the absolute value of the instantaneous amplitude
σ_{af}	Standard deviation of the absolute value of the instantaneous frequency
σ_a	Standard deviation of the instantaneous amplitude
$\mu_{42}{}^{a}$	Kurtosis of the instantaneous amplitude
$\mu_{42}{}^{f}$	Kurtosis of the instantaneous frequency

Note: *Kurtosis* is a measure of the peakedness of a probability density, usually relative to the normal density [4]. The *kurtosis* of a probability density is expressed as $K = \mu_4/\sigma^4$, where μ_4 is the fourth moment about the mean of the random variable and σ is the standard deviation.

The advantage of the decision theoretic approach classifiers with respect to the pattern recognition processing ones is the possibility of operation at lower SNR than >15 dB as requested by the latter ones.

Signal feature extraction, however, is required to perform identification of the emitter, which is accomplished in the same manner as described in Section 3.8 for the radar emitters: by matching each feature with the relevant ones of the likely emitters in the database and assigning a score to each match. The emitter that accumulates the highest total score over the emitter features is considered the emitter identity.

Typical emitter features to be extracted are as follows:

- Signal classification,
- Frequency(ies) of operation,
- RF bandwidth,
- Type of modulation,
- Power level.

3.9.2 COMINT

The task of COMINT equipment in military operations is to record the content of the adversary's communication systems (for subsequent listening and/or analyses) to provide accurate data about their location, their signal structure, and their ECCM features, in order to support mission planning, which may consist of jamming in order to disrupt the adversary's command chain. Analysis of COMINT data is essential for the design of jamming waveforms to be transmitted and for the selection of the CECM equipment to be employed. The main ECCM features of the adversary tactical data links that the COMINT equipment has to measure for the FH emitters are the hop time and the hop bandwidth, and for the SS emitters, the processing gain.

COMINT equipment is also used in civilian applications for spectrum monitoring applications, such as the surveillance of proper broadcast channel widths and received signal quality evaluation in areas of broadcast and wireless services.

COMINT equipment is usually not provided with DF antennas covering 360° azimuth, so their high-gain antennas (to achieve an increased sensitivity and hence a higher SNR for more accurate emitter parameter measurement) are designated in angle by the associated CESM.

The main COMINT functions are:

- Signal acquisition and feature extraction,
- Signal classification,
- Signal demodulation,
- Voice signal demodulation and listening,
- Signal recording,
- Signal reproduction for fine analysis (bit-stream decoding, transmission standard recognition, and speech and speaker recognition).

The COMINT received signal processing flow is shown in Figure 3.50.

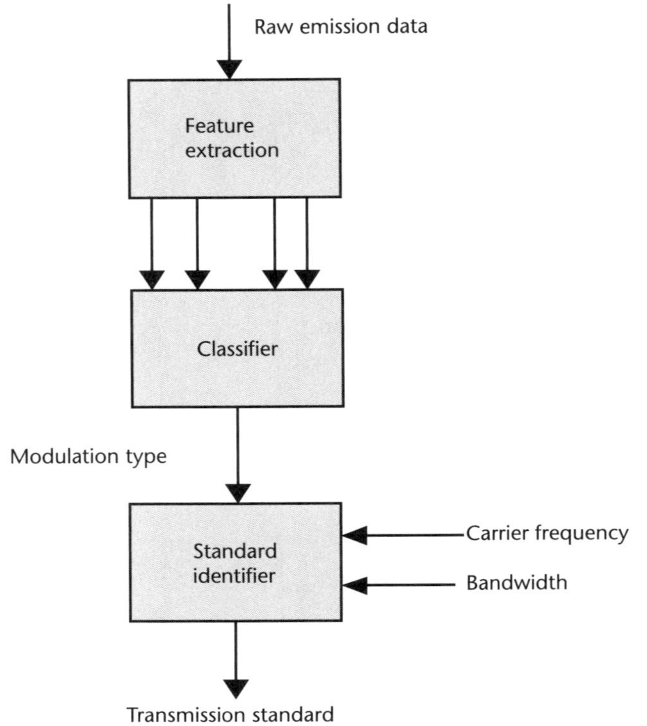

Figure 3.50 COMINT acquired signal processing flow.

The architecture of COMINT equipment is very similar to that of CESM equipment, with the 360° azimuth DF being substituted by a high-gain antenna, and with a large additional number of processors, recorders, and software tools to ease the analysis tasks performed by the operators. The block diagram of a typical COMINT station served by many operators is shown in Figure 3.51. This kind of station is able to operate in the HF band and, thus, thanks to ionospheric refraction, it can intercept and analyze communication emitters located thousands of nautical miles from the station. Typical operator displays showing the capabilities of the available software tools are pictured in Figure 3.52.

3.10 SIGINT

SIGINT (signal intelligence) is currently only a definition that refers to the combination of COMINT and ELINT activities. The current need to introduce modern large-bandwidth communication services pushed these to occupy large available spectral regions in the SHF band. In contrast, the introduction of stealth technologies applied to military platforms has pushed radar designs within the low UHF and VHF bands, where detection is made easier for the resonance of objects of comparable size to the wavelength. The mixing of communication and radar signals within the same bandwidth suggests the use of receiving equipment that exploits common antennas to feed two separate receivers, dedicated to the radar and to the communications signals' interception and classification.

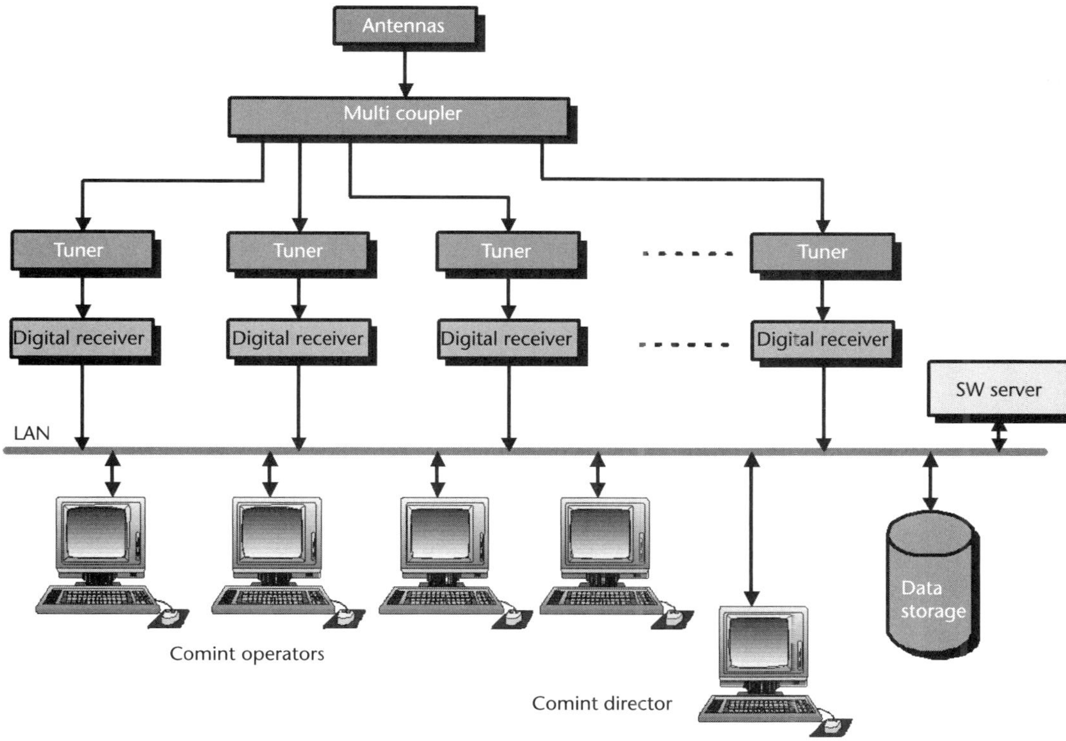

Figure 3.51 Block diagram of a typical COMINT station, operating also in the HF band and served by many operators, each analyzing independently different spectral regions.

This new type of equipment, a real type of SIGINT equipment in that it combines COMINT and ELINT tasks within an integrated piece of equipment, presents substantial advantages in military platforms such as ships and aircraft where reduced volumes are a premium. An example of such equipment is the ELT 819 Airborne SIGINT equipment. Its schematic block diagram is shown in Figure 3.53.

3.11 Conclusion

In this chapter the principles of operation (interception, parameter measurement, and emitter classification) as well as the adopted technologies and techniques of modern passive (i.e., receivers) EW equipment have been described. We have shown how they cope with the challenges provided by the evolution of radar and communications equipment transmitted waveforms, previously illustrated in Chapter 2. No particular emphasis was given to the host platform of the described equipment, although the effort in the development of these technologies was addressed to airborne equipment because the aircraft environment is the most demanding in terms of volume, power dissipation and consumption, vibration levels, and thermal conditions.

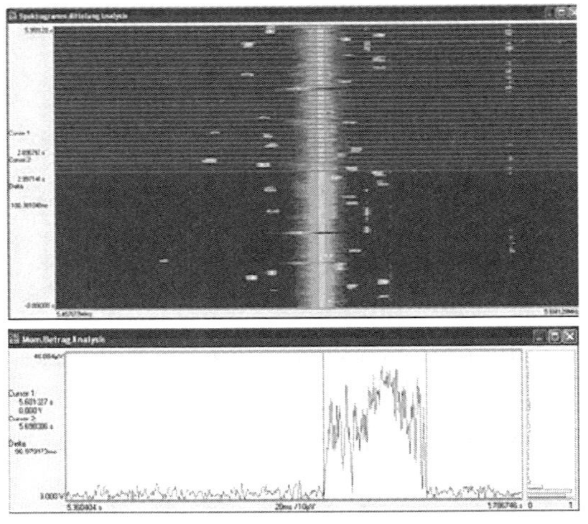

(a)

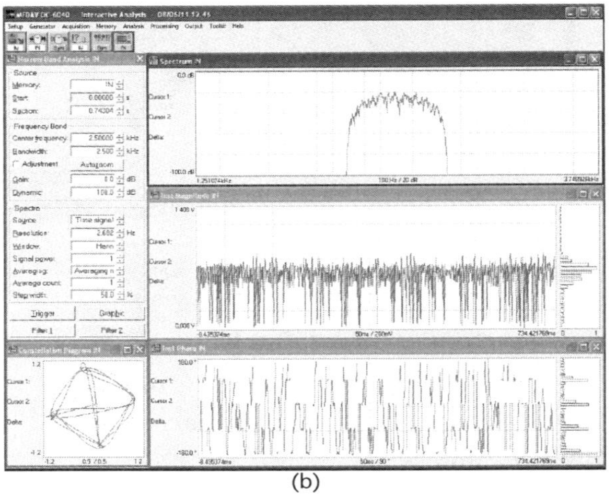

(b)

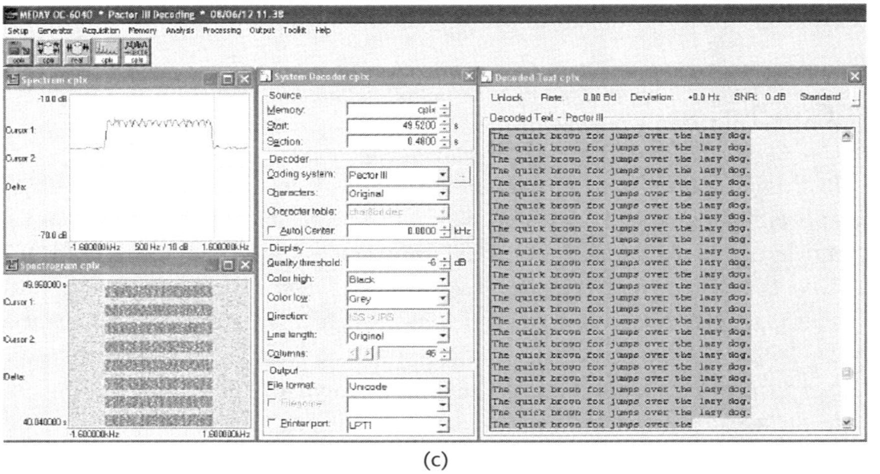

(c)

Figure 3.52 Typical operator displays of a COMINT station operating in the HF band. (a) waterfall and spectrum at a selected time by the operator; (b) simultaneous analysis of three signals; and (c) OFDM spectrum, waterfall, and decoding capability relevant to a digital signal. (Courtesy of MEDAV.)

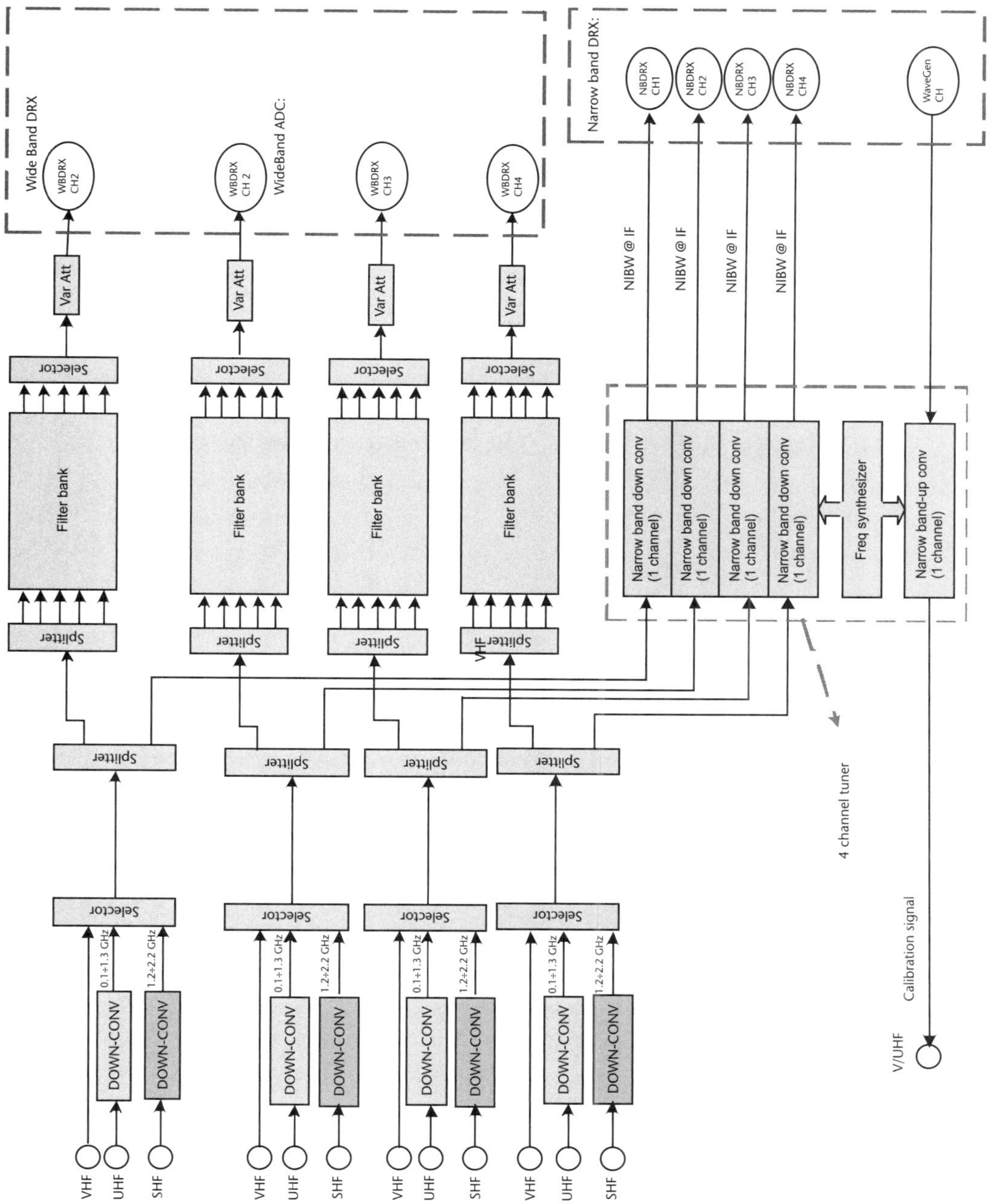

Figure 3.53 SIGINT architecture (bands A to K). (Courtesy of Elettronica S.p.A.)

References

[1] Crochiere, R. E., and Rabiner, L. R., *Multirate Digital Signal Processing*, Englewood Cliffs, NJ: Prentice-Hall, 1983.

[2] Wiley, R., *Electronic Intelligence: The Interception of radar Signals*, Dedham, MA: Artech House, 1985, Chapter 3.

[3] Lynch, D., *Introduction to RF Stealth*, Raleigh, NC: Scitech Publishing, 2004, Chapter 2.

[4] Poisel, R., *Introduction to Communication Electronic Warfare Systems*, Norwood, MA: Artech House, 2003, Chapter 10 and Appendix.

[5] De Martino, A., et al., "Digital Receivers Applications in Electronic Warfare Systems," paper presented at Workshop on Digital Receiver Systems for Defense and Related Commercial Applications, Boston, MA, June 7–12, 2009.

[6] Pace, P., *Detecting and Classifying Low Probability of Intercept Radars*, Norwood, MA: Artech House, 2004.

[7] Ardoino, R., and Megna, A., "LPI Radar Detection: SNR Performances for a Dual Channel Cross-Correlation Based ESM Receiver," paper presented at 39th European Microwave Conference, EuRad session, Rome, September 28–October 2, 2009.

[8] Camuso, P., Foglia, G., and Pistoia,D., "A Comprehensive Analysis on Detection Performances of LPI Signals Filtering Strategies," paper presented at 39th European Microwave Conference, EuRad session, Rome, September 28–October 2, 2009.

[9] Lipsky, S., *Microwave Passive DF*, Raleigh, NC: SciTech Publishing, 2004.

CHAPTER 4

RF Direction-Finding and Emitter Location Techniques

4.1 Introduction

The information about the direction of arrival (DOA) of an emitter signal is of the utmost importance both from an operational aspect (i.e., in what direction with respect to the platform the threat is present and which evasive or engagement manoeuvre the platform can perform) and from a deinterleaving and sorting aspect. Deinterleaving and sorting are performed by EW sensors, and DOA is the only stable emitter signal parameter acquired within the relatively short emitter acquisition period (time frame) of an EW sensor.

EW sensors employ many different direction-finding (DF) techniques, either individually or in combination with similar sensors, to achieve the required DOA and localization measurement accuracy. The basic concepts related to estimation accuracy and its lower bound, the Cramer-Rao lower bound (CRLB), are discussed in Appendix B. Of course, regardless of the DF technique used, the choice of sensor antennas is fundamental to the DF function. A basic introduction to antennas and array of antennas is given in Appendix C.

DF techniques are classified in accordance to the interval of time in which the DOA information is achieved. The measurement can be either *instantaneous* (i.e., within the emitter pulse or a short sample of its waveform in the case of long waveforms, in which case the technique is a *monopulse* technique and requires more than one antenna) or *sequential* in time by means of a movement (usually a rotation) of the sensor's antenna with respect to the DOA of the signal.

The monopulse DF techniques exploit the differences of one instantaneous emitter parameter (such as amplitude or phase or time of arrival) as received by two or more antennas, whereas the sequential DF techniques usually exploit only the emitter signal amplitude as modulated by the movement of the single measuring antenna.

4.2 Amplitude Comparison DF Methods

In principle, the DOA of an RF signal can be determined from amplitude measurements performed during scanning (usually a rotation at high speed in the azimuth plane) of a high-gain parabolic dish antenna with low sidelobes across the direction of the emitter. The signal amplitude, which is supposedly steady within the scanning period ΔT of a high-gain antenna mainlobe, is modulated by the antenna pattern, and the mainlobe peak determines the signal DOA. This method is depicted in Figure 4.1 and a photo of a typical antenna (ELT 888) is shown in Figure 4.2. This method can provide highly accurate DOA measurements (error <1° all over the 2- to 18-GHz band) by associating a RF monopulse beamformer (see later) with the antenna.

This method was applied with antennas scanning in both azimuth and elevation planes to determine bidimensional DOA measurements. Although the approach of a rotating antenna is conceptually simple and can provide DF accuracy on the order of 1°, it is inherently a low POI approach with respect to angle scanning emitters, so it is not widely used in ESM equipment. Currently the only application of this method is in ELINT equipment, in which the high antenna gain, coupled with a high sensitivity receiver (narrow/midband receiver), allows DOA measurements at the antenna sidelobes of the scanning emitter, thus bypassing the low POI relevant to its mainlobe. Further, most of these types of antennas can be stopped or brought to the DOA of the emitter to acquire long samples of the emitter signal for subsequent signal analysis.

The most common approach to amplitude comparison DF is to use four low-gain steady antennas, displaced 90° from each other, in order to provide four angular quadrants of azimuth coverage. This arrangement is commonly used in most RWR equipment. The concept of operation is to compare the amplitudes of the RF emitter signal as received by each antenna. Because amplitude comparison is performed instantaneously, this DF method is often referred to as *amplitude monopulse*. The arrangement of the four antennas and the relevant CVR block diagram

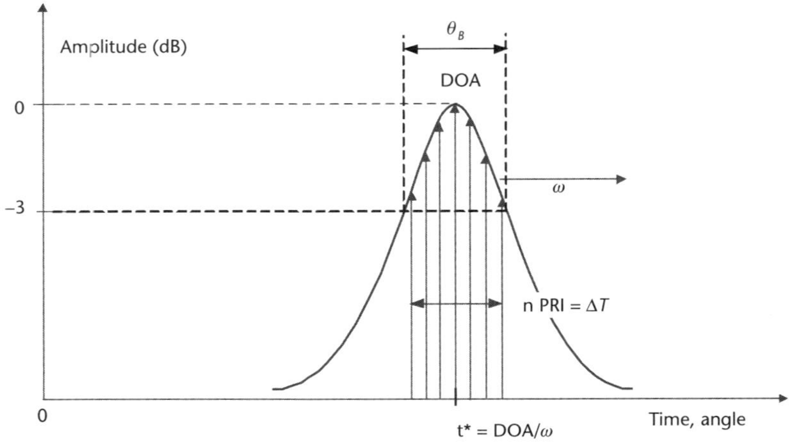

Figure 4.1 DOA measurement with a high-speed ($\omega = 360°/\text{s}$) rotating high-gain antenna ($\theta_B = 1°$; peak amplitude at t^*; DOA $= \omega t^*$; number of measured emitter pulses $n = \Delta T/\text{PRI}$, with $\Delta T = \theta_B/\omega$).

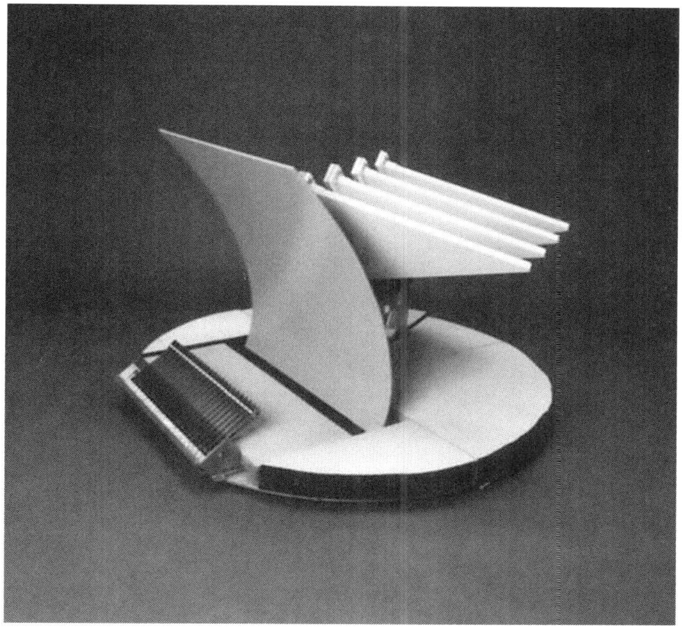

Figure 4.2 ELT 888 full-band (in two sub-bands) high-gain rotating antenna that exploits the monopulse technique within the beamwidth. (Courtesy of Elettronica S.p.A.)

are shown in Figures 4.3 and 4.4, respectively. The antennas usually employed in RWR equipment are planar spiral antennas as shown in Figure 4.5.

The gain pattern of a spiral antenna is reasonably well approximated by a Gaussian function, which in the azimuth plane can be expressed as:

$$G(\theta) = A_i^2 \exp[-k^2(\theta_i - \alpha)^2/\theta_B^2] \tag{4.1}$$

where

A_i is the composite antenna/receiver amplitude gain at peak. [Recall that because the CVR used in the scheme of Figure 4.2.4 uses square-law detectors followed by a lowpass filter, the amplitudes are measured as power levels ($P_i = A_i^2$).]

θ_B is the half power (–3 dB) beamwidth of the antenna pattern (usually for a four-antenna configuration, $\theta_B = 90°$ at the lower frequency end of the covered bandwidth).

$k^2 = -\ln(1/2) = 0.693$.

α is the "squint angle," that is, one-half of the angle between the boresight direction of any two antennas ($\alpha = 45°$ in the four-antenna configuration of Figure 4.3).

The beamwidth θ_B is nominally selected as 90°; however, actual antennas can exhibit beamwidths that vary with frequency from 110° at the lower end to 70° at the upper end of the bandwidth. Furthermore, the squint angle α can vary considerably with frequency. Both of the above-mentioned pattern variations can significantly affect the accuracy of wideband DF equipment. Typical spiral antenna

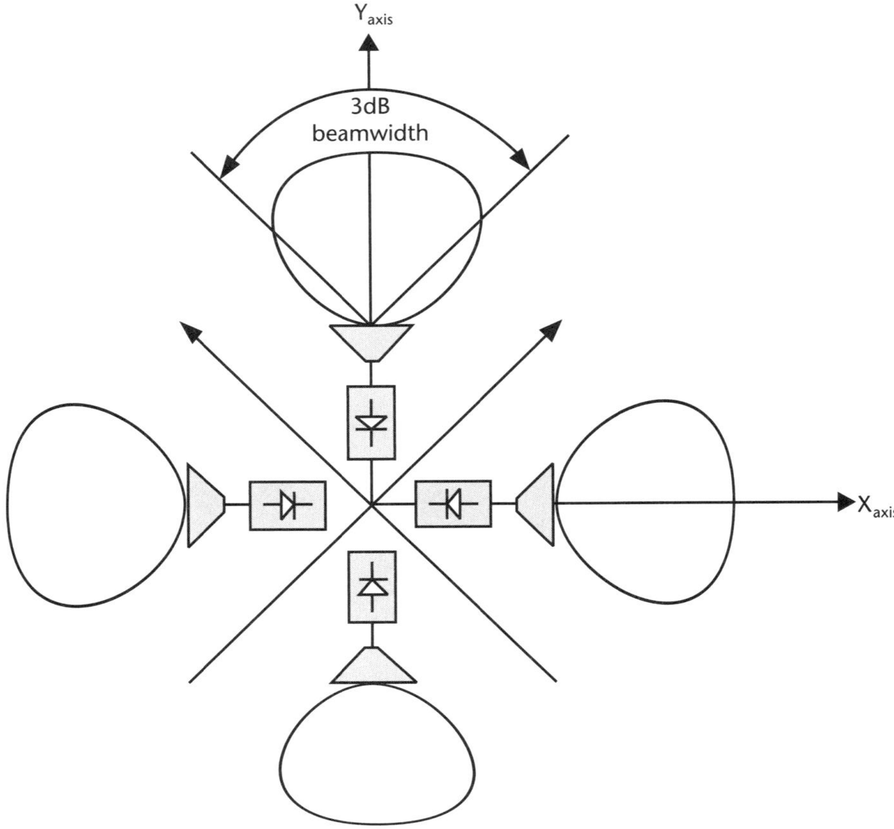

Figure 4.3 Arrangement of the four antennas for the DF amplitude comparison monopulse method.

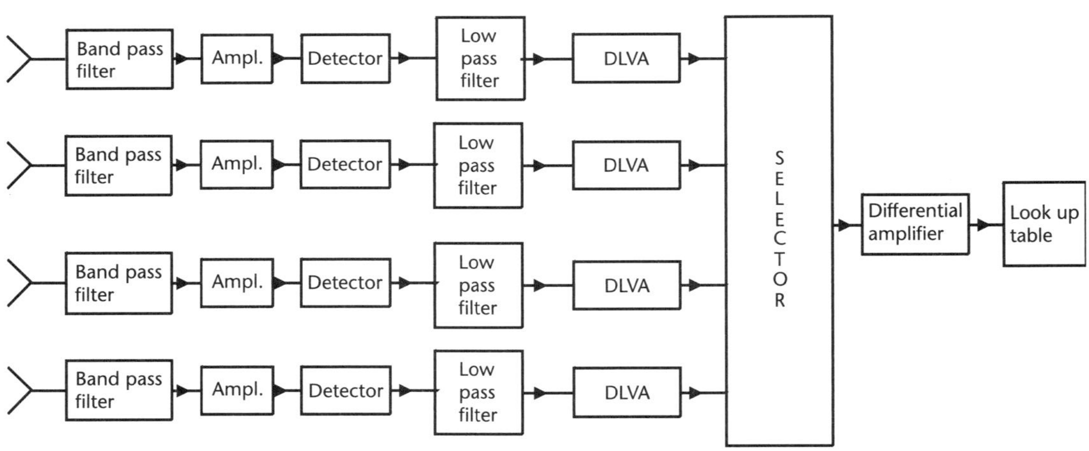

Figure 4.4 Block diagram of a RWR CVR with amplitude comparison monopulse (DLVA = detector logarithmic video amplifier).

beamwidth variations with frequency and polarization are reported in Figures 4.6(a) and (b).

(a) (b)

Figure 4.5 Typical RWR equipment planar antennas: (a) sinuous (both circular polarizations) and (b) spiral (only one circular polarization). (Courtesy of Elettronica S.p.A.)

The measurement of the emitter DOA (i.e., θ) is accomplished in two steps: First, the two antennas that receive the strongest and the next strongest signal amplitudes have to be selected (P_1 and P_2 in Figure 4.3) and thereafter the ratio of these two power levels after the logarithmic amplifiers has to be evaluated. The ratio R (expressed in decibels) of the two power levels after the logarithmic amplifiers is simply the difference between the two outputs:

$$\begin{aligned} R_{dB} &= 10\log\left(\frac{G_1(\theta)}{G_2(\theta)}\right) = 20\log\left(\frac{A_1}{A_2}\right) - 20\log e\cdot\left(\frac{k^2}{\theta_B^2}\right)\cdot[(\theta-\alpha)^2+(\theta+\alpha)^2] \\ &= 20\log\left(\frac{A_1}{A_2}\right) - \left(\frac{1.38}{\theta_B^2}\right)\cdot(4\alpha\theta)\cdot 20\log e = A_{1dB} - A_{2dB} + 48\cdot\frac{\alpha\theta}{\theta_B^2} \end{aligned} \tag{4.2}$$

Therefore, the ratio R_{dB} between the two strongest signals is proportional to the emitter DOA, apart from a bias introduced by the gain mismatch of the two receiving channels as dependent on frequency, which has to be corrected via a LUT. Hence (4.2) becomes:

$$\theta = R_{dB}(\theta_B^2/(48\alpha)) \tag{4.3}$$

The DOA accuracy as dependent from R_{dB} can be evaluated by taking the derivative

$$\theta = (\theta_B^2/(48\alpha))\delta R_{dB} \text{ or } \sigma_\theta = (\theta_B^2/(48\alpha))\sigma_{RdB} \tag{4.4}$$

As can be seen from (4.4) the error in the DOA estimate is constant with the angle θ because $\Delta R_{\text{dB}}/\Delta\theta$ is a constant. In the case of $\alpha = \theta_B/2$ the slope of the logarithmic power ratio is

$$\delta R_{dB}/\delta\theta = 24\,\text{dB}/\theta_B \text{ or } \delta R_{dB}/\delta\theta = 24\,\text{dB/beamwidth} \tag{4.5}$$

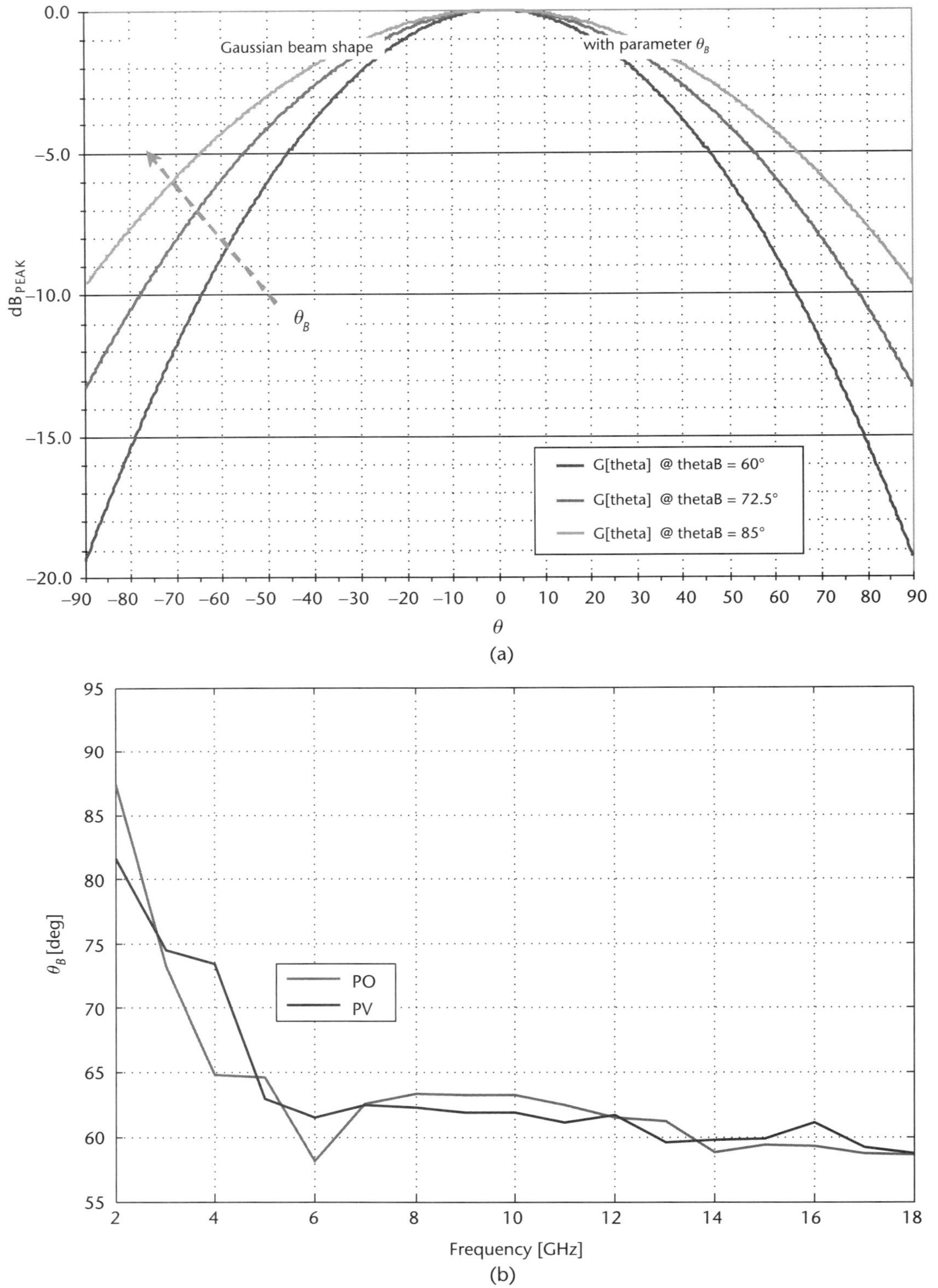

Figure 4.6 (a) Gaussian beam shape relative to beamwidth variations with frequency. (b) Measured spiral antenna beamwidth variations with frequency and polarization. (Courtesy of Elettronica S.p.A.)

There are several potential sources of error in an amplitude comparison DF system. One significant source of error arises from the gain mismatches between the channels that is quickly varying with the frequency and cannot be totally compensated for by a LUT, which corrects the mismatch at fixed frequency steps. Typically, the DF system is designed so that the maximum imbalance is limited to ±1 dB. Therefore, the peak angular error for a system with $\theta_B = 90°$ is, from (4.5), $\Delta\theta = 3.75° \Delta R_{dB} = \pm 3.75°$ or $\sigma_\theta = 2.16°$ rms.

Errors are caused, as mentioned before, by the variations with frequency of the beamwidth and the squint angle of the antennas, but these are generally compensated for with the data provided by the LUT. However the rms error due to further causes can be evaluated similarly as shown before, and these errors add to the one evaluated above.

Assuming no correlation among the various errors, the total error caused by variations in the antenna parameters can be expressed as:

$$\sigma_{\theta,ant} = (\sigma^2_{RdB} + \sigma^2_{\theta,\theta B} + \sigma^2_{\theta,\alpha})^{1/2} \tag{4.6}$$

where $\sigma_{\theta,\theta B}$ is the slope error associated with variations in beamwidth $\sigma_{\theta B}$ and $\sigma_{\theta,\alpha}$ is the slope error associated with the squint angle σ_α, which are, respectively [taking into account (4. 3) and (4.5):

$$\begin{aligned} \sigma_{\theta,\theta B} &= (\theta_B/(24\alpha))R_{dB}\sigma_{\theta B} = \sigma_{\theta B}/\alpha \\ \sigma_{\theta,\alpha} &= \sigma_\alpha/12 \end{aligned} \tag{4.7}$$

A further important error is caused by the receiver noise present in the channels, which affects the measured signal amplitude in each channel, and can be evaluated by expressing the change in the ratio R of (4.2) due to amplitude variations:

$$\delta R_{dB} = 20(\log e)(\delta A_1/A_1 - \delta A_2/A_2) \tag{4.8}$$

Since A_1 and A_2 are the peak signal voltages in the two channels, the rms error σR_{dB} is related to the SNR in each channel, that is:

$$\sigma^2_{A1}/A_1^2 = 1/(2SNR_1) \text{ and } \sigma^2_{A2}/A_1^2 = 1/(2SNR_2) \tag{4.9}$$

Therefore, assuming that the noise in the two channels is uncorrelated we have

$$\sigma_{RdB} = 20(\log e)[1/(2SNR_1) + 1/(2SNR_2)]^{1/2} \tag{4.10}$$

which in turn gives a rms angular error from (4.4) of

$$\sigma_\theta = \sigma_{RdB}/(\delta R_{dB}/\delta\theta) = (\theta_B^2/(48\alpha))20(\log e)[1/(2SNR_1) + 1/(2SNR_2)]^{1/2} \tag{4.11}$$

The SNR of each channel is proportional to the antenna pattern and the two SNR_i are equal only at the crossover point, where $G_1(\alpha) = G_2(\alpha) = 1/2$ and, hence, $SNR_1 = SNR_2 = ½(SNR_M)$, where SNR_M is the SNR at the antenna pattern peak, thus giving:

$$\sigma_\theta = (\theta_B^2/(48\alpha))20(\log e)[2/(SNR_M)]^{1/2} = 0.255(\theta_B^2/\alpha)[2/(SNR_M)]^{1/2} \tag{4.12}$$

which for $\alpha = \theta_B/2$ becomes

$$\sigma_\theta = 0.721\theta_B \,/\, (SNR_M)^{1/2} \tag{4.13}$$

As can be seen from (4.13), the rms angular error is a function of the squint angle α, so it is worth knowing which α is the optimum α that gives the least angular error. For a Gaussian antenna pattern, it can be shown [13] that $\alpha_{opt} = \theta_B/[2(\ln 2)^{1/2}] = 0.6\theta_B$ and $\sigma_{\theta\min} = 0.425/(SNR_M)^{1/2}$. As an example, in the case where we select $\alpha = 45°$, the beamwidth should be $\theta_B = 75°$ and, assuming $SNR_M = 20$ dB $= 100$, we have $\sigma_\theta = 3.75°$.

From (4.5) we can see that by increasing the number of antennas to cover the complete azimuth (i.e., by reducing θ_B and accordingly $\alpha = \theta_B/2$) the DOA accuracy is strongly improved. This is shown in Table 4.1 in which (4.13) is tabulated for two values of SNR_M, having taken into account the further contributions of (4.6).

In early RWR equipment, the amplitude comparison DF was performed by a method derived from the *x-y* display of an oscilloscope. With reference to Figure 4.4, the selector is substituted by two differential amplifiers, each of which is connected to a pair of antenna channels that are looking in opposite directions (refer to Figure 4.3) and that can be called, respectively, according to geographic coordinates north (N) and south (S), east (E) and west (W). Two signals are obtained N-S and E-W that represent the resultant amplitudes along two orthogonal axes (see Figure 4.7).

It can be shown that the measured emitter DOA is obtained as

$$\begin{aligned} DOA &= \tan^{-1}[(E-W)/(N-S)] \\ &= \tan^{-1}[(\log G_E(\theta) - \log G_W(\theta))/(\log G_N(\theta) - \log G_S(\theta))] \end{aligned} \tag{4.14}$$

Usually the two opposite antennas gains are negligible with respect to the gains of the two antennas relevant to the quadrant of the DOA. Therefore,

$$\begin{aligned} DOA &\sim \tan^{-1}[(E)/(N)] = \tan^{-1}[(\log G_E(\theta)/(\log G_N(\theta)] \\ &= \tan^{-1}[R_{dB}(\theta)] = \tan^{-1}[c\theta] \end{aligned} \tag{4.15}$$

Table 4.1 Amplitude Comparison Monopulse Accuracy (°) with an Increasing Number of Antennas Covering 360° Azimuth

Number of Antennas	*3 dB Beam Width (°)*	*Error for Antenna Parameters*	*Error for Noise at SNRM=13 dB*	*Total Accuracy (°)*	*Error for Noise at SNRM=20 dB*	*Total Accuracy (°)*
4	90	6.03	14.51	15.71	6.49	8.86
6	60	4.02	9.67	10.48	4.33	5.91
8	45	3.02	7.26	7.86	3.24	4.43
12	30	2.01	4.84	5.24	2.16	2.95

The errors relevant to variations of antenna patterns have been evaluated with actual data and the error relevant to SNR is evaluated in accordance with (4.13). Note that a well-calibrated LUT can reduce significantly the errors relevant to variations in antenna patterns.

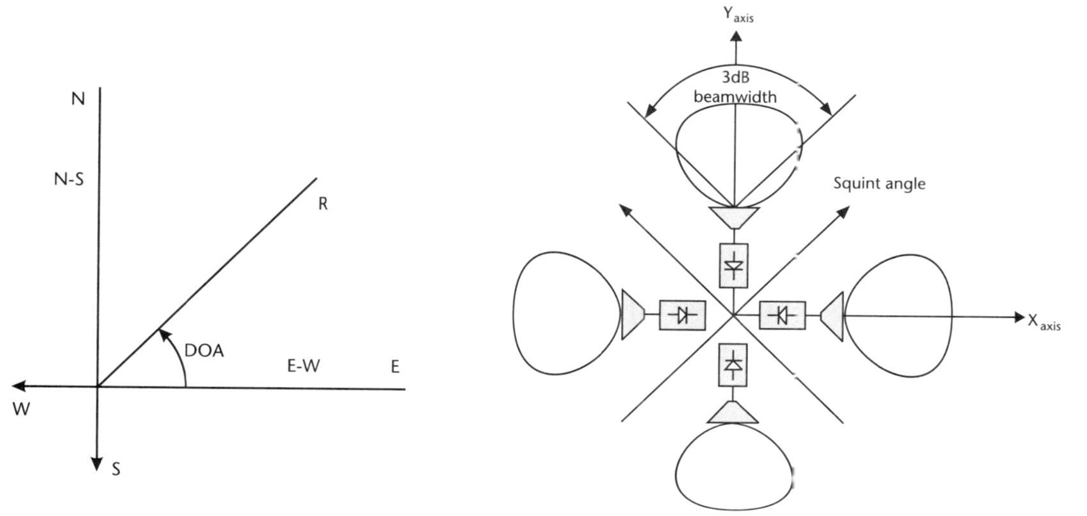

Figure 4.7 Early amplitude comparison monopulse method.

This method, therefore, was able to avoid the selection of the strongest and second strongest amplitudes to perform the ratio R_{dB}.

4.3 Phase Comparison Monopulse DF Measurement Methods

In phase comparison DF the DOA of a single-frequency (or narrow-bandwidth) RF waveform emitted by a source that is a long distance away is determined by measuring the phase differences between the signals received by a set of antennas displaced in various ways (along a line, a circle, a lattice, and so on). With reference to Figure 4.8, a set of $n + 1$ receiving antennas is displaced along a line, thus forming a *linear array*, at distances d_1, d_2, ..., d_n from a reference antenna. The phase differences can be expressed as

$$\Psi_i = (2\pi d_i/\lambda)\sin\theta = (2\pi d_i/c)f\sin\theta \quad \text{with } i = 1,2,\ldots,n \tag{4.16}$$

and can be measured by means of a phase correlator, for which the schematic block diagram is depicted in Figure 4.9.

The accuracy of the phase measurement is directly proportional to the errors in the baseline (Δd_i) (which is in general negligible except at millimeter waves), in the measurement of the carrier frequency (Δf) as well on the effects of the thermal noise. The dependence on the thermal noise can be evaluated from Figure 4.10, which shows the incremental phase change $\Delta\Psi$ due to a noise disturbance.

In fact from the triangle of phasors $\sqrt{S}$, $\sqrt{N}$, and $\sqrt{S'}$ we have (using the Carnot triangle theorem)

$$\cos\delta\psi = (|\underline{S}| + |\underline{S}'| - |\underline{N}|)/2\left|\sqrt{S}\sqrt{S'}\right|$$

For small $\Delta\Psi$ the cosine may be developed with McLaurin's series, which gives

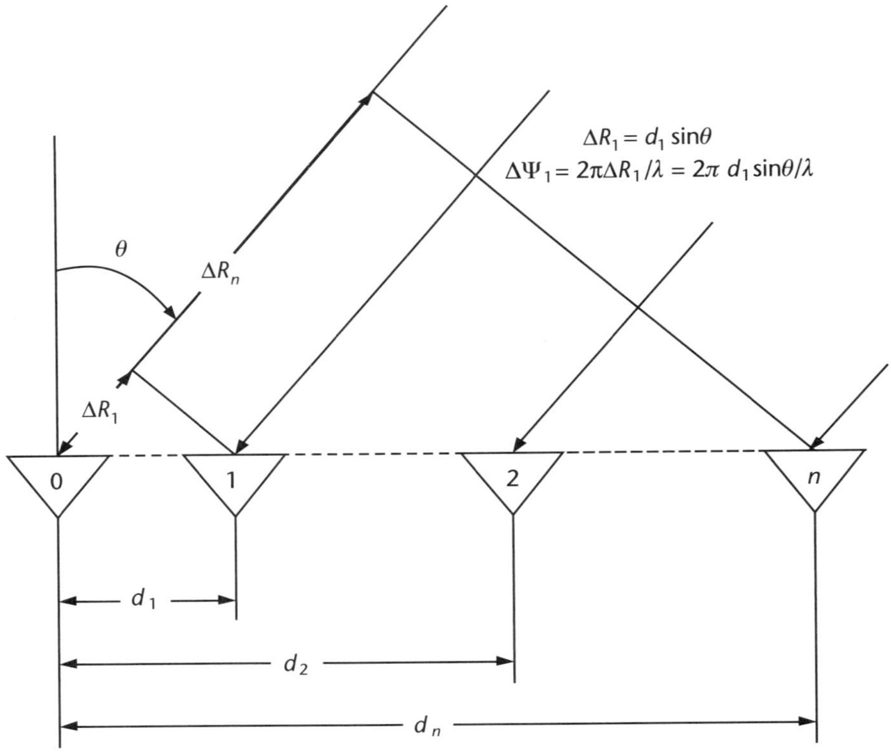

Figure 4.8 Principle of phase DF method (linear array).

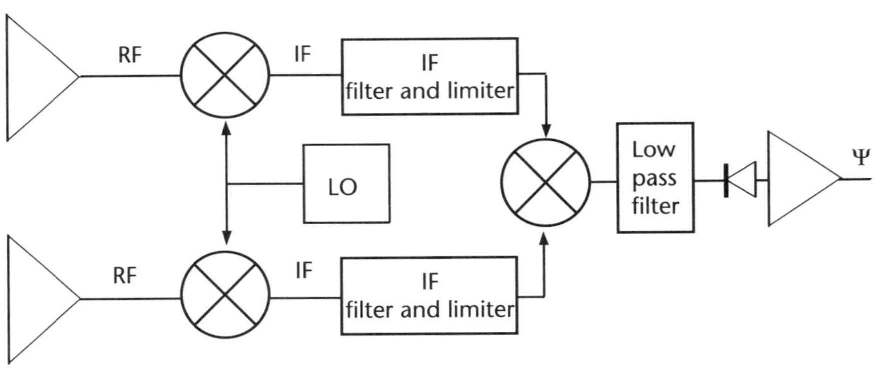

Figure 4.9 Block diagram of the phase correlator.

$$\cos\delta\psi \approx 1 - \delta\psi^2/2 \approx 1 - |\underline{N}|/2|\underline{S}| \quad \text{or } \sigma_\psi^2 = \delta\psi^2 = 1/(2SNR)$$

from which

$$\sigma_\psi = 1/\sqrt{(2SNR)} \tag{4.17}$$

Therefore, the angle accuracy can be expressed as:

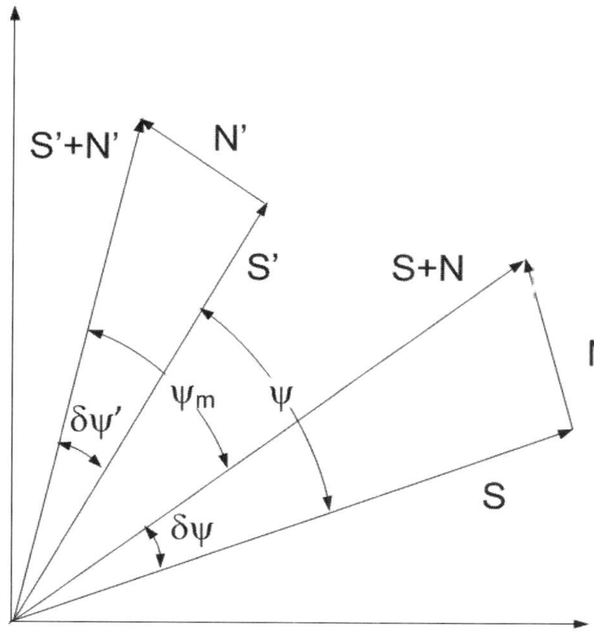

Figure 4.10 Phasor diagram to evaluate the noise error.

$$\sigma_\theta = \frac{c\sigma_\psi}{2\pi d_i f\cos\theta} = \frac{c}{2\pi d_i f\cos\theta\sqrt{2SNR}} \tag{4.18}$$

Phase comparison interferometers are the preferred choice for systems that require high DOA accuracy in that the error can be reduced with large d/λ. However, to reach this goal two important factors have to be taken into consideration: the *coning error* and the *phase difference ambiguity*.

The coning error is present in a linear array in that the phase difference Ψ_i in a 2-D angular world (azimuth θ and elevation ϕ) must be expressed as

$$\Psi_i = (2\pi d_i c) f \sin\theta\cos\varphi \tag{4.19}$$

and because the 1D array measures only $\Psi_i = (2\pi d_i/c) f \sin\vartheta^*$, by equating the two above expressions we have

$$\begin{aligned}&\sin\theta^* = \sin\theta\cos\varphi\\ &\text{or}\\ &\theta^* = \sin^{-1}(\sin\theta\cos\varphi)\end{aligned} \tag{4.20}$$

The coning error is graphed in Figure 4.11 as a function of true azimuth θ with elevation angle ϕ as a parameter.

The coning error limits the selection of a linear array for DOA (azimuth) up to elevation ≤10° (i.e., nearly at the horizon). The coning error can obviously be

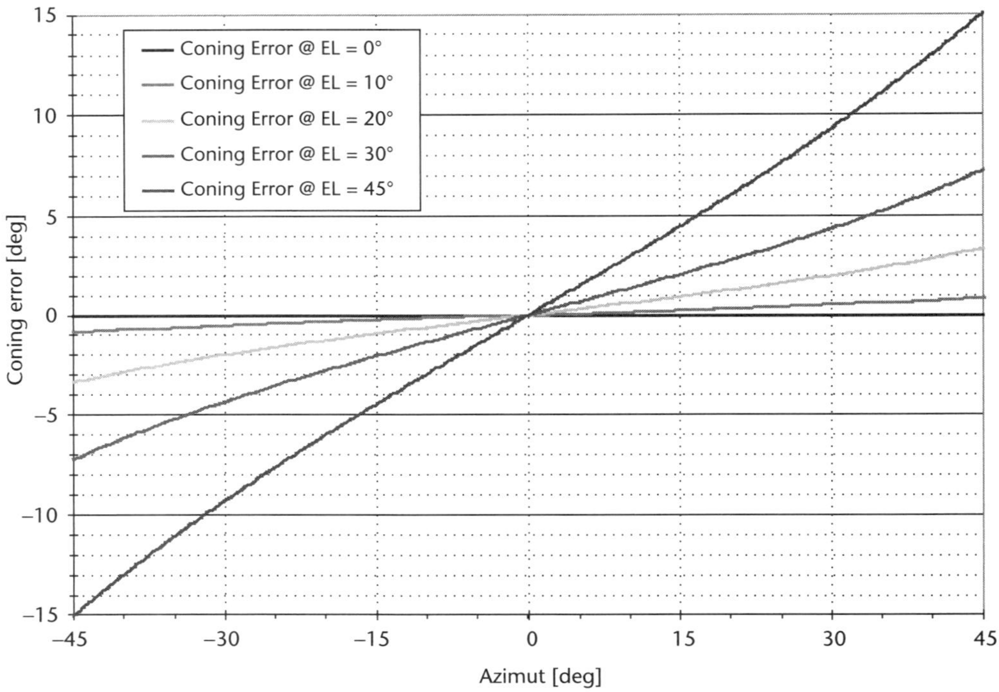

Figure 4.11 Azimuth coning error with elevation as a parameter.

removed by using a planar array antenna or two orthogonal linear arrays capable of measuring the elevation angle ϕ.

The phase difference ambiguity, which comes from the periodicity modulo $[2\pi]$ of the phase $\Psi_{i,}$, can be usually solved for a large-bandwidth (more than one octave) DF measurement either by adopting the smallest base length d_1 (also called the elementary base length) of the array, such that at the upper end of the bandwidth $\lambda_{min} = c/f_{max}$ we have:

$$\Psi i = (2\pi d_i/\lambda_{min})\sin\theta_{max} < \pm\pi \text{ or } d_1 = \lambda_{min}/(2\sin\theta_{max}) \tag{4.21}$$

The elementary base length is exploited either for linear harmonic arrays (i.e., arrays in which the base lengths are $d_i = 2^{i-1} d_1$) or for nonharmonic linear arrays in which the base lengths are prime number multiples of it (as an example, for a two-base array $D_1 = 2\ d_1$ and $D_2 = 3\ d_1$) and the phase ambiguous measurements are solved through the use of the Chinese remainders' method.

Harmonic linear arrays are selected whenever the coverage bandwidth is not larger than one octave because with the multiple bases they can provide highly accurate DF. In contrast, nonharmonic linear arrays are selected whenever the coverage bandwidth is larger than one octave [1, 2].

4.3.1 Correlative Phase DF

Communications signal DF in the VHF/UHF bands over the complete azimuth coverage either on ground/shipboard or in aircraft CESM equipment is usually performed by exploiting phase DF applied between the vertical dipoles of a uniform circular array (UCA). An odd number (3, 5, or 7) of dipoles is selected in order to avoid DOA ambiguities with frequency and thus increasing the measurement bandwidth. In what follows we consider, without loss of generality to a different odd number of dipoles, a five-dipole UCA, which is shown in Figure 4.12 together with its geometrical coordinates.

A circular array can provide both amplitude and phase measurements, which are exploited in the extended phase interferometry (EPI) technique as shown in [3], which offers the potential of applying high-resolution DOA techniques on multiple signals (based on the MUSIC algorithm) although at the expenses of some computational burden at each receiving channel. In this book we consider the results relevant to phase measurements only, which are the most commonly used in CESM equipment.

The signals received by each antenna of the array can be compared in any of several ways. For example, one antenna can be used as a reference and all of the other received signals can be compared to the one received by the reference antenna [3]; or we could simply measure the differential phase between adjacent antennas. Here, we use this second method, which provides the following phase measurements vector:

$$\underline{P}^{(m)T}[p_{1,2}^{(m)}; p_{2,3}^{(m)}; p_{3,4}^{(m)}; p_{4,5}^{(m)}; p_{5,1}^{(m)}] \tag{4.22}$$

where

$$p_{k,k+1}^{(m)} = (2\pi L/\lambda)\cos(\varphi)\sin[\theta + 2(k-1)\pi/5] + n_{k+1} + n_k \tag{4.23}$$

where

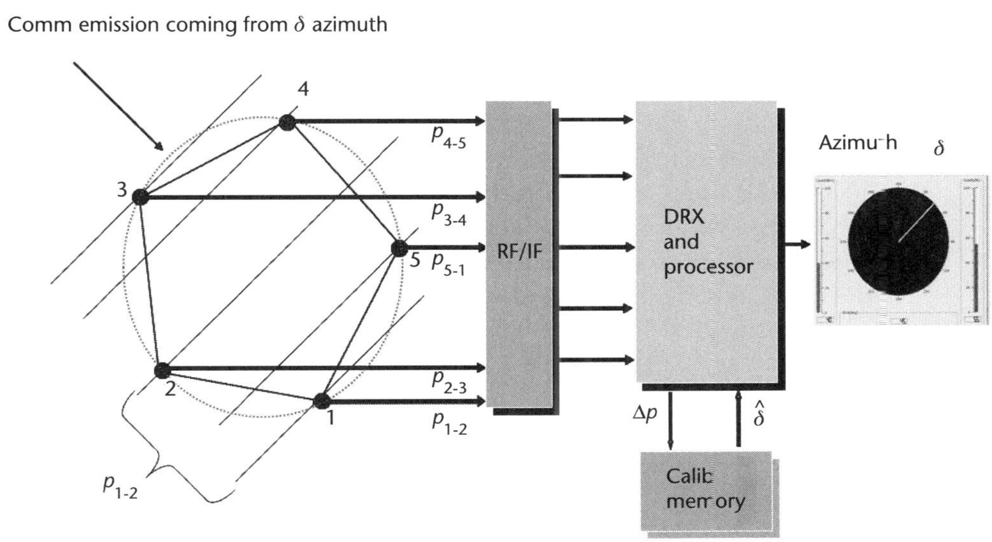

Figure 4.12 Five-dipole UCA and relevant geometry.

θ is the azimuth angle with respect to the radial axis from the center of the UCA to one element (element 1),

ϕ is the elevation angle with respect to the vertical axis passing from the center of the UCA,

n k + 1, n k are the thermal noise contributions to the phase measurements in the relevant receiving channels.

Because the phase responses of the antennas can be very different from the theoretical ones, it is necessary to estimate/measure during the equipment setup the nonideal behavior of the actual antennas and take it into account in the determination of the DOA, through a "correlation" technique, hence the name *correlative DF*. The correlative DF technique consists of two phases: correlation and interpolation.

4.3.1.1 First Phase: Correlation

This phase involves execution of a calibration procedure in which an RF signal is sent to the antenna receivers from a set of positions, displaced at regular steps $\Delta\theta$ in azimuth (and possibly also in elevation at steps $\Delta\phi$). This procedure is repeated at regular steps Δf in frequency within the whole coverage bandwidth of the equipment.

The measured data (i.e., differential phases and amplitudes in case of EPI) are stored in a calibration table, where each entry relevant to a frequency step is a vector $\underline{v}(i\ \Delta\theta;\ k\ \Delta\phi)$ (dimension 5×1) for which the lth component can be represented as $B_l{}^{(i,k)} \exp(jp_{l,l+1}{}^{(i,k)})$ for $l = 1, 2, ..., 5$.

The DOA estimate is performed in two steps during equipment operation. The first step is the comparison (correlation) of actual measured RF signal differential phases (and amplitudes in case of EPI) with the stored data of the calibration table. We represent the RF signal measured amplitudes and differential phases with the vector $\underline{u}$ as follows:

$$u^T = \{A_1 \exp(jp_{1,2}^{(m)}); A_2 \exp(jp_{2,3}^{(m)}); A_3 \exp(jp_{3,4}^{(m)}); A_4 \exp(jp_{4,5}^{(m)}); A_5(jp_{5,1}^{(m)})\} \tag{4.24}$$

Then the first task is the search for the entry that has the data that are most similar to the above vector. Because phase measurement accuracies are dependent on SNR, the measured amplitudes of the received signal must first be normalized (i.e., each amplitude A_i is divided by the modulus of the vector $\|\underline{u}\| = \sqrt{\Sigma_k A^2{}_k}$ to perform a good comparison with the entries of the calibration table).

The normalized received signal vector, $\underline{u}' = u/\|u\|$, whose lth component can be expressed as $C_l \exp(j\ p^{(m)}{}_{l,l+1})$, is then correlated with the likely entries of the calibration table:

$$Corr(u', v(i\Delta\theta; k\Delta\varphi)) = \sum{}_l B_l C_l \cos(p_{l,l+1}^{(m)} - p_{l,l+1}^{(i,k)}) \tag{4.25}$$

The entry $v\ (i\ \Delta\theta;\ k\ \Delta\phi)$, which maximizes Corr $(u',\ v\ (i\ \Delta\theta;\ k\ \Delta\phi))$, determines the estimate DOA* = $(i\ \Delta\theta;\ k\ \Delta\phi)$.

It can be shown that, as the distance of two vectors is defined as

$$d(\underline{u}',\underline{v}) = \sqrt{\left(2 - 2\sum_l B_l C_l \cos(p_{l,l+1}^{(m)} - p_{l,l+1}^{(i,k)})\right)} = \sqrt{\left(2 - 2\,Corr(\underline{u}', \underline{v}(i\Delta\theta; k\Delta\varphi)\right)} \quad (4.26)$$

the entry $\underline{v}(i\ \Delta\theta;\ k\Delta\phi)$ that maximizes the correlation also minimizes the distance between the two vectors (i.e., it is the most likely representation of the measurement vector).

The correlation function becomes increasingly rippled as L/λ increases and, if $\Delta\theta$ and $\Delta\phi$ are not properly selected, its maximization could provide incorrect DOA* estimates. Hence the necessity of the second step discussed next.

4.3.1.2 Second Phase: Interpolation

Once the estimated DOA* = $(i\ \Delta\theta;\ k\ \Delta\phi)$ has been obtained, an interpolation is performed between the adjacent entries of the calibration table along two orthogonal cuts (azimuth and elevation) using polynomial interpolation, as shown in Figure 4.13, in order to minimize the systematic error due to the quantization steps of the calibration table between the true DOA(θ_e, ϕ_e) and the estimated DOA*, that is, to minimize the following errors in azimuth and elevation:

$$\varepsilon_\theta^i = i\Delta\theta - \theta_e \ \text{ or } \ \varepsilon_\varphi^k = k\Delta\varphi - \varphi_e \quad (4.27)$$

It can be shown [3] that in proximity of its maximum (i.e., DOA*) the correlation function of phase interferometry vectors is constituted by two terms: a deterministic term and a statistical term $\zeta^{(i,k)}$ due to the thermal noise in the five DF channels:

$$Corr(\underline{p}^{(m)}, \underline{p}^{(i,k)}) = 5 - 5(\pi L/\lambda)^2 F(i\Delta\theta, \theta_e, \varepsilon_\varphi^l) - \zeta^{(i,k)} \quad (4.28)$$

where the function F in the deterministic term is expressed as:

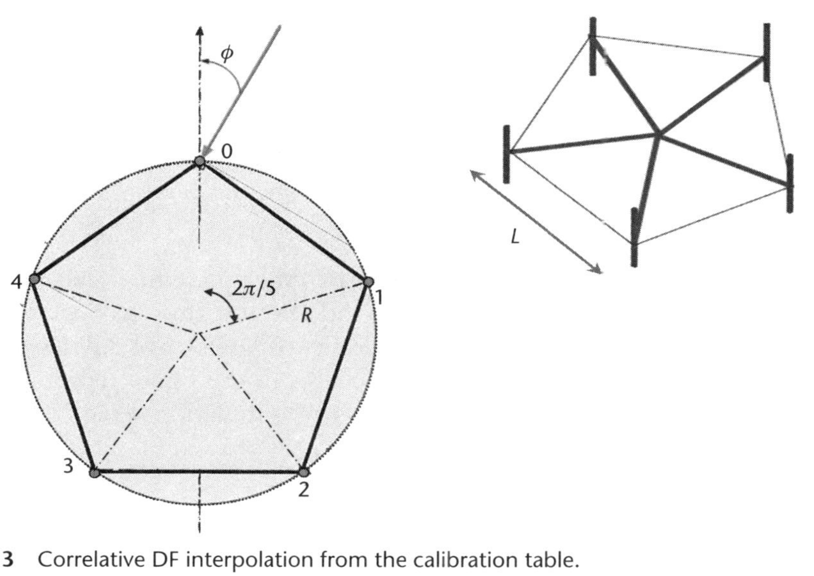

Figure 4.13 Correlative DF interpolation from the calibration table.

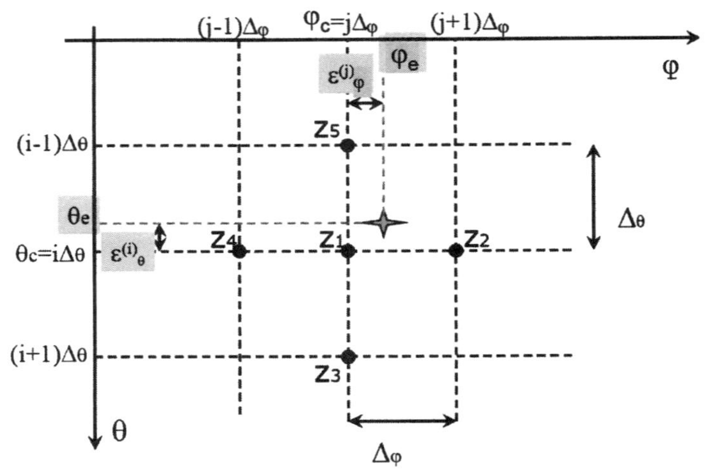

Figure 4.14 Correlative DF interpolation from the calibration table.

$$F(i\Delta\theta,\theta_e,\varepsilon_\varphi^k = (\cos(i\Delta\theta)-\cos(\theta_e))^2 + 2\cos(i\Delta\theta)\cos(\theta_e)(1-\cos(\varepsilon_\varphi^k)) \tag{4.29}$$

and the statistical term $\zeta^{(i,k)}$ is expressed as

$$\zeta^{(i,k)} = \sum\nolimits_l (n_1^2 - n_{1+1}n_1 + n_1(T_{l-1}^{(i,k)} - T_1^{(i,k)})) \tag{4.30}$$

where

$$T_1^{(i,k)} = 2(\pi L/\lambda)(\cos\theta_e)\sin(\varphi_e-(l-1)(2\pi/5)) - \cos(i\Delta\theta)\sin(k\Delta\varphi-(l-1)(2\pi/5)) \tag{4.31}$$

As is evident from (4.28), the correlation function does not depend on ϕ_e and an azimuth error can be produced by the contribution of $\varepsilon^k\phi$, which has to be minimized through the polynomial interpolation in both azimuth and elevation cuts.

The cuts of the function F [4] in azimuth are shown in Figure 4.15 versus the elevation error with a fixed azimuth error of 5° and in Figure 4.16 versus elevation with no elevation error and parameter θ_e.

After the above procedure the DOA estimate is affected mainly by the error contribution due to thermal noise, which can be expressed as

$$\sigma_{\theta rms} = ((\sqrt{10})\sin(\pi/5))\sqrt{(1/(2SNR) + (Q^2/12))} \,/\, ((5\pi L/\lambda)\cos(\varphi_e)\cos^2(\varepsilon_\theta^i)) \tag{4.32}$$

where Q is the phase quantization step in the differential phase measurements.

In Figure 4.17, taken from [4], the azimuth rms error vs. the elevation angle is reported for various SNR values for a UCA with five dipoles operating in the D band with $L/\lambda = 0.75$. The dashed curves in the figure represent the approximation of (4.32), whereas the solid curves are simulation results for a number of SNR values.

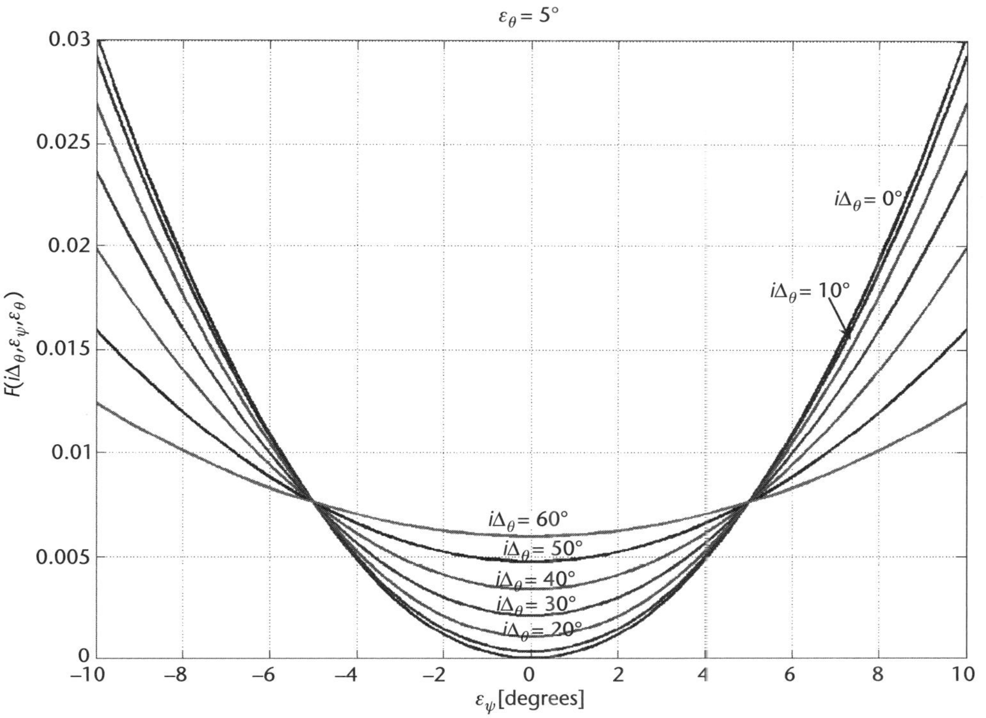

Figure 4.15 *F* function plot vs. elevation with 5° azimuth error.

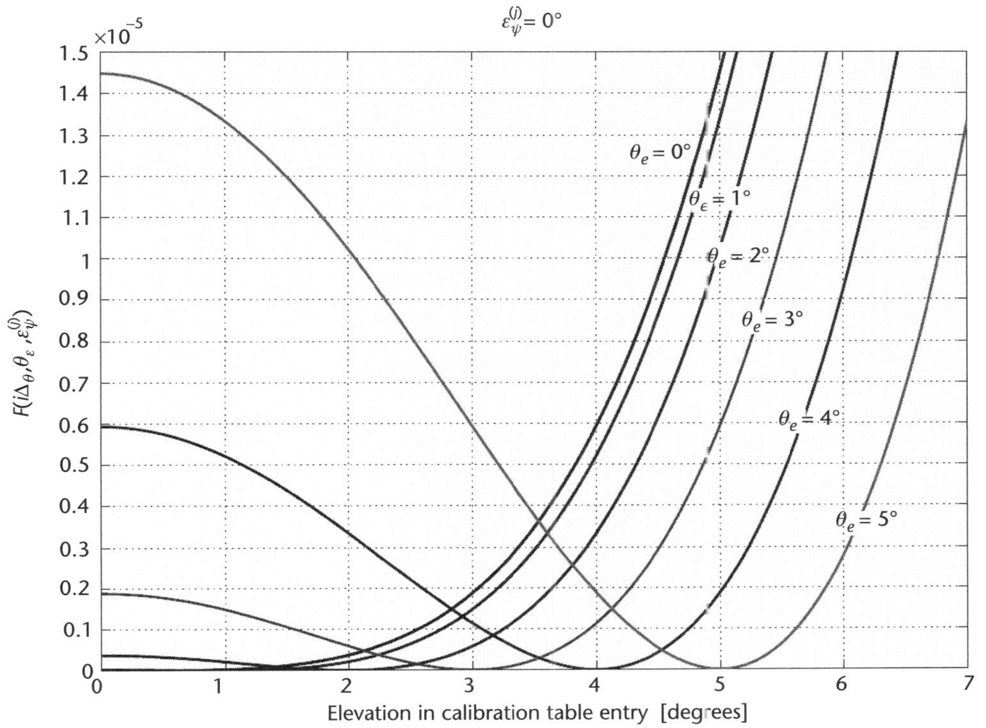

Figure 4.16 *F* function plot for small azimuth angles vs. elevation with no elevation error.

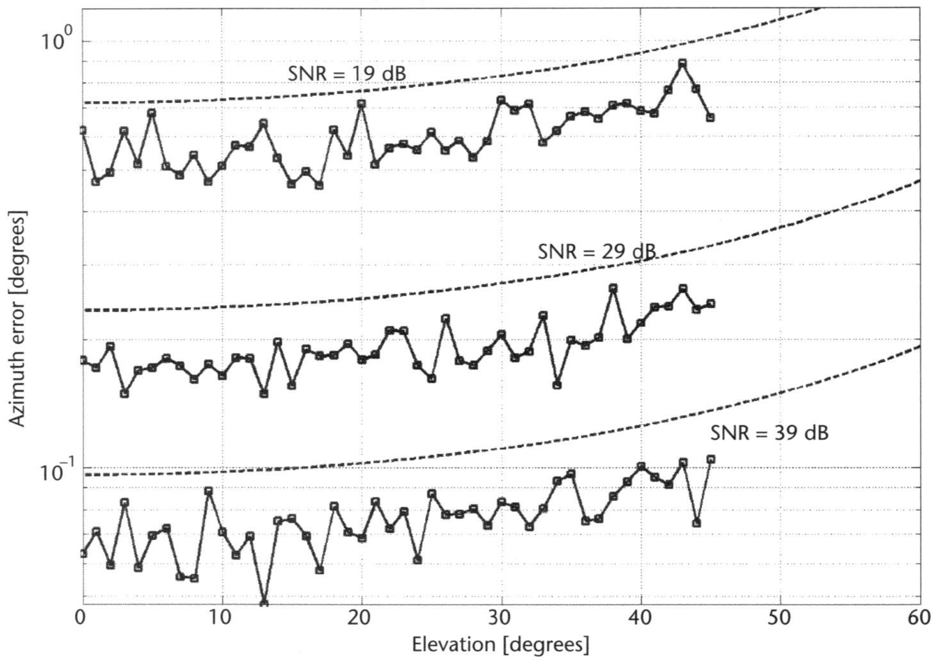

Figure 4.17 Azimuth error (rms) of five-dipole UCA vs. elevation for various SNRs.

4.4 Time-Difference DF

The principle of time difference of arrival (TDOA) measurements is depicted in Figure 4.18 where two strictly synchronized receiving antennas/channels are placed at a distance B, called the base length, from each other. The emitter signal wavefront, with the emitter supposedly at elevation 0° and at a distance R that is much larger

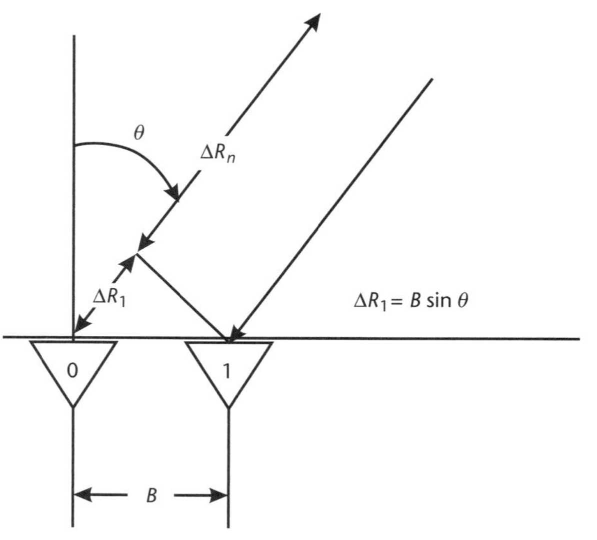

Figure 4.18 Principle of TDOA, $\Delta t = (B/c)\sin(\theta)$.

than the base length, is a plane wave tilted at an angle θ with respect to the baseline of the two receiving channels. In these conditions the delay Δt = TDOA with which the wavefront arrives at the far antenna with respect to the near one is:

$$\Delta t = (B/c)\sin(\theta) \tag{4.33}$$

where c is the light speed ($c = 3 \times 10^8$ m/s). For example, assuming $B = 14$m and $\theta = 45°$ [i.e., sin $(\theta) = 0.707$], we have $\Delta t = 33$ ns. By inverting (4.33) we can immediately derive the angle θ as follows:

$$\theta = \sin^{-1}(\Delta t(c/B)) \tag{4.34}$$

The difference between the arrival time of the same signal at two widely separated (airborne or ground-based) sites has been exploited in many radionavigation aids because the TDOA measurement establishes an isodelay curve (a hyperbola) on the surface of the Earth that passes through the emitter location.

The accuracy of the angle measurement is derived from (4.34) by taking the derivatives of both terms, and indicating with $\Delta(x)$ the error in the measurement of the parameter x:

$$\delta\theta = (c/B\cos(\theta))\delta\Delta t \tag{4.35}$$

As can be seen, the accuracy is inversely proportional to the projected base length B cos(θ) and proportional to the accuracy of the TDOA measurement. In the past, with quite a large $\Delta\Delta t$ error (on the order of hundred of nanoseconds), the technique was used with large base length B.

The current availability of digital receivers, which operate at very high sampling rates (larger than 1 GHz), allows a $\Delta\,\Delta t$ error of less than 1 ns to be achieved. Therefore, the technique can currently be used on board platforms such as ships, large-body A/C, and even ground stations with base length B on the order of 10m. As an example, by assuming $\Delta\Delta t = 1$ ns, $B = 10$m, and $\theta = 22.5°$, we obtain from (4.35) $\Delta\theta \approx 1.9°$. This error can be further reduced by some smoothing algorithm applied on a number of successive TDOA measurements.

The advantage of the TDOA measurement technique in achieving a moderate DOA accuracy is its independence from the signal carrier frequency and hence the absence, when covering the complete RF bandwidth, of the number of antenna arrays required by the phase DF measurement technique.

An example of implementation of the TDOA measurement technique is shown in Figures 4.19(a) and (b), which show, respectively, the block diagram of a double-channel SH receiver and a photograph of a dual DRx board with a common clock. The double-channel architecture is fundamental for differential measurements; such measurements can be carried out in amplitude, phase, or time, depending on the application, as discussed later relevant to further DF techniques. An antenna arrangement on board a ship that provides TDOA measurements over the complete azimuth and frequency coverage is shown in Figure 4.20.

Equation (4.35) is strictly valid in the case where elevation = 0°, which provides a reasonable azimuth estimate for emitters that are very far away from the baseline. Of course, when applied on a single platform that can come near the emitter, the

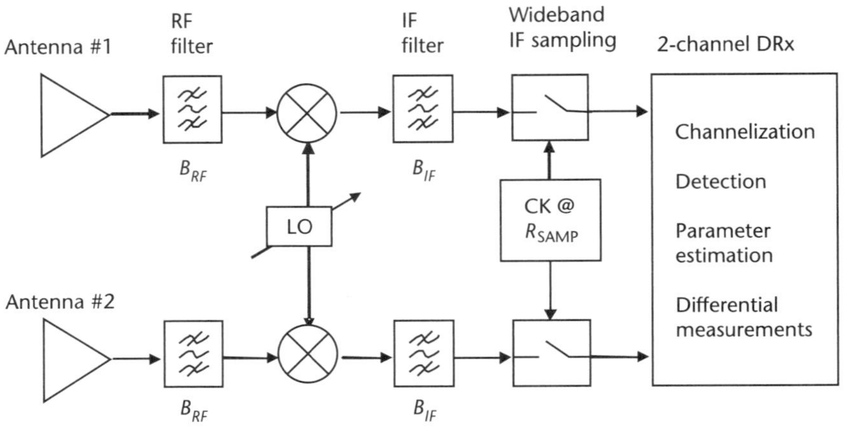

(a)

(b)

Figure 4.19 (a) Block diagram of a double-channel SH receiver. (b) A dual DRx board with common clock. (Courtesy of Elettronica S.p.A.)

Figure 4.20 Antenna arrangement on board a ship. (Courtesy of Elettronica S.p.A.)

coning error, due to emitter elevation ≠ 0°, largely worsens the azimuth measurement accuracy. Reduction of the coning error can be achieved with an additional orthogonal baseline, either vertical (in order to measure the elevation angle) or a further equal horizontal one as shown in Figure 4.20. In this latter case, by performing a TDOA measurement from each baseline we obtain, as shown in Figure 4.21, the following:

$$\Delta t_1 = (B/c)\sin(Az)\cos(El) \text{ and } \Delta t_2 = (B/c)\cos(Az)\cos(El) \tag{4.36}$$

Thus, by taking the ratio of the two above measurements, we get a fair estimate of the azimuth angle:

$$Az^* = \tan^{-1}(\Delta t_1/\Delta t_2) \tag{4.37}$$

with an accuracy of

$$\delta Az^* = (c/B\cos(El))\delta\Delta t \tag{4.38}$$

Thus, in order to achieve a fair DOA accuracy over the whole azimuth coverage, four orthogonal baselines, forming a square of side B, are provided. This arrangement uses also the two diagonals (with baselines $B\sqrt{2}$) and provides good rms azimuth (on the order of 1°) accuracy for the entire frequency coverage.

Two methods can be employed for the TDOA measurement, depending on the type of emitter signal: (1) leading edge TDOA (LE-TDOA) in the case of pulsed emitters, thus exploiting the differential between the pulse rise time; and (2) signal correlation (SC-TDOA) in the case of CW-type emitters.

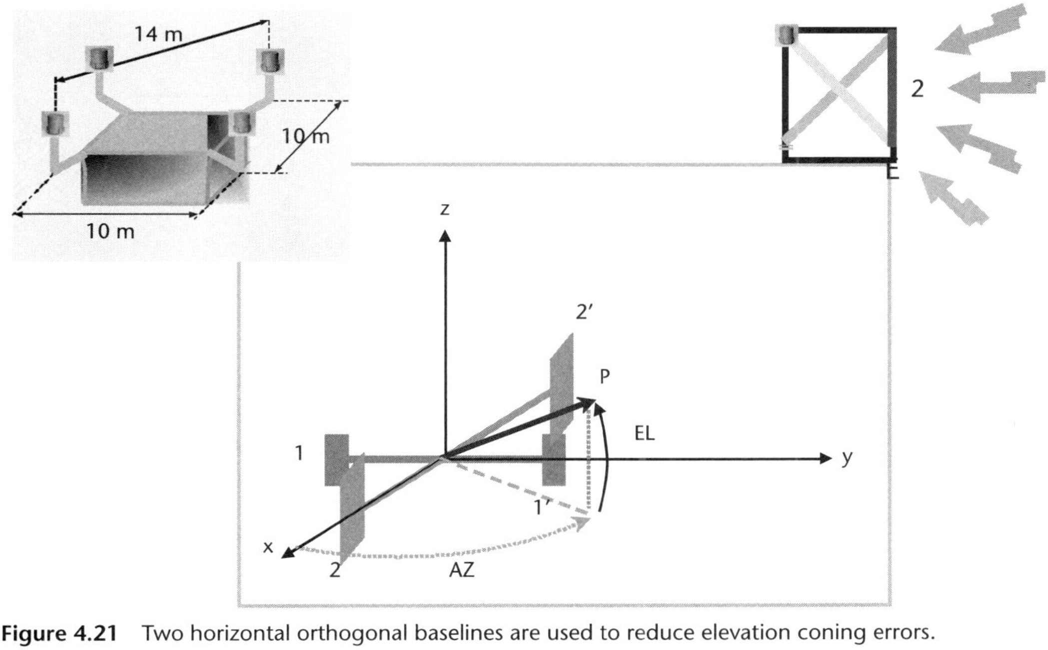

Figure 4.21 Two horizontal orthogonal baselines are used to reduce elevation coning errors.

The principle of the LE-TDOA method is depicted in Figure 4.22, which shows that Δt is measured during the rise time Tr of the pulses received from each antenna at half of the maximum amplitude $A_0/2$.

The accuracy of the TDOA measurement is given by [5] as:

$$\sigma_{\Delta t} = Tr/\sqrt{SNR} \tag{4.39}$$

For conventional pulse radars [6] $Tr = 1/B_{opt}$, where B_{opt} is the bandwidth of the matched filter to the transmitted pulse. In the case of the ESM receiver, the reception bandwidth Br is different from B_{opt} because it has to be appropriately selected for the reception of the pulses from many emitters, as shown in Figure 4.23. The behavior of σ_{TDOA} vs. Br, which can be approximated by assuming

$$Tr' = (1/\sqrt{SNR})\sqrt{(Tr^2 + 1/Br^2} \tag{4.40}$$

is reported in Figure 4.24.

The SC-TDOA measurement functional block diagram is depicted in Figure 4.25. Its measurement accuracy is evaluated as follows [9]:

$$\sigma_{SC-TDOA} \approx \frac{1}{\sqrt{B_N \cdot T_{CORR}} \cdot 2\pi \cdot B_{EQ}} \cdot \sqrt{\frac{2}{SNR} + \frac{1}{SNR^2} \cdot \frac{1}{\left(\sqrt{2} \cdot 2\pi \cdot B_{EQ}\Delta T\right)^2}} \tag{4.41}$$

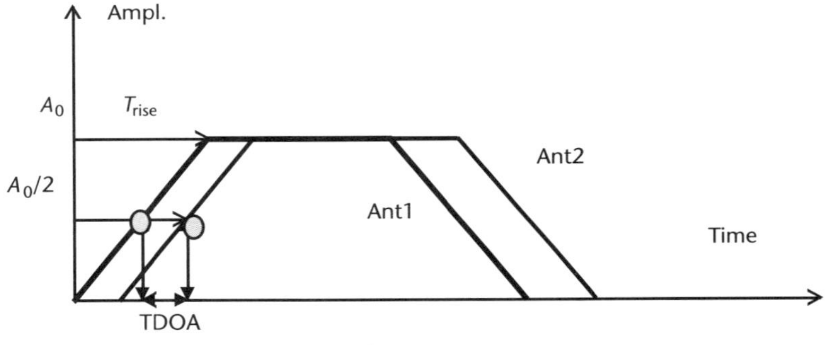

Figure 4.22 LE-TDOA measurement principle.

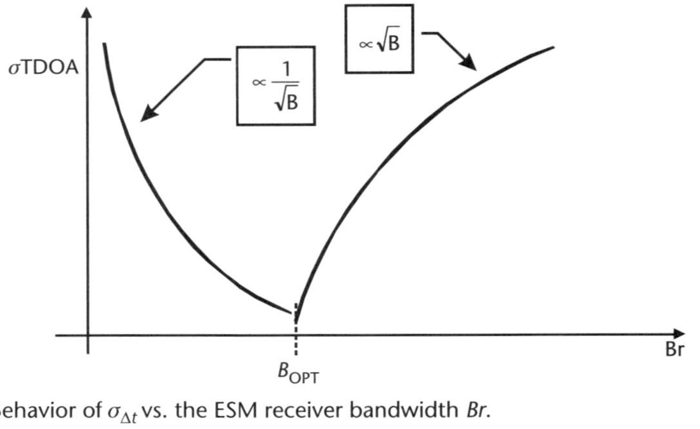

Figure 4.23 Behavior of $\sigma_{\Delta t}$ vs. the ESM receiver bandwidth Br.

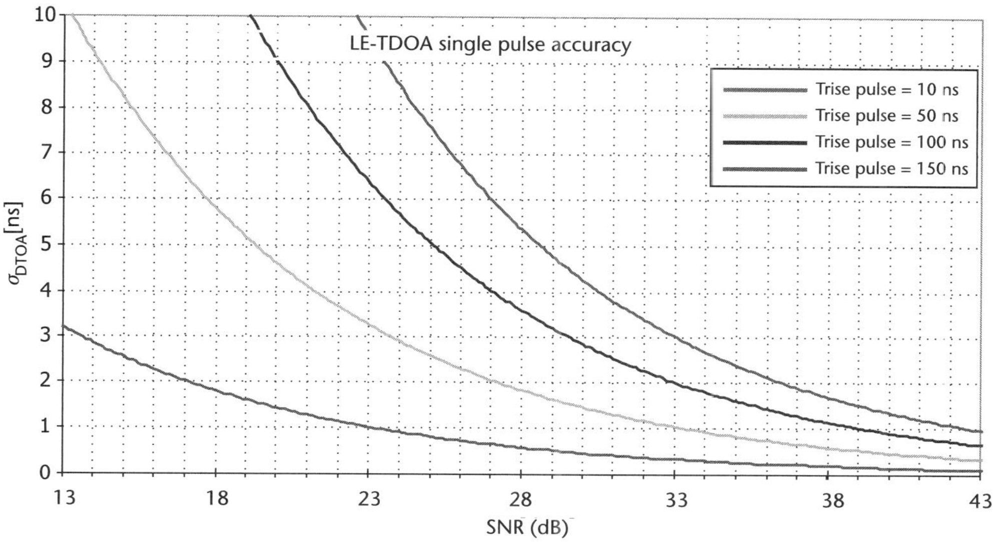

Figure 4.24 Behavior of LE-TDOA accuracy σLE-TDOA vs. SNR.

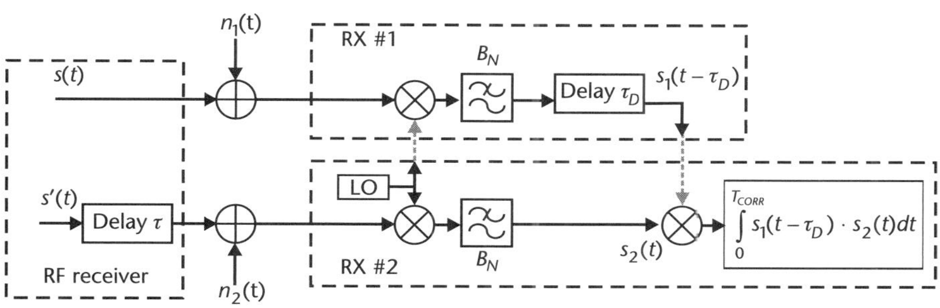

Figure 4.25 Functional block diagram of cross-correlation receiver.

where

B_N = noise bandwidth of the ESM receiver (many MHz);

T_{CORR} = total time of N_{CORR} correlation samples (order of ms);

B_{EQ} = signal bandwidth; $B_{EQ} \approx$ actual bandwidth of CW and PMOP signal modulations; $B_{EQ} \approx 1/\pi\sqrt{(2Tr' \text{ PW})}$ for pulsed signals, where Tr is the rise time and PW is the pulse width (order of MHz);

ΔT period of the sample estimation (order of 100 ns).

The accuracy of the SC-TDOA measurement technique is shown in Figure 4.26.

The advantage of the cross-correlation technique is that it is valid for every type of signal and exploits the large bandwidth B_{EQ}, especially when the signal is either FMOP or PMOP modulated.

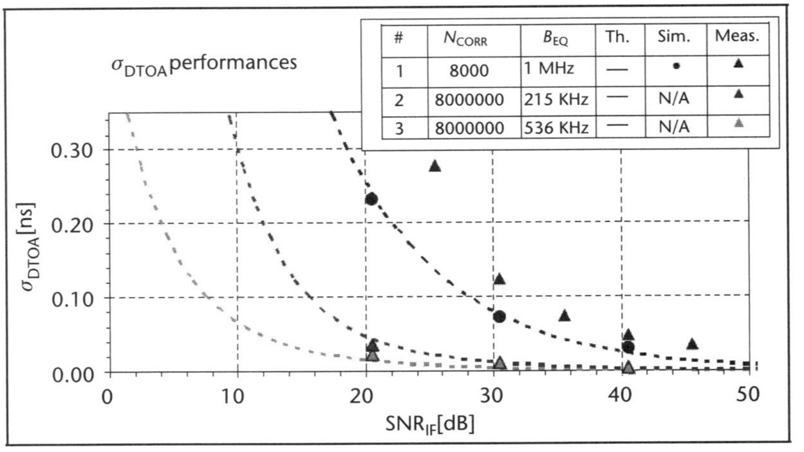

Figure 4.26 Behavior of $\sigma_{SC\text{-}TDOA}$ vs. SNR. (Reprinted with permission of R. Ardoino et al. [9].)

At very low sensitivity levels the SC-TDOA technique is prone to multipath close to the receiving antennas as well as to spurii signals generated by the receiver channels, which can produce erroneous DOA estimates.

Application of super-resolution techniques based on the MUSIC (Multiple Signals Correlation) algorithm allows for discrimination of interfering signals and ensures the SC-TDOA technique's accuracy. An example of this application on a chirped signal with a 2-MHz bandwidth at low SNR is given in Figure 4.27.

4.5 Emitter Location

Determining the geoposition of emitters is a fundamental function of EW systems, because knowledge about their location indicates the disposition of adversarial forces in their territory (electronic order of battle, EOB). In this section we focus on the physics of the measurement methods available with EW systems, following the approach given in [5] and recommending that the interested reader who desires deeper insight into the topic review the new computing methods described in [9].

Determination of the location of an emitter, called a position fix, can be performed using either of the following methods:

1. Measurement of the line of bearing (LOB) of the emitter (in practice, only the azimuth angle of arrival of its signal) from two or more EW intercept sensors. This technique is referred to as *triangulation*.
2. Measurement of the TDOA of the signal at two or more EW intercept sensors. This technique is referred to as *trilateration*. Usually, if the sensors are placed on mobile platforms, in addition to the TDOA, a differential Doppler (DD) measurement can also be performed, providing better location accuracy.

The trilateration and DD methods have been known for a long time and they were executed with intercept sensors placed at very far distances from each other,

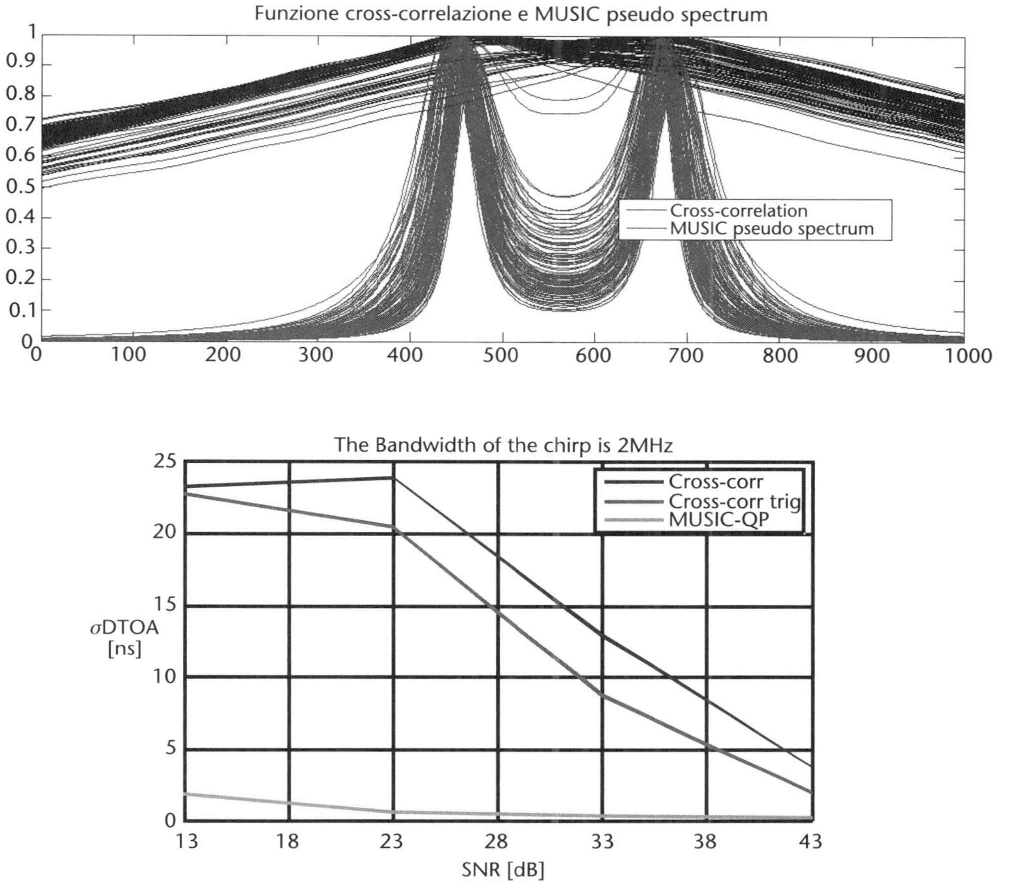

Figure 4.27 Super-resolution technique applied to SC-TDOA. (Courtesy of G. Panci [8] and Elettronica.)

so as to form large baselines that could provide very good location accuracy by overcoming the poor accuracy of the measurements of the time and Doppler frequency differences. Currently, the evolution of electronic technology has allowed highly accurate time and Doppler frequency difference measurement methods to be implemented, which in turn allows for emitter geolocation within a single mobile platform to be provided, as discussed in the following sections. Also discussed is the locating of an airborne emitter from a steady or slow-moving platform.

4.5.1 Triangulation

The DOAs relevant to the same emitter, as provided by two or more EW intercept sensors placed at a distance B (the baseline) from each other, allow for the determination of the geolocation of that emitter through basic triangle geometry (hence, the name triangulation). The principle of the method is shown in Figure 4.28.

The geoposition of emitter E can be evaluated in a straightforward manner as follows:

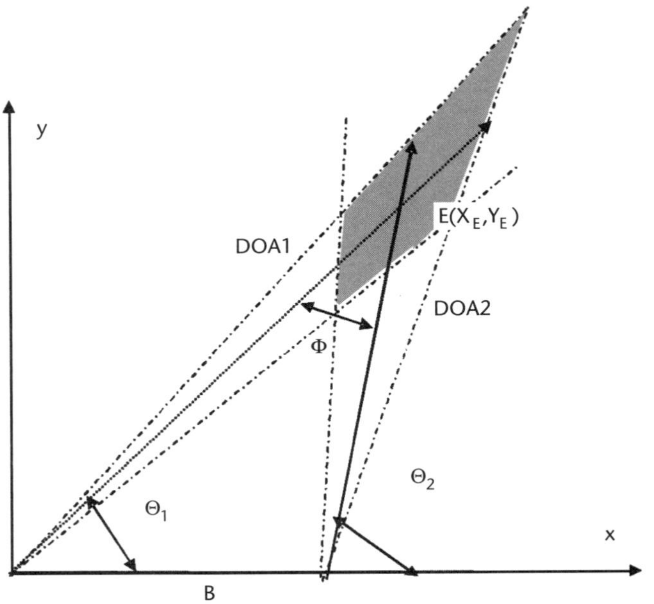

Figure 4.28 Geometry of the triangulation method.

$$\begin{aligned} y_E &= x_E \tan\theta_1 = (B - x_E)\tan\theta_2 \\ x_E(\tan\theta_2 - \tan\theta_1) &= B\tan\theta_2 \end{aligned} \tag{4.42}$$

from which

$$\begin{aligned} x_E &= (B\tan\theta_2)/(\tan\theta_2 - \tan\theta_1) = (B\sin\theta_2\cos\theta_1)/(\sin(\theta_2 - \theta_1)) \\ y_E &= x_E\tan\theta_1 = (B\sin\theta_2\sin\theta_1)/(\sin(\theta_2 - \theta_1)) \end{aligned} \tag{4.43}$$

Because the DOA determinations from the intercept sensors are affected by a DF error, for ease of treatment we assume they have zero mean and variance (or standard deviation) σ_θ. The bearing lines do not determine a single point but rather an area (the hatched area in Figure 4.28). Indeed, if more than two intercept sensors are used, their DOAs do not meet at a single point, but instead form a triangle of error that assumes quite different shapes across the plane xy. Hence, the need for mathematical methods that determine the center point or position fix (PF) within the triangle of error as the location that minimizes the mean squared error from the three DOAs (Stanfield method [10]); some other method could also be used to find the optimum [9].

Coming back to the case of two intercept sensors, it is important to determine the size of the hatched area around PF. To this end, we assume a Gaussian distribution of the bearing error with zero mean and standard deviation σ_θ. The PF will therefore be located within an elliptical area that defines the percentage of location fixes at its interior; usually either 50% (1σ) or 90% (2σ) ellipses are taken into consideration.

With the above assumptions and following the procedure reported in Appendix B of [5], the size of the ellipses are as follows:

$$\sigma_{\min} = (B\sigma_\theta/\sin(\theta_2 - \theta_1))((A - \sqrt{C}/2)^{1/2}$$
$$\sigma_{\max} = (B\sigma_\theta/\sin(\theta_2 - \theta_1))((A + \sqrt{C})/2)^{1/2} \tag{4.44}$$

where

$$A(\theta_2, \theta_1) = \sin^2\theta_2 + \sin^2\theta_1$$

and

$$C(\theta_2, \theta_1) = A^2 + 4A\sin^2\theta_2\sin^2\theta_1 - 8\sin^3\theta_2\sin^3\theta_1\cos(\theta_2 - \theta_1) \tag{4.45}$$

and as depicted in Figures 4.29(a–c), which show, respectively, the geometry of the triangulation method, the error ellipses within a square area of side equal to the base length, and the error ellipses outside the previous area. We can see that the two axes of the ellipse are proportional to $B\sigma_\theta/\sin(\Phi)$, where Φ is the bistatic angle of the triangle formed by the baseline and the position of the emitter (and is shown in Figure 4.28). Further, the ellipses of minimum size at any distance from the baseline are along the axis of the baseline (as can also be intuitively deduced). Indeed, along the axis of the baseline, we have $\theta_2 = \pi - \theta_1$ and $\Phi = \theta_2 - \theta_1 = \pi - 2\theta_1$. Thus by solving (4.44) we have the dimensions of the two axes of the elliptical error probability at 50%:

$$\sigma_{\min} = B\sigma_\theta/\cos^2\theta_1 \quad \sigma_{\max} = B\sigma_\theta/\sin 2\theta_1 = B\sigma_\theta/\sin\Phi \tag{4.46}$$

The normalized location performance (CEP/B) of the triangulation along the axis of the baseline versus the normalized distance (R/B) from the baseline is reported in Figure 4.29 with the angular variance as a parameter.

4.5.2 Trilateration

Emitter location can also be obtained also by measuring the TDOA of the same signal at several sensor sites. As discussed earlier in this chapter, two methods are available: (1) LE-TDOA, with the addition of DD in the case of pulsed emitters, and (2) SC-TDOA, with the addition of DD in the case of CW-type emitters.

The LE-TDOA method is based on the measurement of the TOA of the signal pulse at an intercept sensor and on the computation of its difference (TDOA) with respect to the TOA of the same pulse at a further intercept sensor placed at a distance B (on the order of 10 km) from the first one. The two sensors are obviously synchronized by some GNSS clock and constitute a baseline. The TDOA measurement establishes an isodelay curve (a hyperbola) on the surface of the Earth that passes through the emitter location. By adding a second baseline that is aligned with the previous one, by means of a third intercept sensor at a distance B_2, the TDOA2 computed as the difference between the TOAs of the signal pulse at the second and at the third intercept sensor establishes a second isodelay curve. The

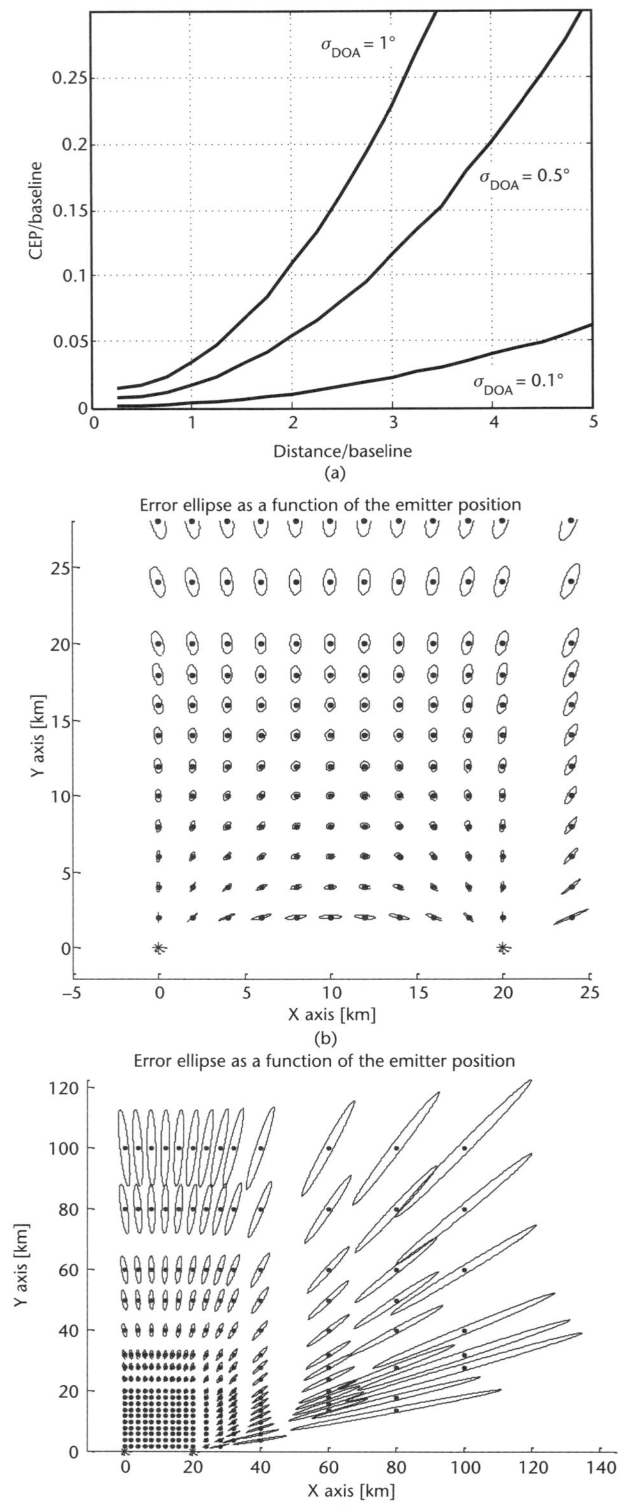

Figure 4.29 (a) Normalized CEP of the triangulation technique vs. distance along the baseline axis. (b) Error ellipse shapes in the sector within the two sensors' base length as a function of the emitter position. (c) Error ellipse shapes in the sector outside the two sensors' base length as a function of the emitter position.

intersection of the two isodelay curves determines the geolocation of the emitter. The mathematical formulation of the method as well its measurement accuracies are described in Appendices B2 and B3 of [5]. In this section we report only the main results; interested readers should refer to [5] for more insight into the method.

The geometry of the trilateration is depicted in Figure 4.30. The expressions of the two isodelay curves are as follows:

$$\text{Iso-delay curve 1} \ ((B_1 - x_E)^2 + y_E^2)^{1/2} - (x_E^2 + y_E^2)^{1/2} = cTDOA_1$$
$$\text{Iso-delay curve 2} \ ((B_2 - x_E)2 + y_E^2)^{1/2} - (x_E^2 + y_E^2)^{1/2} = cTDOA_2$$

The tangent angle to the isodelay curves at the PF (emitter geolocation) is the bisector of their respective bistatic angles (i.e., $\Phi_1/2$ and $\Phi_2/2$, respectively). Further, the minimum position error due to an error Δ_{TDOA} (as an example in the $TDOA_1$ measurement) can be shown to be along the normal to the tangent curve at PF and of amplitude $\Delta s = (c\Delta_{TDOA})/(2\sin(\Phi_1/2))$, that is, inversely proportional to half the relevant bistatic angle.

The performance of the trilateration technique is graphed in Figure 4.31 (taken from [11]), which shows the normalized CEP to the baseline versus the normalized distance from two equal baselines with the TDOA standard deviation as a parameter.

As can be seen by comparing Figures 4.29(a) and 4.31, the latter provides a more accurate PF, albeit at the expense of a more complex ESM equipment architecture and network synchronization and data links requirements. It does, however, cope with the following problems:

- For CW signals, the TDOA measurements have to be performed through signal cross-correlation. This imposes heavier requirements on the data links

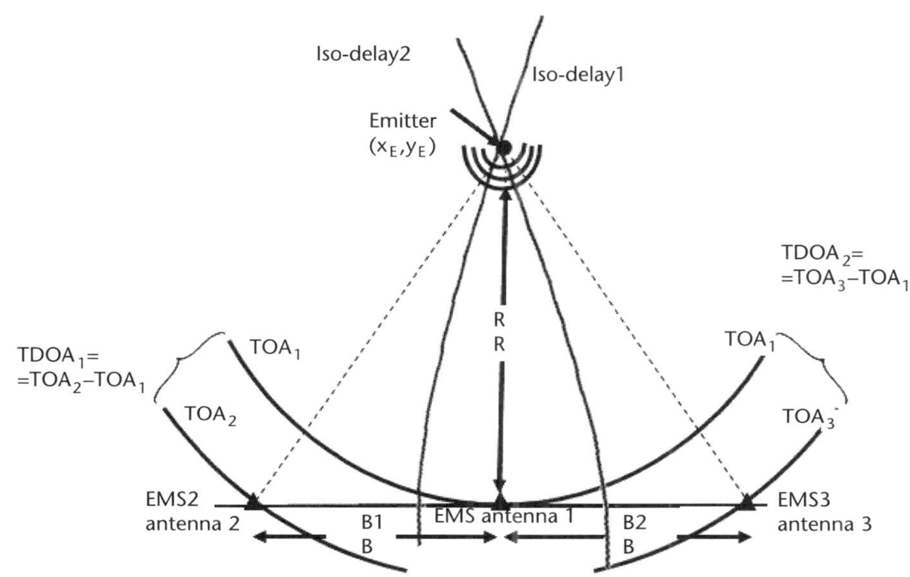

Figure 4.30 Geometry of the trilateration technique.

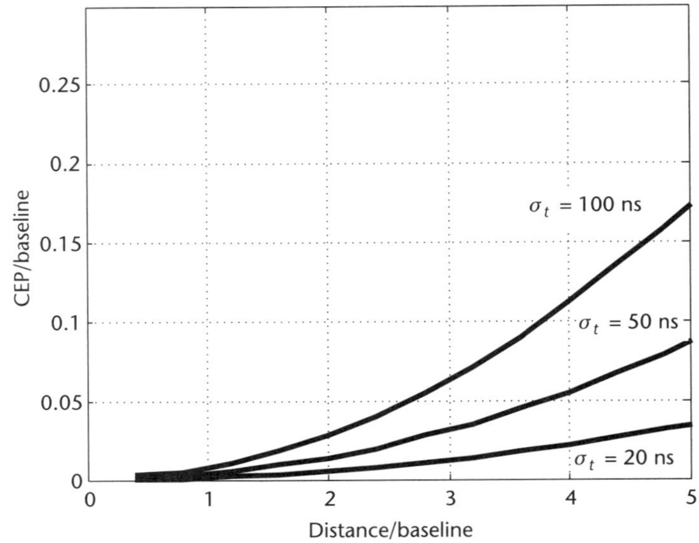

Figure 4.31 Normalized CEP versus normalized distance from the baseline (two equal baselines).

between the stations because they have to exchange the complex signal samples, instead of the synthetic data usually exchanged for pulsed emitters.

- All of the intercept sensors must be simultaneously in sight of the emitter to allow for TDOA measurements on the same signal among the different ESM sensors.
- For short PRI emitters, such as high pulse repetition frequency (HPRF) radar waveforms, some ambiguity problems need to be solved.

The trilateration technique is much requested nowadays, especially for ground applications such as a passive emitter tracker (PET) network, which is becoming an essential complement to the radar network in an IADS for these reasons:

- Passive ESM sensors have a lower cost, due to their receiving-only architecture, with respect to an additional radar in the relevant network.
- The PET network is undetectable, thanks to the passive based detection strategy.
- A higher detection range (RAF) of attacking aircrafts can be obtained (and thus really early warnings) because the PET network exploits the one-way signal attenuation.
- The system has the capability not only to detect, but also to identify targets by means of a priori information stored in the emitter library.
- The system has the ability to operate in very dense environments, using a high number of measured emitter waveform parameters to discriminate and classify airborne emitters.
- The system can exploit the specific emitter identification (SEI) techniques to identify in a unique way the emitters pertaining to a specific platform.

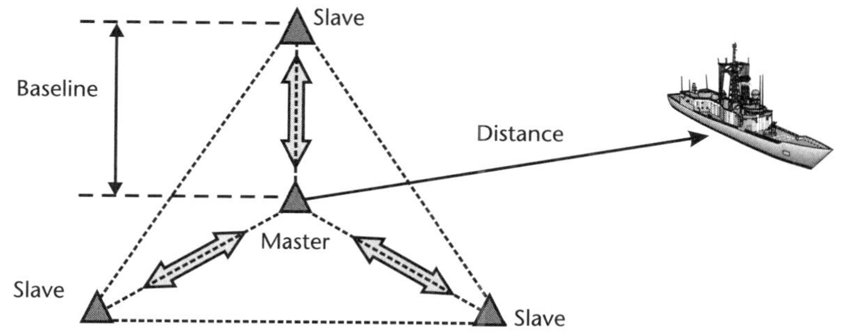

Figure 4.32 Typical layout of a PET network.

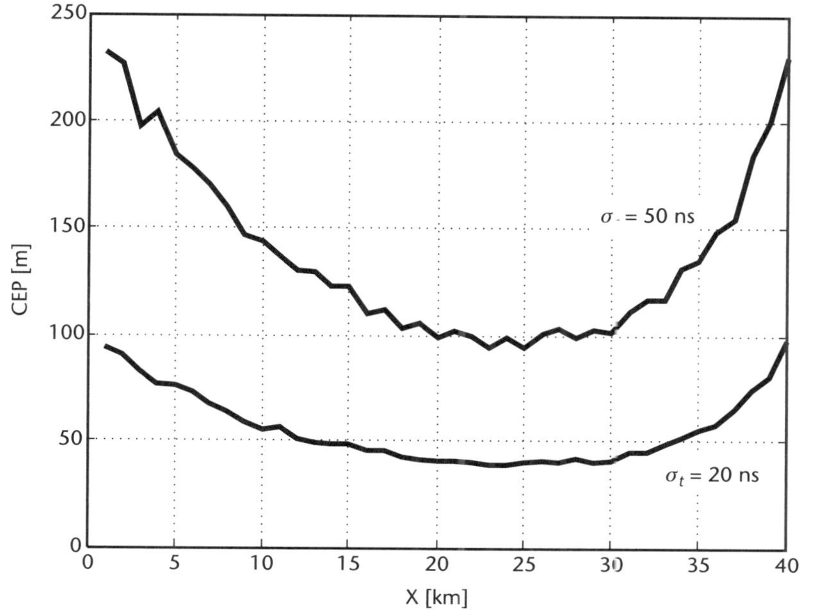

Figure 4.33 PET tracking performance on a moving ship projected along a linear trajectory at constant speed. (Reprinted with permission from D. Benvenuti et al. [11].)

A typical layout of a PET network is shown in Figure 4.32, and its tracking performance is reported in Figure 4.33.

4.5.3 Frequency Difference on Arrival Passive Location Technique

Historically passive location of ground-based emitters from flying platforms was performed using the methods described in the previous sections, such as triangulation and multilateration [5]. Currently, the electronics technology evolution has allowed the development of techniques based on measurement of the Doppler frequency differential within two antenna/receiver groups installed within the same flying platform. This technique provides very accurate location estimates. Because the differential measurements are taken instantaneously on the arrival of the signal wavefront at the platform, the technique is called frequency difference on arrival (FDOA). The advantage of this technique is the short time required to perform

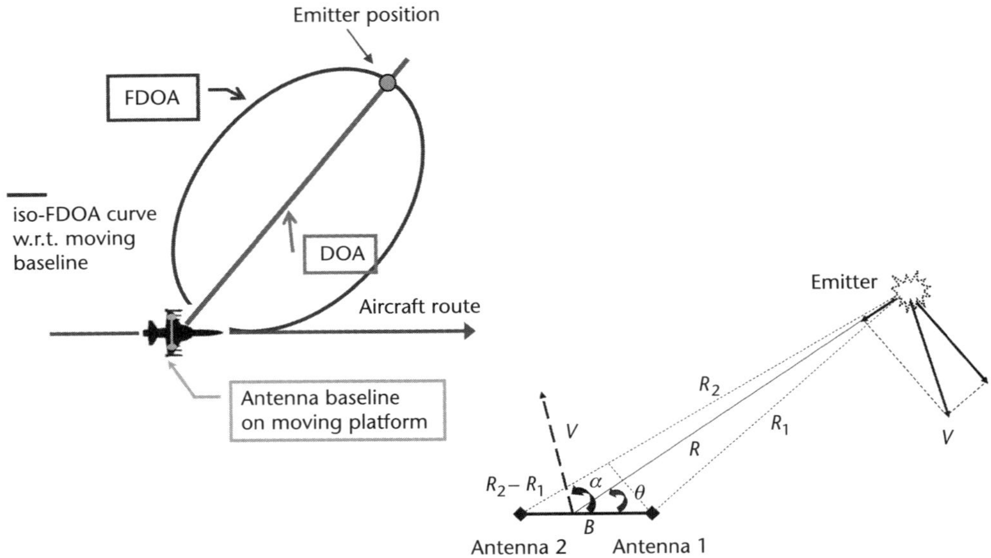

Figure 4.34 Passive location technique.

emitter location, which is on the order of a radar switch-on period when in the EMCON condition.

The location technique combines DOA and FDOA measurements as reported in Figure 4.34, which illustrates the principle of this technique. The motion of an A/C at velocity v with respect to ground induces a rotation of the LOS θ of the radar emitter. The LOS angular velocity ω depends on the tangential component (i.e., orthogonal to the LOS) of the velocity, called v_T, and the distance R between the A/C and the emitter:

$$d(LOS)/dt = \omega = v_T/R = v\sin(\alpha - \theta)/R \ \ [\text{rad/s}] \tag{4.47}$$

where θ = LOS and α is the angle of the A/C velocity vector with respect to the baseline.

Two antennas forming a baseline B on board the A/C can measure through the FDOA of the emitter signal the LOS angular velocity ω as follows:

$$FDOA = (1/\lambda)d(R_2 - R_1)/dt = -(1/\lambda)d(B\cos\theta)/dt = f(\omega B\sin\theta)/c \ \ [\text{Hz}] \tag{4.48}$$

where $f = c/\lambda$ is the carrier frequency of the signal.

The DOA of the emitter can be estimated through the TDOA Δt at the extremes of the same baseline, thus giving

$$DOA = \theta = \cos^{-1}((c\Delta t/B) \ \ [\text{rad}] \tag{4.49}$$

From (4.47) to (4.49) we obtain, through the measurement of TDOA, FDOA, and f from the EW sensor and through the platform navigation data v and α (orientation of the baseline with respect to A/C longitudinal axis):

$$R = v\sin(\alpha - DOA)/\omega = (fBv\sin(DOA)\sin(\alpha - DOA))/(cFDOA) \text{ [m]} \quad (4.50)$$

The normalized variance of the estimated distance R is computed as usual by taking the derivatives of the above equation, and it is composed of three terms relevant to the variances of the measured parameters:

$$\sigma_R/R = [A\sigma^2_{TDOA} + B\sigma^2_{FDOA} + C\sigma^2_f]^{1/2} \quad (4.51)$$

where

$A = [\ (f\ Bv\ \sin(\alpha - 2\theta))/(c\ \mathrm{FDOA}))]^2$; $A/R^2 = [\sin(\alpha - 2\theta)/\sin(\theta)\ \sin(\alpha - \theta)]^2$
$B = [((f\ Bv\ \sin(\theta)\sin(\alpha - \theta))/(c\ (\mathrm{FDOA}^2))]^2$; $B/R^2 = [1/FDOA]^2$
$C \approx [(Bv\ \sin(\theta)\ \sin(\alpha - \theta))/((\mathrm{FDOA}))]^2$; $C/R^2 = [1/f]^2$

In general, in (4.51) the contribution of variance σ_f is not taken into account because C is almost ≈ 0, while:

$\sigma_{\mathrm{LE\_TDOA}} = T_{r/}(\sqrt{2N(\mathrm{SNR})})$ for pulsed emitters, where N is the number of TDOA measurements within the integration time (see Section 4.4);

$\sigma_{SC-TDOA} \approx \dfrac{1}{\sqrt{B_N \cdot T_{CORR}} \cdot 2\pi \cdot B_{EQ}} \cdot \sqrt{\dfrac{2}{SNR} + \dfrac{1}{SNR^2} \cdot \dfrac{1}{\left(\sqrt{2} \cdot 2\pi \cdot B_{EQ}\Delta T\right)^2}}$ for CW-type emitters;

$\sigma_{\mathrm{FDOA}} = (2\sqrt{3}/(2\pi))\ (1/T_p)(\sqrt{\ (1/(M\ \mathrm{SNR})})$ where M is the number of FDOA measurements required to provide an estimate, as shown in Figure 4.35, and T_p is the period at which successive measurements are taken.

As can be seen from (4.50), the accuracy of the range estimate is strongly dependent on the geometry (the angle DOA, the orientation α of the baseline with respect to the A/C velocity vector, and the range of the emitter, which affects the SNR), but it is improved by the length of the baseline, the speed of the A/C, and the frequency of the emitter (i.e., this technique cannot be exploited at the VHF communication frequency band). A sample of the performance of this technique is

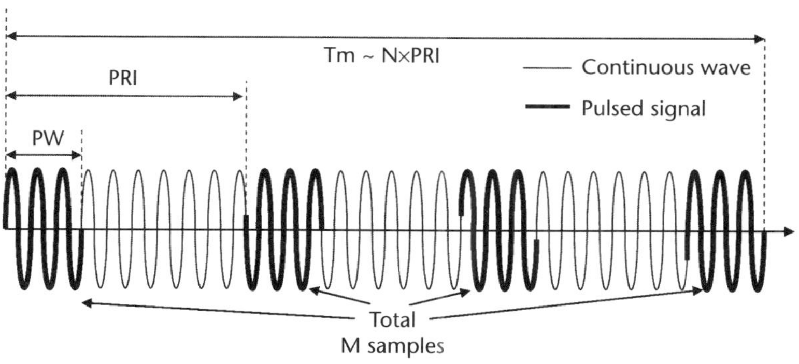

Figure 4.35 Burst of N coherent pulses with total duration Tm.

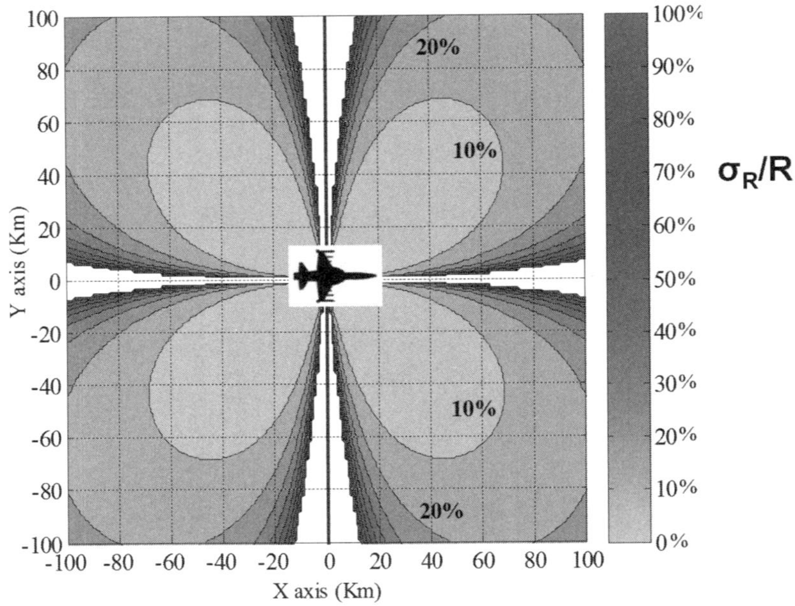

Figure 4.36 Location percentage error diagram: aircraft moving at 200 m/s vs. ground-based emitter with 90 [dBm] ERP. (Reprinted with permission from G. Severino et al. [12].)

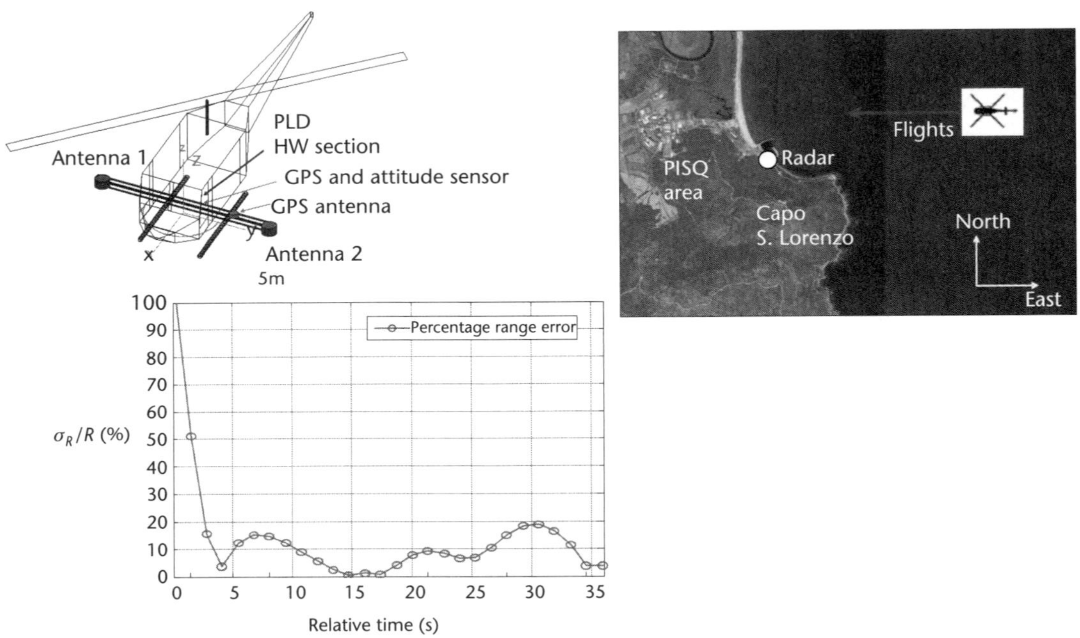

Figure 4.37 Typical location run: percentage error on range estimate during the run. (Reprinted with permission from G. Severino et al. [12].)

shown in Figure 4.36 in which the orientation of the baseline is orthogonal to the A/C velocity vector. The technique has received experimental verification, always with the orientation of the baseline orthogonal to the A/C velocity vector, as shown in Figure 4.37.

4.5.4 Inverse Passive Location

FDOA can be also exploited, similar to the passive location performed by airborne ESM equipment, with a technique developed by Elettronica, called IPL, that provides a passive location estimate of flying emitters from single steady stations on the surface (either ground or sea surface). Obviously, this result may be easily and more accurately obtained by a network of ESM equipment that exploits the TDOA technique described in Section 4.5.2. However, on some occasions stand-alone ships or mobile armored vehicles that are operating in EMCON mode (i.e., radar and communications silence) and without the aid of airborne early warning A/C or surveillance satellites (not all armed forces are provided with those means) need to be warned about the intentions of an approaching A/C emitter.

The principle of operation of the IPL technique is depicted in Figure 4.38. This technique requires the measurement of the following three parameters of an emitter:

- The TDOA or, equivalently, the DOA,
- The Doppler frequency differential in angle FDOA,
- The Doppler frequency difference with time DFD.

The first two parameters mentioned above are obtained as differential measurements between two antenna/receiver channels placed at the extremes of a baseline of length B, while the third parameter, which consists of the derivative of the received frequency with time, is measured by a single antenna/receiver channel designed with stringent requirements that are especially relevant to phase noise.

With reference to the geometry of Figure 4.38 and the considerations already exposed in Section 4.5.2, it is possible to evaluate the above two first parameters as follows:

$$TDOA = (B\cos\theta)/c, \text{ from which } \theta = \cos^{-1}(c\,TDOA/B) \tag{4.52}$$

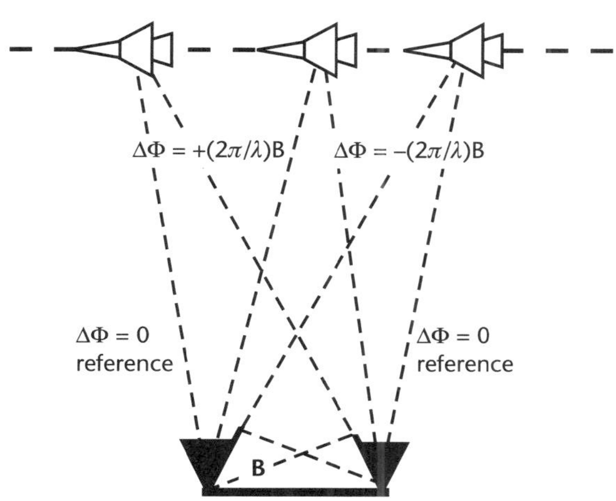

Figure 4.38 Geometry of the IPL technique.

$$FDOA = d(\varphi_{path1} \cdot \varphi_{path2}/dt = (d\pi/\lambda)d(R_1 - R_2)/dt \\ = (2\pi/\lambda)d(B\cos\theta)/dt = (2\pi B/\lambda)\omega\sin\theta \tag{4.53}$$

where $\omega = d\,(\text{LOS})/dt = v_T/R = (v\sin\theta)/R$ is the LOS angular rotation speed.

The FDOA measurement is insufficient to determine the range because ω is dependent on both the transversal component of the velocity of the aircraft v_T as well on the distance R, thus an estimate of the velocity value is needed and can be obtained through the measurement of the time derivative of the radial component $v_r = v\cos\theta$ of the velocity of the aircraft as shown in Figure 4.39. The radial component of the velocity can be measured as the Doppler frequency difference (DFD) with time (provided that the emitter carrier frequency is constant, i.e., a coherent emitter).

The above measurement corresponds in kinematics to the determination of the radial acceleration of the A/C, in that

$$a_r = dv_r/dt = -\omega v\sin\theta = -(v^2\sin^2\theta)/R \tag{4.54}$$

and the variation of the DFD is

$$DFD = df_d/dt = (1/\lambda)dv_r/dt = a_r/\lambda \tag{4.55}$$

So the estimate of the distance $\breve{R}$ is obtained by equating the two following expressions:

$$\breve{R} = v_T/\omega = (v\sin\theta/FDOA)(2\pi B\sin\theta/\lambda) \tag{4.56}$$

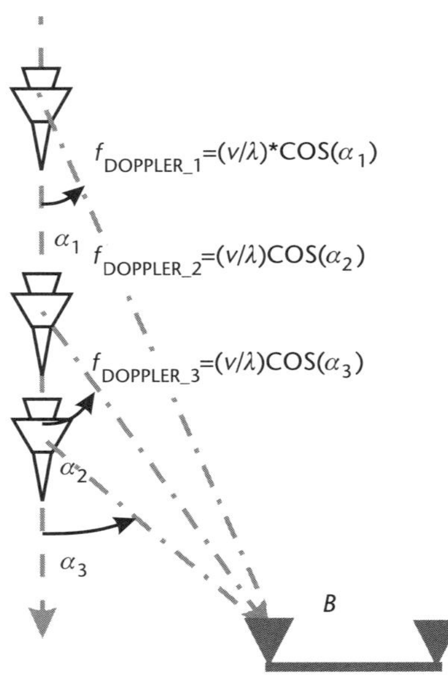

Figure 4.39 Measurement of the DFD.

$$\breve{R} = (v^2 \sin^2 \theta)/a_r = (v^2 \sin^2 \theta)/(\lambda DS) \tag{4.57}$$

from which it can be deduced that $v = (2\pi B)/\text{FDOA}$ and hence

$$\breve{R} = (2\pi B)^2 fDFD \sin^2(\theta))/(c(FDOA)^2) \tag{4.58}$$

The measurement of the Doppler shift is obtained in the case of pulse Doppler airborne radars, which use coherent pulse burst waveforms of duration T_{burst} on the order of 3 to 10 ms, in which carrier frequency, PW, and PRI are kept constant.

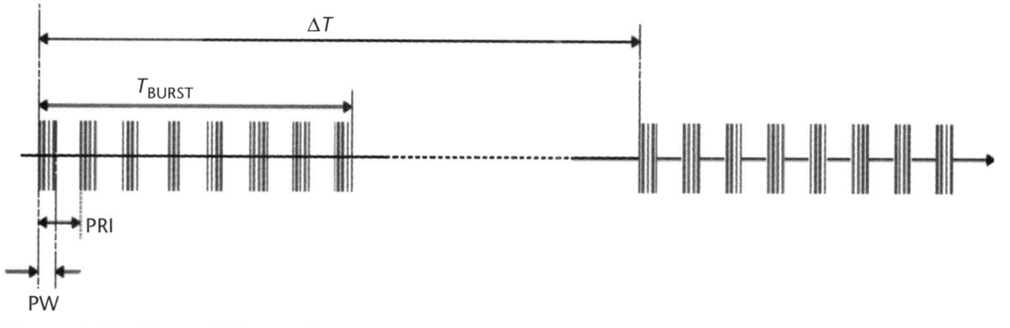

Figure 4.40 Typical PD waveform.

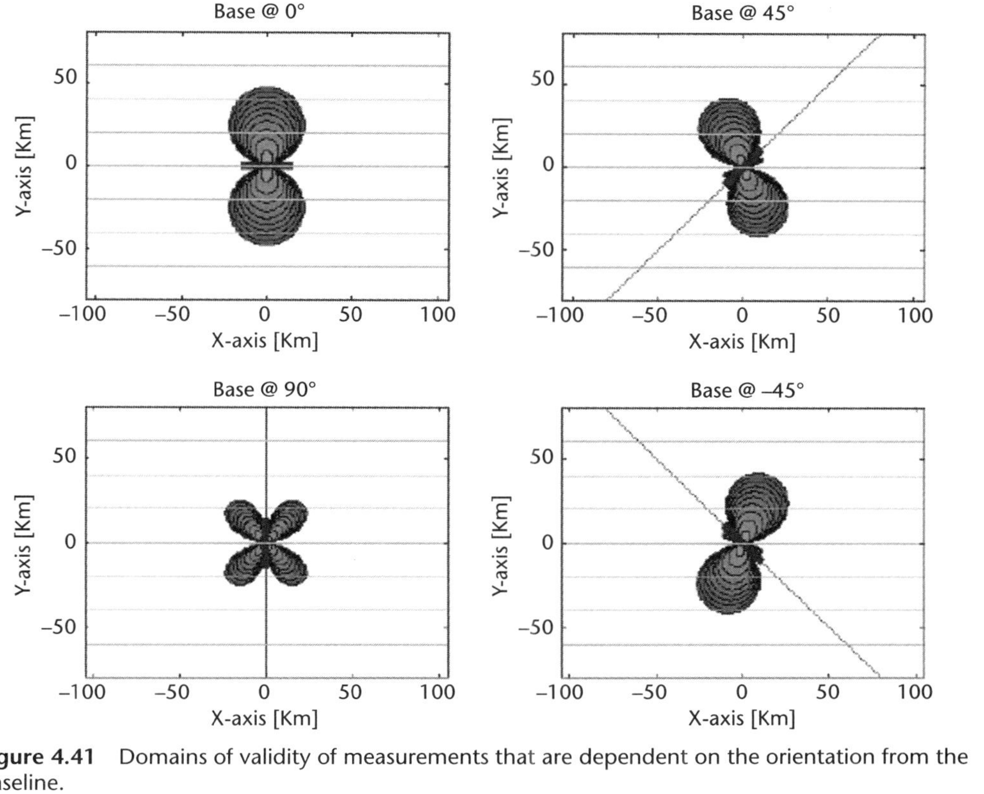

Figure 4.41 Domains of validity of measurements that are dependent on the orientation from the baseline.

The bursts usually show a kind of periodicity ΔT on the order of 1 to 2 sec. A typical PD waveform is depicted in Figure 4.40.

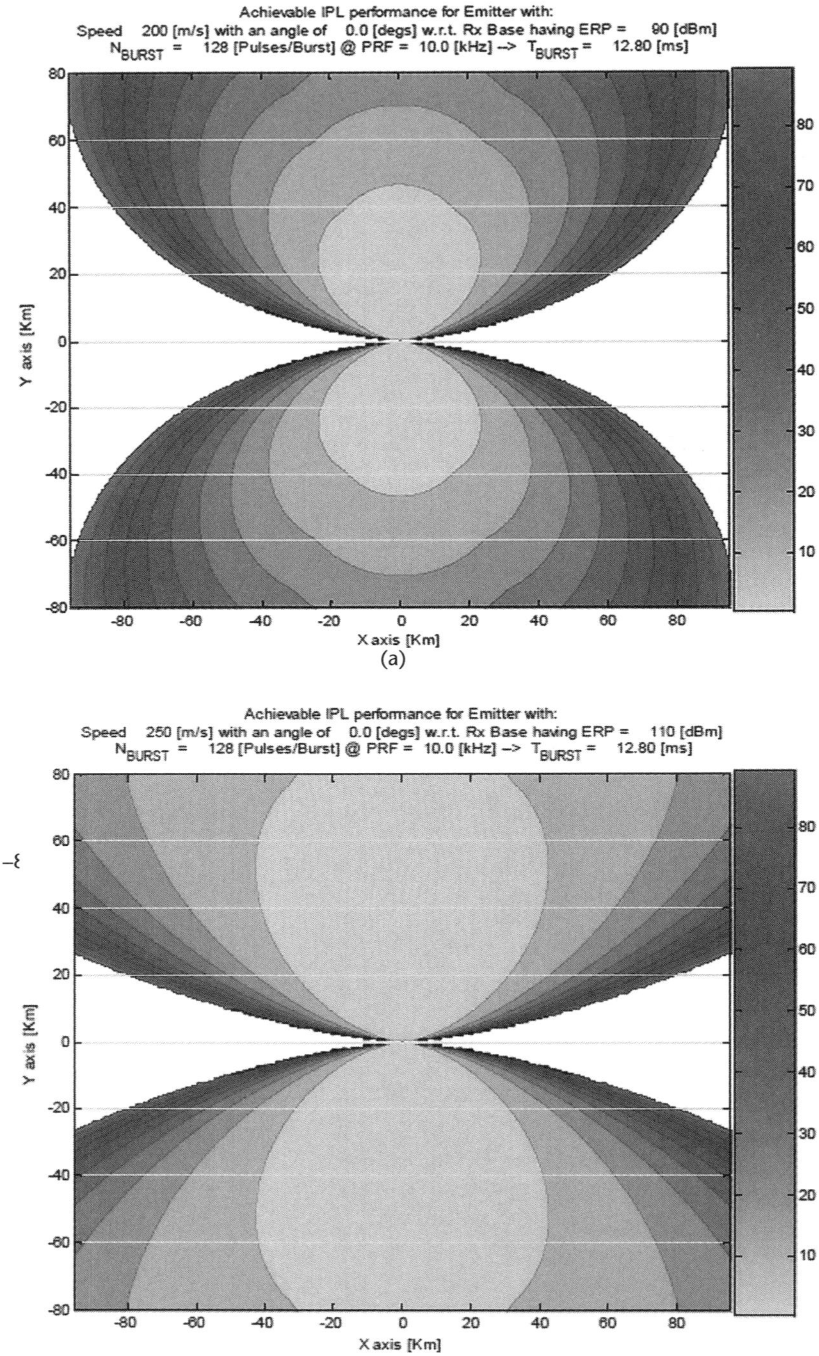

(a)

(b)

Figure 4.42 Normalized IPL accuracy σR/R for an A/C pulse Doppler emitter flying at 250 m/s transverse to the steady platform: (a) A/C emitter ERP = 90 dBm and (b) A/C emitter ERP = 110 dBm. (Reprinted with permission from G. Severino et al. [12].)

The Doppler frequency difference with time DFD can be measured in two modes: (1) as the second derivative of the initial phase of the pulses within each burst (intraburst mode) and (2) as the first derivative of the Doppler frequency measured from the bursts that exhibit the same carrier frequency (interburst mode). The interburst mode, which uses longer observation times with respect to the first mode, provides a better location accuracy. The normalized IPL accuracy σ_R/R, which is dependent on the frequency f, DFD, DOA, and FDOA measurements, can be expressed as:

$$\sigma_R/R = [\sigma_f^2/f^2 + \sigma_{DFD}^2/DFD^2 + 4\sigma_\theta^2 \tan(\theta) + \sigma_{FDOA}^2/FDOA^2]^{1/2} \tag{4.59}$$

The accuracy of the technique, as with all of the measured parameters, is obviously dependent on the input SNR (i.e., on the radar ERP) and the PD waveform (number of pulses in the burst and the number of bursts that have the same carrier frequency within the observation time).

The technique presents some drawbacks:

- No measurement is possible along the direction of the baseline. But this inconvenience can be removed by introducing further baselines at an angle with the first, forming a polygon. The domains of validity of the measurement, which are dependent on direction from the baseline are reported in Figure 4.41.
- The DFD and FDOA measurements are ill conditioned for cases that involve nearly radial trajectories with respect to a fixed platform. In these conditions, however, the ESM equipment provides a warning to the platform to switch on the on-board radars and designate the associated weapon systems.

The normalized IPL accuracy σ_R/R is reported in Figure 4.42 for the case of an A/C emitter flying transverse to the steady platform.

4.6 Conclusion

In this chapter the main passive DF techniques were presented. These techniques exploit the instantaneous signal parameters of RF emitters, such as amplitude, phase, and TOA, to provide warning and/or cueing of ECM systems for subsequent threat countering. The techniques discussed, supported by the evolution of novel electronic technologies, were also shown to be capable, when operating in differential mode either within the same platform or within a group of similar sensors, of providing emitter location either on the ground/sea surface or in the air.

References

[1] Lipsky, S., *Microwave Passive Direction Finding*," Raleigh, NC: SciTech Publishing, 2004, Chapter 6.

[2] Baron, A. D., and Hoffman, C., "Passive DF and Signal Location," *Microwave Journal*, Vol. 25, September 1982, pp. 59–76.

[3] Wu, Y.-W., Rhodes, S., and Satorious, E., "DOA Estimation via Extended Phase Interferometry," *IEEE Trans. Aerospace and Electronic Systems*, Vol. AES-31, No. 1, January 1995, pp. 375–381.

[4] Dinoi, L., Di Vito, A., and Lubello, G., "DF of Ground Based Emitters from Airborne Platforms," paper presented at Radar Conference 2008, IEEE.

[5] Wiley, R., *Electronic Intelligence: The Interception of Radar Signals*, Dedham, MA: Artech House, 1985, Chapter 5.

[6] Skolnik, M., *Radar Handbook*, New York: McGraw Hill, 1990.

[7] Ardoino, R., Capriati, F., and Zaccaron, A., "Performances of a TDOA Estimation Algorithm Based on Cross-Correlation Technique," *Proc. Int. Radar Symposium 2006*, Krakow, Poland, May 24–26, 2006, pp. 189–192.

[8] Panci, G., *TDOA Estimation using Super-resolution Techniques,* Elettronica Technical Report 4073, 2010.

[9] Poisel, R., *EW Target Location Methods*, Norwood, MA: Artech House, 2005.

[10] Stanfield, R. G., "Statistical Theory of DF Fixing," *Proc. IEEE*, Vol. 94, Part 3A, No. 15, 1947.

[11] Benvenuti, D., et al., "A Comprehensive View of TDOA Based Digital Sensors on the Naval Surveillance," Defense Technology Asia (DTA) Int. Conference, Singapore, 2009.

[12] Severino, G., Zaccaron, A., and Ardoino, R., "Airborne Passive Location Based on Combined TDOA and FDOA," paper presented at EW Conference, Bangalore, India, 2010.

CHAPTER 5

Electronic Countermeasure Systems

5.1 Introduction

The objective of ECM systems is to reduce or suppress the effectiveness of enemy defense systems and their relevant weapons systems through soft-kill actions such as confusion, distraction, deception, or seduction. In the absence of electronic countermeasure (ECM) systems, the kill probability of the current generation of weapons systems is very high and could produce a high attrition rate (i.e., large number of casualties and destroyed assets) during a combat mission.

ECM systems cover the whole electromagnetic spectrum and are named in accordance with either the wavelength/bandwidth in which they operate or the equipment function against which they operate. Hence, we have communications ECMs (CECMs) that work against communication systems, radar ECMs (RECMs) that work against radar systems (often simply called ECM), and infrared ECMs (IRCMs) that work against the IR systems.

Although CECM systems are only of the active type (i.e., they provide their countermeasure through appropriate signal transmission), both RECM and IRCM systems can be either of the active or passive type. The latter type, namely, chaff in the RF band and flares in the IR band, provides the countermeasure without any signal transmission, usually a seduction through the concealment of the real target. The classification of the ECM systems is diagrammed in Figure 5.1.

Typical applications of RECM systems are for self-protection and for the mutual protection of mobile platforms. CECMs are primarily used only in the latter application. The block diagram of a typical self-protection jammer (SPJ) is shown in Figure 5.2. A JPS uses on-board countermeasures (i.e., the jammer) along with off-board countermeasures, which are intended to divert the attacking missile from its target. These systems consist of (1) dispensed expendable devices (i.e., chaff and flares), (2) propelled active devices, and (c) towed devices (some of which can be recovered and reused).

The use of off-board and on-board IRCMs implies the use of a missile warning system (MWS) that detects the attacking missile and triggers the countermeasure (CM) in the appropriate direction. If the MWS is constituted by a short-range pulse Doppler radar, called missile approach warner (MAW), then it can also trigger the off-board action at the appropriate time, thus providing a more effective CM.

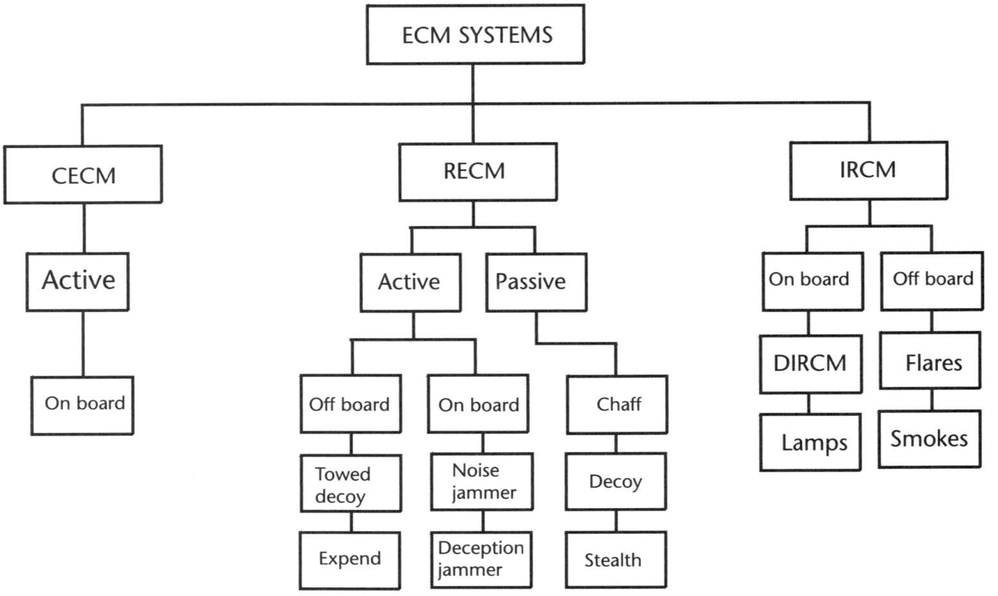

Figure 5.1 Classification of ECM systems.

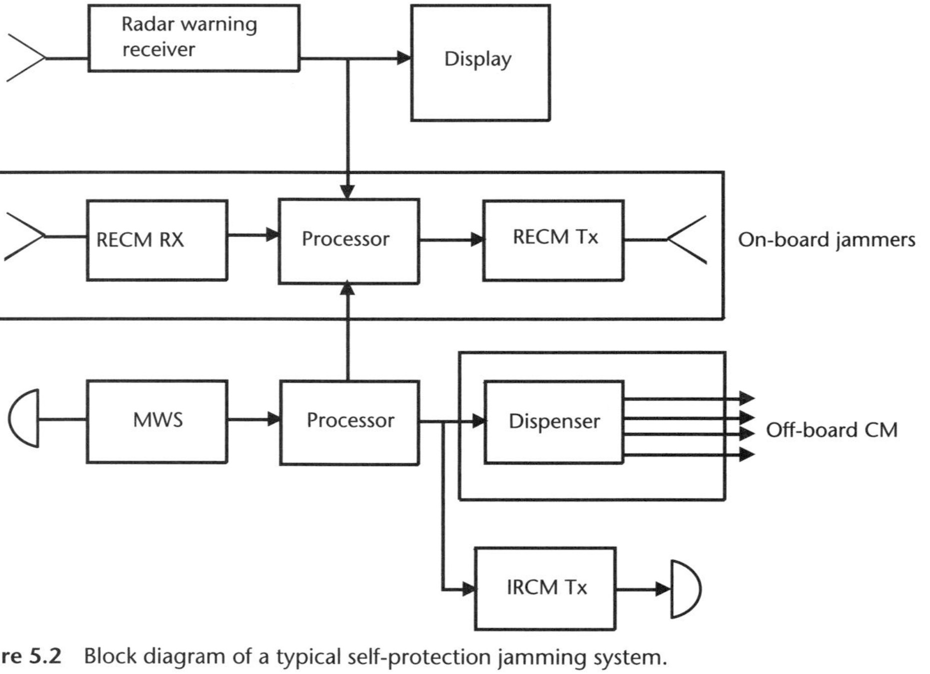

Figure 5.2 Block diagram of a typical self-protection jamming system.

Before describing the architecture of the integrated self-protection system, we recall in Figure 5.3 the functions performed by each element of the system.

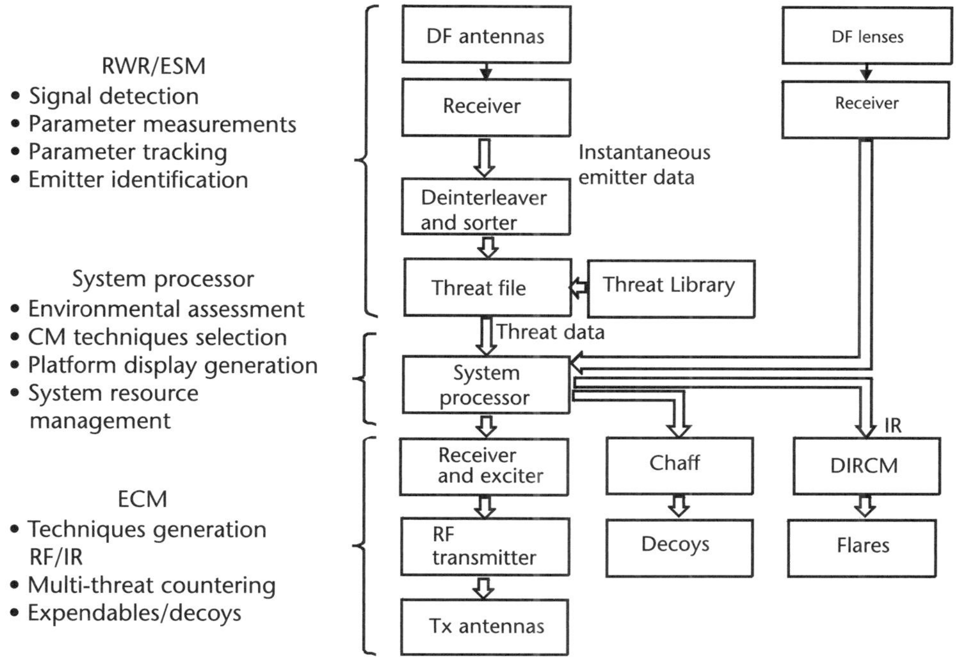

Figure 5.3 Functional block diagram of an integrated self-protection system.

5.1.1 Typical RECM Requirements and Missions

Basically two types of radars are countered by RECM systems: surveillance and tracking radars. The surveillance or search radar functions are designed to locate and automatically track targets within a large volume. This allows the defense centers to perform threat evaluation and weapon assignment (TEWA). Because surveillance radars have to detect targets at long range, they are usually operating at low RF (C to G, according to EW band nomenclature, or VHF to C band according to the standard designation). Low-RF operation is used in order to exploit the low atmospheric attenuation (see Section 2.1) of these bands.

These radars are usually characterized by a rotating antenna, either the parabolic-plus-cosecant-square (cosec^2) type, which provides 2-D target location, or the phased array type, which provides 3-D target location. In both cases, the antennas have a narrow main-beam antenna that scans the search volume and sidelobe beams that are illuminating (and receiving from) all directions. The sidelobe level (SLL) with respect to the peak of the main beam is dependent on the type of employed antenna and is classed as *ordinary* (SLL = –13 to –30 dB_{peak}), *low* (SLL = –30 to –40 dB_{peak}), or *ultra-low* (SLL < –40 dB_{peak}) [1]. The typical transmitted waveforms are pulse-compression signals with 3-D radars showing also some form of waveform agility in the elevation scanning.

The tracking radars are the high-priority threats that need to be countered by RECM equipment because they are associated with the engagement of a weapon system, especially in the terminal phase. When a tracking radar is locked onto a target, the associated weapon is also expected to be directed at the target. The task of the RECM system is to cause the tracking radar to break lock, which in turn removes the guidance information used by the weapon to converge on the target.

The tracking radars are characterized by a narrow beam supported by a relatively small dimension antenna, usually operating at high RF frequencies (I to K_a band, according to EW band nomenclature, or X to K band according to the standard designation). Each aircraft or ship that can be engaged by a weapons system has to be sufficiently protected by RECM systems capable of providing for its self-defense.

In the airborne case it is generally desirable to supplement the strike aircraft's self-protection system with either an escort jammer/close support jammer (EJ/CSJ) or stand-off jammer (SOJ) system. These systems are carried on escort platforms. Escort/close support jamming involves a dedicated aircraft carrying high-power jammers that accompanies the friendly strike force and provides a protective RF jamming shield in support of the entire strike force. Stand-off jamming involves a platform that stands at some distance beyond the effective range of the weapons that defend the target. Sketches of typical EJ/CSJ and SOJ mission profiles are given in Figure 5.4.

The SOJ has the advantage of carrying a large dedicated RECM payload. It has the disadvantage, however, of requiring a very large jammer ERP because it operates far away from the victim radars and because its jamming transmissions are for most of the time into the sidelobe pattern of the victim radar, as discussed next.

5.1.2 EW Radar Jamming Equation

The objective of jamming is to inject an interference signal into the victim radar receiver so as to mask the actual target. The interference signal can be either a random noise waveform or a manipulated replica of the radar waveform. The amount of transmitted jammer power required to achieve this result can be evaluated through the jammer-to-signal power ratio (JSR) at the output of the radar receiver as follows:

The target signal power S received by the radar is (see also Section 2.3) is:

$$S = \frac{P_t G_t G_r \lambda^2 \sigma}{(4\pi)^3 R^4 L_t L_r L_a^2} \tag{5.1}$$

where

P_t = radar transmitter power (W),
G_t = radar antenna gain in transmission,
G_r = radar antenna gain in reception (usually for monostatic radar $G_t = G_r$),
λ = transmitted wavelength (m),
σ = target RCS (m^2),
R = distance of the target from the radar (m),

L_t, L_r, and $L^2{}_a$ are, respectively, the transmitter, receiver, and two-way atmospheric propagation losses.

The receiver noise power is

$$N = kTB_r F_r$$

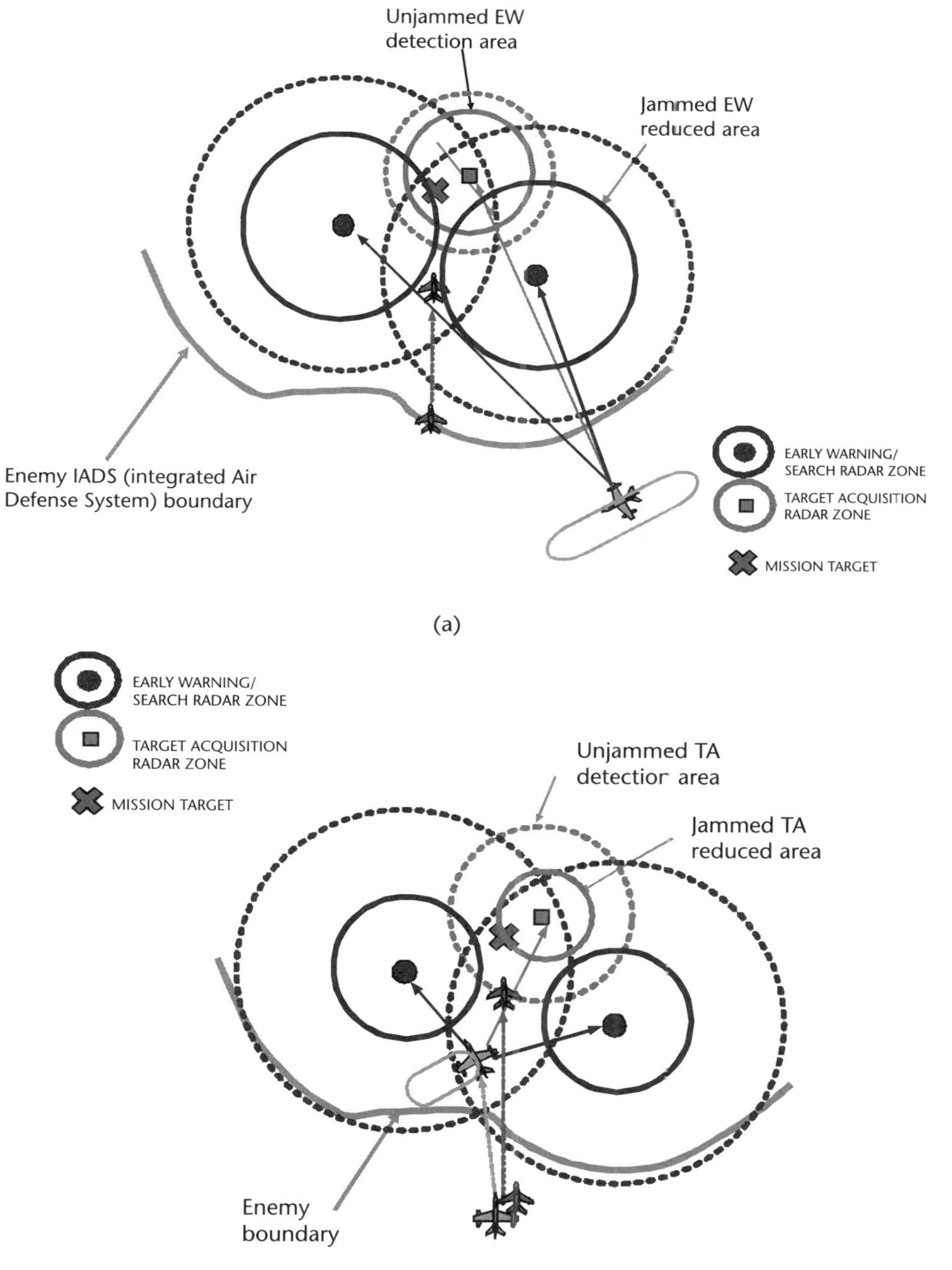

Figure 5.4 Sketches of typical EJ/CSJ and SOJ aircraft missions.

where

B_r = receiver bandwidth at the output of the receiver processing chain (Hz), that is, the ratio of the transmitted signal bandwidth to the receiver processing gain G_p;

kTF_r = noise power density at the receiver output (W/Hz) with $T = 290$ K.

The signal-to-noise ratio at the output of the radar receiver is:

$$S/N = \frac{P_t G_t G_r \lambda^2 \sigma}{(4\pi)^3 R^4 L_t L_r L_a^2 kTF_r B_r} \tag{5.2}$$

and the maximum target detection range from the radar is

$$R_{\max} \left[\frac{P_t G_t G_r \lambda^2 \sigma}{(4\pi)^3 (S/N)_{\min} L_t L_r L_a^2 kTF_r B_r} \right]^{1/4} \tag{5.3}$$

where $(S/N)_{\min}$ is the minimum signal-to-noise power ratio for detection.

The jammer power, spread over a bandwidth B_j, usually larger than the radar receiver bandwidth B_r, is:

$$J = \frac{P_j G_j G_{rj} \lambda^2}{(4\pi)^2 R_j^2 L_j} \frac{B_r}{B_j} \tag{5.4}$$

where

P_j = transmitted jammer power (W),

G_j = jammer antenna gain,

SLL = sidelobe level of the radar antenna (i.e., ratio of mean sidelobe level to peak gain),

G_{rj} = radar antenna gain in the direction of the jammer (i.e., $G_{rj} = G_r$ SLL),

R_j = distance of the jammer from the radar (m),

L_j = jammer transmission losses.

In the presence of a jamming signal the radar detection is impaired by the receiver noise and the jamming signal spread over a bandwidth $B_j > B_r$, thus providing in the receiver bandwidth B_r an equivalent noise power $N' = N(1 + J/N)$. If we assume that $J >> N$ (which is usually the case) we have $N' \approx J$ and hence the cases discussed next.

5.1.2.1 Self-Protection Jammer

For a SPJ, $R = R_j$ and $G_{rj} = G_t = G_r$. By indicating with $\text{ERP}_r = P_t G_t / L_t$ the effective radiated power of the radar, and with $\text{ERP}_j = P_j G_j / L_j$ the effective radiated power of the jammer, we have (by ignoring the term L_a, which is negligible):

$$J/S = \frac{ERP_j}{ERP_r}\frac{4\pi R_j^2}{\sigma}\frac{B_r}{B} \tag{5.5}$$

and hence the self-screening range $R_{j,ss}$; or equivalently the maximum target (i.e., the jammer platform) detection range for the radar is given by the condition $J/S = k$, a constant depending on the jamming technique (or equivalently $S/J = S/N' = 1/k$):

$$R_{j,ss} = \left[\frac{ERP_r}{ERP_j}\frac{B_j}{B_r}\frac{\sigma}{4\pi}\frac{1}{k}\right]^{1/2} \tag{5.6}$$

This equation shows that the larger the target RCS, the larger the ERP_j to achieve a sufficient self-screening range against a given radar.

5.1.2.2 Mutual Protection Jammer

For a mutual protection jammer we have $G_{rj} = G_r$SLL and by assuming $G_r = G_t$ we can determine from (5.3) the burn-through range $R_{t\max,j}$ (i.e., the maximum target detection range in presence of jamming), which is:

$$R_{t\max,j} = \left[\frac{ERP_r}{ERP_j}\frac{B_j}{B_r}\frac{\sigma \cdot R_j^2}{4\pi(S/N')_{\min} \cdot SLL}\right]^{1/4} \tag{5.7}$$

The ratio ρ of the detection range in jamming conditions with respect to clear (i.e., only receiver noise present) conditions is derived from (5.7) and (5.3) and it is:

$$\rho = \frac{R_{t\max,j}}{R_{\max,clear}} = \left[\frac{(4\pi)^2 NR_j^2 B_j}{ERP_j B_r G_r \lambda^2 \cdot SLL}\right]^{1/4} \tag{5.8}$$

This equation shows that the ratio ρ is increasing with short wavelengths, low SLL (i.e., ultra-low sidelobe antennas), and high radar processing gain (proportional to B_j/B_r). The factors that produce a reduction of the ratio ρ are the increase in the effective radiated power of the jammer ERP_j and the decrease of the jammer distance R_j from the radar (i.e., escort jammer) In Figure 5.5 the typical required effective radiated power of the jammer ERP_j vs. frequency for the various jamming roles on aircraft and ships are shown.

5.2 Radar ECM Architecture

Modern military platforms, such as aircraft (whichever their role: strike or air superiority) and ships, operate in high pulse density symmetric conflict scenarios. Pulse density is generated by the large number of radar-controlled weapons systems deployed in any given operational area. Chapter 2 showed that modern radars transmit sophisticated waveforms that constitute a challenge to the interception,

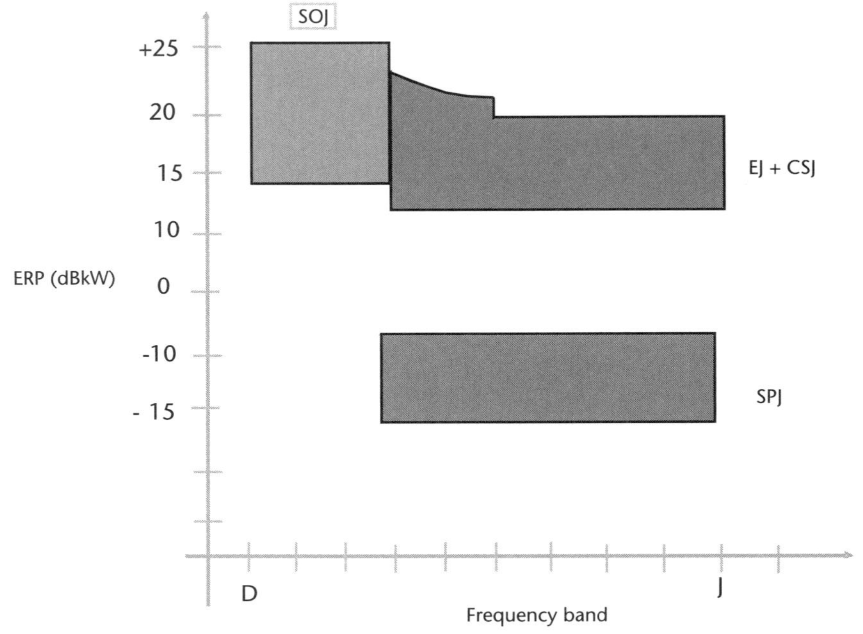

Figure 5.5 Typical required effective radiated power of the jammer ERP*j* vs. frequency for the various jamming roles (SOJ = stand-off jammer; EJ = escort jammer; CSJ = close support jammer; SPJ = self-protection jammer).

detection, classification, and identification of the emitters discussed in Chapter 3 relevant to RESM systems as well as to the selection of appropriate and effective countermeasures to be delivered against them by the RECM equipment.

Further, in such scenarios the platform is most likely attacked by many simultaneous threats of different types, and because each is provided with some specific type of ECCM, the RECM is required to deliver an ECM that is appropriate and effective against each of the simultaneous threats. The tailored ECM against each identified threat is implemented on the basis of intelligence data gathered about the threat. These data are coded and loaded prior to any platform mission into the threat library file of the RECM. The tailored countermeasure to a threat is usually constituted by a sequence or combination of elementary ECM techniques (which will be discussed in Chapter 6) with programmable parameters, which are part of the large inventory of available techniques of the RECM equipment. The sequence/combination of programmable ECM techniques relevant to a given threat is called a *jamming program*.

A RECM system architecture that meets the above requirements [2] is illustrated in Figure 5.6 and has to perform the following functions in sequence:

1. Interception, identification, and prioritization of the simultaneous threats on the basis of their dangerousness,
2. Association of the appropriate jamming program with each threat,
3. Allocation of jammer resources to implement the selected jamming programs,
4. Continuous tracking of DOA and threat signal parameters during engagement with the threat,

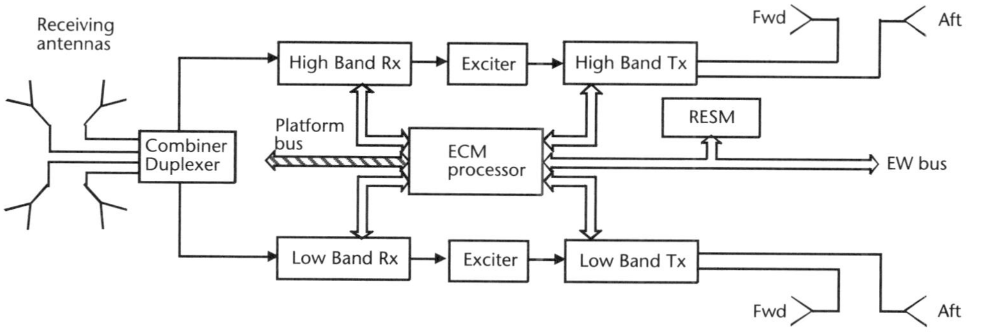

Figure 5.6 Functional block diagram of a typical SPJ.

5. Declaration of a ceased engagement after the threat is no longer locked onto the platform or when a sufficient safe distance from the threat emitter is achieved.

Function 1 is usually performed by ESM equipment, which provides threat designation in terms of identification, DOA, and current threat emitter parameters (i.e., the so-called "radar mode"). Some ESM equipment also provides the threat emitter's distance from the RECM platform (as discussed in Chapter 4) which is very useful to function 5.

Function 2 is provided by the system processor that, upon threat designation, updates the list of engaged threats in accordance with priority rules based on threat dangerousness. It also manages the execution of the appropriate jamming program associated with the threat in the threat library file of the RECM. As will be shown later, this function is performed through a dynamic and adaptive resources management program (called *power management*), which envisages the selection of an alternative jamming program (always coded in the threat library file and associated with the threat emitter) in a case in which some required resources are completely busy with a higher priority threat emitter.

Function 3 from the preceding list is executed by the system resources, which can be either constituted by a number N of identical independent jamming channels with N being equal to the maximum number of simultaneous threats, or by a limited number $n < N$ of highly flexible and powerful jamming channels, which counter the simultaneous threats on a fast-switching time-sharing basis.

A jamming channel is composed of the chain of a wideband receiving antenna, a frequency receiver tuned to the threat emitter's frequency with a bandwidth larger than the radar waveform one, a jamming RF waveform synthesizer called an *exciter* that is capable of producing either noise waveforms or manipulated threat signal replicas, a wideband high-power amplifier, and a wideband transmitting antenna as shown in Figure 5.7.

For most platforms, but especially aircraft, on-board equipment that has limited mass, volume, and power consumption are highly desired and hence any solution that adopts a limited number of highly flexible jamming channels (like that shown in Figure 5.6) is the preferred one. Highly flexible jamming channels capable of swift tuning over the entire threat frequency bandwidth and rapid beam steering over the complete required spatial coverage have been made available as a result

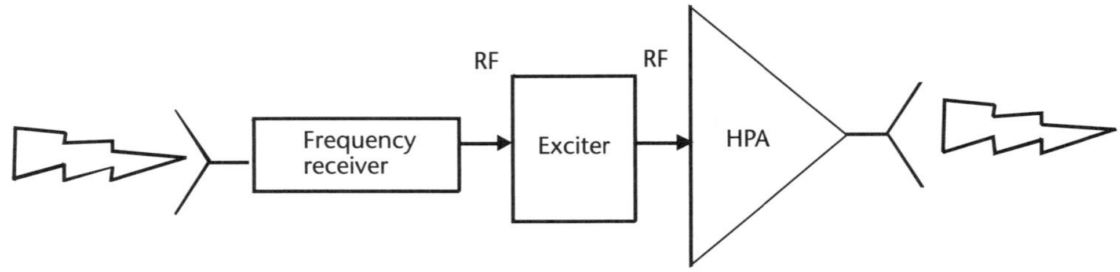

Figure 5.7 Schematic block diagram of a jamming channel.

of the use of modern technologies such as wideband multibit digital RF memories (DRFMs) and wideband solid-state Rx/Tx phased arrays or travelling wave tube (TWT)–based transmitters, as discussed next.

In Figure 5.6 a low-band (LB) section (from D or E to G band) is shown in addition to the high-band (HB) section (usually from G to J band).The LB section may be present only in airborne self-protection jammers (ASPJs), whereas the HB section is used by all of the platforms (aircraft, ships, and armored vehicles) for defense from aircraft attacks.

Because most radar threats are constituted by tracking radars of the single tracking target (STT) type, the architecture of the system is based on adaptive time-shared exploitation of the above-mentioned powerful and quickly switches (from one threat to the next) resources as commanded by a bank of time-of-arrival predictor circuits (PTOA; i.e., predictor of time of arrival), each assigned to a threat. The PTOA operates on the PRI measurements of its assigned threat emitter and generates time-gates (PTOAG) centered at the next pulse TOA with a sufficient width that takes into account the threat emitter PW, the type of PRI changes, and their tolerances and jitter (both measured and checked with the available intelligence data loaded in the threat emitter file of the equipment), as well the duration of the jamming response. This arrangement allows assignment of a complete jamming channel (like that shown in Figure 5.7) to each threat during the corresponding PTOAG as shown in Figure 5.8. A number of best practice system resources management routines are applied by the system processor; their discussion is beyond the scope of this book [3].

Thus, a number of jamming channels (virtual channels) equal to the number of simultaneous threats is made available to counter the threats. As a consequence, the time occupation of one actual channel is basically dependent on the duty cycle of the threat emitter and of the jamming program associated with the threat. The number of actual parallel channels of the RECM is therefore evaluated on the basis of the estimated time occupation of the channel resources during an engagement with the maximum envisaged number of simultaneous threats, of their waveforms, and of the selected jamming responses. A good rule of thumb is to use at least two actual jamming channels in a high-quality RECM. The functional block diagram of a modern ASPJ based on the above-described architecture provided with high-gain linear active phased arrays (covering a large spatial sector by steering a narrow azimuth beam) is shown in Figure 5.9.

In the next sections, after discussing the jamming channel resources mentioned above (namely, the DRFM and the transmitter solutions), we will deal with the

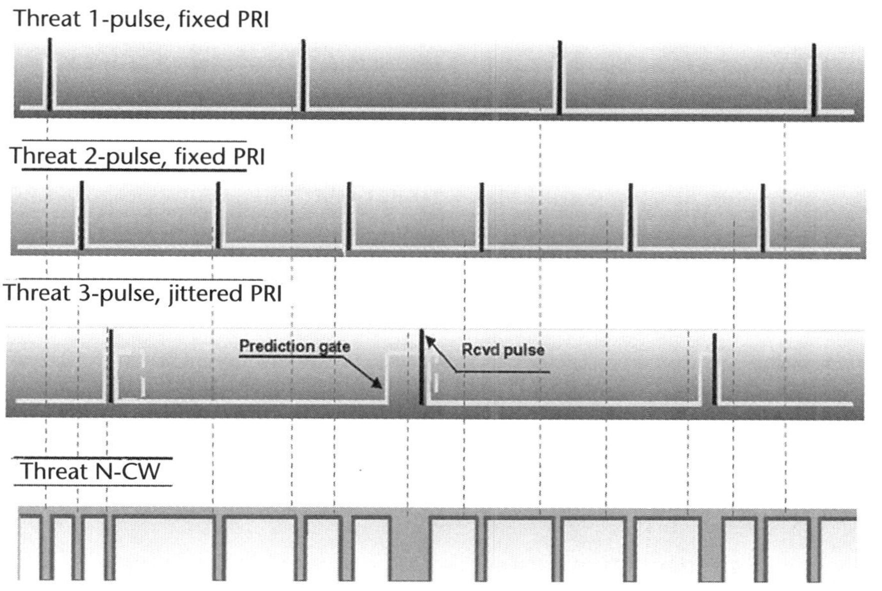

Figure 5.8 Time-shared SPJ resource management through PTOAs in a simultaneous multithreat scenario.

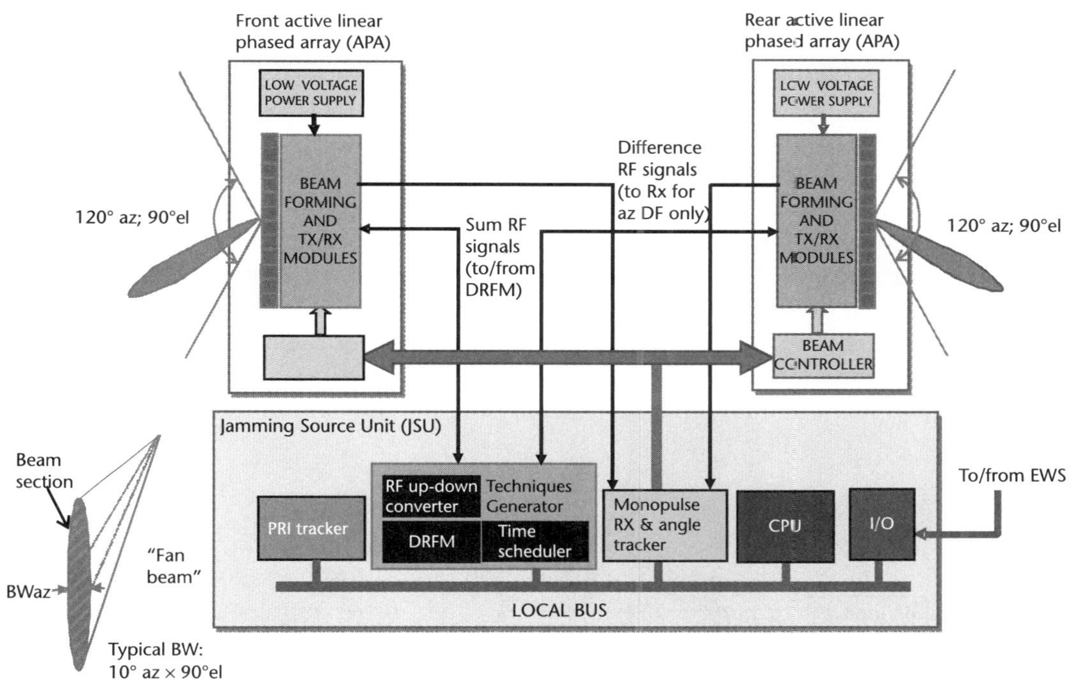

Figure 5.9 Schematic block diagram of a modern ASPJ.

implementation of the various types of RECMs mentioned in the introduction to this chapter.

5.3 Digital Radio-Frequency Memory

As shown in Chapter 2, the vast majority of modern radars employ Doppler and pulse compression techniques, which must be countered by coherent jamming resources that can replicate and manipulate the signals of the radar threats against which they are directed. Only through coherent jamming can RECM equipment overcome the significant advantage of 30 to 50 dB gained by the above techniques being employed in radar signal processing with respect to noise jamming waveforms.

The concept of a DRFM is simple and it began to be applied in the late 1970s, but its implementation required a number of technological advances before it could become an effective device. Its function is the coherent storage (ideally, for an infinite amount of time) of the threat signal so it can be reproduced and manipulated (in time delay and frequency content) by means of digital techniques in order to provide deceptive returns to the threat.

The DRFM has many applications in the EW domain, such as these:

- Radar/communication signal storage for subsequent ELINT/COMINT analyses,
- Deception of covert communications,
- Jamming operations,
- Anti-ARM decoys replicating radar signals,
- Radar signal generation for simulation and test equipment.

The block diagram of a single-sideband (SSB) DRFM is shown in Figure 5.10.

The DRFM receive path is very similar to a conventional superheterodyne receiver with regard to the rejection of spurious signals. At the device RF input, a bandpass filter (BPF) followed by an amplifier is used to pass and amplify only the signals of interest. The RF intercepted signal is downconverted to baseband by a LO. A lowpass filter (LPF) removes the aliasing frequency components above the Nyquist rate and the unwanted mixer intermodulation products. At the output of the LPF, the signal is sampled with a high rate clock and digitized by an ADC with ≥4 to 6 bits. The higher the number of bits, the larger the dynamic range but also the slower the conversion process. The bandwidth of the device is (the Nyquist rate), with $k < 1$, due to the presence of both the BPF and LPF.

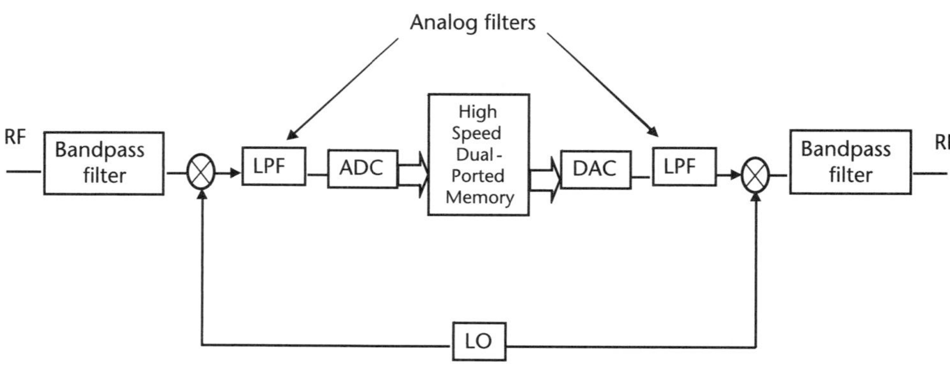

Figure 5.10 Block diagram of a SSB DRFM.

The digital samples are recorded into a high-speed dual-ported memory, so that the stored signal can be replayed simultaneously through memory control. Recording in the dual-ported memory usually requires the use of serial-to-parallel circuitry at input to reduce the conversion data rate, so as to match the dual-ported memory's input bandwidth. Replay of the stored signal vice versa requires parallel-to-serial circuitry at output to match the dual-ported memory's output bandwidth.

The digital signal retrieved from memory is fed to a digital-to-analog converter (DAC) followed by a LPF in order to reproduce the analog signal. The baseband signal is then mixed with the same LO used in the downconversion in order to reconstitute the RF signal. The RF output is then passed through a BPF to provide the transmitter section with only the desired signal spectrum. The retention of the phase of the intercepted RF signal is achieved with a double-sideband (DSB) DRFM, the block diagram for which is shown in Figure 5.11.

The input RF signal phase retention is provided by the digitization of the two channels (I and Q) at the output of the quadrature IF modulator, which also downconverts the RF input signal into baseband by mixing it with the LO frequency.

The block diagram of the DSB DRFM configuration of Figure 5.11 also shows are the memory controller (MC) and high-speed (HS) dual-port memory. The MC manages the storage of the data by assigning the memory address functions as well as the retrieval of the signal data and the relevant delay/recall operations.

The dual-ported memory size determines the maximum duration of the emitter signals that can be stored. Because the memory is nondestructive, the DRFM is capable of reading out a stored signal within a period of 10 to 20 ns (which is an

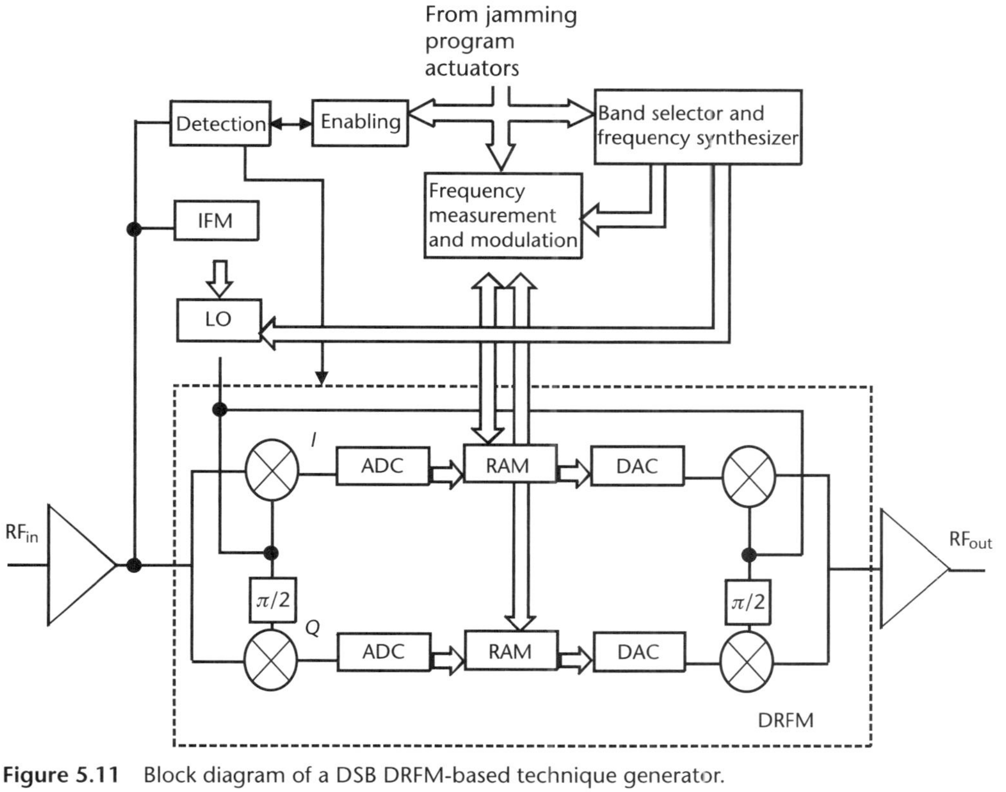

Figure 5.11 Block diagram of a DSB DRFM-based technique generator.

important feature in some CM techniques, as will be shown in Chapter 6). Recall that the signal can be stored for an indefinite period of time and read out as many times as desired or until the threat signal is cancelled from memory.

The operation of the DRFM against coherent radars is limited by the long-term stability of the LOs used in the down-up frequency conversion process. Current LO frequencies are produced by highly stable frequency synthesizers, based on direct digital synthesis (DDS) oscillators and phase-locked control loops (PLLs).

Note that the DSB DRFM has, at the expense of a more complex architecture, the advantages of a bandwidth that is double that of the SSB DRFM, using the same clock rate (i.e., $B_{DSB} = \pm$) and the retention of the intercepted RF signal phase.

The wideband DSB DRFM has become the essential device of RECM systems for its flexibility in replicating any threat signal within its frequency coverage, including wideband frequency and pulse-compression radar signals. Therefore, it is benefiting from serious technological research efforts to improve the high-speed multibit ADC, which is the fundamental device of the DRFM architecture. Multiple bits are required to reduce spurious responses due to the ADC quantization, to prevent small-signal suppression in a simultaneous multithread intercept, and to implement a number of sophisticated ECM techniques. The dynamic range (DR) of the DRFM is dependent on the number n of the ADC bits as is DR $\approx 6n$ (dB). A recent GaAs technology ADC provides a greater than 1-GHz instantaneous bandwidth with up to 6 to 8 bits. Spurious signals in the DRFM are produced for the following causes:

- An imbalance between the quadrature channels, which causes image frequencies as shown in [3, 4].
- Quantization process: a DRFM with n bits has 2^{n-1}amplitude levels that are usually equally divided into positive and negative levels. This arrangement produces only odd harmonics. The spurious levels generated by quantization have been evaluated by Pace [4, Chapter 7, Section 7.7]. Typically a 2-bit DRFM has the worst spurious level of about –18.1 dB below the peak, and a 4-bit DRFM has a worst spurious level of about –34.4 dB below the peak.
- Intermodulation products due to the simultaneous presence of two or more intercepted signals.

5.3.1 Phase-Sampled DRFMs

Phase sampling of the intercepted signal is a much used technique in a DRFM alternative architecture, because the radar signal information is mainly carried in the angle (and hence phase) modulation. In a phase-sampled DRFM, the amplitude information is discarded and the instantaneous phase is sampled and quantized. The block diagram of a 4-bit phase-sampled DRFM is shown in Figure 5.12.

After the LO downconversion, the signal is hard limited so that the phase can be sampled and quantized by the analog comparators. Both I and Q signals are first split and phase shifted by $\pi/4$ rad (45°) forming I_{45} and Q_{45} signals. The four signals—I, Q, I_{45}, and Q_{45}—are then further split and phase shifted by $\pi/8$ rad (22.5°) thus forming the following four further signals: $I_{22.5}$, $Q_{22.5}$, $I_{67.5}$, and $Q_{67.5}$. These eight signals are again split into 16 signals, with 8 of the 16 channels

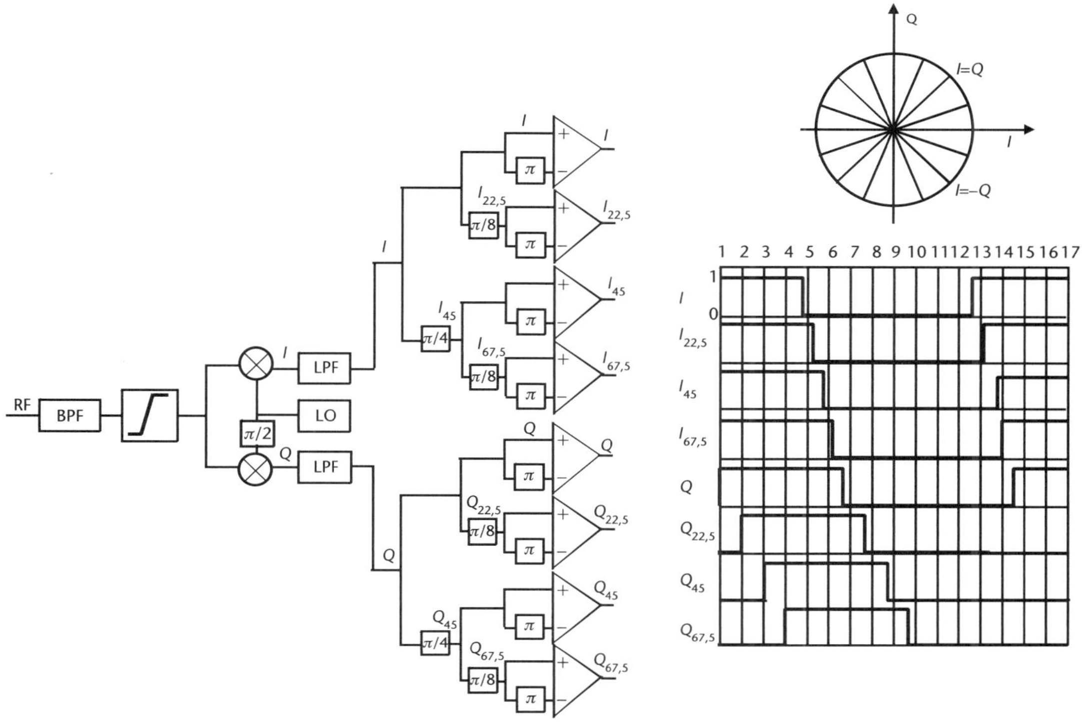

Figure 5.12 Block diagram of a 4-bit phase-sampled DRFM.

further phase shifted by π rad (180°). The 16 channels are then compared two by two to determine the 4-bit binary output representation of the phase. This particular binary code is called an n (n = 4) bit Johnson code, which has 2^n = 16 states, being coded with only 2^{n-1}comparators (C), as shown in Figure 5.13. Note that 4-bit amplitude DRFM requires 15 comparators and a precision resistor string, which is amplitude sensitive. Relevant to the hardware implementation of a phase DRFM, the phase-shifted signals are manipulated using symmetrical resistor rings and ring demodulators [3]. At the output the I and Q components are synthesized from memory and mixed with the LO to reconstruct the signal for retransmission.

The advantages of phase-sampled DRFM include similar performance to the amplitude DSB DRFM, but the storage required is for only half the number of bits (as shown with the Johnson code for n bits). Also, harmonic suppression is improved over amplitude DRFM due to the hard limiting at the I and Q channel inputs. Further, phase modulation can be applied directly at the output signal.

The disadvantages are relevant to the simultaneous signal processing, because small signals are suppressed by up to 6 dB and spurious intermodulation products are generated. Further in applications such as the straight repeater jammer an additional subsystem is required to re-create the missing amplitude information.

The performance comparison between amplitude and phase DRFM reported in [3] shows that a 4-bit amplitude DRFM and a 4-bit phase DRFM have, respectively, worst spurious levels of –35.9 dB_c and 23.5 dB_c and a total amount of spurious signals of 0.27% and 1.28%. Note that the advantage of the phase approach over the amplitude approach diminishes as the number of bits increases.

States-bit	1	2	3	4	5	6	7	8	9	10	11	12	13	14	15	16
I	1	1	1	1	0	0	0	0	0	0	0	0	1	1	1	1
$I_{22.5}$	1	1	1	1	1	0	0	0	0	0	0	0	0	1	1	1
I_{45}	1	1	1	1	1	1	0	0	0	0	0	0	0	0	1	1
$I_{62.5}$	1	1	1	1	1	1	1	0	0	0	0	0	0	0	0	1
Q	1	1	1	1	1	1	1	1	0	0	0	0	0	0	0	0
$Q_{22.5}$	0	1	1	1	1	1	1	1	1	0	0	0	0	0	0	0
Q_{45}	0	0	1	1	1	1	1	1	1	1	0	0	0	0	0	0
$Q_{62.5}$	0	0		1	1	1	1	1	1	1	1	0	0	0	0	0

Figure 5.13 Johnson's code for 4-bit phase-sampled DRFM.

The reconstruction of the I and Q signals for the 4-bit system is performed by phase shifting the eight digital inputs in $\pi/8$ rad (22.5°) increments. The DRFM output can be directly phase modulated by appropriate rotations of the eight inputs before reconstruction of the I and Q channels.

Examples of typical jamming programs implemented by modulators associated with a phase-sampled multibit DRFM and a photograph of a complete DRFM with associated modulators are shown in Figures 5.14 and 5.15, respectively.

5.4 Radar ECM Transmitters

Modern RECM transmitters are composed of an RF chain of a wideband, solid-state, high-gain amplifier (SSA), which drives a wideband, high-power output amplifier (HPA). There are two types of HPAs, based on either vacuum tubes (valves) technology or on SSAs, each one driving an element of a phased array antenna. A RECM wideband vacuum tube HPA is usually made of TWT amplifiers of the helix type and are constructed as shown in Figure 5.16.

The principle of operation of the TWT, which was invented by R. Kompfner at the University of Birmingham, United Kingdom, in 1943 (and later improved by him at the Bell Labs), is the amplification of an electron beam provided by an electromagnetic field. The EM field is produced by current modulated at RF that is circulating in a helical slow-wave circuit wrapped around a tube where electrons are extracted from an anode (also called an *electron gun*) and accelerated along the tube by a high-amplitude electric field applied between the anode and the cathode (also called the *collector*). The electrons deliver a high power that is RF modulated. TWTs, due to their helical structure can be implemented in a practical size only in the microwave bandwidth and are used in many electronic systems beyond EW equipment such as radars and communication links (especially satellite transponders).

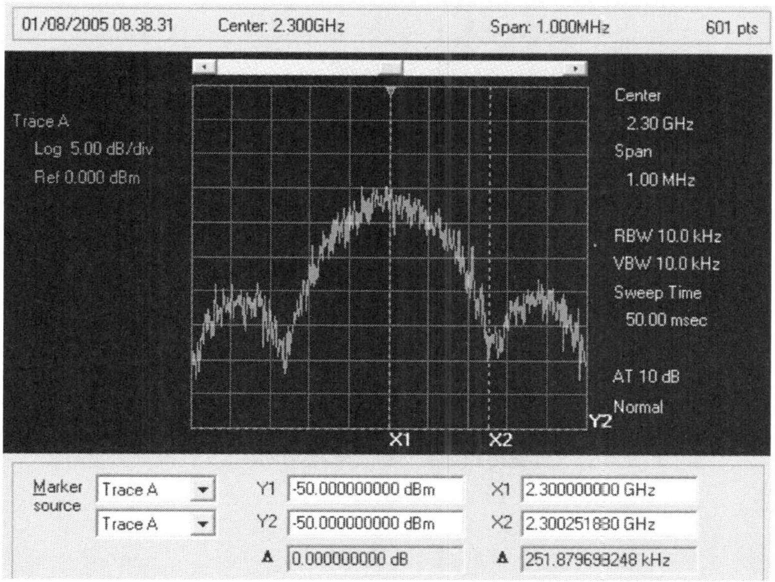

(a)

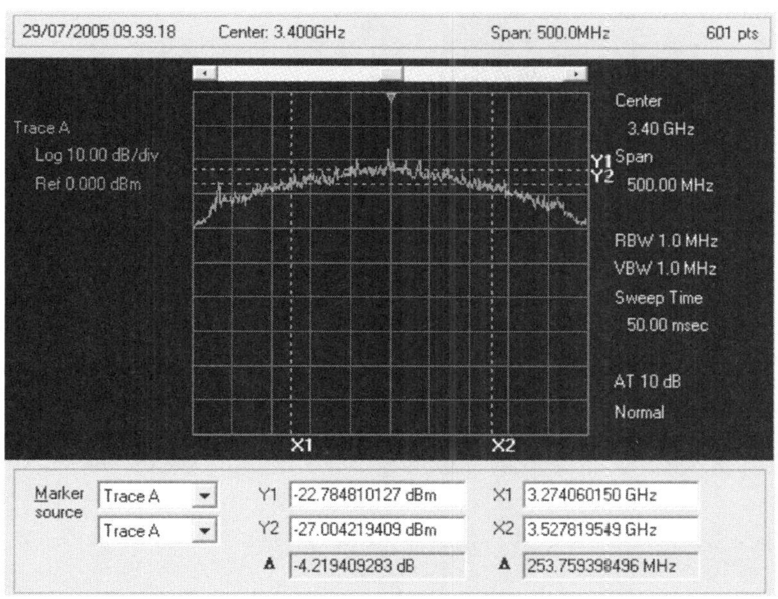

(b)

Figure 5.14 (a) A 250-kHz bandwidth spot noise waveform generated by phase-sampled multibit DRFM. (Courtesy of Elettronica S.p.A.) (b) A 256-MHz bandwidth barrage noise waveform generated by phase-sampled multibit DRFM. (Courtesy of Elettronica S.p.A.)

Wideband TWTs are available in two types: either with a CW power output in the range of 100 to 400W or with pulsed power output in the range of 1 to 2 kW with decreasing duty cycle (from 10% to 4%). The typical power output pattern of a wideband TWT has a parabolic shape with the maximum at midband and nearly equal lower values (on the order of 1 dB) at the two extremes of the bandwidth.

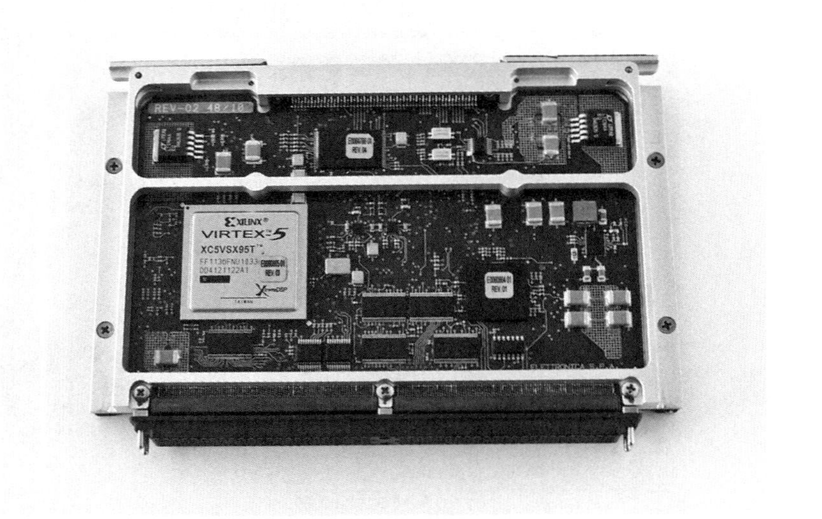

Figure 5.15 A complete DRFM with associated modulators. (Courtesy of Elettronica S.p.A.)

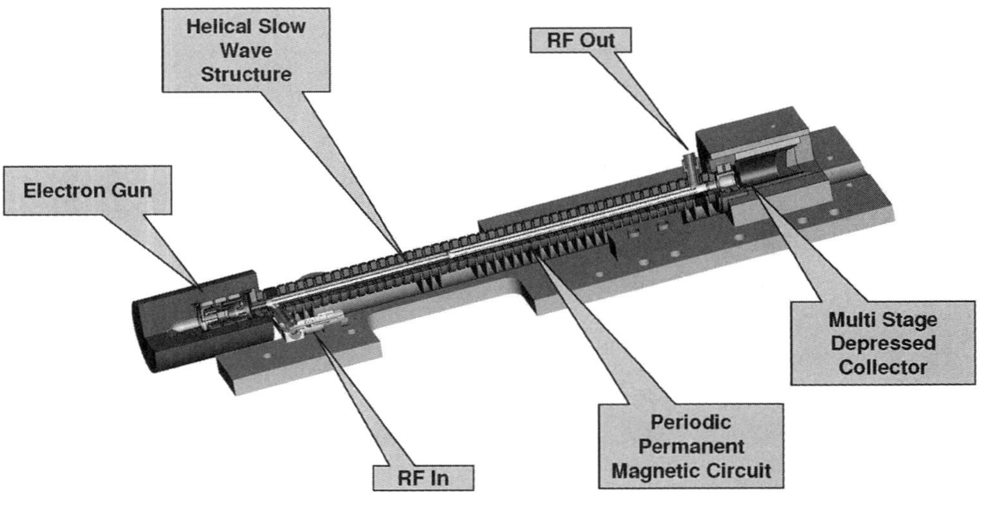

Figure 5.16 Sketch of a TWT. (Courtesy of e2v.)

The efficiency (i.e., the ratio of output power to the power consumption of the device) of wideband TWTs is inversely proportional to the width of the covered band and ranges from 15% to 25%. Of course, one disadvantage is that the high delivered power output is achieved by utilizing additional high-voltage power supplies to provide voltages on the order of 4 to 10 kV between the two electrodes. The higher the voltage, the higher the danger of arcing in the presence of humidity or at high altitudes due to the low air density.

A further disadvantage of wideband TWTs is their limited life (due to cathode poisoning), which is usually less than 2,000 hours of operation. Also, as TWTs become operative, they need a warm-up period (usually on the order of <10 sec), so they must be kept in a constant standby state for the duration of a platform mission, which can rapidly exhaust their operational lifetime. Spare TWTs in storage must be briefly switched on with a periodicity recommended by the manufacturer (the so-called "refresh operation") to prevent the loss of their vacuum due to the permeability of the metal parts used on TWTs.

TWT technology has largely improved over the years with increasing efficiency and decreasing volumes. With regard to volume they have evolved from midi-TWTs (around 1 liter) to mini-TWTs (about 0.2 liter) and finally to microwave power modules (MPMs). MPMs have been developed by a number of U.S. firms under a program started by the Naval Research Laboratory in 1990. A MPM is composed of a mini-TWT that is driven by a high-gain, high-power SSA and enclosed within a small size unit (on the order of 1 to a couple of liters depending on the power output), comprising also a high-density power conditioner (connected to the platform's mains) and a conduction cooling package. Due to their limited size, MPMs have been used as transmitters in towed decoys. A sample MPM is shown in Figure 5.17. Typical TWT performance statistics are listed in Table 5.1.

The power output of wideband TWTs is sufficient to implement the transmitter of a SPJ by connecting the TWT output to a wideband, wide-spatial-coverage horn antenna either with a low-loss power cable if the connection length is on the order of 1m or a waveguide if the length is longer.

When larger power outputs are required, TWTs can be combined through phase shifters at their input and by means of a Butler matrix (which is microwave network composed of 90° and 180° hybrids) at their output as shown in Figure 5.18 for the case of a Butler matrix of order 2. It is used to connect two antennas: depending on the phase setting at the input of the TWTs, the whole power can be concentrated in one of the two antennas. A further combination method is achieved by employing a Rotman lens, instead of the Butler matrix, as a beam-

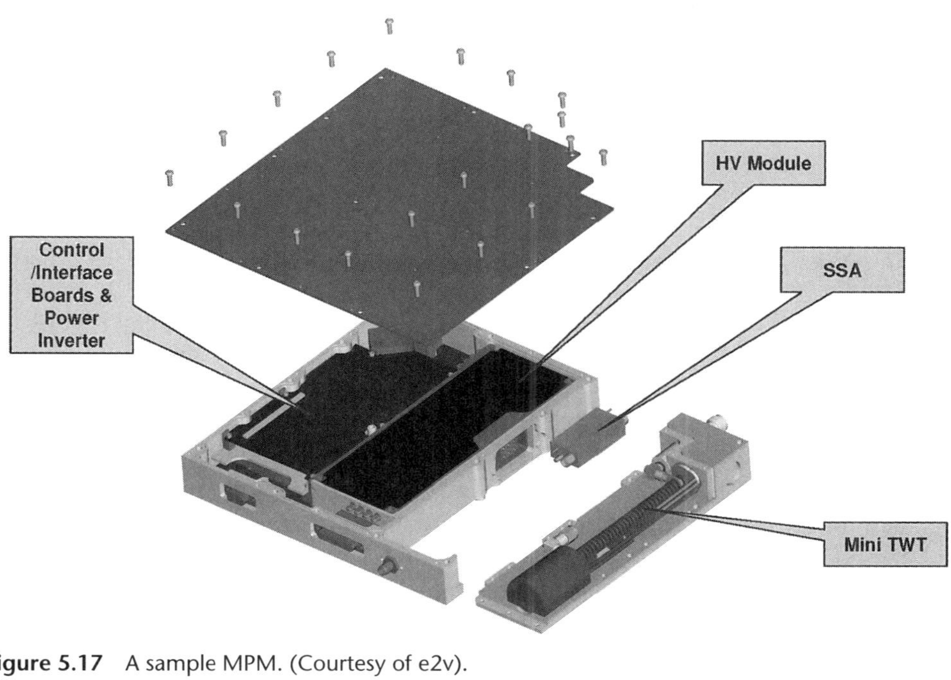

Figure 5.17 A sample MPM. (Courtesy of e2v).

Table 5.1 Typical Performance of CW TWTs (Courtesy of e2v.)

Type	*Product No.*	*Frequency (GHz)*	*Power Minimum (W)*	*Power Maximum (W)*	*Small Signal Gain (dB)*	*Cathode Voltage (kV)*	*Prime Power (W)*	*Size (L × W × H) (mm)*
Midi	N10110	6–18	180	260	45	6.2	1050	329 × 50 × 62
	N10221F	4.5–18	100	210	37	6.2	1050	336 × 50 × 62
Mini	N10173	6–18	75	125	32	4.55	430	220 × 27 × 29
	N20160	4.5–18	50	140	32.5	4.55	470	220 × 27 × 29

a	b	O_1	O_2
0	π	$(2A/\sqrt{2})^2 = 2A^2$	0
0	0	0	$(2A/\sqrt{2})^2 = 2A^2$

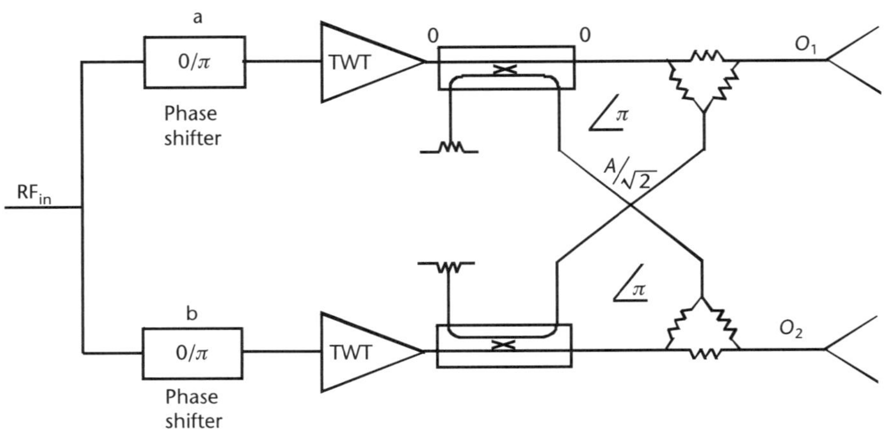

Figure 5.18 Schematic block diagram of a Butler matrix of order 2.

forming network. By operating the TWTs in linear mode it is possible to produce multiple simultaneous beams, overlapped in time.

In recent years advances in semiconductor technology relevant to microwave power generation have allowed for the implementation of small, medium-power-output solid-state (i.e., semiconductor instead of vacuum tubes) amplifiers, that, appropriately driven in phase and coupled directly with the radiating elements of an array antenna, transform the array into an active phased array antenna capable of steering a high-power beam within a large spatial coverage. If in addition to the transmitting path a receiving path is also implemented within the same module connected to the radiating element of the array antenna, the phased array antenna becomes a transmit/receive (T/R) antenna very similar to the ones employed by modern radars.

Active phased array T/R antennas (APAs) constitute a highly efficient solution for integrated EW systems that comprise RECMs compliant with the operational requirement of countering many simultaneous threats. The main advantages of APAs are as follows:

- Maximum ERP is delivered to each threat to be countered, through their angular selection and tracking.
- ERP generation is efficient with reductions in mass, volume, power consumption, and dissipation.
- The use of high-voltage power supplies is avoided.
- If a T/R module fails, system performance degrades gracefully.
- The integrated EW system's sensitivity is increased because of the higher phased array antenna gain, which is useful for countering LPI radars.
- Highly accurate direction finding (HADF) is possible with many techniques (amplitude monopulse and phase interferometry).

The basic characteristic of a phased array antenna is the ability to change the peak direction of its radiation pattern without mechanical displacement of its radiating aperture. The phased array aperture produces a planar wavefront, whose angular direction of propagation can be changed electronically by imposing a suitable phase relationship among the radiating elements that form the array. A conceptual block diagram of a linear phased array antenna is shown in Figure 5.19.

The steering of the beam in a given direction θ is obtained by shifting the phase of the Tx/Rx signal of each T/R module by an amount $\Delta\Psi = (2\pi d/\lambda)\sin\theta$ (where d is the spacing between two elements of the array and λ is the wavelength) with respect to the preceding module in the array. This can be achieved by means of a digitally controlled phase shifter. With reference to Figure 5.19, a signal wavefront can be transmitted in the direction θ_0 with respect to the array by applying the following phase shift to the ith T/R module:

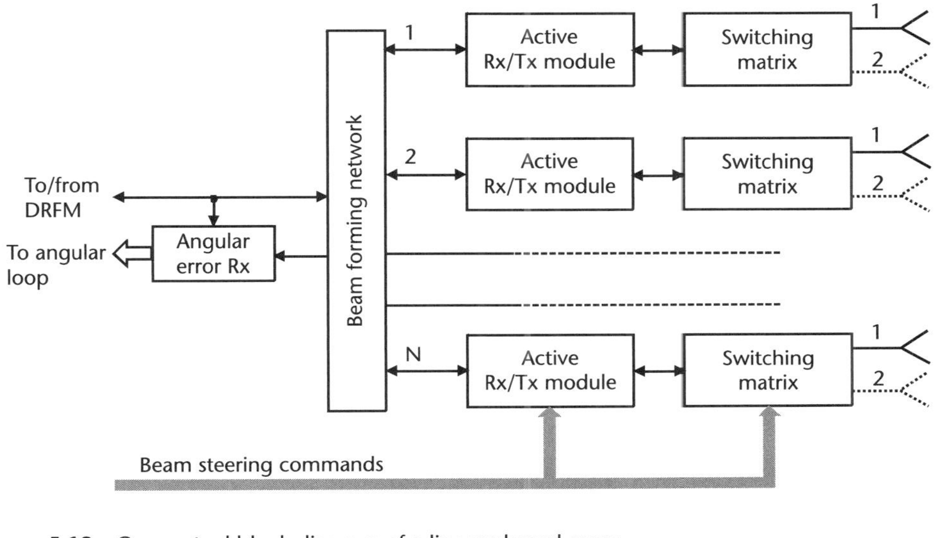

Figure 5.19 Conceptual block diagram of a linear phased array.

$$\Delta\psi_i = (i-1)\frac{2\pi d}{\lambda}\sin\theta_0 \tag{5.9}$$

where $i = 1, 2, \ldots, N$.

Conceptual block diagram of a linear phased array.

As shown in Appendix C, the expression of the array factor (i.e., the far-field radiation pattern of the array) for a uniform illumination of unitary amplitude across the array is:

$$F(\theta,\phi) = g_{el}(\theta,\phi)\sum_{0}^{N-1} e^{-jk((2\pi d/\lambda)\sin\theta-\Delta\phi_k)} = g_{el}(\theta,\phi)\frac{\sin\frac{N\pi d}{\lambda}(\sin\theta-\sin\theta_0)}{\sin\frac{\pi d}{\lambda}(\sin\theta-\sin\theta_0)} \tag{5.10}$$

where $g_{el}(\theta, \phi)$ is the voltage gain of the single radiating element of the array in the direction (θ, ϕ) with respect to the array broadside. The beamwidth of the single element must be equal to the required spatial coverage. The radiating element also provides the polarization feature to the phased array antenna. To meet the requirement of no blind polarization against all type of radars (which use linear or circular polarized antennas), the jammer antenna elements are linearly polarized either at a 45°slant or double linear (horizontal and vertical) with instantaneous selectable polarization to maximize the delivered ERP to the victim radar. Sometimes, due to space limitations, as in the case of towed decoys, circular polarization (either RHCP or LHCP) is also used. The most commonly used radiating elements are cross-printed notch elements (Vivaldi) that provide double linear polarization or sinuous spirals polarized at a 45° slant. Simple vertical-printed notch elements are used with a 45° polarizer in front of the array.

The maximum of (5.10) is obtained for $\theta = \theta_0$ and is:

$$F(\theta_0,\varphi) = Ng_{el}(\theta,\varphi) \tag{5.11}$$

Expression (5.11) implies that the maximum receive gain of the phased array antenna is obtained by giving to each module a phase shift that depends on the knowledge of the direction of arrival and the wavelength of the emission. The beamwidth of the phased array as shown in Appendix C is dependent on the type of complex illumination (amplitude and phase) impressed on the array elements. In case of uniform illumination per (5.10), the beamwidth expressed in degrees is $\theta_B = 51\lambda/Nd$. The beamwidth is inversely proportional to the length of the array Nd. When the beam is scanned to an angle θ_0, the projected length of the array in that direction is reduced by a factor $\cos\theta_0$ and, hence, the beamwidth is increased by the same factor; that is, the scanned beamwidth is $\theta_{B,\text{scan}} = \theta_B/\cos\theta_0$. As reported in Appendix C, one negative characteristic of the phased array in terms of the dependence of the ratio of the distance of the elements on the wavelength d/λ is the appearance of grating lobes in the visible space when the beam is scanned toward its maximum scan angle. The appearance in visible space of grating lobes is negative on transmission because it subtracts power from the main beam and on reception as is can provide an erroneous DOA of the emitter. The avoidance of grating lobes in the visible space is obtained by selecting a spacing between the elements on the order of $d/\lambda_{\min} = 1/2$.

The effective radiated power of a phased array is the product of the antenna gain times the total power delivered by the radiating elements:

$$ERP_{PA} = F(\theta,\phi)g_{el}(\theta,\phi)\cdot NP_{el}/L_a \tag{5.12}$$

whose peak is

$$ERP_{PA,peak} = N^2 g_{peak}P_{el}/L_a \tag{5.13}$$

where P_{el} is the power delivered by each T/R module and L_a represents the transmitting array antenna losses.

The efficiency in ERP generation achievable with phased arrays can be demonstrated by comparison with the ERP generated by a TWT connected to an antenna having the same spatial coverage as the phased array. In this case the antenna gain G_{ant} is the same as the gain of the single radiating element of the array g_{el}, and the ERP of this solution is $\text{ERP}_{\text{TX}} = G_{\text{ant}}P_{\text{TWT}}/L_{\text{Tx}}$. By equating the ERP of the two solutions, we have

$$N^2P_{el}/La - P_{TWT}/L_{Tx} \tag{5.14}$$

By assuming equal transmitting losses from both solutions, we have, noting that the total power transmitted by the phased array is $P_{\text{array}} = NP_{el}$:

$$P_{array} = P_{TWT}/N \tag{5.15}$$

This result shows that, to provide the required ERP, an APA must have an installed power P_{array} that is $1/N$ the required installed power from a TWT. This factor is somewhat reduced taking into account the current lower efficiency of GaAs technology amplifiers (about 10%) with respect to TWTs (about 20%), but in any case a large advantage is gained from the active phased array solution in terms of mass, power consumption, and dissipation.

A typical architecture for a linear APA is shown in Figure 5.20. The APA is composed of a set of radiating elements directly connected to a set of active solid-state T/R modules. The modules are connected to a beamforming network where the signals are processed at RF level to provide both the bearing error signals (in accordance with the monopulse technique) to the receiver unit as well as the emitter signal to a DRFM for the subsequent generation of jamming signals. An interface module, called the *control interface* (shown as *selector*) in the figure, allows for receiving and actuating the commands necessary to swiftly steer the APA beam to the threat direction as well as to manage the APA receiving and transmitting phases.

Due to the narrow antenna beamwidth (in dependence to the array length) and to the platform and threat kinematics, the commands that steer the unit toward each threat have to be constantly updated to ensure that the threat angle tracking is correct. Bearing error signals processed by the beamformer receiver are subsequently processed at the video level by means of a closed-loop control system, as

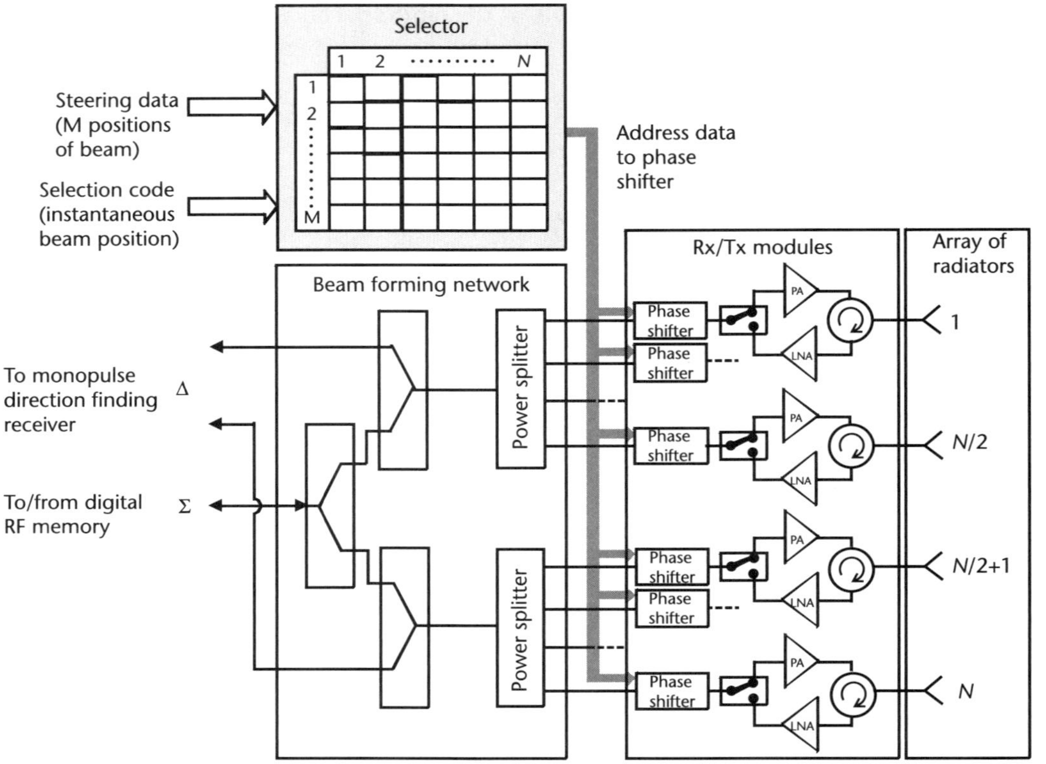

Figure 5.20 Typical architecture of a linear APA.

depicted in Figure 5.21, which compensates for platform maneuvers by receiving the platform navigation data.

The swift beam switching (tens of nanoseconds) among the various threat directions is performed, on issuance of a PTOAG command, by the control interface board in which the updated directions of the *M* tracked threats are memorized. In addition, the relevant phase shift commands (depending on the threat frequency measurement) are evaluated and distributed to each T/R module. A typical implementation of a wideband moderate CW power output T/R module is shown in Figure 5.22.

Generally, a linear phased array is constituted by only one array of radiating elements, but other arrays of radiating elements can be added, as shown in Figure 5.23. They are added either to achieve complete 360° horizontal coverage or to increase the elevation coverage by introducing an equivalent number of switchable outputs to the T/R module as shown in Figure 5.22. A typical implementation of an airborne, fan-cooled linear APA is shown in Figure 5.24.

Whenever very large ERPs are required, planar phased arrays can be employed. As shown in Appendix C, the lattice selected can be either rectangular or triangular (with spacing d_x and d_y, respectively, along the two-axis x and y of the array) and the steering of the beam in two dimensions is achieved by impressing to each element i,j a phase shift $\Delta\Psi_{i,j}$ of the type:

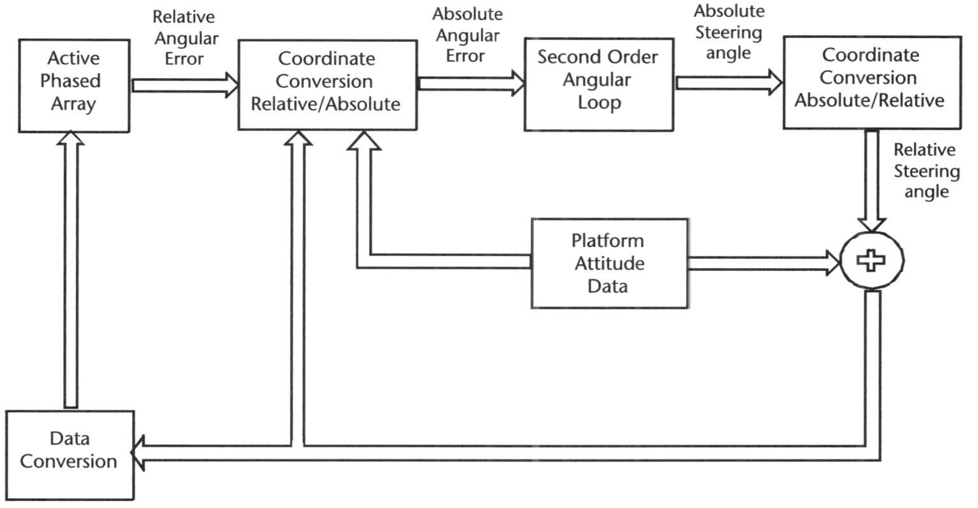

Figure 5.21 Block diagram of an angular closed-loop tracker.

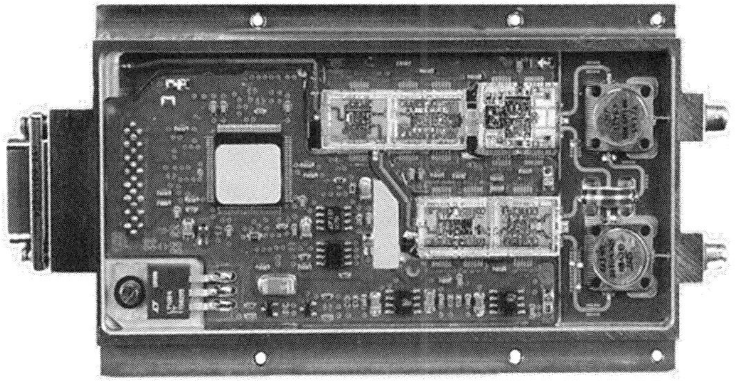

Figure 5.22 Typical wideband moderate CW power output T/R module. (Courtesy of Elettronica S.p.A.)

$$\Delta\varphi_{i,j} = i\frac{2\pi d_x}{\lambda}\sin\theta_0\cos\phi_0 + j\frac{2\pi d_y}{\lambda}\sin\theta_0\sin\varphi_0 \tag{5.16}$$

where (θ_0,ϕ_0) is the direction (elevation, azimuth) with respect to the planar array boresight (0,0) to which the array beam is steered and where $i = 0,1, 2, \ldots, N - 1$ and $j = 0, 1, 2, \ldots, M - 1$ are the coordinates of the individual element in the array of dimension $N \times M$. The peak effective radiated power of such an array is $\mathrm{ERP}_{\mathrm{peak}} = NMg_{\mathrm{peak}}P_{el}/L_a$. This very high ERP is required for ship self-protection systems that have a very large RCS (about 20 to 30 dB above the fighter aircraft one) and for mutual protection systems, both airborne (such as SOJ and EJ systems, as discussed in Section 5.1) and shipboard.

The advantages at the same delivered ERP of the APA solution with respect to a multiple-TWT solution combined with the appropriate Butler matrix and sect oral horns are compared in Table 5.2. The tables cover installation data (i.e., power

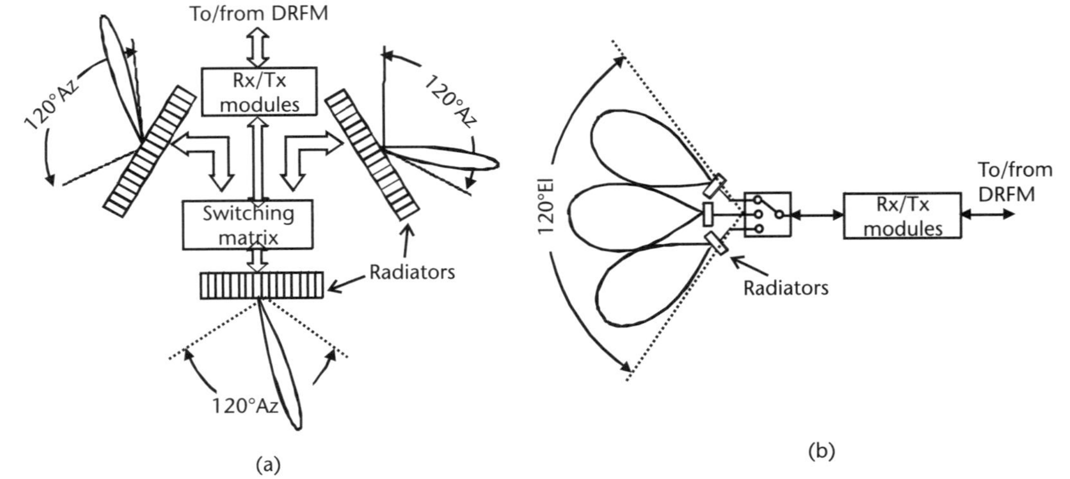

Figure 5.23 Linear phased array configurations for achieving increased spatial coverage: (a) complete horizontal plane coverage and (b) elevation coverage increase by means of three linear phased arrays.

Figure 5.24 Typical implementation of an airborne, fan-cooled linear APA. (Courtesy of Elettronica S.p.A.)

Table 5.2 Comparison Between APA and TWT Solutions for Large Required ERP (Re-adapted from: [5].)

Characteristic	*APA*	*TWTs with Butler matrix*
Type	48 TRM (P_{el} = 4W)	12 CW-TWT (200W)
ERP (Average max boresight) (kW)	40	40
Spatial Coverage	90°Az, ± 30°El	90°Az ± 30°El
Additional Sensitivity (dB)	Up to 20 dB	RX Mode provided by separate passive array and monopulse Rx
Reliability: MTBF (h) 1 flr	300	100
Reliability:MTCBF (h) (2 h mission)	1500	300
Volume (l) (PS Included)	90	117
Mass (Kg) (PS Included)	80	105
Power Consum. (PS Included)	3500 VA	5500 VA
Power Dissipat. (PS Included)	3000W	4000 W

PS = power supply.

consumption, dissipated power, and volume), life cycle costs (mean time between failure, MTBF), and mission criticality. Note that the APA's performance degrades gracefully: the failure of 10% of the T/R modules results in less than a 1-dB loss in ERP. Note also from the table the further APA advantage of an increase in the integrated EW system's (i.e., ESM/ECM) sensitivity and HADF capability due to its monopole beamformer network.

A large ERP wideband APA for ship self-protection is shown in Figure 5.25. Large planar APAs can produce multiple independent beams, both on reception as well on transmission, by using an appropriate complex beamforming network or by partitioning the planar array into subarrays. Overlapped transmitting beams require that the output T/R module amplifier operate in linear mode, and this mode strongly reduces the beam ERP. If a small number of simultaneous, even overlapped in time, beams have to be produced, the partition of the planar array into subarrays is convenient and also reducing the complexity of the beamforming network. Figure 5.26 shows an implementation of four independent beams from a planar APA that uses a monopulse measurement beamforming network.

The architecture of a large ERP RECM, such as the SOJ or the EJ, is very similar to the SPJ one described above. A schematic block diagram of an EJ configuration is shown in Figure 5.27. SOJ equipment can be installed either internally in a wide-body aircraft or within the belly pod of a large fighter aircraft. In the latter case, the pod has to be autonomous; that is, it has to be provided with cooling and an electrical power generator, powered by a ram air turbine generator, because a fighter A/C usually cannot cope with such large incremental requirements. To provide a jammed corridor for the intruder aircraft, the SOJ usually has to loiter about the direction of the intruder corridor. Therefore, the SOJ is composed of two large planar APAs, one on each longitudinal side of the pod and each covering a spatial sector. The spatial section is quite wide in azimuth but narrow in elevation, due to the far distance of operation of the SOJ aircraft with respect to the victim surveillance radars. The two APAs, usually operating in the low-frequency radar band

Two planar APA at 90° for 180° azimuth coverage

(a)

Single planar APA

(b)

Figure 5.25 Large ERP wideband APA for ship self-protection. (Courtesy of Elettronica S.p.A.)

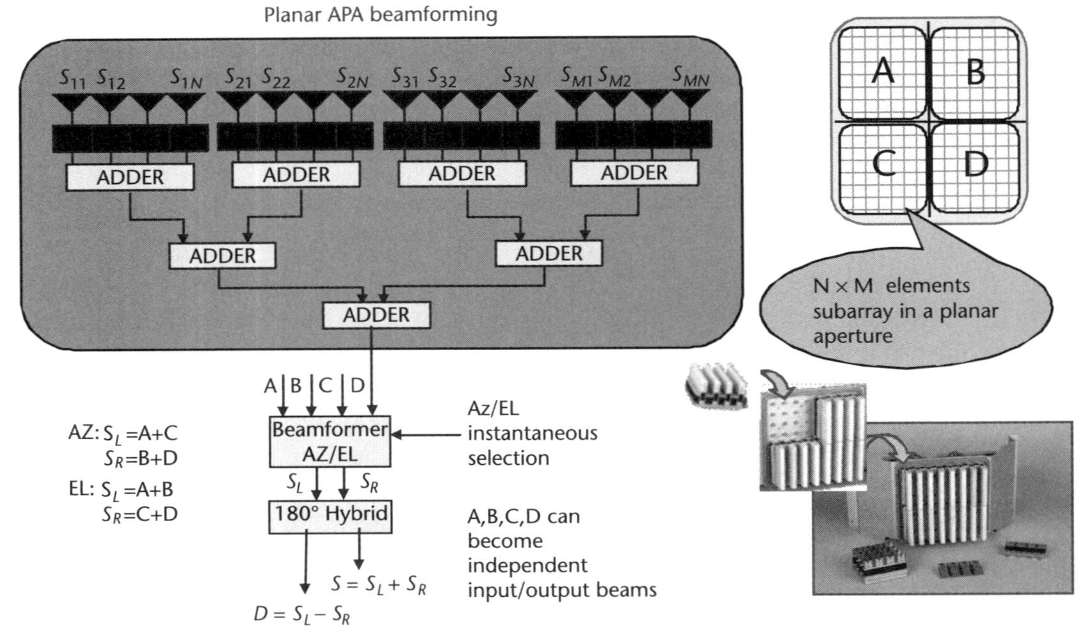

Figure 5.26 Four independent beams from a planar APA that uses a monopulse measurement beamforming network. (Courtesy of Elettronica S.p.A.)

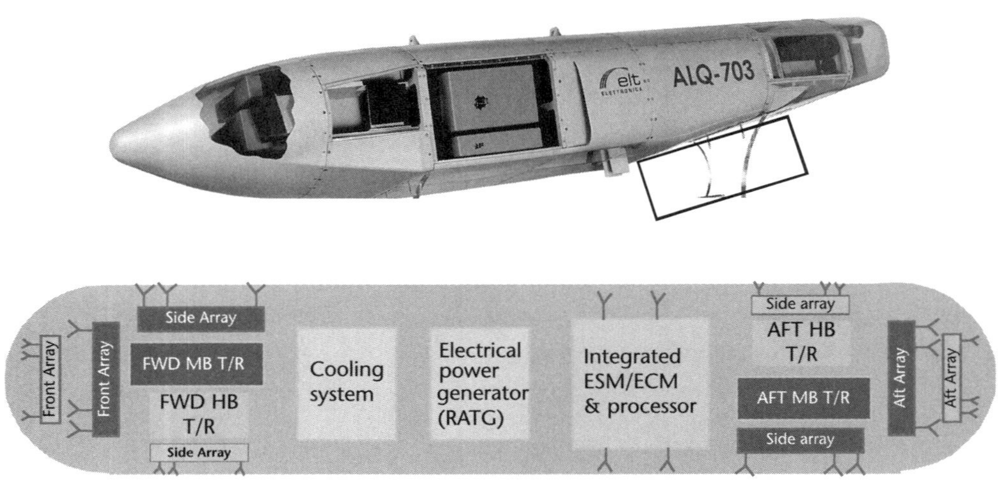

Figure 5.27 Schematic block diagram of an EJ architecture based on high-ERP APAs. (Courtesy of Elettronica S.p.A.)

typical of surveillance radars, are switched on and off alternatively in accordance with the side looking at the intruder corridor. The large, wide azimuth sector and smart types of maneuvers performed by the aircraft while performing 180° turns help to minimize the jamming silence periods.

EJ equipment as well is usually installed in the belly or pylon pods of fighter aircraft that penetrate the adversary's territory together with the intruder aircraft. Because the EJs are within this territory, they usually need complete azimuth and radar frequency band coverage. Due to the large number of threats to be countered during their mission, EJ systems are provided with a large number of jamming resources (DRFMs and simultaneous APA beams) to counter threats.

5.5 Chaff

Chaff is a form of volumetric radar clutter composed of a very large number (up to millions) of small, metallized and shorted tuned dipole antennas. The dipoles are cut at about $\lambda/2$ to achieve the maximum ratio of RCS to weight of used material. These antennas are dispensed into the atmosphere to confuse or disrupt radar operation. Chaff is used either to create a corridor in the air space controlled by a surveillance radar to conceal some intruder aircraft penetrating into enemy territory or to provide self-protection to aircraft or ships when subject to illumination by tracking radar, such as anti-aircraft artillery, SAMs, or AAM seekers.

A complete discussion about the creation of chaff corridors and the relevant chaff characteristics is given in [14], which we recommend to the interested reader. Here it suffices to recall that chaff is dispensed at high altitudes (higher than 10 km) and at a steady rate along quite a long path, so as to fill a large clutter volume that persists for many hours due to the small fall-down rate of the dipoles. Average chaff fall rate is on the order of 0.6 m/s for aluminized glass chaff dipoles, which are the most used type for this application. But the rate can be much higher depending on weather conditions or lower, mainly because of wind conditions. In fact, when chaff is launched from an airborne dispenser, the drag is much greater than gravity so that the chaff foils move at the air mass velocity, and the chaff cloud grows in volume under the effects of prevailing winds and air turbulence.

In the chaff cloud the $\lambda/2$ dipoles are randomly oriented with respect to the illuminating radar polarization, and the average RCS of N randomly oriented dipoles in the radar resolution cell is [6] $\sigma_{av} = 0.15N\lambda^2$. Schleher [6] shows that it is not difficult to produce a sufficient chaff RCS (RCS_c) to conceal a large intruder aircraft of RCS_0, say, $RCS_c \geq 10RCS_0$, within the radar resolution cell of volume $V = R^2\theta_B\phi_B\, c\tau/2$, where R is the distance from the radar; θ_B and ϕ_B are, respectively, the azimuth and elevation –3-dB antenna beamwidths; c is the light speed; and τ is the subpulse duration of the pulse compressed waveform used by the radar. This concealment is not difficult as the required number of dipoles constitutes a low weight and hence are easy to dispense. Note that due to the turbulence produced by the prevailing winds during fall-down of the dipoles, the dipoles can assume any orientation in the chaff cloud. The dipoles that remain vertical (due to gravity) fall more rapidly than the dipoles whose original vertical orientation has changed to a horizontal one, leaving them subject to the forces of the wind. Hence, some time after dispensing, two chaff layers are formed with different prevailing polarizations: an upper layer in horizontal polarization and a lower level in vertical polarization. Because, however, the chaff fall-down rate and prevailing wind speeds are quite low, the chaff cloud can be assimilated to a nearly stationary clutter, which can be easily removed by either medium PRF (MPRF) PD or adaptive MTI Doppler signal processing, both of which are often used in current surveillance radars [1]. Note, however, that cancellation of chaff prevents the use of frequency agility in long-range surveillance radars and this can be exploited by repeater jammers producing false credible targets (see Chapter 6). Even so this type of airborne chaff application is not used as much today[1] and the use of chaff by airborne platforms is almost exclusively for self-protection.

Self-protection chaff is used both by aircraft and ships, although with quite different characteristics with regard to blooming time, useful life, and the required producible RCS. For aircraft the size of a dispenser cartridge is 30 × 30 × 15 mm (i.e., a volume $V = 1.35 \times 10^{-7}$ m^3. The size of a cylindrical chaff foil resonant at $\lambda = 0.03$m is $v = (\lambda/2)(\pi d^2/4)$ where the diameter d is usually 1/1000 of the dipole length. Hence, $v = 3.375 \times 10^{-13}$ m^3. Therefore, the number of chaff dipoles within a cartridge, assuming a high packing density of 50%, is $N = 0.5V/v = 20 \times 10^6$. This produces a $RCS_c = 0.15N\lambda^2 = 765$m^2. Because the aluminum density is 2.5 kg/dm^3, the weight of the dipoles in a cartridge is only about 16.5g.

Chaff payload is usually a mix of different individual dipole lengths. This mix allows for a response over a large range of radar frequencies. A single length responds effectively to its resonant frequency at its second and third harmonics.

The most common method of dispensing chaff rounds from aircraft is by means of a small explosive charge (squib) placed at the base of the cartridge, with the chaff foils contained in a plastic case. The chaff cartridge's squib is electrically initiated, and the gases produced by the explosion eject the chaff out of the aircraft. The initial thrust provides a distance of a few meters and then the chaff dispersion is provided by the airflow conditions and the turbulence within the wake of the aircraft.

Aircraft self-protection chaff is used to produce a break-lock to the STT. The break-lock is achieved by creating a number of false targets in proximity to the actual target within the radar resolution cell (RRC) of volume $V = R^2\theta_B\phi_B\, c\tau/2$. The false targets are produced by a series of chaff bursts (called a salvo) within a short time (tens of microseconds) due to the high aircraft closing speed to the STT and the reduced size of the resolution cell at least in range ($\Delta R = c\tau/2$) due to the radar waveform pulse compression technique. The short times of the salvo are made possible by modern digitally controlled dispensers, which are called chaff and flare dispensers (CFDs) because they are also used to dispense flares, see Section 5.7.1). The dispensing of the chaff salvo is usually accompanied by an aircraft maneuver in order to orient the aircraft speed at 90° with respect to the radar beam boresight, both to reduce the cancellation capabilities of radar Doppler signal processing (the target signal is now competing with the stationary clutter) and to exploit the large cross-range of the RRC [7].

In ship self-protection chaff is dispensed both to distract as well to dilute the RRC. Distraction is used during the acquisition phase of the STT by creating a number of false targets at locations different from the one of the protected platform. Chaff dispensing is coordinated with the action of a noise jammer that produces a range-gate break-lock to the STT. Dilution involves a large chaff RCS (much higher than the ship's RCS) within the same RRC in order to lock the RRC to the chaff cloud, which is moving at the speed of the wind (much closer to the ship's speed) while the ship is maneuvering to get away from the RRC. Chaff for ship

1. In the era after World War II, chaff corridors were used primarily by the Soviet Union, which no longer conducts large-scale defensive or offensive exercises with countermeasures. Extensive attacks in which chaff is combined with noise jamming are quite effective because the available ECCM techniques (see Chapter 6) that are effective against one countermeasure tend to be ineffective against another.

self-protection is dispensed in large quantities (on the order of a few kilograms) by means of rockets that can reach long distances from the ship.

5.6 Communication ECM Systems

The main objective of CECM systems is to deny, over a defined area and by means of jamming, the transportation of information performed in symmetric conflicts through the RF links of an adversary's communication systems, and in asymmetric conflicts through the broadcasting communication services (FM radio and analog and digital TV).

The most important adversarial RF links to be jammed are the ones that are connected to the command and control centers (C2). As discussed in Chapter 2, modern communication systems use mostly digital modulations, but a large number of legacy communication systems that employ analog modulations are still available, especially in the battlefield area. In the descriptions of the architecture and jamming techniques of CECM systems that follow, both types of modulation are taken into account.

The effectiveness of a jamming action against communications (i.e., the denial of the information) is measured by the ratio of jammer (J) to signal (S) power, J/S or JSR, at the input of the victim receiver and, in general, that ratio is required to be JSR > 1. In the case of analog-modulated signals (usually voice communication, either FM or AM), Poisel [8] has shown that to make a voice message unintelligible it is sufficient either to jam at JSR > 1 for about 30% of the duration of the message or to jam continuously along the duration of the message with a JSR ≥ –15 dB for AM voice and JSR ≥ –6 dB for FM voice. Because the number of simultaneous channels in a battlefield is large, it is usually convenient to jam in time sharing three channels.

For digital modulations the unintelligibility of the transmitted message is achieved by increasing the bit error rate (BER) of the channel, which is usually rather low ($<10^{-3}$) to $>10^{-1}$. This can be accomplished either by jamming a continuous broadcasting digital signal for more than 10% of the time or by using some type of modulations (such as BFSK) with JSR ≥ –5dB as discussed in [1, 8].

When discussing JSR it is important to underline the big difference between the ratios of RECM ERP to the victim radar one and of CECM ERP to the victim communication link. This difference is related to the type of propagation used by their victim sensors: Radars use two-way propagation, whereas communication links use one-way propagation. To evaluate this difference, let us consider typical CECM operation in free-space as depicted in Figure 5.28, where the communication transmitter (Tx) and the CECM are, respectively, at a distance R_{TR} and R_{JR} from the receiver. The signal power received at the communication receiver (Rx) $P_{r,t}$ from its transmitter can be expressed as:

$$S = P_{r,t} = \frac{P_t G_t}{4\pi R_{TR}^2} \frac{G_r \lambda^2}{4\pi} \tag{5.17}$$

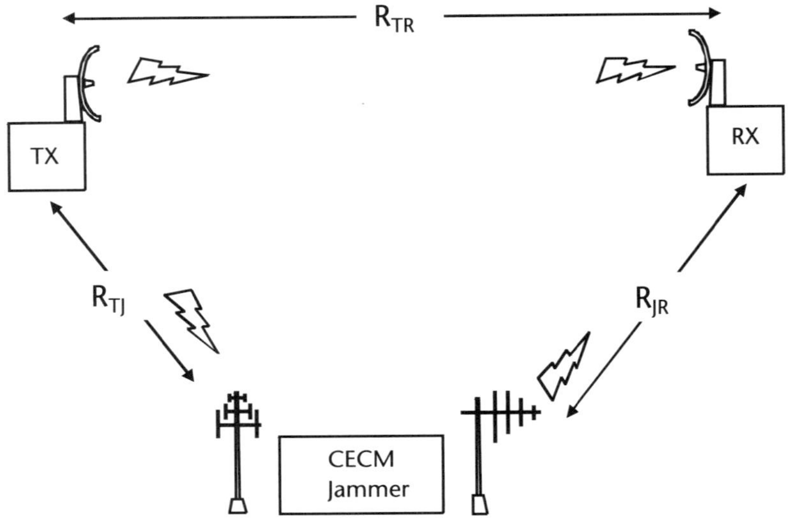

Figure 5.28 Geometry of typical communication jamming operation.

where P_tG_t = ERP$_t$ are, respectively, the transmitted power, antenna gain, and effective radiated power of the transmitter; and G_r is the receiver antenna gain in the direction of the transmitter. Correct operation of the RF link is the reception of the transmitted signal with SNR = $P_{r,t}/N > 20$ dB, where $N = kTB_rF$ is the noise power within the receiver channel bandwidth B_r, which is usually a very low power level.

The signal power $P_{r,j}$ received from the jammer (CECM) can be expressed as:

$$J = P_{r,j} = \frac{P_j G_j}{4\pi R_{JR}^2} \frac{G_{r,j}\lambda^2}{4\pi} \frac{B_r}{B_j} \tag{5.18}$$

where P_jGj = ERPj are, respectively, the transmitted power, antenna gain, and effective radiated power of the CECM; and $G_{r,j}$ is the receiver antenna gain in the direction of the jammer. Thus, in the presence of jamming, the signal power S must compete with the jamming power J, which is much larger than the noise power level N. The condition in which the jamming is effective is expressed by $JSR > k$, depending on the type of modulation, which means:

$$JSR = J/S = \frac{ERP_j}{ERP_t} \frac{G_{r,j}}{G_r} \frac{R_{TR}^2}{R_{JR}^2} \frac{B_r}{B_j} > k \tag{5.19}$$

where $B_j > B_r$ is the jammer signal bandwidth. From this expression it appears that usually $G_{r,j} < G_r$ (that is the case with directive receiver antennas) and the ratio B_r/B_j is always less than 1. Unless the jammer is much closer to the receiver than the transmitter, the ERPj must be larger than ERP$_t$ (that is seldom the case for effective RECM).

The above considerations are further reinforced by taking into account that at low frequencies (up to a few hundred megahertz) and low heights for the Tx,

CECM, and Rx, the propagation model no longer has an attenuation on the order of R^{-2} (i.e., spherical); instead it is on the order of R^{-4} to R^{-6} [8]. This is because of the many reflections that can occur when the transmitter and/or receiver are close to the Earth's surface.

Poisel [8] showed that the change of attenuation law occurs at distances

$$R > D_1 = (4h_r h_t)/\lambda \tag{5.20}$$

where h_t and h_r are, respectively, the Tx and Rx antenna heights, λ is the wavelength, and D_1 is the distance where the first Fresnel zone (an ellipse that is the loci of the points whose path difference is $\lambda/2$ from the two above antennas, which are the foci) first touches a point of reflection on the ground. Indeed, after reflection from the ground, the reflected ray amplitude assumes a constant π radians shift of phase that may provide a destructive contribution to the direct ray at the receiver. At distances beyond D_1, the received power can be expressed as:

$$P_r = \frac{ERP_t \cdot G_r \cdot (h_t h_r)^2}{R^4} \tag{5.21}$$

Although in (5.21) the receiver power is independent of frequency, both the gain G_r and G_t (embedded in ERP_t) are dependent on frequency. Table 5.3 reports distance $D_{1(\mathrm{km})}$ for various antenna heights vs. frequency.

Due to the relative short effective jamming range, as shown earlier, CECMs are mainly employed in ground battlefield applications and are of the following types:

- Ground jammers,
- Expendable jammers,
- Stand-in jammers (SIJs) on board unmanned aerial systems (UASs).

Expendable jammers are usually small units with a limited operational life (that of their batteries) and with a small ERP. They are placed close to the victim receiver by either commando troops or by being launched by artillery and switched on by an RF command.

Table 5.3 Distance D_1 of Change in Attenuation Law vs. Frequency

Antenna Heights $h_t = h_r = h$ *(m)*	*Horizon* $R_{hor} = 4.12 \times 2 \times h^{1/2}$ *(km)*	*Frequency (MHz)*	*Wavelength (m)*	D_1 *(km)*
100	82.4	3	100	0.4
		30	10	4.0
		300	1	40.0
50	58.3	3	100	0.1
		30	10	1.0
		300	1	10.0
25	41.2	3	100	0.025
		30	10	0.25
		300	1	2.5

SIJs installed on board UASs are the most effective CECMs because they are very near the victim receiver and are not affected by the strong R^{-4} attenuation due to their operational height. Further, their proximity to the receiver allows them to use many effective jamming programs (such as a follower jammer; see Chapter 6), which is not possible with CECMs at a far distance from the victim receiver. Airborne SIJs cannot be provided with a large ERP. Usually, the transmitter power is about 100W for power consumption and heat removal constraints and no high-gain antennas appropriate to the VHF/UHF bands can be installed on board aircraft. For these reasons when jamming is required over large areas, ground jammers, even if provided with low-height antennas, are favored because they are capable of being providing a high ERP on the order of 20kW with the employment of high-gain antennas.

CECM operations in the battlefield must be accurately planned and synchronized because the spectrum occupation is so dense that the use of large bandwidth jamming may result in fratricide. For this reason the architecture of the CECM has to envisage the presence of a complete CESM, which can be either integrated with the CECM or functionally connected to it to allow the correct (i.e., not fratricide) jamming operation.

The schematic block diagram of the CECM is depicted in Figure 5.29. It is composed of four main blocks:

1. The *processor* manages the system resources (power management) and is connected to the CESM equipment, which provides continuous monitoring of the environment spectrum (in order to avoid fratricide jamming).
2. The *exciter* is composed of a frequency synthesizer, which provides the carrier frequency of the jamming signal, and technique modulators, which provide jamming modulations (e.g., PRN FM). The exciter operates at a low power level and is sometimes called the *low-power RF section*. Current synthesizers are either based on DDS oscillators or a PLL controlling a voltage-controlled oscillator (VCO).
3. The *power amplifier* has at least two sections, each dedicated to a bandwidth of the CECM. Each section has either a number of parallel solid-state basic power amplifiers, usually operating in Class AB and combined in

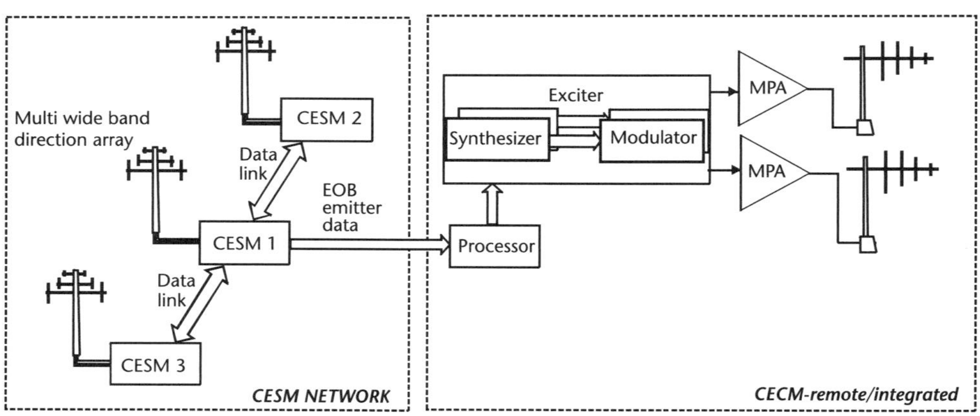

Figure 5.29 Typical architecture of a CECM.

output, or a series of basic power amplifiers, each providing a gain contribution to the chain. Both solutions are capable of providing a large power output.

4. The *antenna* is usually made of a couple of log-periodic antennas capable of covering the complete RF bandwidth from 30 to 3000 MHz. Additional antennas and power amplifiers are required to cover the HF (3- to 30-MHz) band.

5.7 Infrared ECM Systems

The current prime threat to aircraft, also called the "90% threat" since 90% of all aircraft losses in battle in the past 35 years were caused by them, is the IR sensor-guided missile. They have become the current prime threat to aircraft only because, for the past 35 years, most large-scale air operations have been conducted after intensive SEAD operations have neutralized radar-guided SAMs. Before suppression of radar-guided SAMs was achieved, flying low to avoid them made aircraft vulnerable to MANPADS. After they were suppressed, the residual IR weapons continued to take their toll over subsequent protracted operations. IR missiles are much more important in asymmetric warfare, where the enemy has no radar-guided SAMs. But in any major air war, the longer-range radar-guided SAMs and AAMs remain the prime threat. IR missiles are not only employed against aircraft; they also constitute a threat to armored vehicles and ships (see the description of antiship missiles in Chapter 1).

IR missile technology has evolved rapidly over the years and can be classified into three early generations up to the current fourth generation. The vast majority of the IR missiles being deployed belong to the first (spin scan reticle) and second generations (conical scan reticle), which, as discussed in Section 2.6.2, use a relatively simple IR seeker capable of providing target angular tracking through appropriate rotating reticles. The principle of operation is based on the transformation by means of the rotating reticles of the IR energy from the target as collected by the seeker optics into an AM or FM electrical signal at the seeker IR detector. This signal, when compared with the reticle angular reference signals, provides the target angular information with respect to the optics boresight and is provided to the missile guidance system to home in on the target.

The subsequent third generation adopted a pulse position modulation (PPM) through a rosette scan reticle to provide the target's angular information. The fourth generation uses an imaging focal plane array to provide the two targets' angular coordinates to the missile guidance system. For a general description of reticles and IR missiles, further to the discussion Section 2.6.2, we recommend Ref. [9].

The initial period countermeasure (against the first- and second-generation IR missiles) was the use of expendables (flares). These were launched from the defended platform in order to seduce the IR seeker toward a more radiation intense and appealing target. The development of more intelligent IR seekers, such as the last seekers of the second generation and third- and fourth-generation seekers, that can discriminate the targets from the flares has motivated the development of active jamming systems, such as IRCMs.

Because IR seekers are purely passive, it is not possible to determine in which IR bandwidth its receiver operates. However, because IR seekers are primarily sensitive in the 2- to 5-μm (mid-IR) wavelength band (i.e., the wavelength regions of aircraft engines' and exhaust gases' energy radiation), the active IRCMs have to be provided with simultaneous multiband high-intensity flash lamps or laser sources operating in the same bandwidth.

Two jamming principles are employed by an active IRCM: deception and saturation/destruction of the IR missile seeker detector, both of which provide the IR missile's angular tracking break-lock. Table 5.4 reports on the efficiency of the two jamming techniques vs. the various IR missile generations.

Deception is achieved by generation of an IR signal of higher intensity than the target IR signature (i.e., a jamming to signal ratio >>1), but with different AM or FM at the same scanning rate of the reticle to provide an angular error to the IR detector. This deception technique (a wobbled waveform) is very similar to the angular pull-off technique applied by the RECM against the COSRO tracking radar, as described in Chapter 6. Deception is provided either by a number of high-intensity flash lamps that instantaneously cover a wide spatial coverage (the flash lamps are usually preventively activated during the take-off and landing phases of the defended aircraft) or by a lower intensity directionally radiating flash lamp or laser source aimed against the incoming IR missile through an angular closed loop tracking that is guided by an IR detector very similar to the IR missile one. This latter type of jammer, called a directional IR countermeasure (DIRCM), has to be activated upon the angular designation provided by a missile warning or approach system (see Section 2.6.3) about the approach of the attacking IR missile.

Saturation of the IR missile detector and its AGC loop (also called *dazzling*) is achieved through a high-power laser DIRCM. Damage to or destruction of the IR missile sensor detector or even the frosting of the IR dome that encloses the seeker optics is achieved by a very-high-power laser DIRCM. The required laser power levels are reported in Table 5.5

Antisensor laser DIRCMs exhibit advantages over flash lamps due to their laser-directed energy (small laser beam divergence) and the laser source wavelength tunability in the IR missile seeker bandwidth. The latter is obtained by combining a laser with wavelength conversion devices, usually exploiting the second-order nonlinearity of a crystalline material, such as lithium niobate (Li NbO_3), to accomplish optical parametric generation (OPG), that is, the generation of two signals at larger wavelengths than the input or laser pump signal. Fiber lasers are the most used source due to their high efficiency, absence of water cooling, compact and robust package, and human eye–safe radiation levels.

Table 5.4 Efficiency of Active IRCM Techniques

Missile Name	*Type of Seeker*	*Deception*	*Dazzling*	*Damaging (Laser Power >100 mJ/pulse)*
Sidewinder (SA7/9, SA13)	Spin or conical scan reticle	Very efficient	Very efficient	Not efficient
Stinger (SA18) Chaparral (AA10/11)	PPM reticle	Not efficient	Very efficient	Very efficient
IRIS-T, MISTRAL	Imaging	Not efficient	Efficient	Very efficient

Table 5.5 Optronic Threat and Laser Power Requirements (Adapted from: [11].)

Type of System	*Type of Sensor*	*Wavelength Range (μm)*	*Laser Power Requirement*
Sight/ reconnaissance	TV/CCD, image intensifier	0.4–1.1	*Deception:* 0.1–100W (CW)
			Damage: >1 J/pulse
			IR-dome destruction: 1 kJ/pulse
	FLIR, IRST	1.5–5	*Deception:* 1–1,000W (CW)
		3–5	*Damage:* >1 J/pulse
		8–12	*IR-dome destruction:* 1 kJ/pulse
Missile systems	Reticle seeker	1.6–2.3	*Deception/dazzling:* 1–20W (CW)
		2.7–3.6	*Damage:*>0.1 J/pulse
		3.7–4.8	*IR-dome destruction:* 1 kJ/pulse
	Imaging	0.4–1.1	*Dazzling:* 0.1–50W (CW)
		3–5	*Damage:* >1 J/pulse
		8–12	*IR-dome destruction:* 1 kJ/pulse
	Beam rider	0.8–1	*Dazzling:* 0.1–100W (CW)
		1.06	*Damage:* >1 J/pulse
		10	*IR-dome destruction:* 1 kJ/pulse
	Semi-active laser	1.06	*Dazzling:* 0.1–100W (CW)
			Damage:>1 J/pulse
			IR-dome destruction: 1 kJ/pulse

The jammer-to-signal ratio JSR required for deception from a DIRCM at the IR missile optics can be evaluated as:

$$JSR = \frac{H_t T_s(\lambda_t) f(\theta)}{\Delta H_a T_S(\lambda_a)} \tag{5.22}$$

where

H_t = laser DIRCM irradiance at the IR missile optics (W/cm^2);

$T_S(\cdot)$ = atmospheric transmittance factors, respectively, at the laser λ_t and target emission λ_a wavelengths;

$f(\theta)$ = off-axis attenuation of the optics at an angle θ from the optics boresight;

ΔH_a = target irradiance contrast (see Section 2.6) at the IR missile optics (W/cm^2).

The laser DIRCM irradiance and the target irradiance contrast at the IR missile optics at distance R from the DIRCM /target are evaluated as follows:

$$H_t = \frac{P_t T_S(\lambda_t, R)}{\Omega_t R^2} \tag{5.23}$$

$$\Delta H_a = \frac{\Delta I_a T_s(\lambda_a, R)}{R^2} \tag{5.24}$$

where

P_t = laser DIRCM transmitted power (W),
Ω_t = laser beam divergence (sr),
ΔJ = target radiation contrast (W/sr).

Typical values of the target radiation contrast ΔI are:

- *Helicopters:* ΔJ_h = 100 W/sr,
- *Turboprop aircraft:* ΔJ_p = 1000 W/sr,
- *Fighter aircraft:* ΔJ_f = 10000 W/sr.

If we assume that $\lambda_t = \lambda_a$ for a tunable laser, we have:

$$JSR = \frac{P_t f(\theta)}{\Omega_t \Delta J} \tag{5.25}$$

From this equation we can see that the required laser power for deception is independent of distance R and dependent only on the target radiation contrast, provided that the missile optics off-axis rejection is moderate; that is, the DIRCM angular tracking loop is quite accurate.

The schematic block diagram of a laser DIRCM is shown in Figure 5.30. The laser/flash lamp DIRCM system is composed of three main functional blocks: the missile warning system (MWS), the system processor, and the laser/flash lamp turret. Some DIRCM systems also have flare dispenser that is activated by the system processor.

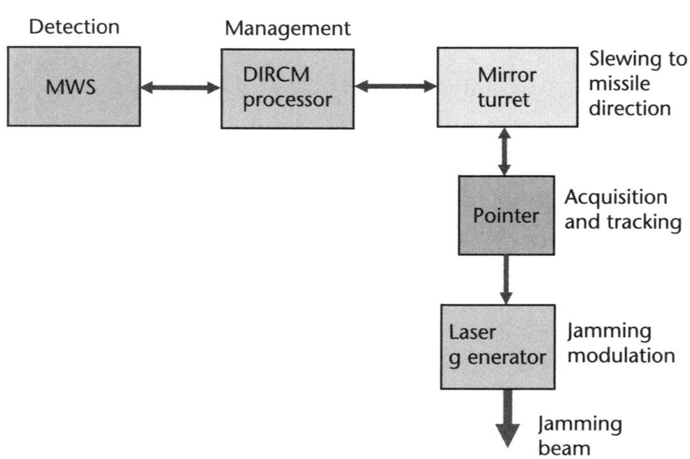

Figure 5.30 Schematic block diagram of a laser DIRCM.

The angular threat designation to the laser/flash lamp DIRCM is provided by the MWS (see Section 2.6), which is composed of either an IR sensor (which detects the heat energy produced by the missile plume) or an UV sensor (which detects the energy of the spectral lines of missile propellant). On receipt of the angular designation, the processor addresses the laser/flash lamp turret to the designated angular direction. The turret has pointer/transmitter optics that track and jam the incoming IR missile. The pointer is provided with an IR detector focal plane array that performs very fine DF measurements of the heat energy produced by the missile plume. These measurements serve as the inputs to the servo-loop of the turret driven by the processor. The laser/flash lamp generator is provided with a multiband energy source that produces the AM jamming IR energy (usually a wobbulated AM high prf waveform), which is delivered to the transmitter optics.

The missile engagement by a DIRCM lasts a very short period—just a few seconds, as shown in the graph of Figure 5.31—so all DIRM functions have to be swiftly accomplished, especially the slewing of the turret to the missile DOA and the transmission of the wobbled jamming waveforms. An ELT 572 laser DIRCM turret is shown in Figure 5.32.

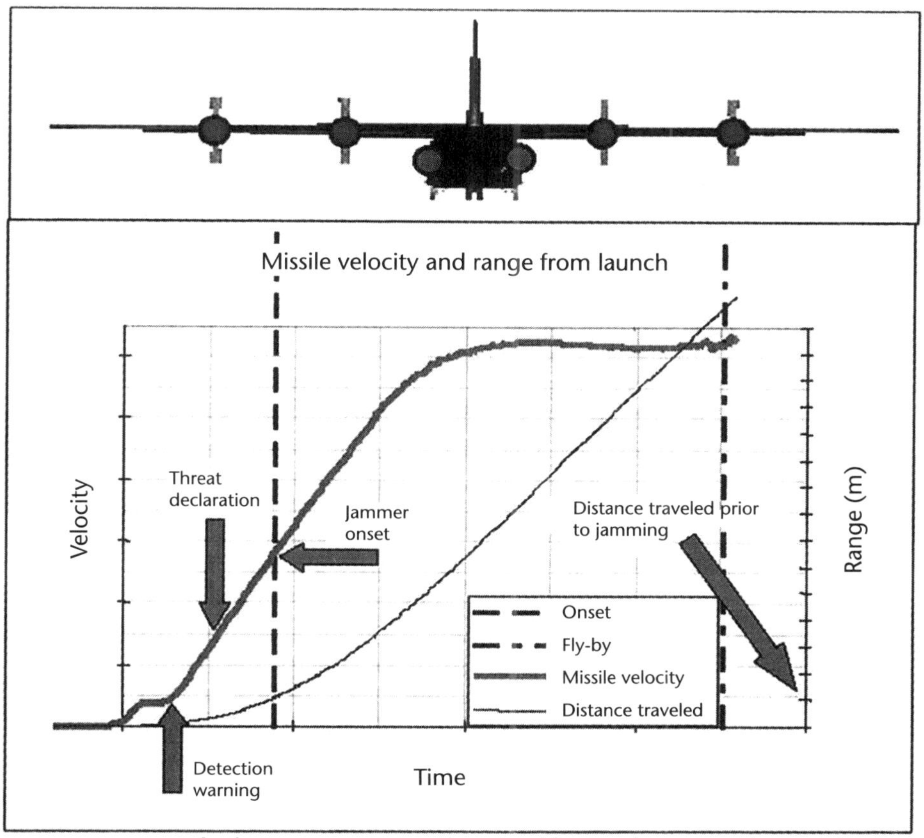

Figure 5.31 IR missile engagement duration.

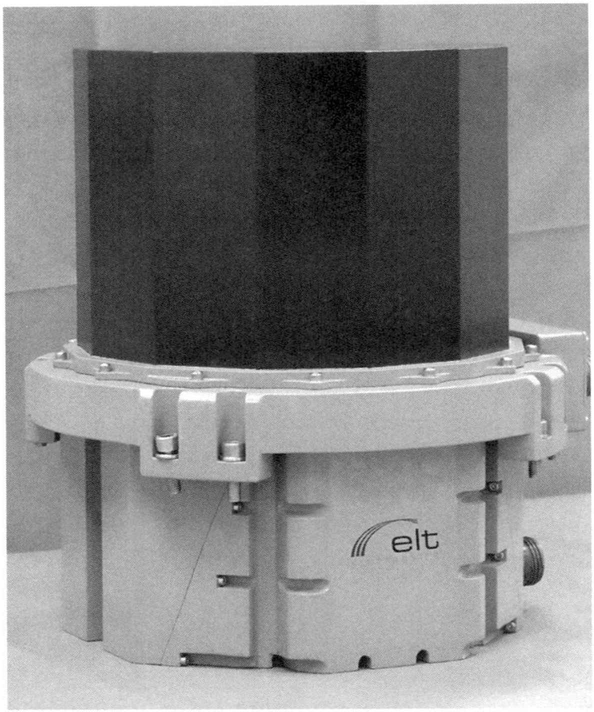

Figure 5.32 ELT 572 laser DIRCM turret. (Courtesy of Elettronica S.p.A.)

5.7.1 Flares

Flares are expendable IR decoys that are deployed by aircraft and ships to seduce the attacking IR missiles to a position safely distant from the defended platform. On-board aircraft flares are deployed from the same cartridge dispenser used for aircraft self-protection chaff. The flares have a cartridge size (50 × 25 × 200 mm) known as MJU7B in the United States or 218 in the United Kingdome that is identical to that of chaff cartridges. Other nations use different sizes for their cartridges, but the dispenser for both expendables is the same and is called a chaff and flare dispenser (CFD). Due to their low relative cost with respect to the on-board active ECMs, CFDs are always present on board aircraft and frequently they constitute the only aircraft ECM.

Flares employ the same type of electrically initiated squib as is used in chaff cartridges, but because the flare pellet is mainly composed of a magnesium (which immediately ignites in presence of oxygen in the atmosphere) and aluminum/Teflon powder bonded together by the binder material Viton (hence, the acronym MTV for magnesium-Teflon-Viton), the ignition mechanism is quite different from the chaff mechanism in order to prevent inadvertent or precocious ignition of the flare within the aircraft. In fact, the flare pellet is contained in a can and is activated by a sequencer/shutter mechanism that is composed of a piston placed at the base of the pellet. Upon squib ignition, the expanding gases move the piston, which drives the pellet and the sequencer out of the can. When out of the can, the shutter is removed, allowing the magnesium to ignite.

Because most IR seekers can distinguish magnesium flares, the composition of modern flares is being changed to mimic the spectrum of the aircraft they are defending. The main characteristics of aircraft flares are as follows:

- Radiation intensity (W/sr) in the seeker bandwidth (ranging from hundreds of watts per steradian to kilowatts per steradian depending on the aircraft to be defended—helicopters or fighter aircraft),
- Activation or rise time (usually a few tens of milliseconds to reach the peak radiation intensity),
- Persistence of radiation (usually 3 to 6 sec).

The IR threat is currently also present in some antiship RF missiles, which are starting to incorporate IR homing seekers in order to provide a more accurate final phase of flight to the target. The principal sources of IR energy in a ship are the engine room and the stack, which in most warship designs are placed at the center of the ship and thus the center of the ship constitutes the most appealing position for missile impact. The seduction of IR missile homing to a position far from the ship is provided [10] either by a series of flare subrounds that detonate progressively at increasing range from the ship (a sort of walk-off technique, which is discussed in Chapter 6) or in addition to the first technique by a number of subround floating smoke pots (on the sea surface) deployed from the same previous subrounds, which allow the buoying of decoys above the sea surface. Both techniques are performed by the NATO decoy launching system Sea Gnat. The Sea Gnat launches rounds from an Mk 36 deck-mounted multibarrel launcher. A pictorial representation of the countermeasure is shown in Figure 5.33.

Naval IR rounds differ from aircraft ones in the use of phosphorous-based materials because of their size when they bloom, which needs to nearly equate to the defended ship's size, and for their persistence, which is on the order of 30 to 60 sec (larger than the period of a missile engagement). Further the naval IR rounds are required to radiate in both the mid-IR band (3 to 5 μm) as well as the far-IR band (8 to 14 μm) with the proper radiant intensity ratio between the two bands to counter the bicolor seeker discrimination technique [10].

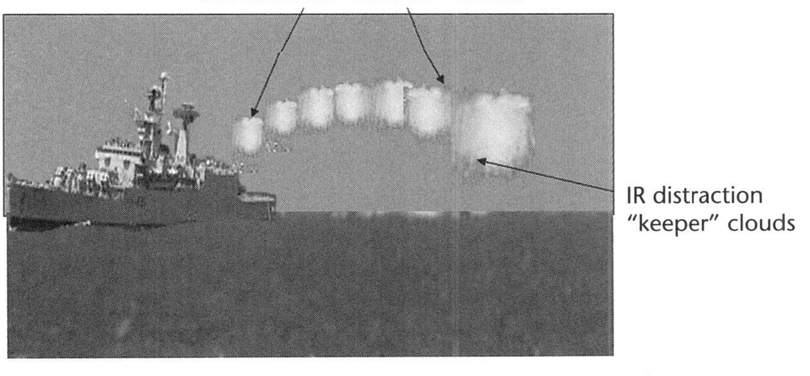

Figure 5.33 Pictorial representation of a series of flare subrounds that detonate progressively at increasing range from the ship.

5.8 Conclusion

In this chapter we discussed about the architecture and composition of the different types of ECM equipment, each devoted to one region of the electromagnetic spectrum. We illustrated their capabilities, which are expanded on further in Chapter 6 with a discussion of the implemented jamming and deception techniques against the various sensors and weapons systems discussed in Chapters 1 and 2.

References

[1] Farina, A., "*Electronic Counter Measures*," Chapter 24 in M. Skolnik (Ed.), *Radar Handbook*, New York: McGraw Hill, 2008.

[2] Schneider, W. J., "Digital Countermeasure Memories: New Techniques Possible," in *International Countermeasures Handbook*, Dedham, MA: Artech House, 1986.

[3] De Martino, A., Nanni, S., and Orsini, M., "Adaptive Multithreat Jamming Efficiency," paper presented at EW International Conference, Pratica di Mare, Rome, Italy, May 11, 1995.

[4] Pace, P., *Advanced Techniques for Digital Receivers*, Norwood, MA: Artech House, 2000, Chapter 7.

[5] Kerins, W., "Spur Levels in Multiple-Bit DRFMs," *Journal of Electronic Defense*, Vol. 14, No. 1, January 1991, pp. 49–54.

[6] Schleher, C., *EW in the Information Age*, Norwood, MA: Artech House, 1999, Chapter 7, Section 7.2.

[7] Galati, G., and Lombardi, P., "Design and Evaluation of an Adaptive MTI," *IEEE Trans. on Aerospace and Electronic Systems*, Vol. AES-14, November 1978, pp. 899–905.

[8] Poisel, R. A., *Introduction to Communication EW Systems*, Norwood, MA: Artech House, 2006, Chapters 2, 7, 8, 11, and 17.

[9] Torrieri, D., *Principles of Secure Communication Systems*, Second Edition, Norwood, MA: Artech House, 1992.

[10] Neri, F., *Introduction to Electronic Defense Systems,* 2nd Ed., Raleigh, NC: SciTech Publishing, 2006, Chapter 5.

CHAPTER 6

ECM Techniques and Sensor ECCMs

6.1 Introduction

The active ECM systems discussed in the previous chapter are placed on board military platforms (aircraft, naval surface craft, ground vehicles, and even buildings) and, through the use of RF or IR transmissions, are meant to disrupt the operation of hostile threat radars and IR guided missile systems and to inhibit communications among forces.

The techniques used by active ECM systems can be classified as either jamming or deception techniques. Jamming is defined as "the intentional and deliberate transmission of signals with the purpose of disturbing, masking or otherwise degrading the reception of the useful signal by the sensor system" [1]. Deception is defined as "the intentional and deliberate transmission of signals with the purpose of misleading in the interpretation or use of the information by the sensor system" [1]. Active ECM systems generally have a set of jamming and deception techniques that are fixed or set on a mission basis. In some modern systems sophisticated DSP can even make selections and set parameters in real time in accordance with the threat/platform engagement situation.

In this book ECM techniques are described to the extent needed for an understanding of active ECM effectiveness, and the parameters of the waveforms associated with the techniques are based on the general physics of the engagement rather than on countering a specific sensor (radar or IR seeker or communication receiver). In this chapter we first deal with the ECM techniques applicable against radar, which is the key sensor in most modern weapons systems employed on land, sea surface, and air platforms. We begin with a discussion of the ECM techniques used against surveillance radars and follow that up with the techniques used against the tracking radars and RF-guided missile seekers. We also discuss the effects of the many ECCM techniques employed in the two above-mentioned radar types and the applicable modifications to the ECM techniques to cope with them.

In the second part of the chapter, the discussion covers the ECM techniques that are used against modern communication systems, beginning with digital tactical data links and their relevant ECCM techniques and following with RF-initiated improvised explosive devices.

6.2 ECM Principles and Techniques Used Against Surveillance Radars and Related ECCMs

As discussed in Chapter 2, a surveillance radar's functions are to detect and locate the position of a target within a large coverage volume. In Chapter 2 we discussed the basic structures of such devices, ranging from the basic 2-D search radar type to advanced 3-D and phased array radar types. We also discussed a surveillance radar's signal processing techniques, including coherent transmission modes (and relevant Doppler signal processing, up to the airborne PD technique) and pulse compression techniques (up to the use of LPI waveforms). Coverage of this material was preparatory to the discussions about the interception, classification, identification, and direction finding by ESM systems, as discussed in Chapters 3 and 4. In this chapter we discuss the ECCM techniques introduced in these radars to resist the ECM techniques employed by the RECM systems discussed in Chapter 5.

Conceptually, we can model a surveillance radar as a rotating antenna with either a wide beam (2-D) or multiple beams (3-D) in elevation, whose narrow main beam sequentially scans the complete azimuth (or a sector) while its sidelobes receive responses from all directions. In accordance with this model, jamming can be produced by RECM systems through the continuous transmission of noise waveforms into the radar antenna that increase the receiver noise level, thus strongly reducing the target SNR and possibly even completely masking the target (depending on its distance from the radar as shown by the jamming equations of Chapter 5). Similarly, deception can be performed by RECM systems by using replicas produced by DRFMs of the actual radar waveform. This produces a large number of false targets that saturate the radar automatic detection and tracking (ADT) processing [2]. Deceptive false targets are competing only against the radar receiver noise and hence can be generated at a relatively low jamming power.

Modern surveillance radars, in addition to the signal processing techniques discussed in Chapter 2, employ a number of ECCM techniques to reduce the effects of the above-mentioned ECM techniques. They are discussed in the following subsections.

6.2.1 Frequency Agility in Transmission

When applied in pulse-to-pulse mode, the jamming noise bandwidth can be compelled to equate the complete radar frequency agility bandwidth (usually on the order of about 10% of the central frequency), thus spreading the jammer power into a barrage noise bandwidth and reducing the target masking effectiveness, as shown in Figure 6.1. Similarly, it compels the deceptive false targets (the replicas) to be produced only at distances beyond than that of the jammer, identified as RJ in Figure 6.2.

6.2.2 PRI Agility

PRI agility has the same effect as frequency agility with respect to the deceptive false targets.

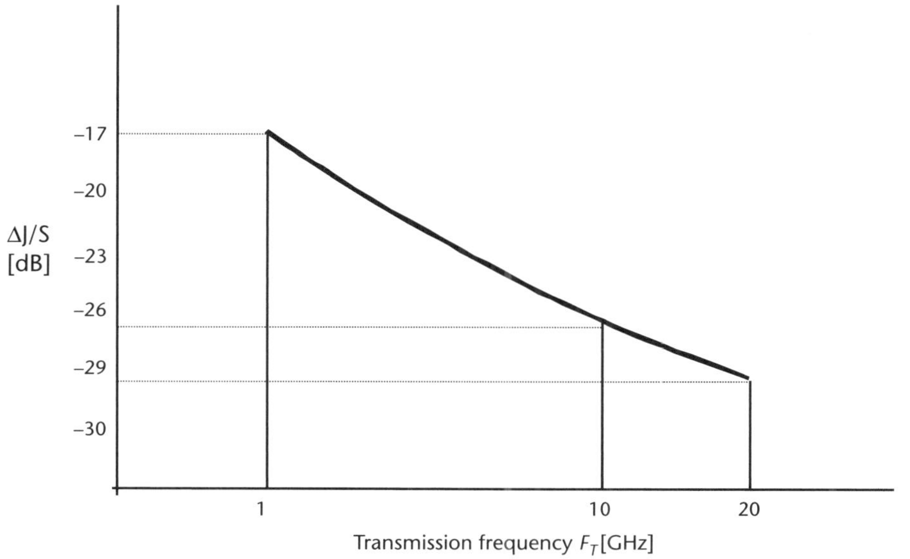

Figure 6.1 JSR reduction factor, that is, the ratio of the radar pulse bandwidth (2 MHz) to the jammer bandwidth for a pulse-to-pulse radar frequency agility of 10% of central frequency.

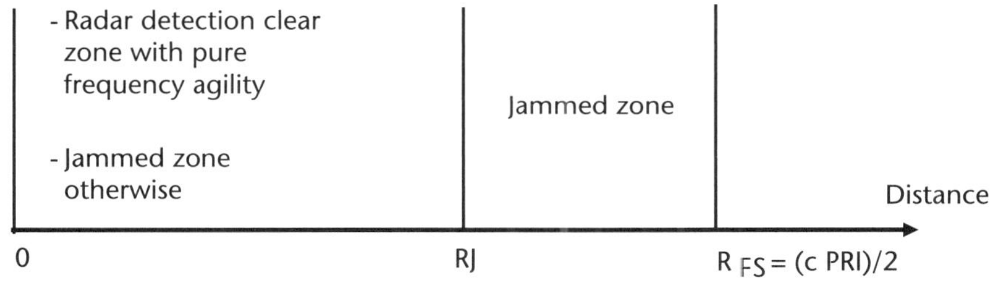

Figure 6.2 Jammer effectiveness in range when dependent on the frequency agility mode.

6.2.3 Ultralow Sidelobes

Ultralow sidelobes are the sidelobe levels far from the mainlobe (on the order of –20 dB with respect to the isotropic level). They strongly reduces the noise jamming level coming from the sidelobes and compels the RECM equipment to strongly increase its ERP to produce effective deceptive false targets.

6.2.4 Multiple-Sidelobe Cancellers

The sidelobe canceller (SLC) is an additional receiving adaptive antenna system capable of strongly reducing the continuous noise jamming power received in the sidelobes of the radar antenna. This reduction is achieved by controlling the overall (i.e., radar antenna and additional antenna system) receiving antenna pattern and placing a null in the direction of the jammer(s). Usually an additional antenna system such as this is implemented with a linear array of elementary radiators (dipoles), each connected to a receiving channel in which both the gain and the phase shift are adaptively controlled, as shown in Figure 6.3. Suboptimal results can be achieved with phase shift only control. The basic principle of the adaptive array can

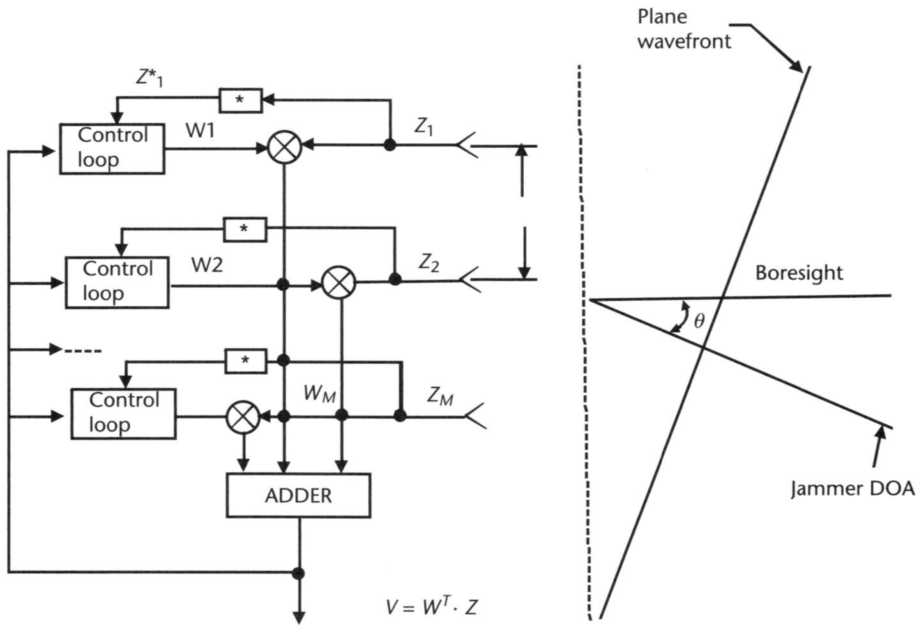

Figure 6.3 Block diagram of an adaptive array.

be illustrated considering a linear array that receives a planar wavefront, inclined at an angle θ with respect to the boresight as in Figure 6.3.

Taking the phase of V_1 as the reference, the set of the complex received signals can be expressed as a complex vector $\underline{v}^T$ as follows:

$$v^T = (v_1, v_2, \ldots, v_m) = A(1, e^{-ja}, e^{-2ja}, \ldots, e^{-(m-1)ja}) \tag{6.1}$$

where $a = 2\pi(d/\lambda)\sin\theta$ is the phase shift between two adjacent elements of the array.

Each complex signal is weighed by a complex weight of the type $w_k = B_k e^{jb}$ and then coherently summed to produce the signal $g(\theta)$ at the input of the receiver:

$$g(\theta) = \underline{w^T} \cdot \underline{v} = \sum_1^m w_i v_i(\theta) \tag{6.2}$$

If the weights are uniform and real, that is, if $\underline{w}^T = (1, 1, 1)$, the linear array has its maximum along the boresight and the total received signal becomes

$$g_0(\theta) = A \frac{\sin\left(\pi m \frac{d}{\lambda} \sin\theta\right)}{\sin\left(\pi \frac{d}{\lambda} \sin\theta\right)} \tag{6.3}$$

The simultaneous presence of a strong noise jamming signal $\underline{v}_d$ of amplitude D coming from direction θ_d,

$$\underline{v}_d^T = (v_{d1}, v_{d2}, \ldots, v_{dm}) = D(1, e^{-j\varphi}, e^{-2j\varphi}, \ldots, e^{-(m-1)j\varphi} \tag{6.4}$$

where

$$\phi = 2\pi(d/\lambda)\sin\theta_d$$

produces a total received signal:

$$\begin{aligned} g_0'(\theta) &= \underline{w}^T \cdot (\underline{v} + \underline{v_d}) \\ &=_0 \underline{w}^T \cdot \underline{v} + \underline{w}^T \cdot \underline{v_d} = g_0(0) + g_0(\theta_d) \\ &= A + D \frac{\sin\left(\pi m \frac{d}{\lambda} \sin\theta_d\right)}{\sin\left(\pi \frac{d}{\lambda} \sin\theta_d\right)} \end{aligned} \tag{6.5}$$

Because D can be much larger than A, it can completely mask the target signal A. The reduction of the received jamming power may be achieved by controlling the weights $\underline{w}$ in such a way as to nullify the second term in (6.5). This condition [3] is obtained when the weights $\underline{w}$ are:

$$\underline{w}_{opt} = kM^{-1}\underline{s}* \tag{6.6}$$

where

k = gain factor;

$\underline{s}$ = versor (i.e., unit vector) of the steering direction θ_s to which the main antenna (or the array) is pointing and at which we want to achieve the maximum antenna gain; $\underline{s} = (1, e^{-j\psi}, e^{-2j\psi}, \ldots, e^{-(m-1)j\psi})$ with $\psi = 2\pi(d/\lambda)\sin\theta_s$;

* denotes the *complex conjugate* operator, which, when applied to $\underline{s}$, simply means a change of sign in the exponential;

M $(m \times m)$ is the covariance matrix of the array whose generic kth, rth element is expressed (in the case of $D >> A$) as

$$m_{k,r} = E(v_k v_r^*) \cong E(D^2 e^{-j(k-r)\Psi} + n_k n_r^*) \tag{6.7}$$

where $E(\cdot)$ denotes the *expected value* operator (or time average in the case of stochastic processes) and n_k and n_r are the receiver noise samples taken at the same time in the kth and rth receiving channels. The matrix element $m_{k,r}$ is therefore an index of the spatial correlation of the noise jamming samples received by the corresponding receiver channels, and M^{-1} is the inverse of such a matrix.

Hence, the procedure to determine the optimal weights for the cancellation of the received noise jamming power consists of three steps:

1. Estimation of the spatial correlation of the signals received by any two elements of the array. The estimation process is performed in a period longer than the receiver noise correlation period and the time duration of the transmitted waveform, in order to avoid the cancellation of the desired signal.
2. Computation of the inverse covariance matrix M, which can be quite time consuming depending on the size m of the array.
3. Multiplication of the matrix M^{-1} by the vector $\underline{s}^*$.

In the past, the most often used methods for the determination of optimal weights were time iterative methods, also called closed-loop methods such as the Applebaum-Howel-Brennan (AHB) method [3]. Currently, due to the increase in computing capabilities, methods for computing the inverse of the covariance matrix directly are being employed. These methods are called open-loop methods.

In any case the evaluation of the optimum weights for noise jamming cancellation requires a transient time for the estimation of the spatial correlation of the signal received at the elements. This transient time lasts a number of durations of the transmitted waveform. Reference [4] reports typical noise jammer cancellation ratio (JCR) values on the order of 30 dB, seldom up to 40 dB.

Examples of typical performances for the AHB method, whose basic principle is shown in Figure 6.4, are reported [2] in Figures 6.5 and 6.6.

6.2.5 Sidelobe Blanker

The sidelobe blanker (SLB) is an additional receiving channel. It is provided with an auxiliary antenna (usually a semi-omni antenna) whose function is to prevent the detection of strong targets and interfering jammer pulses (as they might appear after pulse compression) entering the radar receiver through the antenna sidelobes. The schematic block diagram of a SLB is shown in Figure 6.7.

By properly selecting the auxiliary channel gain G_{aux}, any pulse coming from the sidelobe region is received by the auxiliary channel with a level exceeding that of the same signal as received by the main antenna and thus can be suppressed. The SLB acts on a single-pulse basis and for each radar range bin. Only signals entering through the radar's main beam are allowed to proceed along the further radar signal processing. The target detection performance of the SLB in terms of P_D and P_{FA} vs. the SNR and JNR have been analyzed in [4], to which we address the interested reader.

A further fix implementation in modern search radars is the use of a range filter map (RFM) associated with each channel (radar and auxiliary). The RFM is a 2D map that collects the radar echoes of all range cells (after pulse compression) and all Doppler filter outputs of a radar burst. The two independently collected RFMs are then compared in order to compare, in the case of overlapping range cells, the Doppler contents and thus discriminate the target from the false target produced by a RECM.

The joint cascaded operation of a SLB and SLC, as shown in Figure 6.8, is indeed a very powerful ECCM and cannot be defeated with the employment of a single ESJ. A number of ESJs (at least three) operating in synchronous mode through the use of a proper triggering link among the aircraft is required. Each ESJ

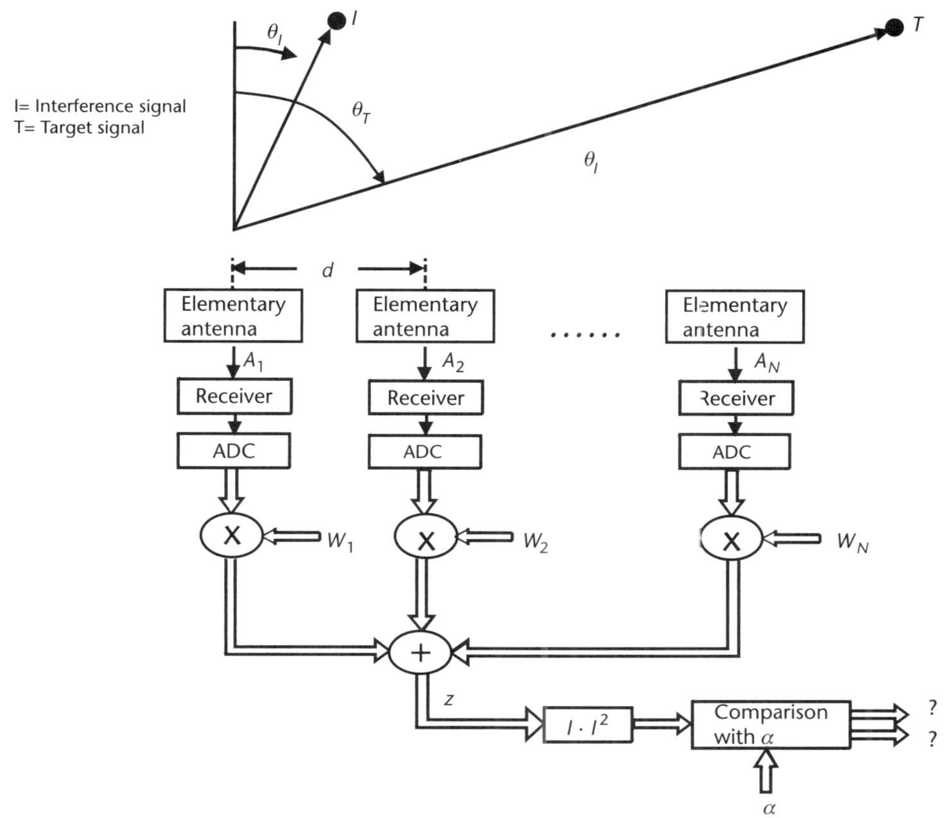

Figure 6.4 Basic principle of the AHB closed-loop method for determination of the optimal weights of an adaptive array.

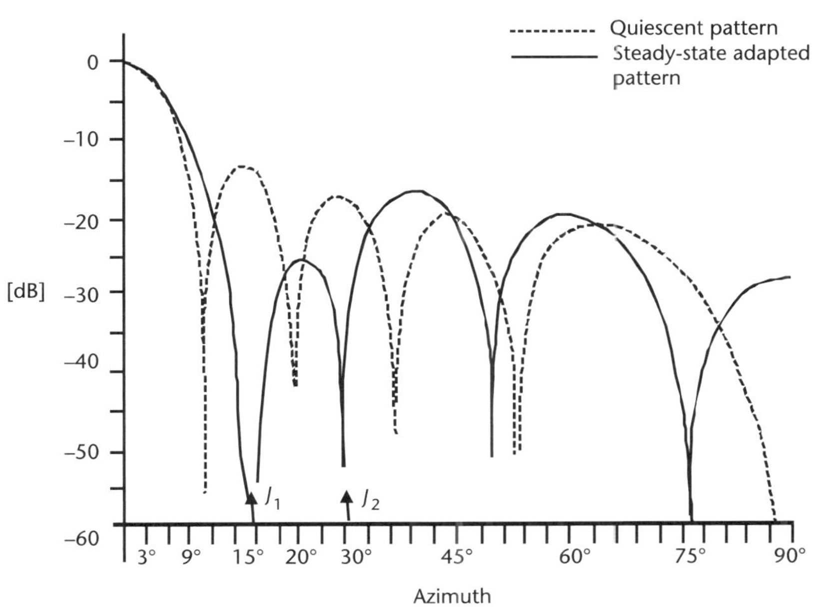

Figure 6.5 Quiescent and steady-state pattern of a 10-element adaptive array with two noise jammer sources present: first jammer J_1 with $J_1/N = 30$ dB at azimuth = 17.4° and second jammer J_2 with $J_2/N = 10$ dB at azimuth = 30°.

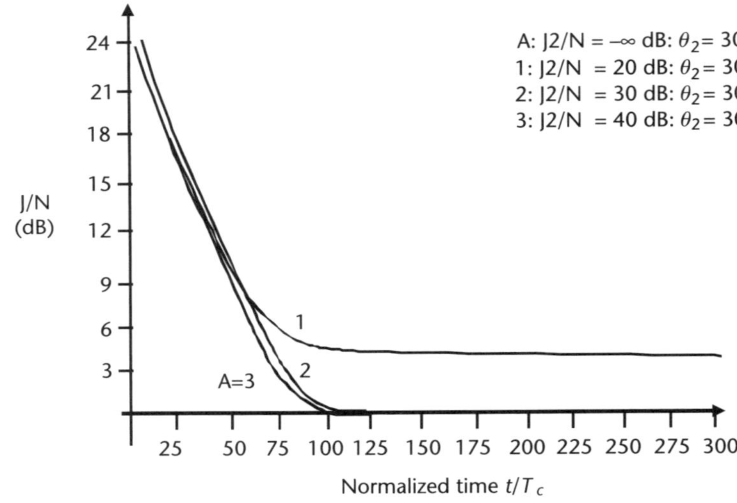

Figure 6.6 Transient behavior of a 10-element adaptive array in the presence of two jammers: J_1 at azimuth = 17.4° and J_1/N = 30 dB; J_2 at azimuth = 30° with variable J_2/N; T_c is the transmitted waveform duration (in this case, a pulse).

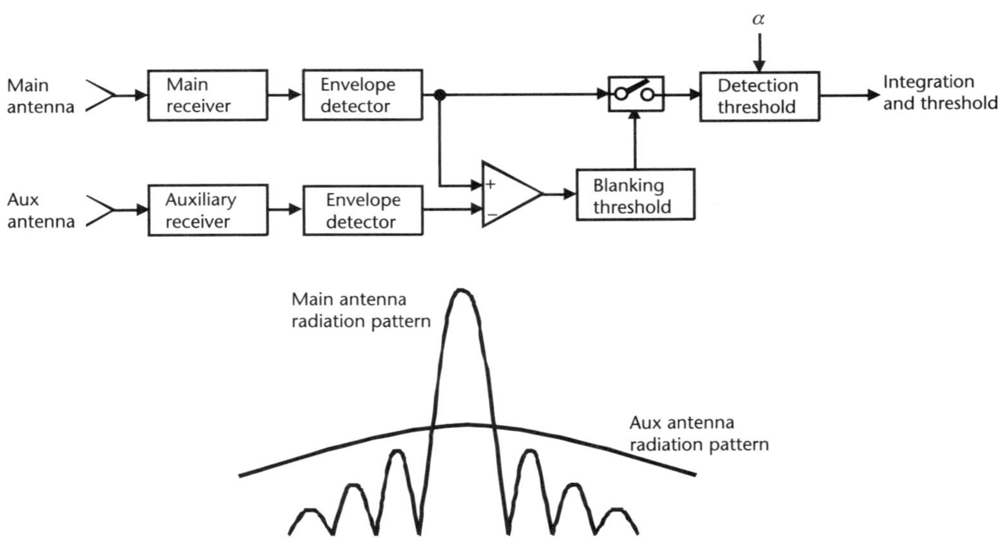

Figure 6.7 Schematic block diagram of the SLB principle.

upon synch-command should alternately transmit a noise waveform of sufficient length (in terms of radar waveform width τ) while the other(s) should transmit large-amplitude false credible targets (i.e., replicas of the radar waveform with the correct Doppler shift, so as to deceive the above-mentioned RFM).

6.2.6 Adaptive Arrays

In this type of radar the principle of jammer cancellation is similar to that of multiple SLC operation, but, with respect to the latter, multiple SLC offer the capability of enhancing the target signal while canceling the disturbance. To achieve this

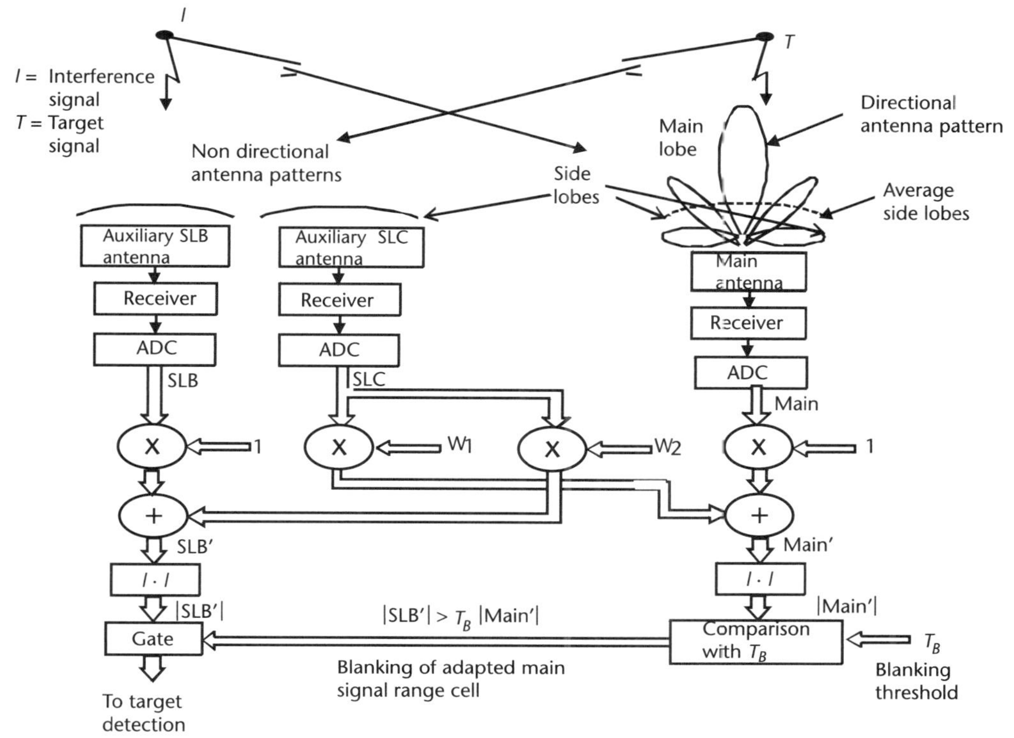

Figure 6.8 Functional block diagram of the cascade operation of a SLB and SLC.

capability, space-time adaptive processing (STAP; see also Chapter 2) is employed that combines receive beamforming and Doppler filtering.

The operation of STAP involves the use of N antennas in the beamforming (which theoretically provides $N - 1$ degrees of freedom; i.e., theoretically it can suppress up to $N - 1$ displaced jammers) and M transmitted coherent pulses to allow Doppler filtering. Fully adaptive phased arrays are not practically possible because a phased array radar needs a large number of receiving elements (on the order of thousands), so usually the elements are grouped in a reasonable number of subarrays, which feed the adaptive processor. The scheme and description of a phased array radar constituted by $N = 24$ subarrays is shown in [4].

The various ECCMs potentially available in surveillance radars and the functions they enhance in order to remove or reduce the jamming and deception actions are shown in Figure 6.9 [2].

6.2.7 Noise Jamming

The objective of noise jamming is to produce an interference signal that masks the target signal completely. In principle, the optimal jamming signal should have the characteristics of receiver noise, that is, it should be as close as possible to a white Gaussian noise (WGN).

Noise jamming is classified by the ratio of jamming bandwidth B_j to the radar waveform bandwidth B_r (which pulse compression techniques, if any, further reduce in accordance to the radar processing gain G_p). If the ratio B_j/B_r is large, the

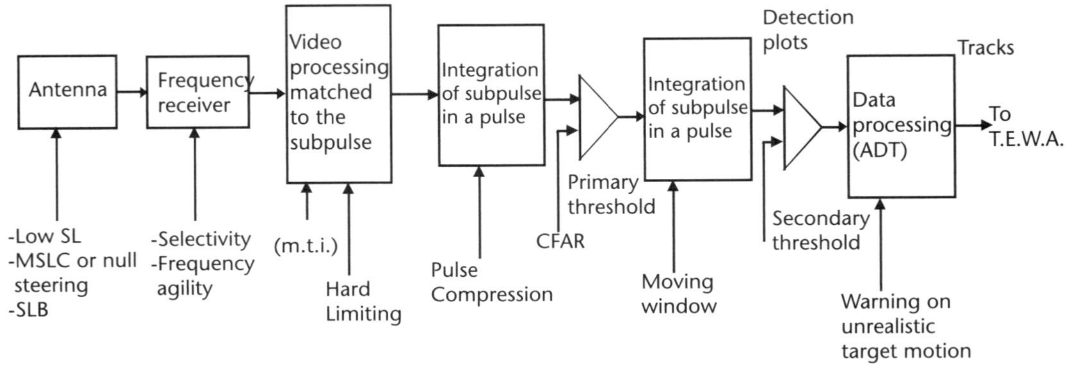

Figure 6.9 Schematic block diagram of search radar signal and data processes and relevant applicable ECCM techniques.

noise jamming is called *barrage noise* jamming; otherwise, if the ratio is small (on the order of 3 to 5) the noise jamming is called *spot noise* jamming.

In theory, a large power WGN could be produced by power amplification of a low-power WGN produced by a bandpass-filtered electronic device (this technique is called *direct noise amplification*, DINA). Unfortunately, due to the deeply saturated operation of power RF amplifiers, the large noise fluctuations of WGN will be clipped by the saturation level, thus eliminating the noise peaks that produce the false alarms in the victim radar receiver.

The usual way to produce a nearly WGN with large spiky fluctuations is to use a random frequency or phase-modulated waveforms that are not affected by the power amplifier saturation level. Note, however, that this is less effective in JNR because the radar azimuth correlation processing, such as the moving window integrator, strongly reduces its false target effects as shown in Figure 6.10. If this technique is used with the addition of a digital noise generated by regenerative shift register techniques, it can potentially be synchronized to the victim's radar waveform, thus removing the above-mentioned advantage of the radar azimuth correlation processing. This technique, when coupled to the generation of false random targets, produces a jamming technique, called *smart jamming*, that is resistant to the rejection provided by the SLB and SLC ECCM techniques [1]. Indeed the pulse-type signals cannot be rejected by the SLC due to its short duration with respect to the SLC time constant (loop transient time), and the excessive number of random false targets would excessively blank the radar main-beam target returns as operated by the SLB.

6.2.8 False Targets

This technique is applicable against search radars with constant antenna scan period (ASP) and type (AST). It consists of replicas of the radar waveform sent within each PRI of the main beam (the so-called time-on-target, TOT) with two time delays, τ_1 and T_1, both variable in time, that are applied, respectively, to:

- The received radar waveform in order to simulate a radial displacement of the false target,

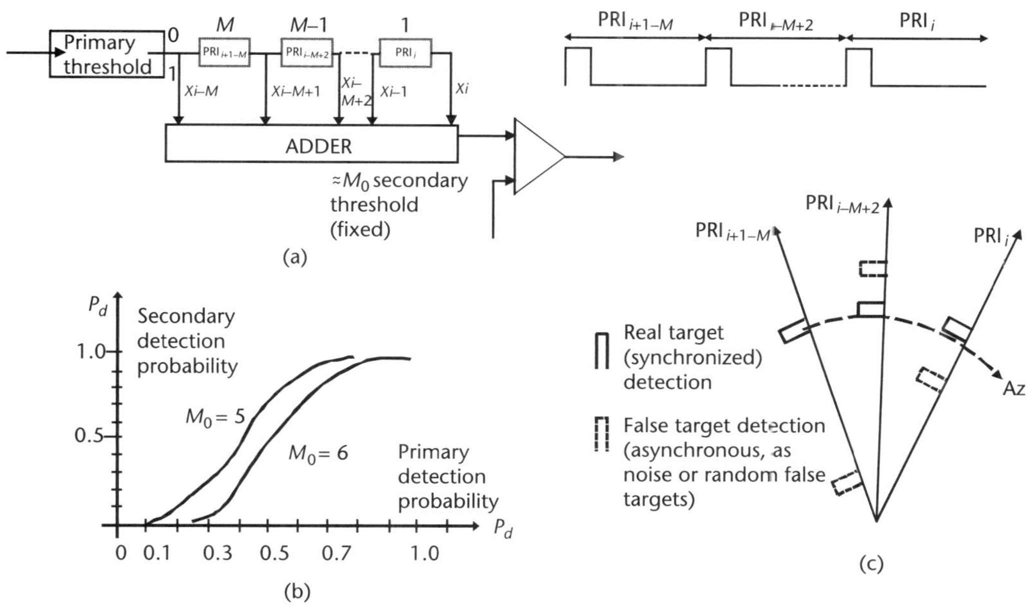

Figure 6.10 (a) Moving window integrator (MWI) schematic block diagram and PRI law. (b) Operational characteristics of the MWI with $M = 10$ integrated synchronous samples and M0 secondary threshold. (c) MWI cancellation capability of asynchronous detections.

- The initial time of the replica in order to simulate an azimuth displacement within the main beam (until the SLB does not blank the replica).

This technique is meant to produce a number of credible false targets to the radar ADT function and is enhanced by the combined synchronized operation of a couple of jammers slightly displaced in angle (on the order of one radar beamwidth).

6.3 ECM Principles and Techniques Used Against Tracking Radars and Related ECCMs

Tracking radars, such as the single target tracker (STT) or multifunctional phased array radar (MFR), that perform multiple simultaneous target tracking, as shown in Chapter 2, are used to determine the polar coordinates [range R and both angular (θ, φ)] of a target of interest with respect to the radar position, which is assumed to be the reference. The relevant measurements, which are performed at a high data rate (nearly continuously), have to be accurate to a fraction of its pulse width τ (transformed in length) and of its beamwidth θ_B and are used to plot the target movement and accurately derive its rate of change with respect to time in order to guide the weapons systems to the successful interception of the targets.

Some types of tracking radars, such as those used for the command guidance of missiles or as illuminators for semiactive homing guidance, need only measure angular coordinates in order to maintain the target within the beamwidth of the radar antenna. Tracking radars achieve these required accuracies by using narrow beamwidths ($\theta_B < 2°$) and pulse widths (or compressed pulse widths) on the order of a fraction of a microsecond (i.e., range accuracies on the order of 10m). Further they improve the resolution and the accuracy of the future target positions by measuring continuously or with a very high data renewal rate the position and motion of the tracked target.

Range tracking is performed through the early-late (E-L) gate tracking loop, which drives the early and late gates to a position where the energy levels in both gates are equal (this position corresponds to the peak amplitude of the matched filter to a rectangular pulse). Such tracking loops are usually designed to operate also with memory, which allows them to coast at the last measured range rate during periods of lost data (either due to jamming or signal fading).

During the tracking phase the range tracking loop bandwidth is on the order of a few tens of hertz (i.e., time constants on the order of tens of microseconds) in order to achieve the required measurement accuracies and protect the tracking from sudden interferences (e.g., such as that produced by chaff as discussed in Chapter 5).

Further, recall that to prevent any interference (even other targets) located at ranges different from the tracked range, the tracking gate generator must provide a tracking gate that has a width that is slightly larger than the total width of the E-L gates so as to isolate the return of the target of interest from any other signal. This tracking gate also determines which signal enters both the angle tracking gate as well as the AGC loop, which adjusts the gain of the receiver's amplifiers to a level commensurate with the received power level of the tracked signal.

Therefore, the RECM jamming program must be capable of entering the radar receiver at the time dictated by the position of the radar range tracking gate in order to be effective.

From the above discussion it is evident that the jamming programs used by the RECM of the defended platform are different in the case of self-protection from those used for mutual protection. In the case of self-protection, the jamming signal has to be transmitted with a negligible delay after the interception of the radar signal at the jammer antenna.

The usual self-protection jamming program is a sequence of techniques performed in order to deceive first the radar range tracking loop and in a second step the angular tracking loop, with the aim of producing a radar break-lock and hence delaying or avoiding the guidance of the weapons system associated with the STT or MFR.

6.3.1 Range Tracking Loop Deception

Range tracking can be deceived by the range-gate pull-off (RGPO) technique, also called the range-gate walk-off (RGWO) technique. The objective of the RGPO technique is to move the range tracking gate away from the true target position to a position at which only the jamming signal is present. This implies that the RECM has to initially transmit the jamming signal as coincident as possible in time with the target return signal. This is accomplished within the RECM by placing the DRFM repeater as near as possible to its receiving antenna and by synthesizing a replica of the incoming pulse to be transmitted either by a separate transmitting antenna (quite isolated from the receiving one) or by the same receiving antenna switched to transmission. In the latter case the DRFM reconstructs the radar RF pulse from a small slice of the incoming pulse by transmitting (after the reception switch-off) a number of contiguous slices, to which a head-to-tail phase correction is applied in order to avoid the so called *moding* effect, which may provide severe JSR losses if the retransmitted slices are out of phase between them. This technique is called *slice repeater mode*. The schematic block diagrams of the two possible RECM configurations are shown in Figure 6.11.

The immediate RECM replicas are transmitted for a sufficient period to capture the radar AGC to the level of the replicas (progressively higher than the target returns). Thereafter, the replicas are moved according to a pull-off rate from the actual target position (i.e., a progressively increasing range from the radar; or a less decreasing range, mimicking a decelerating target maneuver).

The pull-off rate has to be within the design capabilities of the tracking loop; that is, it has to be similar to a possible deceleration of the target. Otherwise, if it is larger, the tracking loop is unable to follow the walk-off and the tracking gate will return to its memorized position, which is the position of the actual target. The pull-off rate for an airborne RECM usually mimics a deceleration <3g (30 m/s^2).

The range difference with respect to the actual target range is expressed as:

$$\Delta r(t) = \frac{1}{2} a t^2 \text{ for } 0 < t < T \tag{6.8}$$

where $t = 0$ is the instant at which the pull-off rate is impressed after the capture of the AGC level.

The delayed replicas are synchronized with the arrival of the incoming radar pulse, and the false target movement is provided with elementary delay time steps (with a resolution on the order of <2 ns). Also the phase shift of the transmitted signal is corrected in accordance with the delay rate.

The time evolution of the RGPO program is shown in Figure 6.12. In Figure 6.12(b) the impressed delay to the replica is shown. Typically, the maximum delay is designed so that $\Delta r_{max}/(c\tau/2) \geq 4$, i.e., $aT^2 \geq 4c\tau$, which for a pulse width of $\tau = 0.5$ µs and an acceleration $a = 3g$ provides a total time of the program $T = 6.3$ sec and a distance $\Delta r_{max} \approx 600$m.

Once the range gate is sufficiently distant from the actual target (i.e., a delay of more than one pulse width), angular deception techniques can be applied to the RECM signal. If the RGPO program were unable, for any reason, to capture the

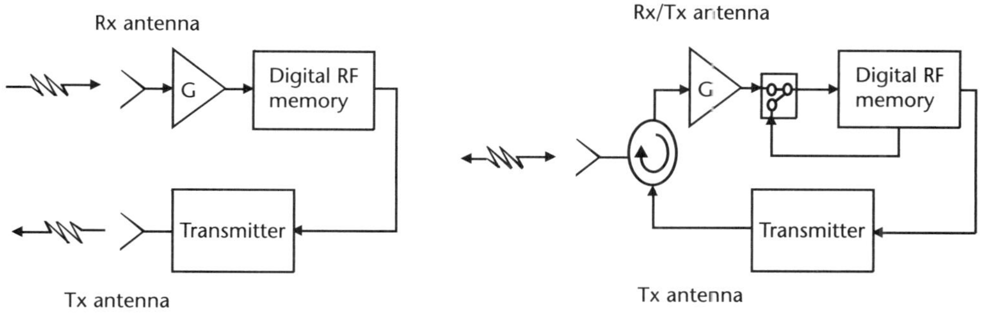

Figure 6.11 Schematic block diagrams of the two possible RECM configurations for the RGPO program.

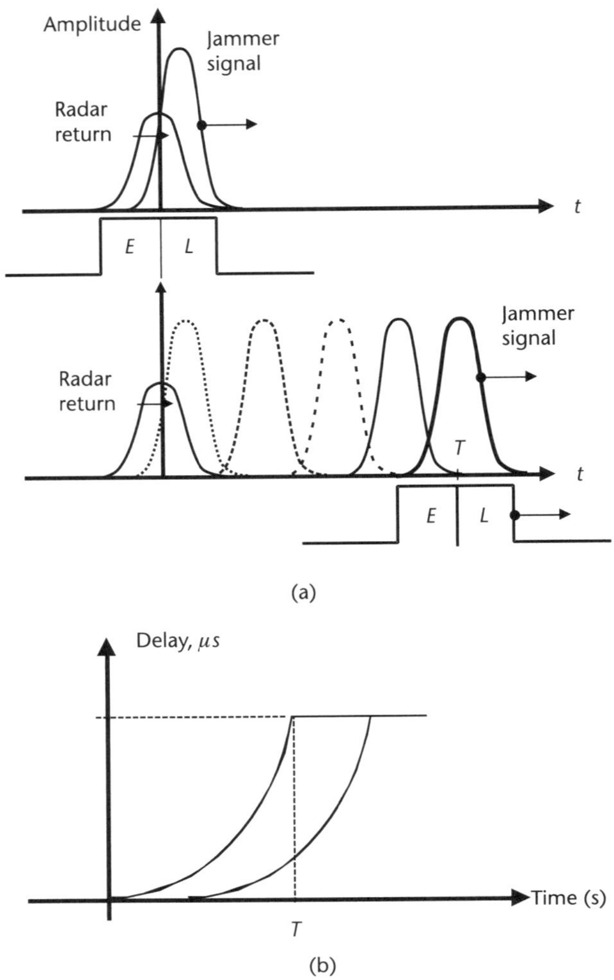

Figure 6.12 Time evolution of the RGPO program.

radar range tracking gate, the subsequent RECM transmissions and modulations would be outside that gate and hence ineffective.

Because there is no means for the RECM system to determine that the RGPO program was successful in capturing the radar range tracking gate, the RGPO program has to be recycled from the true target position to make another attempt at gate capture. The recycling of the program may be initiated as soon as the delay Δ of the first cycle is $\Delta \geq 2\tau$. The first RGPO cycle, when it has reached its maximum delay, continues to transmit a replica from that position. Therefore, the RECM is transmitting two pulses for each intercepted pulse from the victim radar: One is in accordance with the RGPO recycling and the other is continuously positioned at the outer range of the RGPO program. This latter false target is referred as a *hook* or *hold-out target*. Without the hook target, the pull-off recycling may in some cases make useless angle jamming modulations, which are instead continuously applied to the hook target. This means that when the radar range tracking gate hooks onto it, the radar angle tracking loop is continuously exposed to the erroneous angle modulations.

A number of ECCM techniques are employed by the tracking radars to counter the RGPO technique. The most used is leading-edge (LE) range tracking, which is performed either by increasing the weight assigned to the early gate received power with respect to the late gate's received power or by differentiating[1] the incoming radar pulse and applying the E-L gate technique to the differentiated LE of the pulse. Both techniques track the target at the beginning of the early gate, thus strongly reducing the received power from the jammer due to its inherent retransmission delay, which causes it to appear only at the late gate. To increase ECCM effectiveness, the tracking radar may also use *jittered* PRIs and, if possible, *frequency agility* or *diversity* in transmission, in order to prevent the prediction of the next pulse TOA and its carrier frequency.

Another much used ECCM technique is the addition of range guard gates on either side of the range tracking gate. The early and late gates are meant to assess whether a large noise cover pulse is present and the late gate only whether a RGPO program is in place. In both cases, for a limited period, the radar can keep track of the previous position of the true target and then pass to track-on-jam if the jamming persists.

These ECCM enhancements are countered by the RECM with the transmission of a *dual-mode* jamming technique consisting of the overlapped transmission of a barrage noise cover pulse of duration equal to the PRI maximum jitter and of a pulse radar replica at the same radar carrier frequency. The jammer noise signal prevents the operation of any form of E-L gate tracking (because the E-L gate is floating within the cover pulse), while the second signal appears as an appealing target. To this pulse is applied the RGPO program, while the noise cover pulse is stopped as soon as the delayed replica is at a distance of 2τ from the actual target position. At the same time, the angle jamming modulations are applied to the radar pulse replicas.

Some PD fire control radars employ range tracking to provide a more accurate measurement of target range. They use a modified range tracker operation at the narrowband detection Doppler filter and thus the RGPO technique must be applied in conjunction with Doppler signal compensation. This technique is called velocity and range-gate pull-off (VRGPO).

In the case where the RECM has to counter an STT for mutual protection purposes, the only applicable jamming technique is noise cover pulse. Noise jamming has to be applied for a short period—the required time for the protected platform to perform a rapid maneuver; otherwise, the jamming platform may be subject to track-on-jam by a weapons system or home-on-jam by a missile.

1. The use of an E-L gate applied to differentiated pulses compromises the performance of the radar range tracking because the received energy in each processed pulse is much lower with regard to the regular pulse due to the reduced target signal energy received and to the increase in the noise power level. Furthermore, as shown in Chapter 2, the position of the effective center of the target RCS is fluctuating due to the range glint and thus the LE can jitter in position by more than the transmitted delay of the RECM. In general, this technique is applied only in the presence of a very large RCS as is the case for an antiship missile seeker that is attacking a large ship. Simple calculations can be used to show that the decrease in the jammer power processed at the radar receiver, due to the RECM use of a barrage noise cover pulse, is partially compensated for by the above incurred losses to the target radar signal.

6.3.2 RECM Techniques Used Against Radar Doppler Tracking

Doppler radars, either CW or PD, use Doppler filtering of the incoming target returns to eliminate all other signals that may be at the same angle or even within the same range interval and to reduce any interference, intentional or nonintentional, that lies outside the target Doppler filter bandwidth. Target tracking is also performed through Doppler frequency tracking by using a similar EL mechanism (now in the Doppler frequency domain) for the range tracking based on TOA as shown in Figure 6.13.

In the above scheme the VCO is properly set by the control loop at the frequency that provides a zero output from the amplitude (i.e., the energy content) difference of the E-L Doppler filters. The Doppler tracking must be initialized after having measured in the target acquisition phase the Doppler frequency relevant to the target's velocity and acceleration relative to the radar (f_0 in Figure 6.13).

The extraction of the Doppler frequency information in PD/CW radars requires the use of coherent processing intervals (CPIs) on the order of a few to many milliseconds, which typically include many pulse interval times of the radar transmission. RECM techniques exploit the CPI duration to generate signals at Doppler frequencies by phase modulating the incoming signal and retransmitting it either with the content of a number of Doppler frequencies (multiple false Doppler targets, MFDT) in order to confuse the radar's target acquisition or with the VGPO technique.

The objective of the VGPO technique is to capture the Doppler tracking gate away from the true position up to a position several filter bandwidths away (which is achieved with a false Doppler target of energy providing a JSR > 3 dB) and then apply the angle modulation pull-off techniques. As for the case of the range tracking gate, the pull-off rate of the false target must be within the design capabilities of the Doppler tracking loop, because the pull-off program represents either an accelerating or decelerating target with respect to the true target. Like the RGPO technique, due to the lack of knowledge, if the first cycle of the technique was successful, the VGPO program is recycled, while a hook target is kept at the maximum

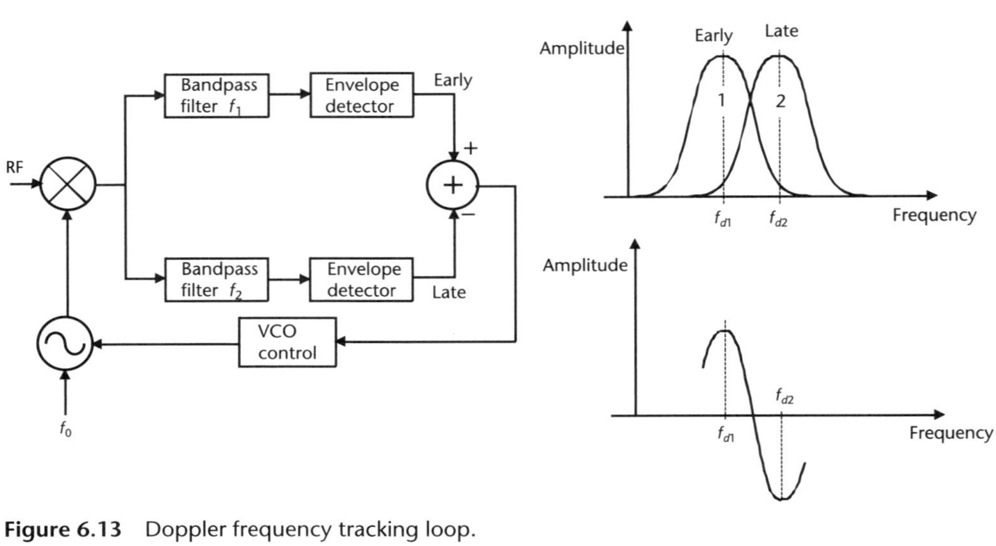

Figure 6.13 Doppler frequency tracking loop.

(or minimum) reached closing speed. The time evolution of the VGPO program is shown in Figure 6.14.

Because the rate of change of the Doppler frequency represents an acceleration [i.e., $df_d/dt = k \cdot dv_c/dt = k\ a(t)$[and the closing speed is expressed as $v_c(t) = a(t) \cdot t + v_{c0}$, care must be taken that the final value of acceleration $a(T)$ does not exceed the maximum acceleration value that can be developed with a target maneuver (i.e., $3g$ for an A/C with pilot, $15g$ for an unmanned A/C such as an AAM). As an example, consider a false target mimicking a manned aircraft maneuver characterized by a parabolic increase of the closing speed v_c. With a final acceleration value of $3g$ we can express v_c as:

$$v_c(t) = v_{c0} + kgt^2 \tag{6.9}$$

which reaches its maximum after a period T equal to:

$$T = 3/k \tag{6.10}$$

Assuming $T = 3$ sec, we have $k = 1$ (s^{-1}) and, hence, the increase in the closing speed is

$$\Delta v_c = v_c(T) - v_{c0} = 9g \approx 90(m/s) \tag{6.11}$$

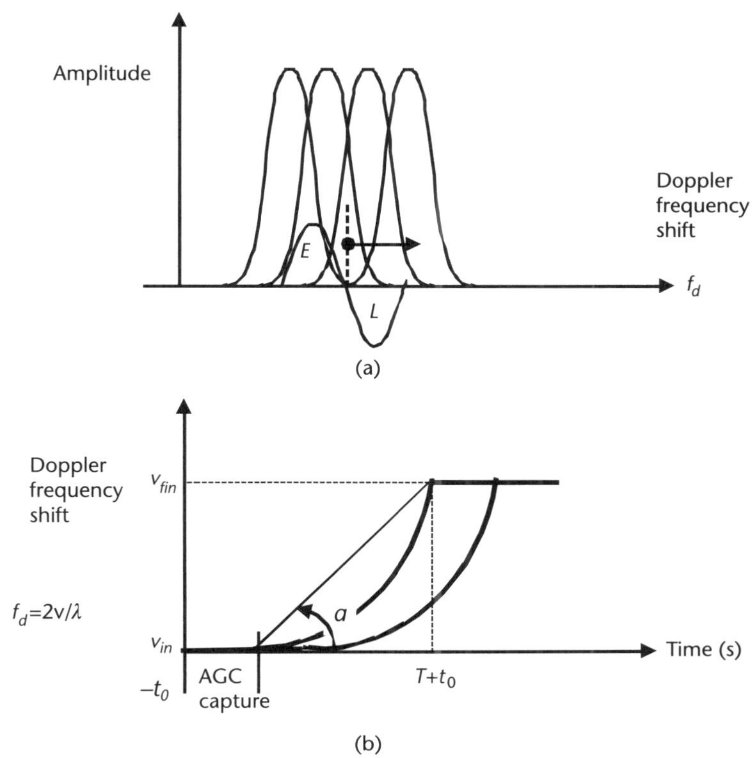

Figure 6.14 Time evolution of the VGPO program.

which corresponds, assuming the radar carrier frequency to be f_t = 10 GHz (or λ = 0.03m), to a Doppler frequency increase $\Delta f_d = 2 \cdot \Delta v_c / \lambda$ = 6 kHz. Because radar Doppler tracking bandwidths are on the order of 1 kHz, the VGPO program would have produced a pull-off of almost six filter widths.

Note that the Doppler frequency change impressed by the VGPO program is in addition to the Doppler frequency already provided by the closing speed and acceleration of the defended platform, so the RECM is required to select the additional acceleration of the VGPO program, taking into account the instantaneous kinematics parameters of the platform.

The ECCM techniques relevant to the VGPO program that the radars can put in place are as follows:

- *Coherency check*. The phase modulation impressed to the replica also produces relatively strong spurious signals that may be identified by detectors placed in the Doppler filters close to the filter containing the true target Doppler frequency. In this case the radar may coast on the true target Doppler filter, ignoring the new more appealing signals.
- *Velocity guard gates*. The gates are placed on either side of the true target Doppler filter to monitor the insurgency of false targets and make the radar coast on the true target Doppler filter (i.e., keep the tracking on the previous position of the true target for a limited period) and then pass to track-on-jam if the jamming persists.

As recalled at the beginning of this discussion, the objective of both RGPO and VGPO techniques is to provide a larger JSR for the subsequent angle deception techniques (which may require a JSR > 30 dB when competing with the true target signal). This is achieved by removing the target signal from the tracking gate.

Because most modern tracking radars employ some Doppler signal processing (MTI, LPRF PD), it is necessary to add to the delayed replicas of the RGPO program (which is a time modulation technique) the relevant Doppler frequency shifts due to the false target deceleration (usually a few kilohertz when mimicking an A/C). This VRGPO program is efficiently produced by modern DRFMs.

6.3.3 RECM Techniques Used Against Radar Angle Measurement

Angle measurement and tracking provided by STT or MFAR radars are the essential target position parameters required to achieve effective interception of intruders into defended territories (including military ships), because target angle position alone can in many cases provide interception from radar-guided weapons (i.e., through track-on-jam or home-on-jam techniques).

Because of the importance of STT angle measurement and tracking, much effort has been spent by radar designers on protecting this capability and by RECM designers on denying it through the development of effective systems and techniques. As discussed in the previous paragraphs, the RECM techniques used against radar range tracking are only the initial part of a jamming program meant to ease the success of the following part of the program in which techniques against radar angle tracking are employed.

As discussed in Chapter 2, old tracking radars (some of which are still present in many of the legacy systems of some nations) used sequential lobbing techniques to provide angle measurement, among which the most effective was the conical scan technique. In this technique angle measurement is achieved by nutating the antenna main-beam pattern around the boresight of the antenna parabolic dish reflector and by measuring the amplitude modulation of the target returns. This index becomes a measure of the target angular offset from the antenna boresight. The angular tracking loop of the STT is able to completely correct the angular offset, which is achieved when the AM index is zero, that is, when the radar antenna boresight is aligned to the target direction, as shown in Figure 6.15.

The RECM applied technique is an amplitude modulation as near as possible to the nutation frequency, so as to be accepted within the angular tracking loop bandwidth (centered at the scanning frequency f_s). Typically this band is a few hertz.

Even though the target return signal may be overwhelmed by the RECM signal, unless the AM frequency is within the above-mentioned bandwidth, the radar angle tracking loop will coast on its prior data or continue to extract the correct target position from the RECM signal (which is modulated at f_s by the nutation of the receiving feed). The self-protection RECM, therefore, first measures the frequency of the amplitude modulation (which is available until the radar antenna boresight is not aligned with the target direction) and then locks onto it through a PLL. As soon as the scanning frequency is locked, the RECM starts transmitting a signal at the radar carrier frequency that is modulated with a square wave (which produces the maximum index of modulation) at the scanning frequency but with a phase shift of π radians with respect to the received radar amplitude modulation. This technique is called inverse gain and is shown in Figure 6.16.

As can be noted from Figure 6.16(a), which displays the combined target plus jammer responses (i.e. in the case that RGPO or VGPO programs were not successful), the target modulation is still present and a large JSR would be required to pull

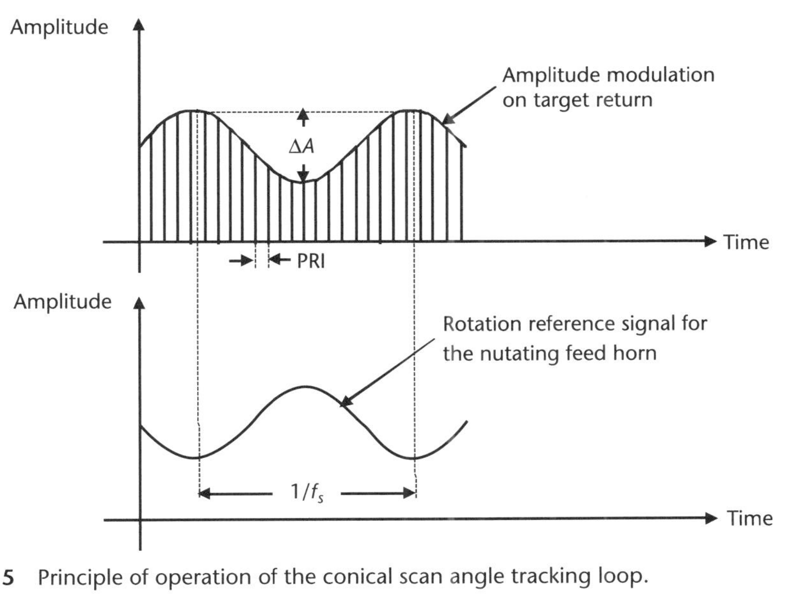

Figure 6.15 Principle of operation of the conical scan angle tracking loop.

off the target angular tracking loop. In Figure 6.16(b) only the jammer response is shown, which would produce the pull-off of the radar angle tracking from the target position in the case of successful RGPO or VGPO programs.

The first ECCM technique employed by the conical scan tracking radars was to adopt a two-antenna system: one transmitting antenna with a fixed beamwidth, centered at the parabolic dish boresight, and one receiving antenna with a nutating beamwidth at frequency f_s around the boresight of the transmitting antenna. This technique is called *conical scan on receive only* (COSRO) and is also applied to lobe switching angle measurement schemes (see Chapter 2), where the technique is called *lobe on receive only* (LORO).

Because with the COSRO technique the radar transmitting antenna is not nutated any more, the RECM is unable to measure the angular scanning modulation f_s. A different RECM technique is therefore employed, called *swept-square-wave* (SSW), in which the frequency of the amplitude modulation is slowly but continuously varied, as shown in Figure 6.17, over the expected range ($f_{s,\min}$ to $f_{s,\max}$) of the nutation frequencies that may be used by the radar. The above range is usually comprised from 20 to 300 Hz. The sweep rate has to be slow enough that the

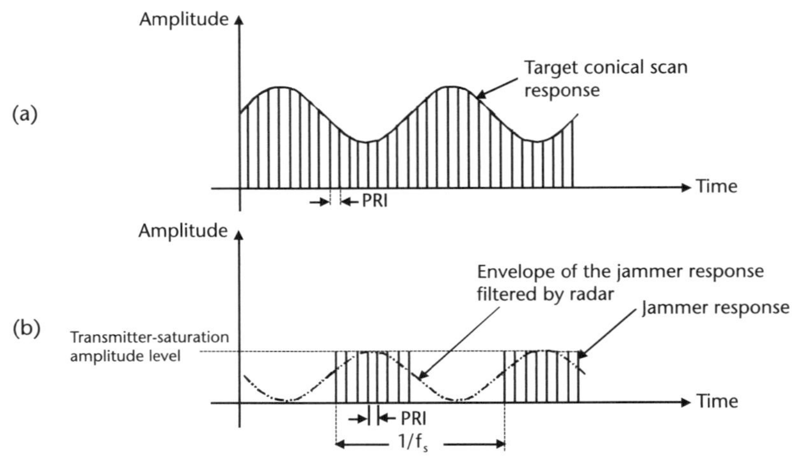

Figure 6.16 Inverse gain technique applied against conical scan tracking radars.

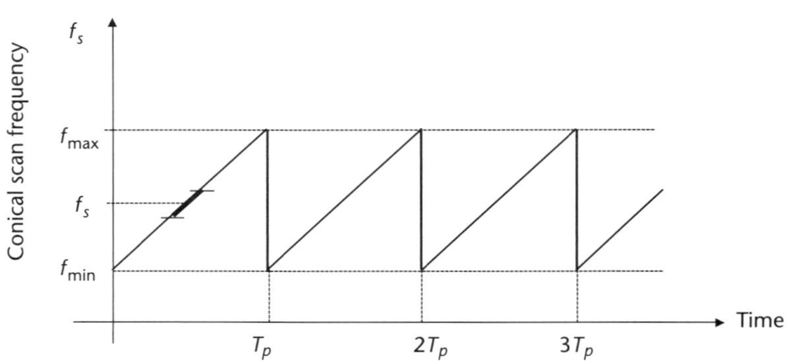

Figure 6.17 SSW technique.

amplitude modulation at the radar scanning frequency f_s persists for a sufficient period T in order to produce the pull-off of the angle tracking.

Because the angle tracking loop bandwidth is typically very narrow (a few hertz), the required period T is a fraction of a second and hence the coverage of the complete conical scan frequency interval (about 280 Hz) would require more than 50 sec. This sweeping frequency period is too large, especially in a missile engagement. However, because the employed amplitude modulation (the on-off square wave) is rich in odd harmonics (especially the third and fifth harmonic, which are attenuated, respectively, by 10 and 14 dB with respect to the fundamental), it is sufficient to sweep the fundamental square wave from 20 to 60 Hz in order to cover the complete conical scan frequency range. In this case the sweeping period can be 4 to 5 sec and the effective jamming period is $T = 0.25$ sec, sufficient to induce the angle tracking pull-off.

The use of an inverse gain or SSW technique requires [5] a JSR of about 10 to 20 dB when the signal has to compete with the true target return. Therefore, the technique is always preceded by the RGPO or VGPO program, which usually requires only a JSR of > 3 dB.

To reduce the total frequency sweep period, modern RECMs employ the *jog detection* technique, which is based on monitoring the amplitude variations of the tracking radar along the frequency sweep. In fact, during the frequency sweep, if the dwell time at a certain frequency is insufficient to disrupt the radar angle tracking, it may yet be sufficient to induce momentary perturbations to the angular tracking loop. These small angular movements of the transmitting antenna (and hence the relevant variations of received amplitude at the RECM) can be detected,

and the relevant RECM transmitted frequency provides an indication of the radar's conical scan frequency on reception only. With this indication the RECM stops the SSW frequency sweep and begins a fine SSW frequency sweep around the above indicated frequency.

The SSW technique is also applied in IR jamming (by flash lamps or DIRCMs) of the rotating reticles of the IR missile seekers, albeit with a different scale of rotation speeds and tracking loop bandwidth due to the smaller moments of inertia of the small IR reticle with respect to the nutating RF antenna feed.

Modern tracking radars employ the monopulse technique (either amplitude or phase version) for angle measurement. This technique avoids the need for amplitude modulation, because it can extract the needed angular information from a single return by measuring its amplitude differences or phase difference as received at two or more displaced elements of the radar antenna (see Chapter 2).

The RECM techniques used against monopulse angular tracking are cross-polarization, cross-eye, countdown, ground bounce, chaff illumination, and expendable and towed decoys. These techniques are discussed in the following subsections.

6.3.3.1 Cross-Polarization

Some radar parabolic reflector antennas show one potential vulnerability to the amplitude monopulse technique for the presence of an antenna pattern in orthogonal (cross) polarization although at a level of less than –25 dB with respect to the design polarization. In the cross-polarized antenna pattern, the nulls occur where the design polarization pattern shows the maxima; that is, where the two polarized beams show interchanged patterns as shown in Figure 6.18. Therefore, when the amplitude difference of the two beams is performed (as required by the amplitude monopulse technique), the two error signal patterns are out of phase.

The net result is that for the same position in angle relative to the boresight, the cross-polarized pattern drives the tracking loop to the opposite direction with respect to the one relevant to the design polarization. Thus, the RECM has to transmit at orthogonal polarization with respect to the received one in order to produce a steady angular error. The transmitted polarization is required to have an accuracy of about 1° with respect to the cross-polarization in order to be effective as shown with the following consideration. Let us assume that the tracking radar is designed with vertical polarization and its cross-polarized (i.e., horizontal polarization) response is –25 dB with respect to the isopolarized gain. An error of ε deg in the transmitted signal with regard to horizontal polarization will produce a vertically polarized jamming component in the response with a power

$$P_{j,v} = P_j \sin^2 \varepsilon \cong P_{j,h} \sin^2 \varepsilon \tag{6.12}$$

If $\varepsilon \approx 3°$, $\sin^2\varepsilon = -25$ dB and this means that the transmitted signal is producing a vertical component of the same level as the cross-polarized one. Thus the cross-polarized signal should be accurate to <1° with regard to horizontal polarization in order for its vertical component to be less than –30 dB and for the technique to be effective.

Due to the difficulty of reaching such an accuracy, the RECM employs a feed of a parabolic dish reflector that is rotated at 90° with regard to the received

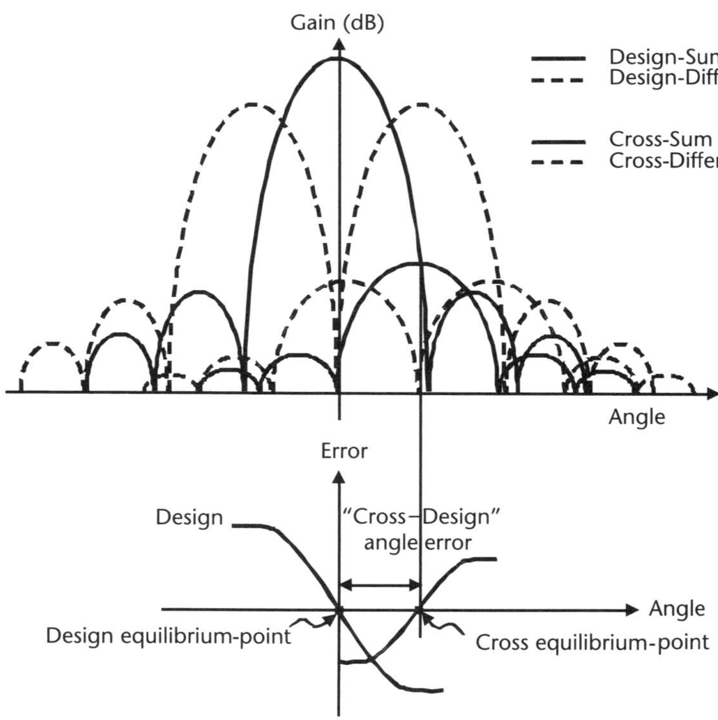

Figure 6.18 Orthogonal sum and difference patterns and relevant amplitude monopulse gradients.

polarization, and then slowly swept across this position of ±2° in order to provide a kick to the angular tracking loop when the transmitted signal is at the actual cross-polarization.

Another RECM method is to use a large active phased array antenna with N radiating elements in double polarization (H and V). Once the received polarization is measured with the same accuracy as above, the transmitting elements are selected in the proper ratio n_H/n_V (where $n_H + n_V = N$) in order to produce the cross-polarized jamming signal with the required accuracy.

The cross-polarization technique is not used much for the following reasons:

- It requires many antenna fixes, such as the application of polarization filters to the receiving feeds and application of a polarization twist to the Cassegrain antenna subreflector, which drastically reduces its effectiveness (see the pictures of the tracking radars NA30 and NA25 reported in Chapter 2).
- There is no inherent reason why the radar receiving antenna must be designed at the same polarization as the transmitting one because the target's scattering power is uniformly distributed at all polarizations.

The behavior of the phase monopulse radar technique relative to cross-polarization is very similar to that of the amplitude monopulse radar technique in that the cross-polarized RECM signal produces a π radians phase shift between the two monopulse receiving antennas (or the two semi-arrays of a phased array antenna) of this technique as shown in [5], thus providing a distortion in the error signal pattern that feeds the radar angular tracking loop.

6.3.3.2 Cross-Eye

The cross-eye technique exploits the angular glint error produced in the angular tracking loop by a target composed of a large number of independent reflectors as discussed in Chapter 2, where the case of two reflectors (the dumbbell model, analyzed by Howard [6]) is reported.

In practical applications of this technique, the two independent reflectors, spaced at a distance L, called the base length, are exchanged for two coherent jammers, which transmit a power ERP_2 and $ERP_1 = a^2 ERP_2$ where $a \leq 1$ is an attenuation factor. Its signals arrive at the radar victim antenna at a distance R from the base length, with a phase shift ϕ. Under these conditions, for which the geometry is shown in Figure 6.19, it can be shown [6–9] that the radar monopulse network is subject to an angular error ε as follows:

$$\varepsilon = \frac{\alpha}{2} \cdot \frac{1 - a^2}{1 + a^2 + 2a\cos\phi} \tag{6.13}$$

where $\alpha = (L \cdot \cos\gamma)/R$ is the angular extent of the base length as seen by the radar from distance R, and γ is the angle between the normal direction is the ortthogonal direction to a segment to the base length and the radar antenna boresight.

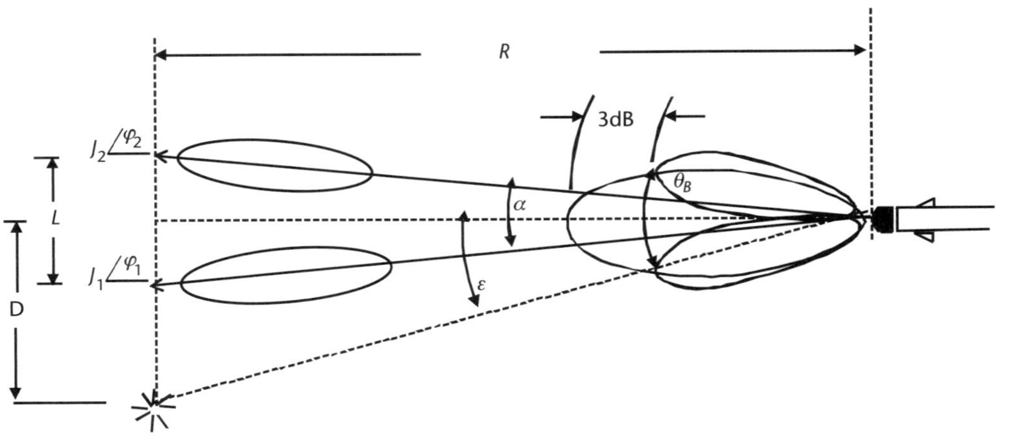

Figure 6.19 Geometry of the cross-eye technique.

Equation (6.13) is valid as long as the angular error ε is within $\theta_B/2$(i.e., $\varepsilon \leq \theta_B/2$), where θ_B is the radar –3-dB sum beamwidth. Beamwidth vs. phase shift ϕ is plotted in Figure 6.20 using the power ratio $a^2 = J_1/J_2$, where J_1 and J_2 are the jammer powers transmitted by the two jammers and received by the victim radar at a distance R. Jammer powers J_1 and J_2 are proportional to the respective transmitted jammer ERP through the same proportionality factor $k = (\mathrm{ERP}/4\pi R^2)(G\lambda^2/4\pi)$, where G is the radar antenna receiving gain.

As shown in Figure 6.20, the maximum of the error is achieved when the two signals reach the radar antenna out of phase ($\phi = \pi$ radians) and the peak error is expressed as

$$\frac{\varepsilon}{\alpha} = \frac{D}{L} = \frac{1}{2} \cdot \frac{1+a}{1-a} \tag{6.14}$$

where D is the apparent position from the center of the base length and is aligned with the base length of the "ghost" jammer (the combination of the two jammers).

To implement practical jammer systems, some important considerations have to be made relevant to the preceding angular error expression:

- The phase shift of nearly π radians between the two coherent jammer signals is the one *arriving at the radar antenna*. To achieve this relationship in most engagement geometries, a *retrodirective jammer architecture* like that shown in Figure 6.21 must be used.
- The phase shift accuracy becomes more stringent the lower the ratio of a relevant to the amplitude between the two signals that provides the larger distance of the "ghost" target from the defended platform.
- The lower the ratio a, the higher the required jammer ERPj of the two coherent sources to compensate for the jammer power loss ΔJ at the radar receiver due to the out-of-phase condition for the effectiveness of the technique. Jammer loss is expressed as follows:

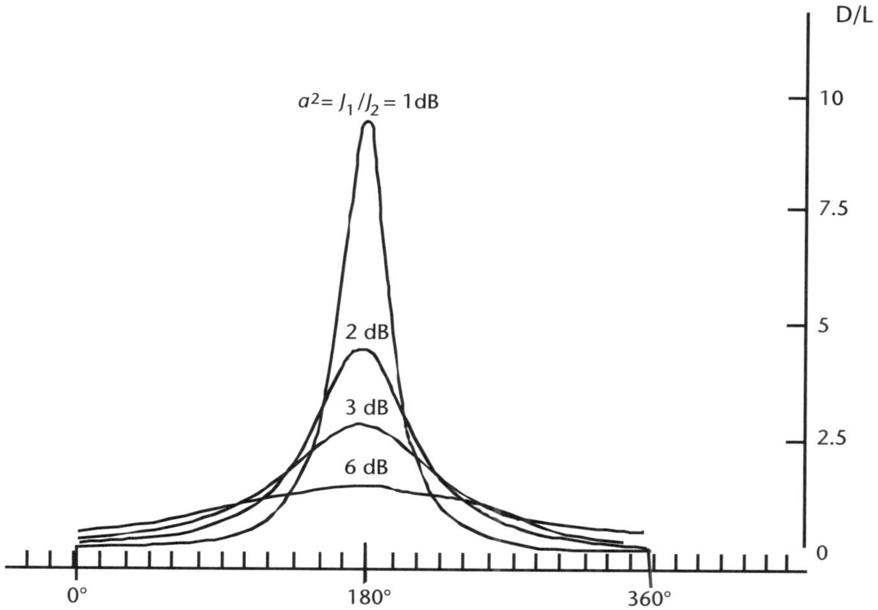

Figure 6.20 Cross-eye induced error to a monopulse receiving antenna.

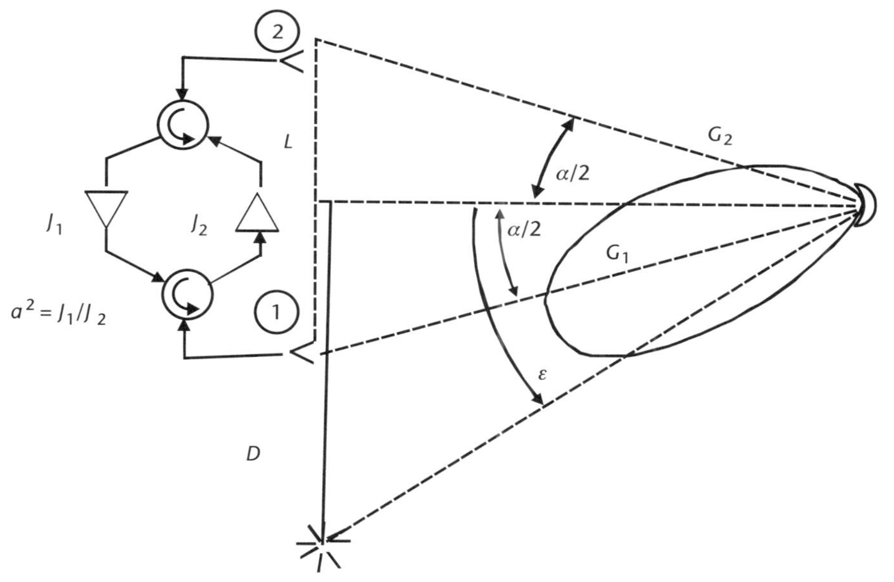

Figure 6.21 Retrodirective system architecture of a cross-eye jammer.

$$\Delta j = (1 - a)^2 \tag{6.15}$$

and is plotted in Figure 6.22.

From theory and experimental trials [8, 9], it has been shown that the required residual JSR (i.e., after removing the loss ΔJ) must be JSR $\geq$ 6 dB for a ratio $a = -2$ dB.

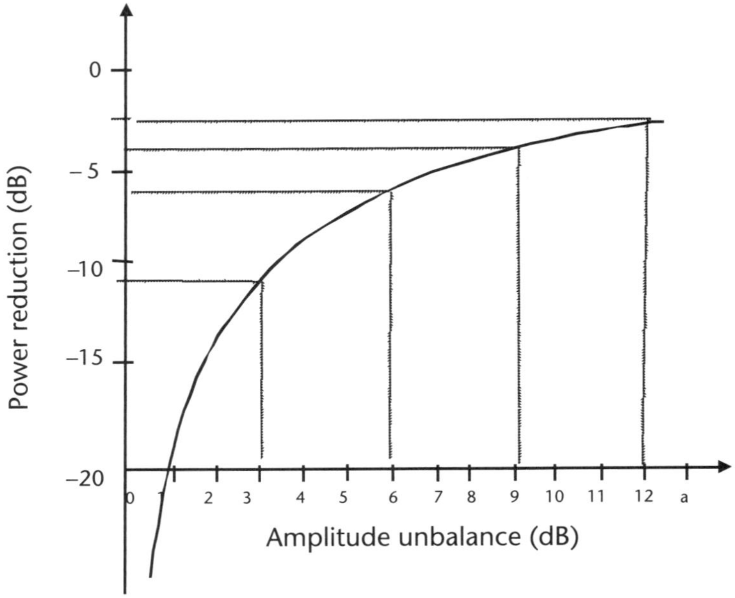

Figure 6.22 Cross-eye jammer power loss.

The preceding considerations imply that practical jammers should implement amplitude ratios $a < -1$ dB in the presence of the target return or perform the cross-eye technique after a RGPO/VGPO program has succeeded in capturing the range tracking gate as for the previously discussed angular gate deception techniques. Notwithstanding the expression of (6.13) is evaluated in terms of an angular error to the radar angular tracking, the induced glint error, as shown in (6.14), is produced by an apparent source of the return signal at a distance D from the center of the base length.

The plot of the maximum angular cross-eye error vs. the inverse of the distance R from the radar is shown in Figure 6.23 with the ratio a as a parameter. From this figure note that at large distances [i.e., where $\alpha_B = \alpha/(\theta_B/2) = L/(R\theta_B/2) < 0.01$,

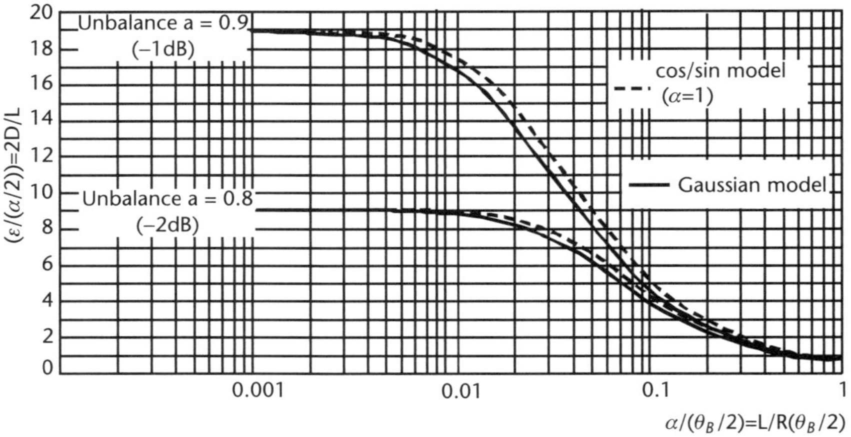

Figure 6.23 Constant power cross-eye (CPCE) angular error vs. the inverse of the distance R to the radar.

small angular extent of the base length with regard to the radar beam] where large JSRs are available, the angular error is negligible. At shorter distances of $\alpha_B > 0.05$, as the angular extent of the base length is increased with respect to the beamwidth, the angular cross-eye error decreases as the imbalance *a* of the jammer signals is increased by the different antenna gain values at which the two jammer signals are received. This means that as the jammer approaches the radar, the closer it gets, the larger is the imbalance applied to the jammer signals by the antenna gain and the smaller is the distance *D* of the apparent source from the base length.

Large angular errors at short distances can be achieved by keeping constant the jammer signals imbalance as seen by the radar. The additional imbalance produced by the radar beamwidth is compensated via a tracking closed loop performed by the retrodirective jammer architecture. This second method is called *constant gain cross-eye* (CGCE) and is the one implemented by modern RECMs employing active Rx/Tx phased array antennas. This implementation has the further advantage (as for the jog detection) of being able to assess the effectiveness of the jamming technique. As the jammer is in a closed loop with the radar beamwidth, then, through the measurement of the received unbalance increase produced when the latter aims at the ghost target, the jammer is able to increase the transmitted power unbalance from its two sources in order to keep a constant unbalance between the signal powers received by the radar (as weighed by the radar beamwidth).

The normalized cross-eye gain of the CGCE method vs. the inverse of the distance *R*, with the initially impressed imbalance *a* as a parameter, is shown in Figure 6.24. The same figure also shows the normalized cross-eye gain of the CPCE method for comparison. Note from the figure that when the distance of the radar from the target is on the order of $\alpha_B = \alpha/\theta_B/2 > 0.1$ (i.e., $L/R\theta_B \geq 0.05$), the CGCE method induces a resonance of the angular tracking error dependent on the monopulse characteristics of the radar antenna. This refers to the angular distance of the Δ beamwidth peaks from the antenna boresight or Σ beamwidth peak. This resonance has been verified in experimental trials. Due to the kinematics of the missile engagements with a target, this resonance error produces a large miss distance to the missile in the last phase of the engagement. Some simulation results of active missile engagements against fighter aircraft defended by a CGCE jammer

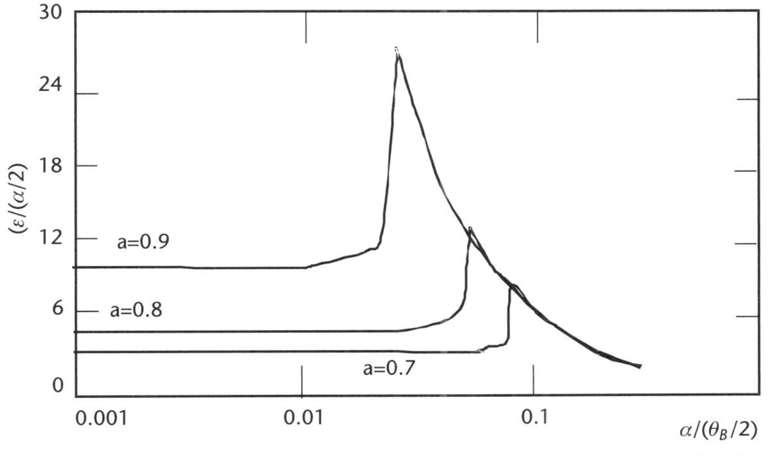

Figure 6.24 Normalized cross-eye gain of the CGCE method vs. the inverse of the distance *R*.

are graphed in Figure 6.25. On the bottom left side of this figure, the model of the aircraft (with four independent scatterers to simulate its RCS and glint because the jammer technique is performed without any RGPO) and the model of the jammers, with the initial differential amplitude of the two jammers (placed at the wingtips in order to achieve the maximum base length of L = 10m in the forward and rear sectors of the aircraft) are set to achieve a miss distance on the right side of the aircraft. On the top left side of the figure a pictorial view of the AAM/aircraft engagement is shown with two distances reported: $R_{\text{switch-on}}$, representing the start of the engagement from the radar, and $R_{\text{CE-on}}$, representing the start of the CGCE jamming technique. On the right side of the figure, two engagements are reported with the aircraft and missile kinematics data and the achieved miss distances calculated from the extremes of the wingtip.

The experimental verification of the monopulse characteristics of the radar antenna have removed the limitation attached to (6.13) of the maximum cross-eye angular error being less than half the radar –3-dB beamwidth (i.e., $\varepsilon < \theta_B/2$). Indeed, whenever the cross-eye produced distance D is sufficiently large, either because the base length L is large (so that $\alpha = L/R > \theta_B/2$) or the imbalance a is very small, a break-lock is produced. This condition can be exploited in many engagements. An example is given later when discussing the ground bounce technique.

The cross-eye technique described here produces a warped phase front just at the victim radar antenna. This warped phase front introduces an angular error in all types of angular tracking. This error is applied to antennas that tend to align their boresight normal to the incoming (supposedly) planar phase front (single target within the tracking gate).

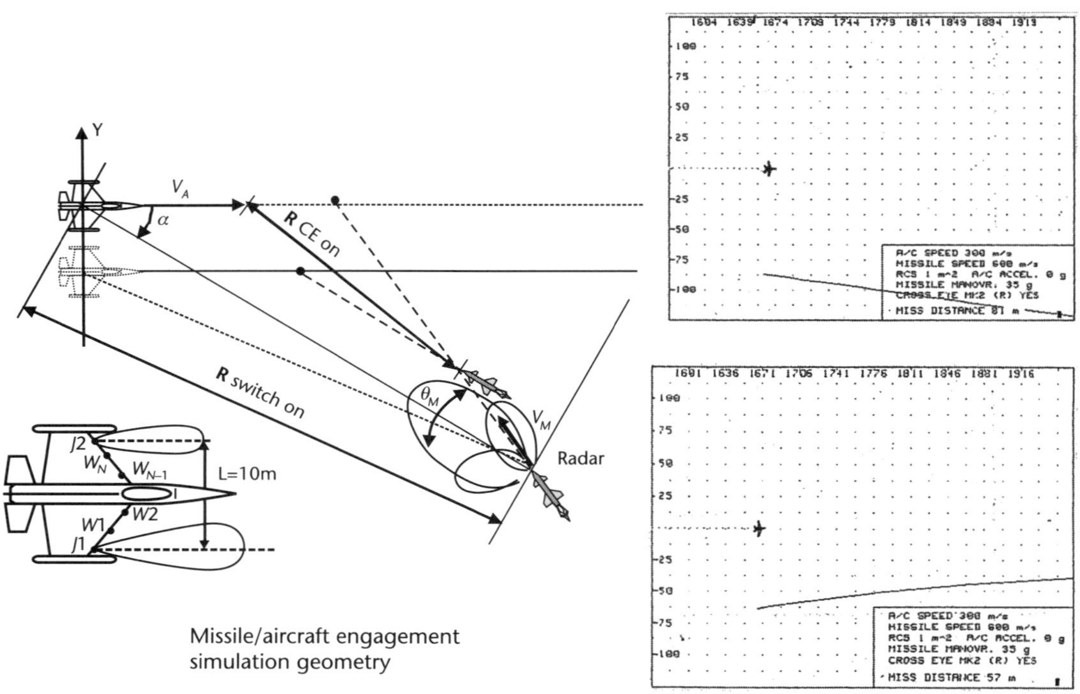

Figure 6.25 Some simulation results for active missile engagements against fighter aircraft defended by a CGCE jammer.

Two important considerations must be mentioned that are relevant to possible ECCM techniques used against the cross-eye technique as applied to active missiles:

- The cross-eye SPJ is in most cases not affected by the burn-through range of most SPJs (as discussed in Chapter 5 when dealing with the jammer equation). This is because the missile seeker beamwidth is pointed toward the ghost target. It illuminates the actual target and receives target returns only from the skirts of the main beam. This two-way antenna-gain reduction produces an increase in JSR that partially compensates for the R^{-2} target signal S increase due to the approaching distance. This compensation is larger, the narrower the radar antenna beamwidth, as shown in Figure 6.26.
- If the cross-eye technique is applied in the presence of the target return (i.e., within the range tracking gate, which results in quite short base lengths for achieving the minimum retransmission delay by the jammer), only radars employing super-resolution techniques (and hence array antennas) would have the ability to continue with the angular tracking of the actual target.

Simple methods applicable to monopulse antennas such as the complex indicated angle method [10, 11], which computes the complex monopulse ratio Δ/Σ (both the real and imaginary parts of the ratio, whereas the usual monopulse method computes only the real part of the ratio), can only indicate the presence of jamming without providing any angular indication relevant to the target [12].

6.3.3.3 Countdown

Angle tracking loops operate with a narrow servo bandwidth, that is, they require a large number of pulses to extract the angular error. After application of the RGPO

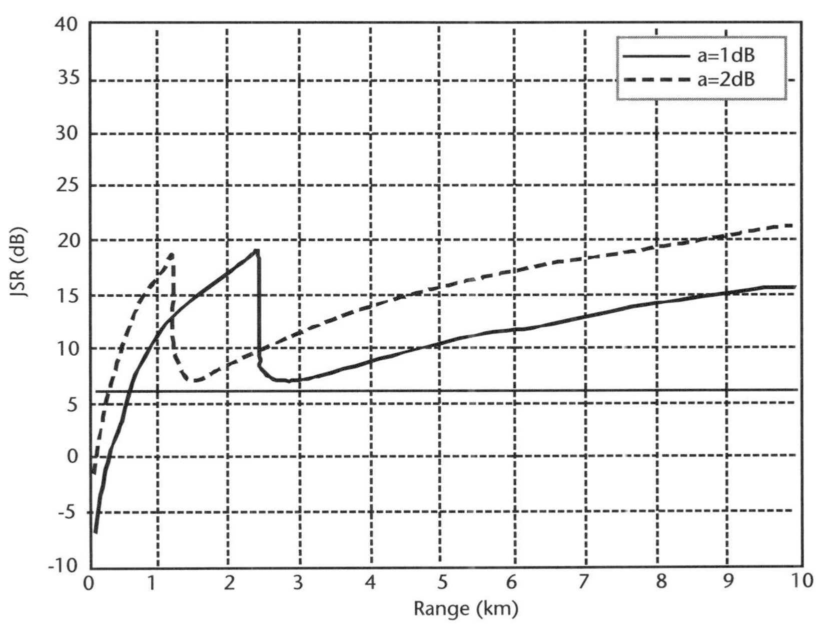

Figure 6.26 JSR increase in the terminal phase of the engagement.

or VGPO program and shifting of the tracking gate to the hook false target position, it is possible to reduce progressively the fraction of pulses to which the repeater responds (hence, the name *countdown* for this technique) and are within the tracking gate. This reduction continues until the number of pulses is insufficient for extracting an accurate angle error. This technique also affects the AGC loop, which has an amplification level that tends to keep constant the operational level of the successive circuits. A reduction in the number of integrated pulses decreases this level, so the AGC increases the amplification level toward the saturation level of the amplifiers, thus introducing nonlinearities that further worsen the angular tracking accuracy.

6.3.3.4 Ground Bounce

The ground bounce technique can be used by the RECM of a low-flying aircraft against an AAM provided with a PD or semiactive CW tracking radar seeker. It transmits replicas of the radar waveform toward the underlying terrain in a manner such that the transmitted energy is scattered toward the tracking radar. Two necessary conditions have to be met in order for this technique to be effective:

- The direct jammer energy aimed toward the seeker must be lower than that reflected from the terrain. This poses a sidelobe level limit to the jammer antenna or the use of a dedicated jammer antenna, pointed to ground, to perform such a technique.
- In the case of an attack from a PD radar, the geometrical conditions of the reflection should allow a minimum delay with respect to the target echo; otherwise, the radar will continue to track the aircraft. This implies that the technique must be activated when the missile is quite far away so that both the target signal and the jamming signal reflected by the terrain are within the same resolution cell of the radar.

Suppose an aircraft is flying at constant speed v_a and height h_a, while the missile is diving from a height h_m as shown in Figure 6.27. The initial distance between the missile and the aircraft is R_0. To allow simple calculations, let us suppose that the kinematics of the engagement is contained in the elevation plane and we have the following expressions for the aircraft and the missile:

$$\begin{aligned} x_a(t) &= G_0 - v_a t \\ y_a(t) &= h_a \end{aligned} \quad \text{and} \quad \begin{aligned} x_m(t) &= v_m t \sin\theta \\ y_m(t) &= -v_m t \cos\theta \end{aligned} \tag{6.16}$$

In these expressions G_0 is the slant range on the terrain at the beginning of this technique and the successive expressions can be evaluated as:

$$G = \sqrt{R^2 - \Delta h^2} = r_a + r_m \tag{6.17}$$

where r_a and r_m are, respectively, the slant distances of the point of the terrain reflection of the aircraft and the missile signals. From inspection of Figure 6.27, we have

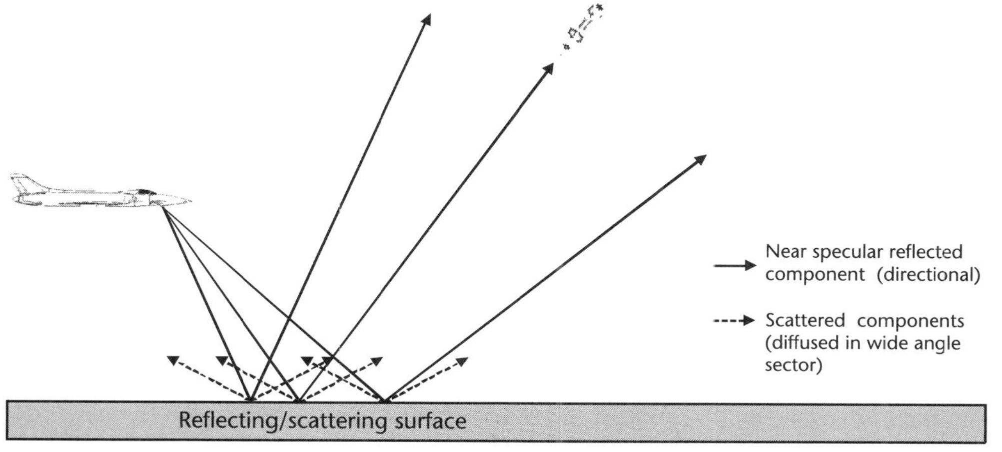

Figure 6.27 Geometry of a ground bounce engagement.

$$r_a = \frac{h_a}{h_m} r_m \quad \text{and hence} \quad G = r_m(1 + h_a/h_m) \tag{6.18}$$

The jammer power J_t received by the missile from the signal reflected by the terrain is:

$$J_t = \frac{ERP_j}{4\pi R_a^2} \cdot \frac{\sigma_g}{4\pi R_m^2} \cdot \frac{G_m \lambda^2}{4\pi} \tag{6.19}$$

while the target return power as illuminated by the missile seeker radar is:

$$S = \frac{ERP_m}{4\pi R^2} \cdot \frac{\sigma_a}{4\pi R^2} \cdot \frac{G_m \lambda^2}{4\pi} \tag{6.20}$$

where

$\sigma_g = \sigma_0 R_a^2 \theta_a^2$ is the bistatic RCS of the terrain patch illuminated by the jammer, which has an antenna with a –3-dB beamwidth θ_a and σ_0(dB$_{m^2/m^2}$) is the specific RCS of the terrain;

$R_a = (h_a^2 + r_a^2)^{1/2}$ and $R_m = (h_m^2 + r_m^2)^{1/2}$ are, respectively, the distances from the reflecting terrain patch of the jammer and the missile;

$R = (G^2 + \Delta h^2)^{1/2}$ is the distance of the target from the missile;

σ_a is the target RCS;

G_m is the missile antenna gain;

ERP$_j = P_j G_j$ and ERP$_m = P_m G_m$ are, respectively, the effective radiated power of the jammer (with power P_j and antenna gain G_j) and of the radar on board the missile (with power P_m and antenna gain G_m).

Under these conditions the JSR is:

$$JSR = J/S = \frac{ERP_j}{ERP_m} \cdot \frac{\sigma_g}{\sigma_a} \cdot \frac{R^4}{R_a^2 R_m^2} = \frac{ERP_j}{ERP_m} \cdot \frac{\sigma_g}{\sigma_a} \cdot f(R) \tag{6.21}$$

By assuming with good approximation that $R \approx R_m$ we have:

$$JSR = \frac{ERP_j}{ERP_m} \cdot \frac{\sigma_g}{\sigma_a} \cdot \frac{h_m^2}{h_a^2} = \frac{ERP_j}{ERP_m} \cdot \frac{\sigma_0 \theta_j^2 R_m^2 (h_a^2/h_m^2)}{\sigma_a} \cdot \frac{h_m^2}{h_a^2} = \frac{ERP_j}{ERP_m} \cdot \frac{\sigma_0 \theta_j^2 R_m^2}{\sigma_a} \tag{6.22}$$

The JSR at the missile receiver from the direct path is:

$$JSR_d = \frac{ERP_j (G'/G_j)}{ERP_m} \cdot \frac{4\pi R_m^2}{\sigma_a} \tag{6.23}$$

Therefore, the ground bounce technique can be effective if $JSR/JSR_d > 1$ or:

$$\frac{JSR}{JSR_d} = \frac{ERP_j/ERP_m}{ERP_j/ERP_m (G'/G_j)} \cdot \frac{\sigma_0 \theta_j^2 R_m^2}{4\pi R_m^2/\sigma_a} = \frac{\sigma_a \sigma_0 \theta_j^2}{4\pi} \cdot \frac{G_j}{G'} > 1 \tag{6.24}$$

From (6.24), using typical maximum values for the terrain reflectivity ($\sigma_0 \approx -20\ dB_{m^2/m^2}$) and the usual target RCS and large jammer beamwidth, the required jammer antenna gain G' in the direction of the missile has to be more than 40 dB below the jammer antenna gain G_j, practically a null in that direction. Therefore, the employment of the ground bounce technique requires the adoption of a separate jammer antenna looking at ground and isolated toward the upper hemisphere by the aircraft structure.

A more effective jamming technique that exploits the multipath produced by the terrain is the low altitude cross-eye (LACE) technique [13]. This technique can be exploited by aircraft flying at low altitude against radars provided with angular tracking loops in both polar angular dimensions (θ, ϕ) (i.e., typically AAM or AAA against a low-flying attack helicopter). The technique uses two jammers in the vertical plane (although separated by a moderate base length) and exploits the terrain reflection, which produces images of the two cross-eye jammers within the missile beamwidth. The large virtual base length (double the flight height) produced by the flight generates break-locks to the angular tracking of the radar. The geometry of the engagement is shown in Figure 6.28 and the simulation results of a LACE engagement relevant to a low-level flight by a helicopter against a tracking radar with 2° beamwidth are shown in Figure 6.29. The technique has been experimentally verified.

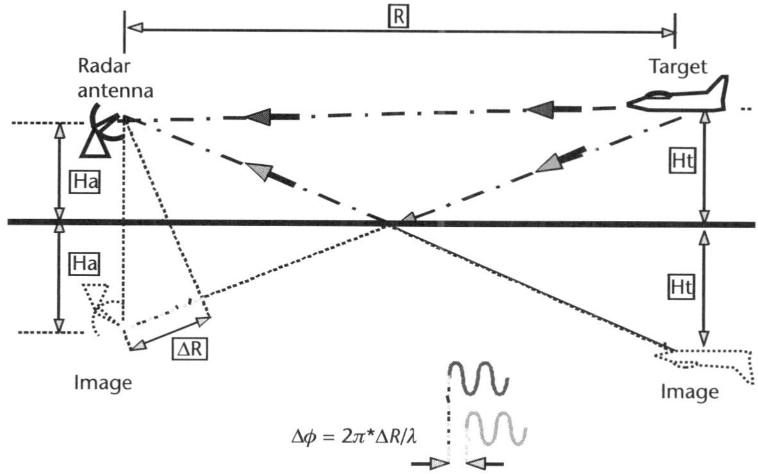

Figure 6.28 Geometry of the LACE engagement.

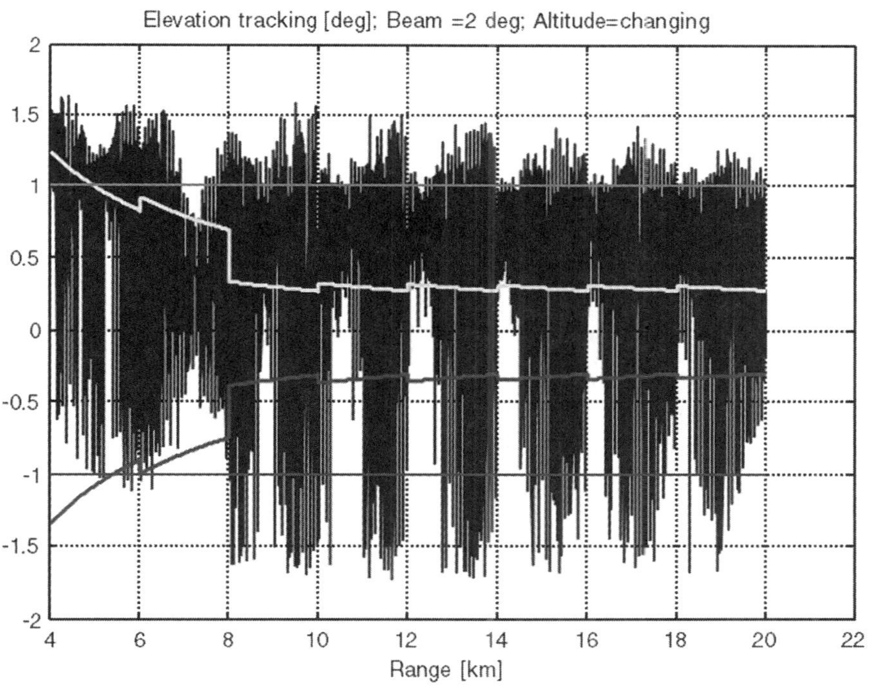

Figure 6.29 Simulation results relevant to the elevation tracking error during a low-flying helicopter LACE engagement. Helicopter height is decreasing from 100 to 50m. The two curves (yellow and red) show the limits of the radar beamwidth (2°). The passing of the two limits indicate regions of break-locks.

6.3.3.5 Chaff Illumination (CHILL)

In Chapter 5, when discussing chaff, we described a technique capable of producing a radar break-lock. This technique involved a maneuver that consisted of flying an aircraft at nearly 90° with respect to the radar boresight in order to exploit the launch of chaff, which immediately produces a large RCS clutter at nearly zero Doppler and cannot be discriminated by radar Doppler signal processing.

Break-lock can also be achieved without any aircraft maneuvering by the illumination of the chaff launched at the back of the aircraft through a proper replica by the on-board RECM of the radar waveform. The replica is transmitted according to a VRGPO program (or simply VGPO in case of CW radar). This technique is very effective against coherent radars that exploit Doppler information to discard static targets and is very convenient because it exploits a relatively cheap off-board decoy.

6.3.3.6 Expendable and Towed Decoys

With respect to passive decoys, such as chaff and corner reflectors (the latter used in the defence of ships), the ejection of an active decoy from a defended platform provides more flexibility against an approaching radar-guided missile because its transmitted signal can be modulated to produce more realistic or confusing target parameters to the victim radar.

In the airborne case, *expendable decoys* are most often used against semiactive CW missiles because a low-power RECM is sufficient for their countering. The expendable decoy takes the shape of a minimissile, with a payload very similar to that of a repeater jammer. In addition to being angularly separated from the defended target, the expendable decoy transmits a deceptive VGPO.

After having been launched in the direction of the incoming missile, the decoy rapidly decelerates and begins to fall (slowly if a small parachute has been deployed). Therefore, its transmitted signal has to be modulated with a VGPO program in order to produce a credible false target for a long period of time (usually the life of its battery).

Shipboard expendable decoys are launched by an on-board decoy launching system with a ballistic trajectory in the direction of the incoming missile but at some angular offset with respect to the ship to missile direction. In this way the decoys can lure the missile leading-edge range tracking loop. Their operational life is increased by being suspended in the air by a small parachute and by a large size battery.

The *towed decoy* (TD) is used to defend both ships and aircraft. Naval TDs normally float above the water at long distances from the defended ship and are of the repeater type. Airborne TDs can be either of the repeater type or RF fed at low power via a fiber-optic cable. The advantage of the airborne TD is that its jamming signal is inherently provided with the same Doppler content of the aircraft target, thus it is sufficient for its transmitted energy to exceed the target return in order to lure the CW semiactive missiles' tracking gates.

Fiber-optic TDs have a remote transmitter (usually a mini-TWT power amplifier) for which both RF modulating signals (via fiber-optic cables) and a high-voltage power supply (via the inner core of the cable; generated internally to the towing aircraft) are provided via the same towing cable. As such they are able to exploit the large variety of jamming programs available from an on-board RECM technique generator (see Chapter 5) and are able to counter more advanced threats than just those from CW semiactive missiles. An illustration of the deployment from aircraft of expendable and towed decoys is shown in Figure 6.30.

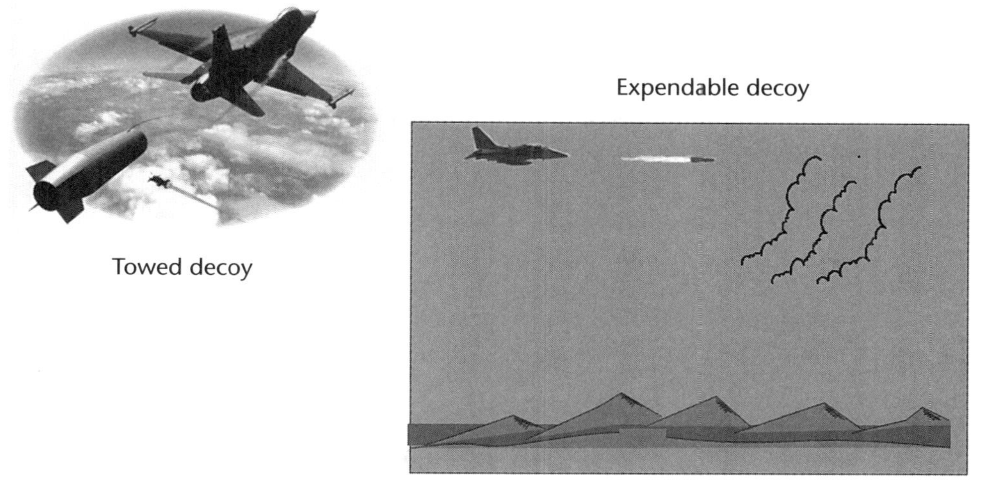

Figure 6.30 Airborne SPJ: effective off-board active decoys.

The TD technique presents two conical regions within which an approaching missile will fuse on and destroy an aircraft. They result from the fact that the missile, while tracking the decoy signal, can pass close enough to the aircraft to fuse on and destroy it, either before or after passing the decoy. Both cones have the vertex on the decoy and an aperture $\alpha = 2\ \tan^{-1}((r_l + W)/L)$, where r_l is the lethality range of the fuse, W is the aircraft fuselage width (usually negligible with regard to r_l), and L is the towing cable length. The cone aperture is usually small in dependence of the large towing cable length. Indeed, if the aircraft and the TD are aligned with respect to the direction of arrival of the missile in the forward sector of the aircraft, the missile will prevalently home on the aircraft and its Doppler-based fuse system will explode in proximity of it. In the aircraft rear sector, if the missile is homing onto the TD, but for any reason is not exploding in its proximity, it may continue its course and explode in proximity to the aircraft. The geometry of the above situation is shown in Figure 6.31.

To avoid such inconvenience, as soon as a missile lock-on is detected the pilot performs a "gentle weave" maneuver (so as not to dramatically increase the aircraft RCS as viewed by the missile) to prevent the aircraft and the TD being aligned with the DOA of the missile. Because the missile's fuse is activated by the TD body, it is possible that the missile explosion may damage or destroy the TD. In such a case the aircraft is usually provided with at least one further decoy, which is immediately deployed to protect the aircraft against a second attack.

As discussed above the airborne TD is very effective against CW semiactive missiles but is less effective (unless provided with a much larger ERP) against AAA tracking radars, which can exploit frequency agility and LE range gate tracking.

6.4 Conclusions About RECM Techniques

In the preceding sections both main ECM and ECCM techniques employed by surveillance radars and tracking radars, including the multifunctional active

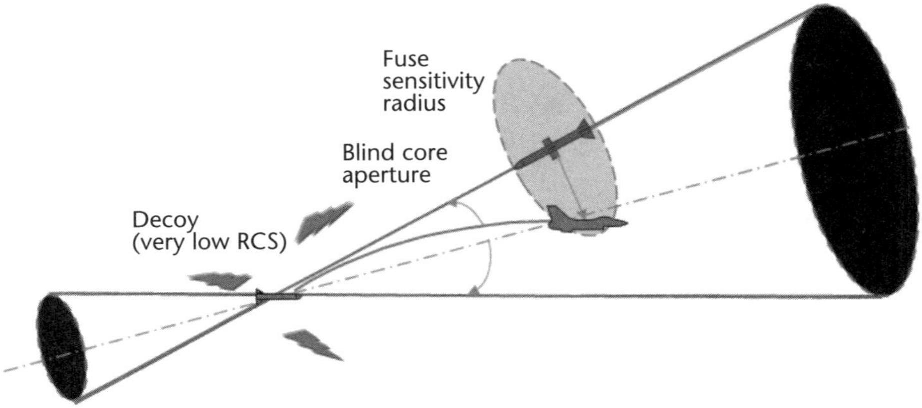

Come aperture (°) VS cable-length and fuse sensitivity			
Fuse sens. radius	Towing cable length		
	50m	100m	200m
10m	22.5°	11.4°	5.7°
20m	44°	22.5°	11.4°

Figure 6.31 Towed decoy limitations: blind cone.

electronically scanning array radars, were discussed. The most effective ECM technique against surveillance radars is a combination of short noise cover pulses (phase modulated at the pulse compression chip rate) capable of masking the jamming aircraft and use of a large number of credible false targets (within the radar main beam, possibly performed by a small number of synchronized ESJs) at ranges behind the ESJ in order to overcome the PRI jitter and frequency agility exploited by the radar. In most cases, it is necessary to first apply an ECM technique that pulls the tracking gate off the target before attempting angle deception. Among the techniques used, the most effective against active missile seekers is the cross-eye technique, which can provide a number of break-locks. When applied in the presence of the target return (i.e., within the range tracking gate), it is difficult to discriminate. The most effective technique against semiactive missiles is constituted by the towed decoy.

Many ECM techniques have been reported in the literature against SAR and bistatic/multistatic radars, but until now no experimental results have been available to assess their effectiveness.

6.5 ECM Principles and Techniques Used Against Communication Systems

In earlier times militarily, more value was associated with listening to the adversary's conversations and exchanging data via radio links than with jamming its transmissions. The widespread use of modern digital and encryption techniques in communication systems has greatly reduced the ability to quickly extract information from an adversary's exchanged messages. So jamming the nodes of the adversary's communication network has become of primary importance because it allows the commander to be separated from his forces and limits the enemy's situation awareness by disrupting the exchange of information within its network. A further important aspect of jamming to communication networks in military operations is the following consideration by W. Perry, reported in [14]: "If we jam an enemy radar, we remove one weapon from our attack, but if we jam a C2 system we disrupt an arsenal of weapons."

As discussed in Section 2.4, military communication systems are of three types:

- Combat net radios, which provide short- and medium-range voice (digitally modulated) communication;
- Tactical data links, which are used to transfer data within a C2 system from one node to another;
- Satellite communications, which provide very-long-range data exchange between users.

The main ECCM (or antijam) characteristics of modern digitally modulated communication systems are given in Figure 6.32. The two types of FHSS communication waveforms—slow (SFH, more data bits per hop) and fast (FFH, more frequency hops per data bit)—for the data bit BFSK modulation are shown, respectively, in the Figures 6.33 and 6.34.

Modern communication waveforms require two types of circuitry in their receivers: demodulation circuitry and synchronization circuitry. Demodulation circuitry operates the correlation of the received waveform with the code available at the receiver matched to the transmitted one, thus decoding the information content of the transmitted message. Synchronization circuitry allows the above-mentioned correlation function by aligning the two codes. ECM techniques can therefore be aimed at either causing errors to the content of the received messages or preventing

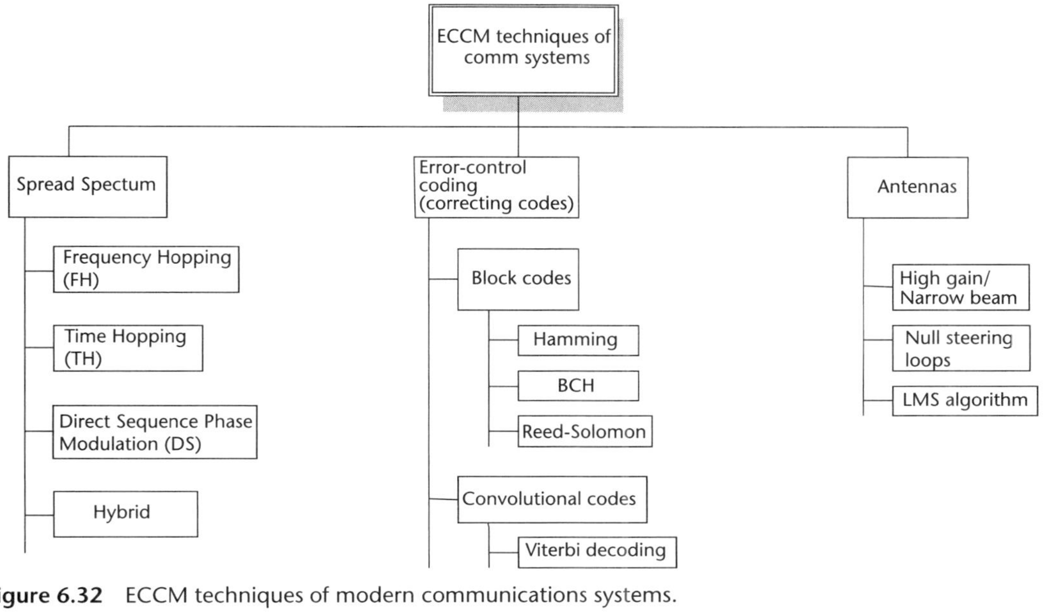

Figure 6.32 ECCM techniques of modern communications systems.

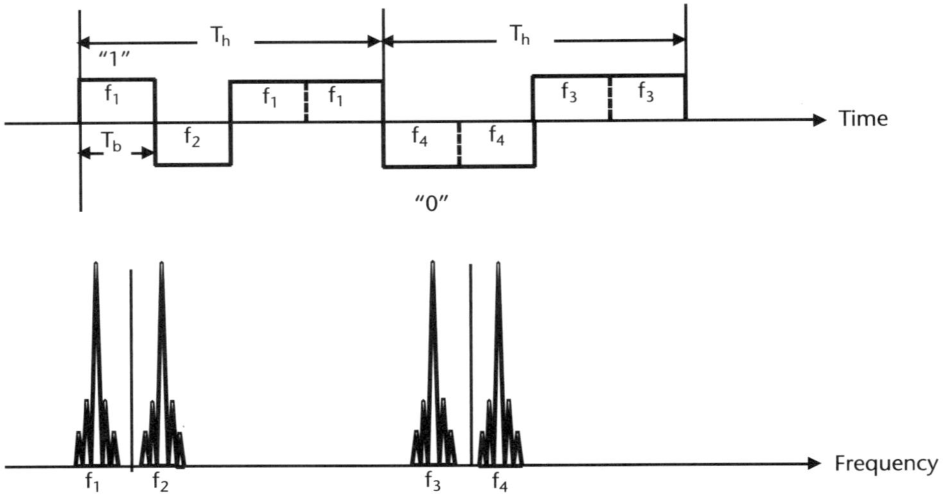

Figure 6.33 SFH BFSK transmission.

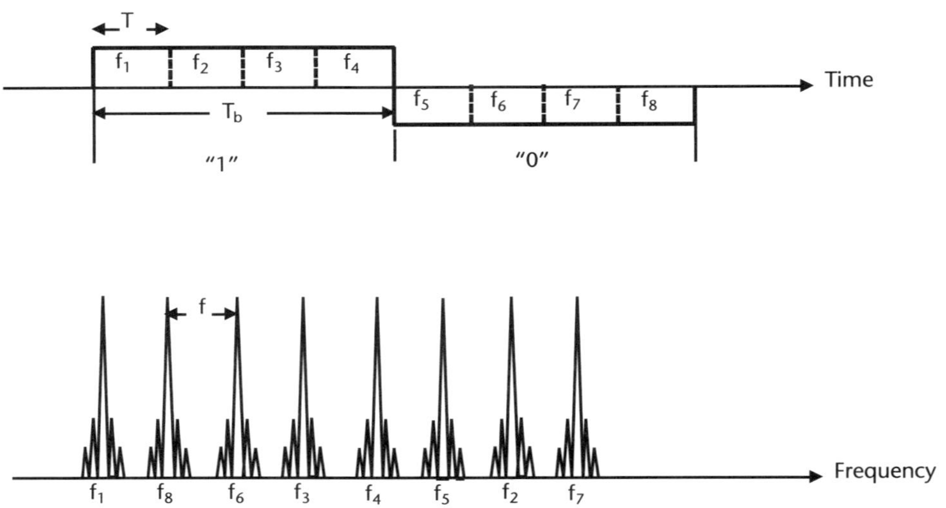

Figure 6.34 FFH BFSK transmission.

the synchronization of those messages. Both options have to be considered when selecting a jamming technique. The idea is to determine which technique provides the higher vulnerability to the victim communication system. The main ECM techniques discussed in this section are listed in Figure 6.35 [14]. The discussion will follow the path indicated by Poisel [15], who analyzed the susceptibility of the communication techniques reported in Figure 6.32 with respect to the ECM techniques of Figure 6.35.

ECM techniques against modern comm systems

- Waveforms
  - Narrowband noise (NBN,spot)
  - Wideband noise
    - Partial Babd Noise (PBN)
    - Broad Band Noise (BBN, barrage)
  - Swept Noise
  - CW tones
    - Single tone
    - Multi-tone
  - Smart jamming
- Strategies
  - Follower or Repeat back jammer
  - Pulsed jammer

Figure 6.35 ECM techniques and strategies used against modern communication systems. (*From:* [14].)

6.5.1 Noise Jamming

The noise jamming technique modulates the jamming carrier frequency with a random Gaussian noise waveform with a bandwidth that may vary from the one relevant to a single communication channel (narrowband or spot noise) to the entire spectrum occupied by the digital SS communication [wideband or barrage (or partial band) noise].The objective of this technique is the insertion of noise into the receiver to increase the BER and deny the information content of the communication, as discussed in Appendix E.

Broadband noise (BBN) is useful against all types of SS communication techniques, but it requires a very large jammer power with a limited effect, proportional to the jamming power received by the bandwidth of a single channel; that is, a ratio equal to the processing gain of the SS techniques, $G_p = B_{ss}/B_{ch}$ for DSSS, where B_{ss} is the entire communication bandwidth and B_{ch} is the bandwidth of a single channel, or $G_p = N_T$ for FHSS, where N_T is the total number of channels of the communication.

BBN is also effective in jamming the synchronization phase of a SS communication when the receiver is trying to align with the transmitter PN code sequence. The 50% probability of no synchronization of a FHSS communication (where coarse acquisition is accomplished with the matched filter approach or serial search code acquisition approaches) under BBN jamming is evaluated in Chapter 6 of [15] and shown in Table 6.1. Note that such a JSR level for a BBN jammer operating in the full VHF range (30 to 90 MHz, with a channel bandwidth $B_{ch} = 25$ kHz) relevant to a FH combat net radio (such as Jaguar-V) would require a jammer transmitting power on the order of a few kilowatts, which is possible only with a large ground-based jammer.

Partial band noise (PBN) places noise energy across a number (but not all) of the channels of the victim communication system. PBN performed with an optimal bandwidth fraction $\gamma > 0.3$ (where $\gamma = B_j/B_{ss}$ is the fraction of the spread-spectrum communication bandwidth B_{ss} occupied by the jamming waveform of bandwidth B_j) can be even more effective than BBN for some SS jamming techniques.

Narrowband noise (NBN) jamming places the jamming energy within a single channel. As in the case of BFSK jamming shown in Appendix E, the jamming

Table 6.1 Required BBN JSR (dB) for > 30% Probability of No Synchronization of a FHSS Communication Operating at SNR = 10 dB

P*	10	20	30
JSR (dB)	–3	–0.5	1

* p = minimum number of hops for synchronization.

bandwidth can be either the data signal one, the complementary signal one, or both. This type of jamming can be applied in a jammer follower against a SFH communication system, taking into account that the BER produced by the JSR has to be modified by the fraction $\alpha < 1$ of the presence of the jammer within the hop time at the receiver:

$$P_{error} = (1-\alpha)\cdot p_{noise} + \alpha \cdot p_{jammer+noise}$$

where p_{noise} is the probability of error in the presence of receiver noise only and $p_{\text{jammer+noise}}$ is the probability of error in the presence of combined jammer and receiver noise.

The *tone jamming* technique involves the placement of one or more jamming tones in appropriate channels of the communication system's spectrum. Single-tone jamming, which is ineffective against FHSS, is somewhat effective against DSSS when the tone is close to the carrier frequency [15, Chap. 8]. Multitone jamming can be performed by distributing m tones across the spectrum against FHSS and DSSS. Against FHSS the m tones have to hop randomly to attempt to hit as many FH channels as possible. Against DSSS more effective jamming is achieved if the m tones are placed at integer multiples of the data rate [15, Chap. 6]. In both cases the achievable BER is lower than 10^{-1}. Multitone jamming can be effectively applied against FHSS synchronization with an appropriate number of tones (at JSR = 0 dB) larger than the number of correlators of the acquisition receiver of the FH communication system. The spectrum coverage of these is shown in Figure 6.36.

Swept noise jamming is achieved by sweeping a relatively large band B_{sj}, comprising many contiguous FH channels, across the entire FH bandwidth with a dwell time and JSR in each transmitting channel that are high enough to generate the required BER. Let us assume that the bandwidth to be swept is B_{ss} and the instantaneous jamming bandwidth is $B_j = n\Delta f$, where Δf is the bandwidth of a communication channel. Assume further that the victim communication system is an SFHSS with h hop/s and data rate R kbps to achieve a BER ≥ 0.1 and that we want **in order** to jam $M = R \cdot \text{BER}$ bps. Because each hop contains R/h bits, the number of hops k that has to be hit in $T = 1$ sec is $k \cdot R/h = M = R \cdot \text{BER}$ (i.e., $k = h \cdot \text{BER}$). So the number of sweeps that B_j has to perform within the bandwidth B_{ss} in $T = 1$ sec is k and the sweep rate is $\omega = kB_{ss}$ (MHz/s) with a sweep period of $T_1 = T/k = 1/k$. The dwell time in each channel of width Δf is $\tau = Bj/\omega = B_j/k\ B_{ss} = B_j/(h \cdot \text{BER} \cdot B_{ss})$. Because the dwell time must be a fraction $\gamma \geq 0.3$ of the hop time, we have $\tau = \gamma/h$ and hence $B_j = \gamma\text{BER}\ B_{ss}$.

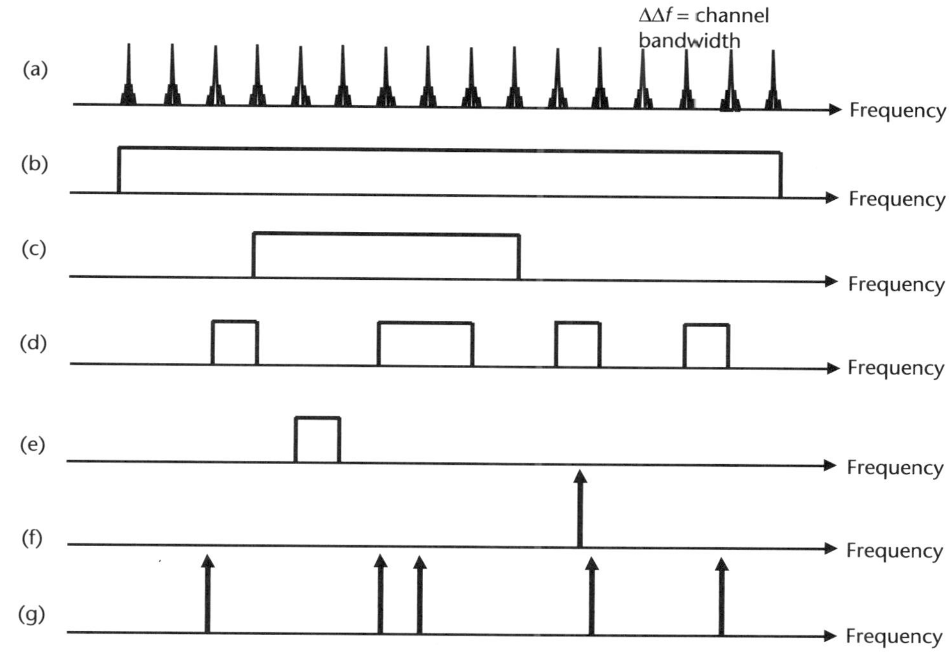

Figure 6.36 Spectrum coverage of the principal CECM techniques: (a) FFH spectrum, (b) barrage noise, (c, d) partial band noise, (e) spot noise, (f) single-tone jamming, and (g) multitone jamming.

As an example if we assume B_{ss} = 60 MHz (the entire VHF combat net radio communication range) and a data rate R = 20 kbps spread over h = 100 hop/s, we have to jam 2,000 bps. Therefore, we would need to sweep a noise bandwidth $B_j \geq 0.3 \times 0.1 \times 60\ MHz = 1.8$ MHz for 10 time/s in order to provide a jamming time $\tau = Bj/\omega = 1.8/600 = 3 \times 10^{-3} = 3$ ms in each channel of width Δf.

The time representation of the swept jamming technique is shown in Figure 6.37.

Recall that jamming a voice transmission for more than 30 percent of its duration significantly degrades the intelligibility of the message so as to deny the transfer of information. For digital communications the criterion to determine the effectiveness of a jamming waveform is to consider changing the BER from the design value of 10^{-2} to 10^{-4} in the presence of noise only (the baseline case) to values above 10^{-1} [15]. The reader should consider the basic BER evaluation of BFSK modulation when affected by a jamming noise or CW tone as discussed in the Appendix E then review Table 6.2, which lists the main results of an analysis performed by Poisel [15] on the effectiveness of the jamming techniques listed in Figure 6.35.

6.5.2 Follower Jammer

As discussed in Chapter 5, a follower jammer, also called a repeater or repeat-back jammer, is composed of a CESM, which measures the angle and frequency at which the SFHSS system is currently transmitting and, after some processing of duration T_j performed to recognize the emitter, commands the jammer to transmit the new received frequency. This jammer response has to meet a necessary condition in order to be effective, which is constituted by the fraction of hop period αT_h (where α

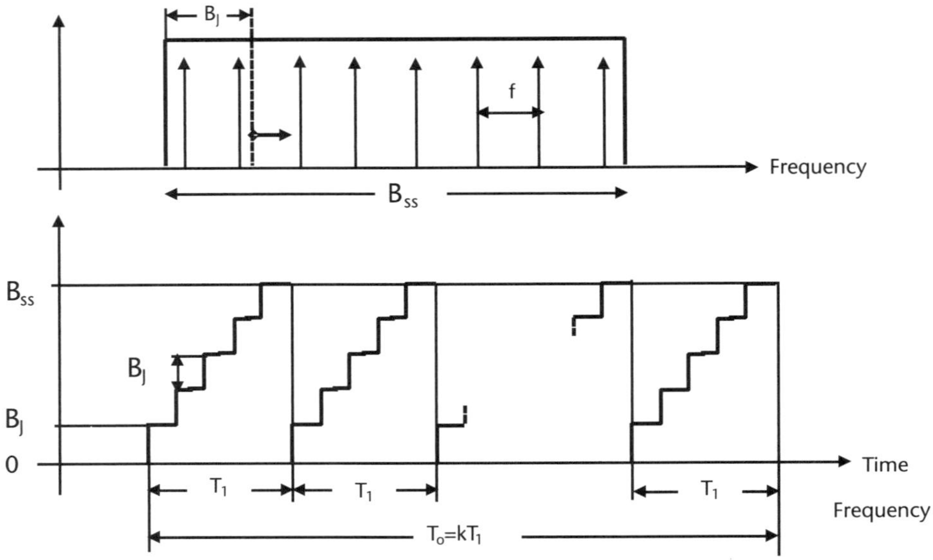

Figure 6.37 Time representation of the swept jamming technique.

> 0.3 is the fraction and T_h is the hop period) left to the jamming waveform by the geometry of the propagation and by the processing time at the jammer to provide the narrowband jamming waveform centered at the hop frequency. This condition can be expressed with reference to the geometry and the timing of the engagement shown in Figure 6.38 as follows:

$$(a+b-d) \leq c\left[(1-\alpha)T_h - T_j\right] \tag{6.25}$$

where

a = distance from the transmitter to the jammer,
b = distance from the jammer to the receiver,
d = distance from the transmitter to the receiver,
c = speed of light (3×10^8 m/s),
T_j = processing time at the jammer
$(1-\alpha)T_h - Tj$ = allotted fraction of time for the jammer response to be effective during the residual time αT_h.

Equation (6.25) can be shown to represent an ellipse of a canonical equation by considering a Cartesian coordinate system with its origin in the middle of distance d and the abscissa axis along the line connecting the communication transmitter and the receiver:

$$\frac{x^2}{(k+d)^2/4} + \frac{y^2}{k(k+2d)/4} = 1 \tag{6.26}$$

Table 6.2 Effectiveness (i.e., BER > 10^{-1}) of Jamming Techniques Used Against Modern Digital Communications

Jamming Technique	*Digital Communication Type*					
	DSSS BPSK/QPSK	*Notes*	*SFHSS*	*Notes*	*FFHSS*	*Notes*
Noise Jamming						
BBN	Effective, but not practical	JSR > $G + \text{SNR}_{in}$	Effective, but not practical	JSR > −5 dB (uncoded) Coding reduces effectiveness	Effective, but not practical	JSR > −3 dB (uncoded) Coding reduces effectiveness
PBN	Effective if centered	$\gamma > 0.1$ and JSR > G_p	Effective for $\gamma > 0.3$	Effective also with coding, but frequency diversity requires larger γ	Marginal for $\gamma > 0.3$	—
NBN	Ineffective	—	Follower effective for $\gamma > 0.3$	—	Marginal for $\gamma > 0.3$	(Also follower)
Swept jamming	Ineffective	—	Effective	—	NA	—
Tone Jamming	—	—	—	—		—
Single tone	Ineffective	—	Follower effective for $\gamma > 0.3$	—	Marginal for $\gamma > 0.3$	(Also follower)
Multitone	Ineffective	—	Effective for $\gamma > 0.3$	—	Marginal for $\gamma > 0.3$	(Also follower)
Pulse Jamming	Ineffective	—	NA	—	NA	—
Smart Jamming	Effective	—	Effective	—	Effective	—

where and the two ellipse semiaxes are $A = (k+d)/2$ and $B = (1/2) \cdot [k(k + 2d)]^{1/2}$. From this equation note that for the usual modern jammer processing times on the order of 1 to 2 ms, the follower jammer (either on ground or on board an UAS) constitutes the preferential jamming technique to be employed against SFHSS systems. For FFHSS systems the follower jammer technique can be applied only by an UAS close to the receiver.

6.6 Conclusions About ECM Techniques

In the first part of this chapter the most effective ECM techniques used against modern surveillance, tracking, and multifunction radars were discussed by illustrating their operating concepts and describing their effects.

The last part of the chapter, namely, Section 6.5, was devoted to the ECM techniques used against the modern spread-spectrum communication systems. The

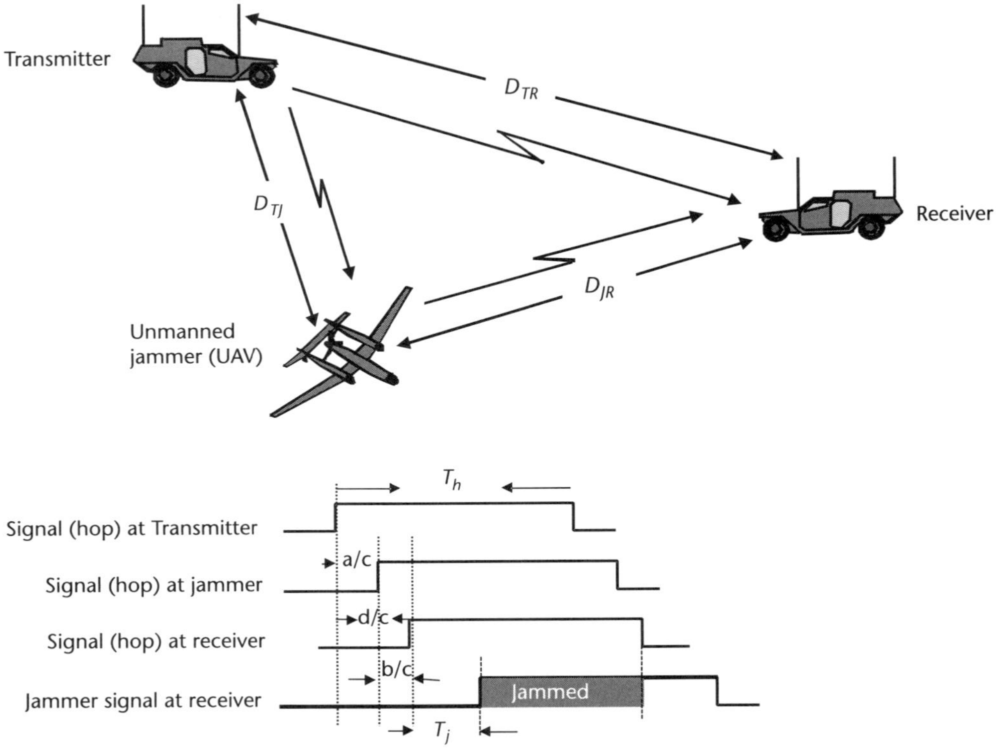

Figure 6.38 Propagation geometry and timing of the follower jammer technique.

discussion of these techniques was mainly addressed at increasing the BER to a level that denies the exchange of reliable information among the communication receivers.

References

[1] Schleher, C., *EW in the Information Age*, Norwood, MA: Artech House, 1999.

[2] De Martino, A., "Jamming Techniques Against Modern Search Radars," paper presented at EW International Conference, The Hague, The Netherlands, July 1985.

[3] Gabriel, W., "Adaptive Arrays—An Introduction," *IEEE Proc.*, Vol. 64, No. 2, February 1976, pp. 239–272.

[4] Farina, A., "Electronic Counter-Countermeasures," Chapter 24 in M. Skolnik (ed.), *Radar Handbook*, New York: McGraw Hill, 2008.

[5] Chrzanowski, E., *Active Radar Electronic Countermeasures*, Norwood, MA: Artech House, 1990.

[6] Howard, D., "Tracking Radar," Chapter 9 in M. Skolnik (ed.), *Radar Handbook*, New York: McGraw Hill, 2008.

[7] Neri, F., et al., "Experimental and Theoretical Results of an Airborne Cross-Eye Based Angular Deception Jammer," *AGARD Proc.*, Florence, Italy, February 1986.

[8] De Martino, A., "Cross-Eye Based Jammers for Airborne and Shipboard Applications," paper presented at NATO Conference, Brussels, Belgium, 1988.

[9] Faulks, L., "The Principle of Reprocity and Cross-Eye," paper presented at European AOC Conference, Interlaken, Switzerland, 2008.

[10] Asseo, S., "Detection of Target Multiplicity Using Monopulse Quadrature Angle," *IEEE Trans. on Aerospace and Electronic Systems*, Vol. AES-17, No. 2, March 1981, pp. 271–280.

[11] Bogler, P., "Detecting the Presence of Target Multiplicity," IEEE Trans. *on Aerospace and Electronic Systems*, Vol. AES-22, No. 2, March 1986, pp. 197–203.

[12] Grazzini, M., "Technical Note on Complex Indicated Angle as a Method of Cross-Eye and Towed Decoy Jamming Detection," Elettronica Internal Report 2004.

[13] De Martino, A., and Rossi, V., "Method and Apparatus for Generating Angular Deception Signals," Italian Patent No. IT2009RM00330, June 26, 2009.

[14] Pettit, R., *ECM and ECCM Techniques for Digital Communication Systems*, Belmont, CA: Wadsworth, 1982.

[15] Poisel, R. A., *Introduction to Communication EW Systems*, Norwood, MA: Artech House, 2002, Chapters 2, 7, 8, 11, and 17.

Appendix A
Signal Detection in Sensor Receivers

The main factor limiting the sensitivity of an electronic receiver is the *noise*, which is unwanted electromagnetic energy that impairs the receiver's ability to detect the desired signal. Noise may originate within the receiver itself, as is the case with *thermal noise* generated by the thermal motion of the conduction electrons in the receiver input stages, or it may enter via the receiving antenna along with the desired signal. Except for very sensitive receivers, like those applied in radioastronomy or in long-range surveillance radars, the dominant noise component in receivers is a result of thermal noise.

The thermal noise power N (W) generated by a receiver of bandwidth B (Hz) at temperature T (K) is equal to:

$$N = kTB \tag{A.1}$$

where k is Boltzmann's constant ($k = 1.38 \times 10^{-23}$ J/K), T is the temperature (usually assumed to be 290 K), and B is, with good approximation for the purposes of this book, the 3-dB bandwidth of the receiver. Equation (A.1) is usually expressed in logarithmic form through the decibel [i.e., $X_{dB} = 10 \log(x/x_{ref})$] with the reference power expressed in milliwatts (mW = 10^{-3}W) and the reference bandwidth expressed in megahertz (MHz = 10^6 Hz). Hence, it becomes:

$$N_{dBm} = -114(dB_{m/MHz}) + B(dB_{MHz}) \tag{A.2}$$

In the EW domain two basic types of receiver are considered: the superheterodyne and the amplified crystal video. Both can be schematized with regards to the mechanics of signal detection by an *envelope detector* as depicted in Figure A.1, where the amplifier/filter boxes are ideally considered to provide a gain to the input signal and noise within a rectangular filter of the same bandwidth as the receiver.

The output noise power N_{out} of practical RF receivers is always larger than that for the ideal case GN_{in} because noise is introduced from each component of the receiver. The additional noise power is taken into account by the noise figure F, which is defined as:

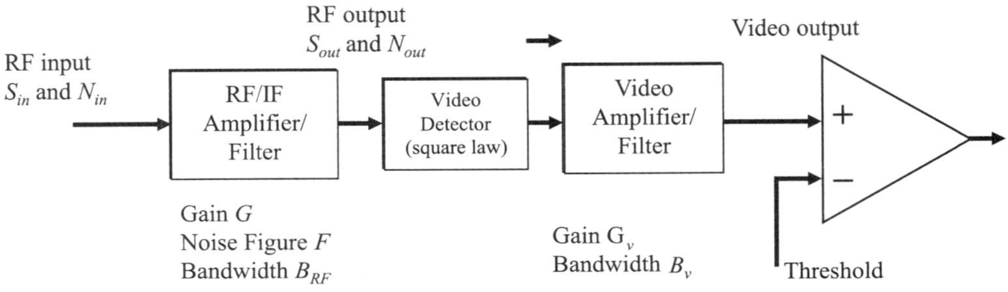

Figure A.1 Envelope detector.

$$F = N_{out}/GN_{in} = \frac{actual \cdot amplifier \cdot noise \cdot output}{noise \cdot output \cdot of \cdot ideal \cdot amplifier \cdot at \cdot Std \cdot Temperature} \tag{A.3}$$

The noise figure takes into account the degradation of the SNR in the transit through an actual amplifier as can be shown by considering the SNR at input and output:

$$\frac{SNR_{out}}{SNR_{in}} = \frac{S_{out}/N_{out}}{S_{in}/N_{in}} = \frac{GS_{in}/GFN_{in}}{S_{in}/N_{in}} = \frac{1}{F} \quad \text{with} \quad F > 1 \tag{A.4}$$

If the receiver input end is constituted by many amplifier stages connected in a cascade as in Figure A.2, it can be shown from the definition of the noise figure in (A.3) that the overall noise figure of the chain[2] is expressed as:

$$F_{chain} = F_1 + (f_2 - 1)/G_1 + (F_3 - 1)/(G_1 G_2) \tag{A.5}$$

The difference between a superheterodyne receiver and an amplified video receiver is that the first type of receiver converts the RF signal into a lower frequency f_{IF}, because band B_{IF} is much narrower than f_{IF} (i.e., $f_{IF} >> B_{IF}$). The radar sensors are usually of the superheterodyne type with $B_{IF} \approx 2 \cdot B_v$.

The input noise power at the video detector of Figure A.1 can be expressed as:

$$N_{in} = kTB_X F \tag{A.6}$$

where the bandwidth B_X can be either B_{RF} or B_{IF}.

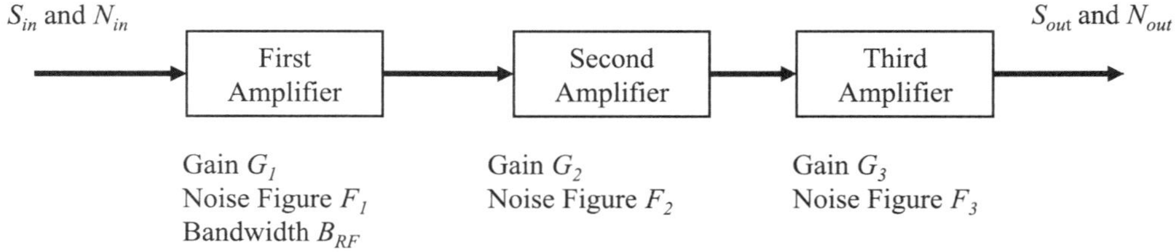

Figure A.2 Overall noise figure of a three-stage RF amplifier chain.

The video detector converts the noise within the RF bandwidth B_X to baseband bandwidth B_v. The video noise is computed by the convolution of a bandpass filter of width $2B_v$ with the noise of the RF bandpass having a width B_X. This process is schematically illustrated in Figure A.3 for the case of $B_X/2B_v = 3$ where three independent noise samples (of Rayleigh distribution with $2B_v$ bandwidth) are added to the baseband.

The *effective* or *equivalent noise bandwidth* of the receiver chain of Figure A.1 can thus be expressed as:

$$B_{eq} = \sqrt{2B_v B_{RF}} \tag{A.7}$$

thus providing a postdetection SNR integration gain of

$$\frac{SNR_{out}}{SNR_{in}} = \frac{GS_{in}}{GkTB_{eq}F} \Big/ \frac{S_{in}}{ktB_{RF}F} = \sqrt{\frac{B_{RF}}{2B_v}} = \sqrt{n} \tag{A.8}$$

The distributions of noise alone (Rayleigh distribution) and signal plus noise (Rice distribution) at the output of the Rx chain are shown in Figures A.4 and A.5 [1].

The detection of a signal in the receiver chain of Figure A.1 is based on establishing an amplitude threshold V_T at the output of the envelope detector. If the signal amplitude is strong enough to cross the threshold, a detection occurs; if it is not strong enough, no detection occurs.

The video output of the envelope detector can be represented as shown in Figure A.6. Note that its fluctuating appearance is due to the random nature of the receiver noise. With reference to Figure A.4, if the threshold is lowered, there would be more threshold crossings, most of them due to large fluctuations of noise, which produce false signal detections or *false alarms*.

The receiver noise at the input of the detector is described by a Gaussian pdf with zero mean value and variance σ_n^2 that is equal to N_{in} of (A.6):

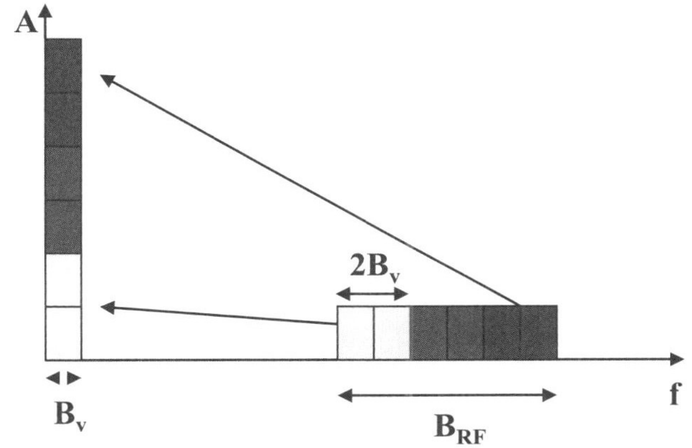

Figure A.3 RF bandwidth conversion to baseband operated by the video detector.

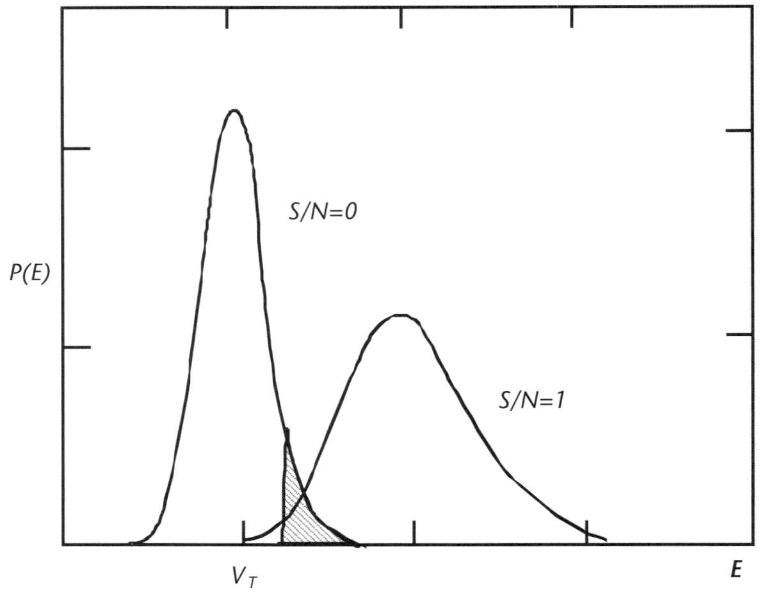

Figure A.4 Probability density functions for noise alone and for singal plus noise [1].

$$p(n) = \frac{1}{\sqrt{2\pi\sigma_n^2}} \exp\left(-\frac{n^2}{2\sigma_n^2}\right) \tag{A.9}$$

The noise $r(t)$ at the output of the envelope detector is described by a Rayleigh pdf of the form:

$$p(r) = \frac{r}{\sigma_n^2} \exp\left(-\frac{r^2}{2\sigma_n^2}\right) \tag{A.10}$$

Therefore, the *probability of false alarm* P_{fa}—that the envelope of the noise exceeds the amplitude threshold Th and thus provides a false alarm—can be evaluated as

$$P_{fa} = \int_{Th}^{\infty} \frac{r}{\sigma_n^2} \exp\left(-\frac{r^2}{2\sigma_n^2}\right) \cdot dr = \exp\left(-\frac{Th^2}{2\sigma_n^2}\right) \tag{A.11}$$

By itself, the probability of false alarms does not give a good indication of the problems to the receiver caused by an excessive number of false signal detections. A better indication is provided by the *false alarm time* T_{fa}, which can be shown [2] to be:

$$T_{fa} = \frac{1/Beq}{P_{fa}} = \frac{1}{B_{eq}} \exp\left(\frac{Th^2}{2\sigma_n^2}\right) \tag{A.12}$$

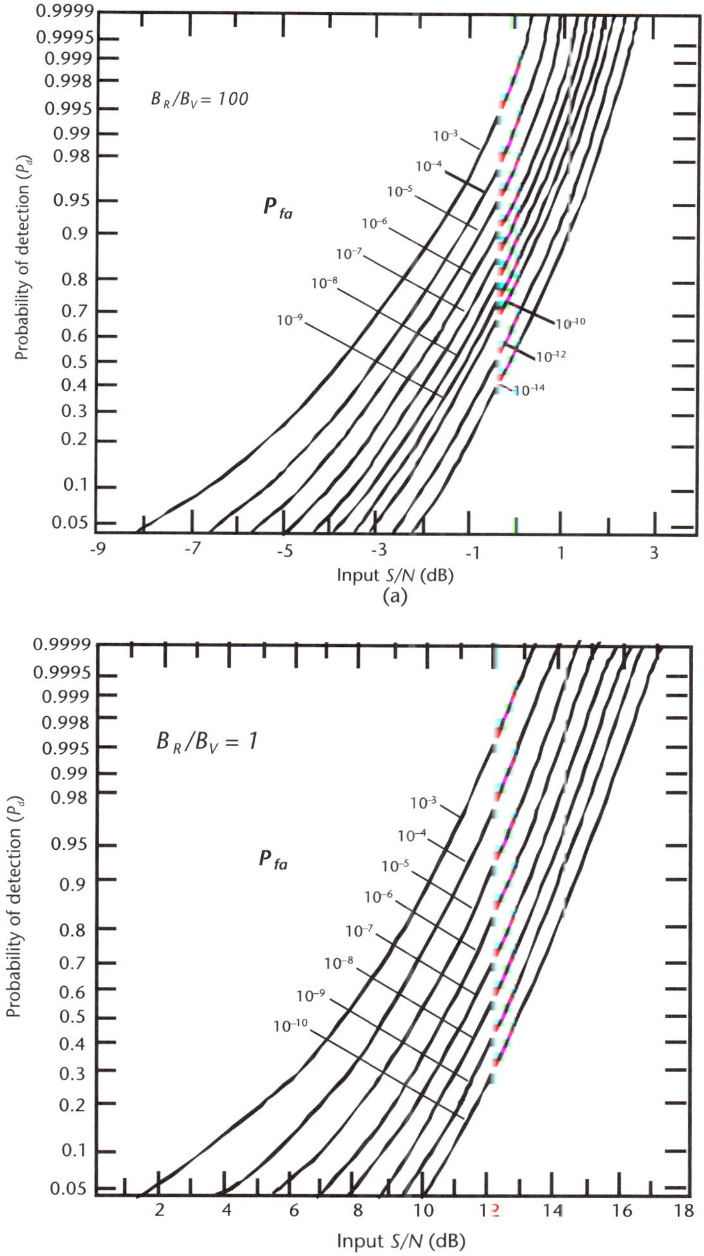

Figure A.5 Probability of detection for a sine wave in noise vs SNR with a parameter P_{fa}, (a) Bv/BRF=100, (b) $B_v/B_{RF} = 1$ [1].

As a numerical example let us assume that we want an average false alarm time $T_{fa} \geq 15$ min = 900 sec and that the equivalent bandwidth $B_{eq} = 2$ MHz. Then from the last equation we have $P_{fa} = 5.55 \times 10^{-10}$ and $Th/\sigma_n = \sqrt{-2\ln P_{fa}} = 6.52 = 16.3$ dB

In the case of radar, where the bandwidth $B_{IF} << f_{IF}$, we can represent the echo signal at the input of the detector as:

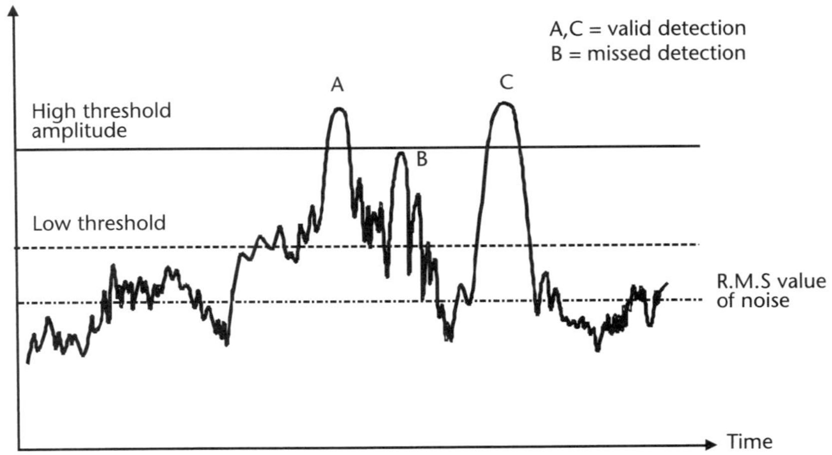

Figure A.6 Video output of the envelope detector.

$$x(t) = A \cdot \sin 2\pi f_{IF} t \tag{A.13}$$

Hence, the input to the detector becomes and the pdf of the envelope $r(t)$at the video band output is given by the Rice pdf:

$$p_s(r) = \frac{r}{\sigma_n^2} \exp\left(-\frac{r^2 + A^2}{2\sigma_n^2}\right) \cdot I_0\left(\frac{rA}{\sigma_n^2}\right) \tag{A.14}$$

where $I_0(z)$ is the modified Bessel function of zero order and argument z.

The probability of detecting the echo signal is expressed as:

$$P_d = \int_{Th}^{\infty} p_s(r) \cdot dr \tag{A.15}$$

This probability is not easy to calculate, but there are many graphs available in the literature [1]. As an example in Figure A.5(a) the probability of detection P_d is reported vs. SNR = $A^2/2\sigma_n^2$ with P_{fa} as a parameter.

Note that in the case of a radar that on reception uses a matched filter (i.e., a filter that maximizes the SNR) as its bandwidth, B_{RF} is equal to that of the transmitted signal: $B_{RF}/2B_v = 1$. The postdetection SNR gain is achieved by integrating along time n successive video detected pulses [2]. This procedure encompasses less demanding P_{fa} and P_d (and hence a lower SNR) than for a single pulse detection.

A.1 Integration of Successive Radar Pulses

A target, even if illuminated by a scanning antenna, provides a number n of echo signals. The process of summing all of the radar echoes from a target is called *pulse integration.* Integration can be performed in a *coherent* way, by preserving the phase of the echo pulse signals so as to add in phase their phasors without any loss. This can be accomplished only before the detector, hence, coherent integration is also called *predetection integration.*

Integration can also be performed in a *noncoherent* way, by adding the pulse amplitudes after the envelope detector (hence, noncoherent integration is called also *postdetection integration*). The operation of both types of pulse integration in shown in Figure A.7.

The result of coherent integration of n pulses having the same SNR_{in} is

$$SNR_{out} = n\ SNR_{in}$$

Because SNR_{out} is the required value to meet the final P_d and P_{fa} conditions, the single pulse required SNR is n times less $SNR_{in} = SNR_{out}/n$. The result of noncoherent integration can be expressed as:

$$SNR_{out} = nE_i(n)SNR_{in} > \sqrt{n} \cdot SNR_{in}$$

where the factor efficiency $E_i(n) < 1$ takes into account the envelope detector losses.

The integration improvement factor $I_i(n) = nE_i(n)$ as a function of n, P_d, and P_{fa} was evaluated by Marcum shows and reported in Figure A.7. Inspection of Figure A.7 indicates that when only a few pulses are integrated (and thus they need to have a large SNR) the postdetection improvement is nearly equal to the predetection one, whereas when a large number of pulses is integrated (and thus their SNR is not large) the postdetection improvement differs significantly from the predetection one.

A.2 Coherent Detection

Most radars apply coherent processing to the received echo signals in order to take into account the Doppler phase variation from a pulse to the next induced by the target motion. Coherent detection is applied in modern digital signal processing with an I,Q detector (see Figure A.7). In this detector the narrowband IF signal

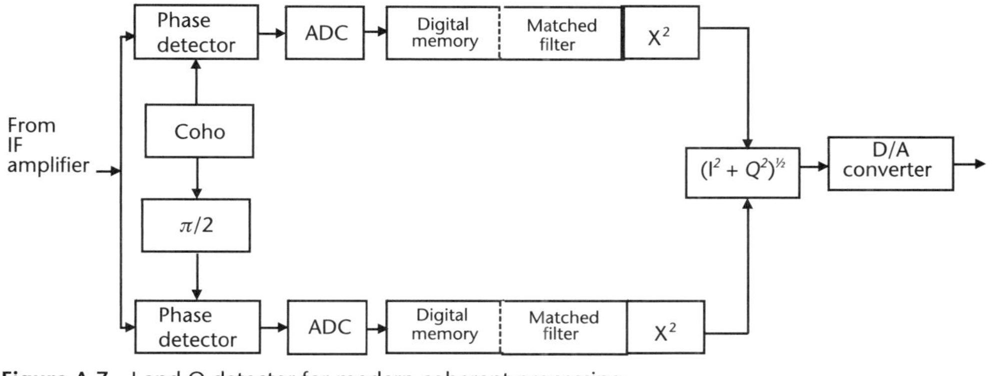

Figure A.7 I and Q detector for modern coherent processing.

$(t) = A(t)\ \sin(2\pi f_0 t + \phi(t))$ is decomposed into two components: in-phase $I(t)$ and quadrature $Q(t)$, which are expressed, respectively, as

$$s(t) = I(t)\sin(2\pi f_0 t) + Q(t)\cos(2\pi f_0 t)$$

with

$$I(t) = A(t)\cos\varphi(t) \text{ and } Q(t) = A(t)\sin\varphi(t)$$

Thus, the two channels together provide the amplitude and phase modulations of the input signal. If the outputs of the I and Q channels are squared and summed, then the square root of the sum is the envelope $A(t)$ of the envelope detector shown previously.

The digitized signals are obtained through sampling and A/D conversion of the two components to produce a series of complex digital samples $I + jQ$. Sampling must be performed at least at the Nyquist rate (i.e., twice the input signal bandwidth B) in order to faithfully reproduce the signal. Because there are two channels in the I,Q detector, the ADC in each of the I and Q channels needs only to sample at the rate B sample/s. With this rate there is a SNR loss of about 0.6 dB compared to continuous sampling, since the sampling is not guaranteed to occur at the peak output of the matched filter. In practice, by sampling at a rate $2B$ samples/s much of the above SNR loss is recovered.

References

[1] Tsuin, J. B., *Microwave Receivers with Electronic Warfare Applications,* Raleigh, NC: Sci-Tech Publ. Inc, 2006.

[2] Skolnik, M., *Introduction to Radar Systems,* New York City, NY: McGraw-Hill, 1962.

[3] Marcum, J., "A Statistical Theory of Target Detection by Pulsed Radar, Mathematical Appendix," *IRE Trans.,* 1960.

Appendix B
Introductory Concepts of Estimation Theory

The detection of a signal in a sensor is generally followed by the estimation of signal parameters that are related to some interesting information about the emitter such as its amplitude, time, angle of arrival, and carrier and modulating frequencies. Because the received signal is corrupted by disturbances of various types, we often desire to obtain a valid estimate $\tilde{u}$ of the unknown parameter u starting from statistical information about the signal and its disturbances (e.g., noise).

In this appendix, we briefly describe two widely used methods of parameter estimation: maximum likelihood estimation and the least-squares method. Before we do so, however, we have to introduce a very important boundary in defining the performance of any estimating method: the Cramer-Rao lower bound (CRLB). A demonstration of the CRLB is beyond the scope of this book. Interested readers are referred to [1] for deeper insight into the topic.

Let us assume that we collected N measurements of a random variable y, which is related to the unknown parameter u. The measurements can be represented by a vector $\underline{y}^T = [y_1, y_2, \ldots, y_N]$ and we define an estimate of the parameter u as $\tilde{u} = \phi(\underline{y})$, where $\phi(\underline{.})$ is a suitable function that represents mapping from the samples to the estimate. Let us further assume that the vector of measurements $\underline{y}$ provides a *sufficient statistic* (i.e., it provides enough information about the observations to make a proper decision) and has an exponential pdf of the form:

$$p(\underline{y}|u) = \exp[A(u)q(\underline{y}) + B(u)]h(\underline{y}) \tag{B.1}$$

As an example the Gaussian pdf with mean μ and variance σ^2 is of the exponential type with respect to both parameters μ and σ^2.

The probability density function (pdf) of the estimate $\tilde{u}$ depends on the unknown parameter u and is expressed as $p(\tilde{u}|u)$ (which can be either the Bayesian conditioned pdf (also called a *likelihood function*) or a pdf depending on the parameter u.

The efficiency of an estimator is given by the distance of the estimate $\tilde{u}$ from the unknown parameter u, that is, by how much the relevant pdf $p(\tilde{u}|u)$ is clustered around the value u. This distance is defined in the theory of measurements by two parameters:

- *Estimator accuracy*, that is, the ability of the estimator to provide an estimate $\tilde{u}$ whose expected value $E(\tilde{u}|u)$ is as close as possible to the parameter u , where:

$$E(\tilde{u}|u) = \int \hat{u}p(\hat{u}|u)d\hat{u}|$$

- *Estimator precision*, that is, an indication of the conditional variance of the estimator accuracy:

$$\text{var}(\tilde{u}|u) = \int |\hat{u} - E(\hat{u}|u)|^2 \, p(\hat{u}|u) \cdot d\hat{u}$$

A sufficient estimator that satisfies the relationship $E(\tilde{u}|u) = u$ is called *unbiased* and is surely an accurate one. However, it is not necessarily the best estimator, because the efficiency of the estimator is also dependent on its precision, that is, the minimum variance around the mean value.

The theorem of Cramer-Rao states that there is a lower bound, the CRLB, to the conditional variance of any scalar estimator. Therefore, if we can find an unbiased estimator whose variance is equal to the CRLB, it will be the optimum one and it is called an *efficient estimator.*

The CRLB for a biased estimator, that is, an estimate that is expressed as $E(\tilde{u}|u) - u = b(u)$,is

$$CRLB(\tilde{u}) = \text{var}(\tilde{u}|u) \geq (1 + b'(u))^2 / I(u)$$

Where $I(u)$ is the so-called Fisher's information in the sample, defined as

$$I(u) = \int_{-\infty}^{+\infty} N \cdot \left[\frac{1}{p(y|u)} \cdot \frac{\partial p(y|u)}{\partial u} \right]^2 p(y|u) \cdot dy = N \cdot E\left[\left(\frac{p'}{p} \right)^2 \right]$$

and

$$b'(u) = \int \left[E(\hat{u}|u) - u \right] \cdot p'(y|u) \cdot dy$$

$$p'(y|u) = \frac{\partial p(y|u)|}{\partial u}$$

The CRLB can be easily computed in the case of a parameter z, which is a univocal differentiable function of the parameter u, that is, $z = g(u)$. In this case:

$$CRLB(\breve{z}) = \text{var}(\hat{z}|u) = \left[\frac{dg(u)}{du} \right]^2 \cdot CRLB(\hat{u})$$

B.1 Maximum Likelihood Function Estimator

The maximum likelihood estimator (MLE) method is quite simple and powerful and requires a minimal amount of a priori information about the parameter to be estimated. The property for which $\tilde{u}_{\text{MLE}}$ is said to be the maximum likelihood estimation of an observation $\underline{y}$ is that:

$$p(\underline{y}|\hat{u}_{MLE}) \geq p(\underline{y}|\hat{u}) \quad \forall \hat{u}$$

That is, $\tilde{u}_{\text{MLE}}$ maximizes the likelihood function $p\ (\underline{y}|\tilde{u})$ or, in other words, $\tilde{u}_{\text{MLE}}$ is the value of $\tilde{u}$ that most likely caused the observed values $\underline{y}$ to occur. As an example let us consider a case where we have N noisy measurements of a parameter u:

$$x_i = u + n_i \quad i = 1, 2, \ldots, N$$

where n_i are zero mean independent Gaussian distributed random variables with variance σ^2.

The conditional pdf of $\underline{x}$ is:

$$p(\underline{x}|u) = (2\pi\sigma^2)^{-N/2} \exp\left(-\frac{1}{2\sigma^2} \sum_1^N (x_i - u)^2 \right)$$

To find $\tilde{u}_{\text{MLE}}$ let us maximize $p\ (\underline{x}|u)$ by equating to zero the derivative of its log-likelihood function:

$$L(x|u) = \ln(p(x|u)) = -\frac{N}{2}\ln(2\pi\phi^2) + \left(-\frac{1}{2\sigma^2} \sum_1^N (x_i - u)^2 \right)$$

$$i.e., \quad 0 = \frac{\partial L(\underline{x}|u)}{\partial u} = \frac{1}{\sigma^2} \sum (x_i - \hat{u}_{MLE})^2$$

from which

$$\hat{u}_{MLE} = \frac{1}{N} \sum x_i$$

So the estimate $\tilde{u}_{\text{MLE}}$ does not depend on the noise variance and it is unbiased because the expected value of $\tilde{u}_{\text{MLE}}$ is $E(\tilde{u}_{\text{MLE}}|u) = u$, which is also the expected value of the observation $\underline{x}$.

The conditional variance of the estimate $\tilde{u}_{\text{MLE}}$ is:

$$\text{var}(\hat{u}_{MLE}|u) = E\left[(\frac{1}{N} \sum_1^N (x_i - u)^2 \cdot |u \right] = \frac{N\sigma^2}{N^2} = \frac{\sigma^2}{N}$$

The preceding result is intuitive because the precision of the estimate is improved as N increases.

The CRLB for this case of an unbiased estimate is

$$\sigma^2_{CRLB} = \frac{1}{I(u)} = \frac{\sigma^2}{N}$$

(as demonstrated in [1]). In this case we have shown that the MLE is efficient because it equates the CRLB.

The example above about how to apply the MLE method has led to a solution in closed form. However, in most practical cases the application of the method involves much more complicated and nonlinear mathematical relationships that can be solved either numerically with iterative methods or by linearization around some working point.

B.2 Least-Squares Method of Estimation

The least-squares method (LSM) of estimation is derived from the general least-squares method, which has the goal of estimating a number N of parameters from a larger number $M > N$ of measurements. The method assumes that there is a linear dependency among the parameters measured.

Let $\underline{z}(M \times 1)$ be the vector of the measured values, $\underline{x}(N \times 1)$ the vector of the parameters, and $\mathbf{A}(M \times N)$ the matrix that represents the linear dependency between the measured values and the parameters. The relationship among them is expressed as:

$$z = A\underline{x} \tag{B.2}$$

The system of equations represented by (B.2), being $M > N$, does not have a unique solution. We are interested in finding a particular solution $\underline{x}_{\text{LS}}$ that minimizes the square norm of the error

$$\varepsilon = \left| A \cdot \underline{x} - \underline{z} \right|$$

which is the norm that intrinsically weighs more the large errors and less the smaller ones. The square norm of the error is expressed as:

$$q = \underline{\varepsilon}^T \cdot \underline{\varepsilon} = (A\underline{x} - \underline{z})^T \cdot (A\underline{x} - \underline{z}) = \underline{x}^T A^T A\underline{x} + \underline{z}^T \underline{z} - \underline{x}^T A^T \underline{z} - \underline{z}^T A\underline{x} \tag{B.3}$$

where the symbol T denotes the transposition. The minimization of q is achieved by equating to zero the derivative of q with respect to the vector $\underline{x}$, taking into account the following identities valid for any square matrix $\mathbf{B}$:

$$\frac{\partial}{\partial \underline{x}} \underline{x}^T B \underline{x} = (B + B^T)\underline{x}$$

$$\frac{\partial}{\partial \underline{x}} \underline{x}^T B \underline{z} = B\underline{z}$$

$$\frac{\partial}{\partial \underline{x}} \underline{z}^T B \underline{x} = B^T \underline{z}$$

Thus,

$$0 = \left. \frac{\partial q}{\partial \underline{x}} \right|_{\underline{\underline{x}}_{LS}} = 2A^T(A\underline{x}_{LS} - \underline{z})$$

which gives the solution

$$\underline{x_{LS}} = (A^T A)^{-1} A^T \underline{z}$$

The matrix $\mathbf{P} = (\mathbf{A}^T\mathbf{A})^{-1}\mathbf{A}^T$ is called the *pseudoinverse* of the matrix **A**. If **A** is square we immediately recognize that $\mathbf{P} = \mathbf{A}^{-1}$, the *inverse* of matrix **A**.

In the complex field the square norm is expressed as

$$q = \underline{\varepsilon}^H \varepsilon = (A\underline{x - z})^H (A\underline{x - z})$$

where the superscript H (Hermitian) denotes the complex conjugated transposition [i.e., $\mathbf{A}^H = (\mathbf{A}^*)^T = (\mathbf{A}^T)^*$] and the pseudoinverse matrix is $\mathbf{P} = (\mathbf{A}^H\mathbf{A})^{-1}\mathbf{A}^H$.

In the LSM we are interested in estimating the parameter when the measurements are affected by noise. In this case the relationship is expressed as

$$\underline{z} = A\underline{x} + \underline{n}$$

and the quadratic error norm is the expectation of it:

$$q = E\left[(A\underline{x} + \underline{n - z})^T (Ax + \underline{n - z})\right] = (A\underline{x - z})^T (A\underline{x - z}) + E\left[\underline{n}^T \underline{n}\right]$$

After equating to zero, the partial derivative with respect to $\underline{x}$ provides the same estimate as in the case of the absence of noise: $\hat{x}_{LSE} = (\mathbf{A}^T\mathbf{A})^{-1}\mathbf{A}^T\underline{z}$ but with a variance $\text{var}(\hat{x}_{LSE}) = \sigma_n^2 = \text{CRLB}$.

Hence, we see that the minimum least-squares estimator is an efficient estimator.

Reference

[1] Crescimbeni, R., and G. Galati, "Basic Concepts on Detection, Estimation, and Optimum Filtering," in *Advanced Radar Techniques and Systems,* G. Galanti (ed.), London, UK: Petere Peregrinus, 1993.

Appendix C
Antennas and Phased Array Antennas

The antenna is a reciprocal transducer between free-space propagation and the transmission line internal to the EW equipment. The term *reciprocal* refers to the fact that it can operate in either the Tx or Rx mode. The function of the antenna during transmission is to concentrate the radiated energy within a shaped beam in the direction of its boresight. On reception, the antenna collects the energy coming within its beam from the free space and delivers it to the receiver.

Antennas normally used in the radar bands have shaped, highly directive patterns that can be mechanically scanned in one or two angular directions (azimuth and elevation) through the movements of one or two pedestals on which they are installed or by electronic means such as in the phased array type of antenna. The vast majority of past generation radars used parabolic reflectors in various forms. Most current generation designs employ planar phased arrays. Antennas used in the communication bands are usually designed for omnidirectional coverage, with the exception of those used for point-to-point transmission (i.e., radio links), which are similar to those used in the radar bands.

The ability of an antenna to concentrate energy in a particular direction is called *directive gain* G_D. The directive gain is defined as the ratio between the maximum radiation intensity per unit solid angle, which occurs within the main beam having dimensions θ_B and ϕ_B (the half power beamwidth in radians, respectively, in the two orthogonal planes), to the average radiation intensity over the whole solid angle (4π steradians). Assuming P to be the total power radiated by the antenna, this definition can be expressed as follows:

$$G_D = (P/\theta_B\phi_B)/(P/4\pi) = 4\pi/\theta_B\phi_B \quad \text{(C.1)}$$

or by expressing θ_B and ϕ_B in degrees we have $G_D = 41253/\theta_B^{(\circ)}\phi_B^{(\circ)}$.

The definition given above does not provide the effective power radiated by the antenna but merely its ability to concentrate the radiated energy within the main beam. To take into account the effect of the antenna losses, a *power gain* G is defined that provides a measure of the effective power radiated within the main beam and is proportional to the directive gain through the *antenna efficiency factor* η (typically $\eta = 0.55$ to 0.65 for good antennas):

$$G = \eta G_D \tag{C.2}$$

From antenna electromagnetic theory [1], an antenna can be considered a planar aperture of different shapes provided with an illumination (or current distribution) $I(x,y)$ over the aperture that in the far-field region of the antenna (i.e., at distances $R > R_{FF} = 2D^2/\lambda$, where D is the maximum size of the aperture and λ is the wavelength) produces an electric field vector intensity pattern $\underline{E}(R,\theta,\phi)$:

$$\underline{E}(R,\theta,\phi) = \underline{a}\frac{e^{-jkR}}{R}E(\theta,\phi)\cdot\exp(j\underline{k}'\cdot\underline{\rho}) \tag{C.3}$$

where

$\underline{a}$ = polarization versor (i.e., unit vector),

$E(\theta,\phi)\exp(j\underline{k}\cdot\underline{\rho})$ = radiation pattern at distance $R = (x^2+y^2+z^2)^{1/2}$ from the origin,

$k = 2\pi/\lambda$ and $\underline{k}'^T = k(u\,\underline{x}_0;\ v\,\underline{y}_0;\ \underline{z}_0\cos\theta)$,

$\underline{\rho}'^T = (x'x_0;\ y'\,\underline{y}_0;)$ = vector of the point coordinates in the planar aperture and

$\underline{x}_0; \underline{y}_0$ = versors of the coordinates in the plane of the aperture and $\underline{z}_0$ the versor of the aperture boresight,

$u = \sin\theta\cos\phi$ and $v = \sin\theta\sin\phi$,

$\underline{k}'\cdot\rho'$ is the scalar product of the two vectors.

The radiation pattern $E(\theta,\phi)$can be evaluated as

$$E(\theta,\phi) = E(u,v) = \iint dx'dy'I(x',y')\exp(-jk(x'u + y'v)) \tag{C.4}$$

where the double integral is extended over the area of the aperture as shown in Figure C.1.

Because the antenna is a reciprocal device its *capture area* A_e relevant to an incident plane wave is:

$$A_e = \frac{G\lambda^2}{4\pi} = \rho_a A \tag{C.5}$$

where A (m^2) is the physical area of the antenna and ρ_a is the antenna aperture efficiency, which arises from the aperture illumination distribution $I(x',y')$. The plot of the square of the field intensity $|E(u,v)|^2$ when its maximum is normalized to unity is equivalent to the normalized power radiation pattern $P(\theta,\phi)$ of the antenna.

Usually the aperture illumination $I(x',y')$ can be assumed [2] to be of the type $I(x',y') = I_1(x')I_2(y')$, that is, the product of two independent distributions along the two aperture coordinates. Equation (C.5) then becomes:

$$E(\theta,\phi) = \int_{-a/2}^{a/2} dx'I_1(x')\exp(-jkx'u)\cdot\int_{-b/2}^{b/2} dy'I_2(y')\exp(-jky'v) \tag{C.6}$$

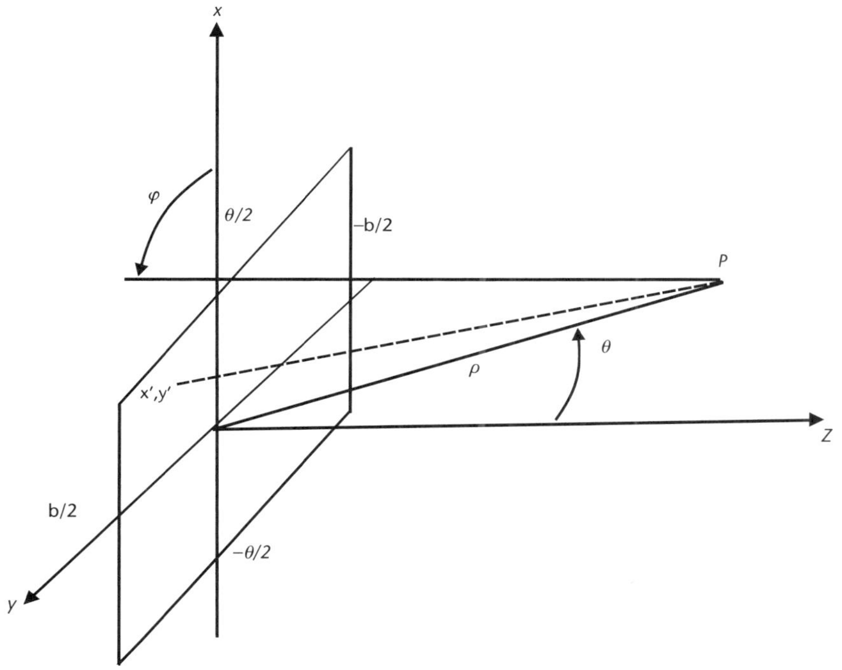

Figure C.1 Aperture reference coordinates.

A special case is the uniform illumination along both coordinates: $I_1(\cdot) = I_2(\cdot) = 1$. In this case the radiation patterns along the two principal planes $(\theta,0)$ and $(\theta,\pi/2)$ become:

$$E(\theta,0) = \frac{\sin\left[(ka/2)\sin\theta\right]}{(ka/2)\sin\theta} \text{ and } E(\theta,\pi/2) = \frac{\sin\left[(kb/2)\sin\theta\right]}{(kb/2)\sin\theta} \tag{C.7}$$

These patterns are both of the form sinx/x as shown in Figure C.2.

Uniform illumination is the most efficient aperture distribution (so $\rho_a = 1$ for this illumination), but presents a first sidelobe amplitude of only –13.2 dB with respect to the peak amplitude. The beamwidth, as measured between the half-power points, is $\theta_B = 0.88\lambda/a$ (rad) or $\theta_B = 51\lambda/a$ (deg). The reduction of the first sidelobe level is achieved by tapering the illumination at the extremes of the aperture at the expense of aperture efficiency ($\rho_a < 1$) as reported in [3].

A special case is constituted by a circular aperture whose radiation pattern, due to the symmetry in the aperture plane, is a solid of rotation only dependent from the angle θ with respect to the boresight. The radiation pattern for a uniformly illuminated circular aperture is evaluated as:

$$E(\theta) = \int_0^{2\pi} d\phi \int_0^{\rho_0} d\rho \cdot \exp(-jk\rho u) = 2\pi\rho_0^2 J_1(\delta)/\delta \tag{C.8}$$

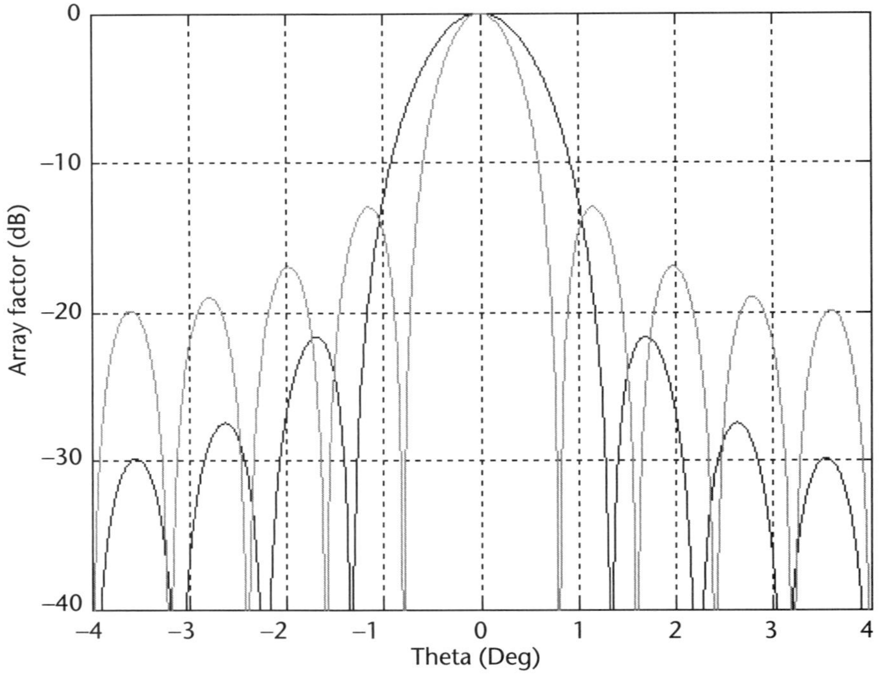

Figure C.2 Radiation pattern of uniform (light grey) and cosine (black) aperture illumination.

where $\Delta = k\rho_0 \sin\theta$ and $J_1(\cdot)$ is the first-order Bessel function. The first sidelobe is –17.6 dB with respect to the peak, and the beamwidth is $\theta_B = 58.5\lambda/2\rho_0$ (deg).

C.1 Antenna Types

One of the most widely used radar antennas is the parabolic reflector. The geometrical properties of the parabola provide that the path lengths of any ray coming from the focus, where the feed is placed, and reflected from the parabolic surface, are the same to a plane perpendicular the parabola axis. Thus, the point energy source placed at the feed is transformed into a plane wavefront at the output of the parabola.

Among the various reflector antenna types used in the radars we mention here only two: the Cassegrain reflector, which is often used in tracking radars, and the cosecant-squared reflector, which is largely used in 2D surveillance radars.

The principle of the Cassegrain antenna is shown in Figure C.3. The feed is placed at the vertex of the parabolic reflector in front of a hyperbolic subreflector. The plane wavefront coming from a target at long distance is transformed into a convergent beam at the feed (which is the real focus of the antenna) by the subreflector (which acts as a secondary focus). The Cassegrain reflector antenna has the advantages of reducing the axial dimension of the antenna and providing an easy connection of the feed to the radar Tx/Rx lines. A photograph of the Selex SI NA30 tracking radar, which uses a twisted polarization Cassegrain reflector, is shown in Figure 2.59 of Chapter 2.

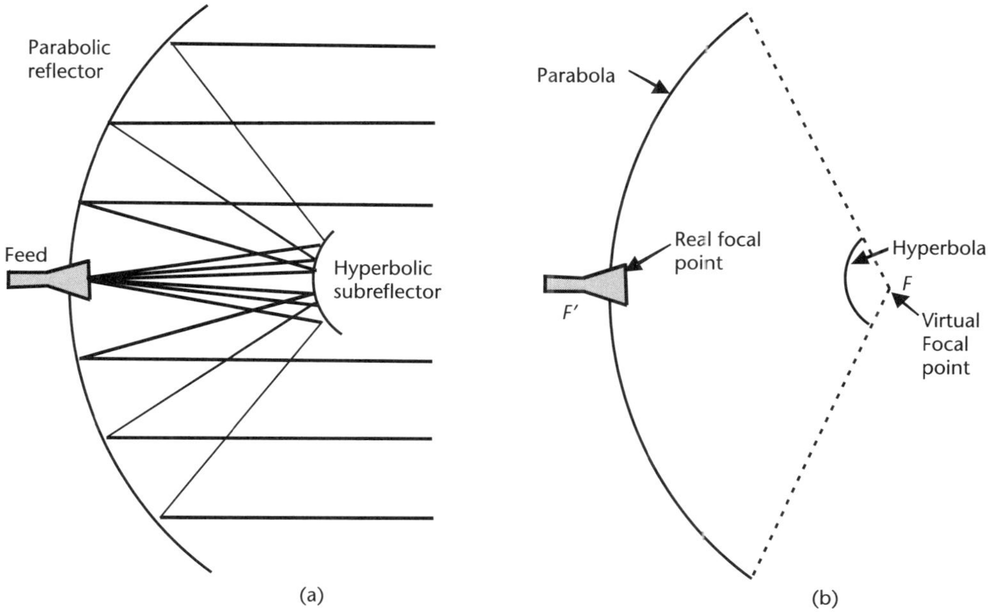

Figure C.3 Cassegrain parabolic reflector antenna: (a) principle of operation and (b) optics schematization.

The cosecant-squared reflector antenna is used in surveillance radars in order to shape the large elevation beam in accordance with a pattern proportional to cosec$^2\varepsilon$, where ε is the elevation angle. This produces a constant power echo signal from a target flying at constant altitude H as shown in Figure C.4.

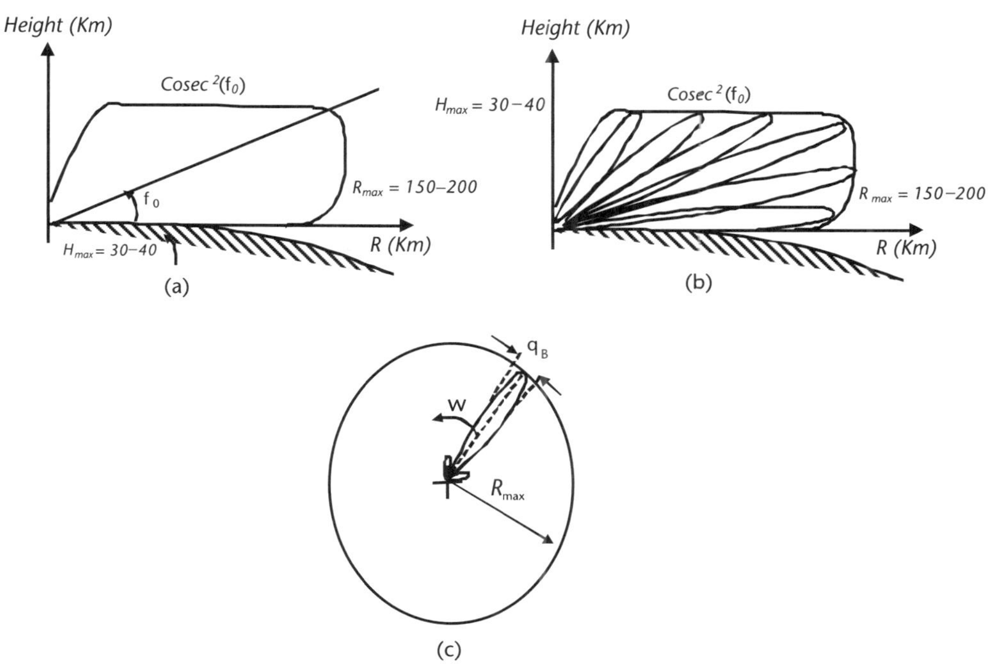

Figure C.4 Cosecant-squared antenna in surveillance radars: (a) free-space elevation coverage, (b) actual elevation coverage in the presence of reflecting terrain, and (c) azimuth coverage.

C.2 Array Antennas

An array antenna consists of a number n ($n \geq 2$) of individual radiating elements, suitably spaced with respect to one another in a lattice. The relative amplitude and phase of the signals applied to each of the elements are controlled to obtain the desired radiation pattern from the combined action of all (or part of) the elements. The array can be considered to be an aperture with a discrete number of radiators instead of an infinite number as is the case for a full radiating surface. As such, the array is characterized by the same parameters as an antenna, such as gain, polarization, beamwidth, and sidelobe levels, but with the exception that the parameters need to be determined as a function of the scan angle of the beam.

The array elements themselves dictate many of the array properties such as the array polarization, spatial coverage, operating bandwidth, and mutual coupling of elements within the array lattice. The array elements can consist of dipoles, slots, spirals, and wideband flared notch (Vivaldi) radiators. Three geometrical forms of array antennas are of interest in the EW domain: the linear array, the planar array, and the conformal array. A *linear array* consists of elements arranged along one straight line. A *planar array* is a 2D configuration of elements arranged to lie in a plane. A *conformal array* is a 1D or 2D configuration of elements arranged to lie along a curved line or a curved surface.

Relevant to the direction of the maximum radiation, the array can be classified as a *broadside array* or an *endfire array* when the direction of the maximum radiation is, respectively, perpendicular or parallel to the line or plane where the elements lie. Endfire arrays have not been widely used in the radar bands. Endfire linear arrays are more commonly used in the communication bandwidth when medium gain and large beamwidth are required.

The performance parameters of a phased array [1] can be directly derived from (C.3) by considering the far-field radiation pattern of the nth element of the array shown in Figure C.5, which is:

$$\underline{E}_n(R,\theta,\phi) = \underline{E}_n(u,v) = a_n \frac{e^{-jkR}}{R} \underline{g}_n(u,v) e^{j\underline{k}\cdot\underline{\rho}_n} \tag{C.9}$$

where

$g_{\underline{n}}(u,v)$ = a vector quantity, normalized to unity at its peak, that represents the far-field pattern of the radiating element, including polarization and taking into consideration the effects of mutual coupling with the other array elements;

a_n = complex weight (i.e., amplitude and phase); also called *excitation* of the nth element in the array;

the other symbols are the same as those defined for (C.3).

In the determination of the array radiation pattern, we can omit the factor e^{-jkR}/R, which is a factor common to all of the elements of the array, and follow the approach of (C.3), which in this case in lieu of the integral extended to the

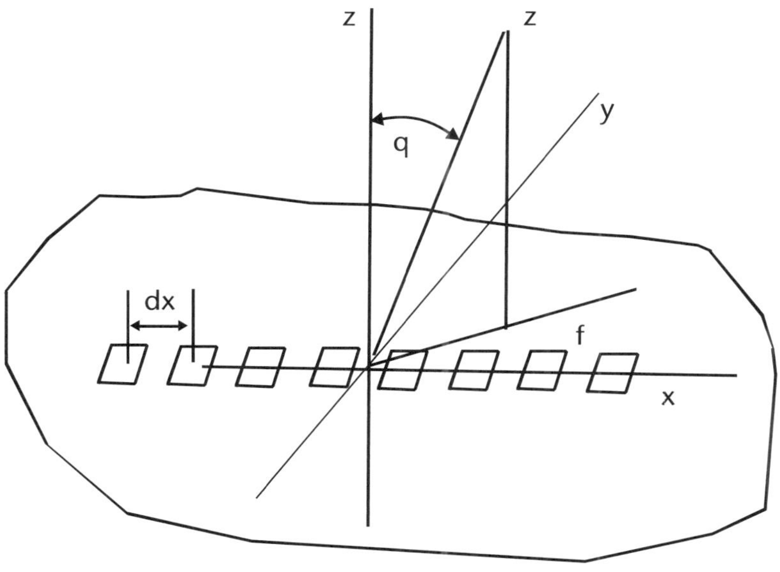

Figure C.5 Coordinate system for the evaluation of the radiation pattern of a linear array.

illumination over the aperture becomes the summation of the contributions of N elements of the array:

$$\underline{E}_a(u,v) = \sum_{1}^{N} a_n \underline{g}_n(u,v) \cdot e^{j\underline{k}\cdot\underline{\rho}_n} \tag{C.10}$$

This general expression is valid for any type of array. Let us now consider some more simple expressions relevant to regular linear and planar arrays, that is, arrays that show a periodicity in one or two dimensions.

In the case of a regular linear array, the elements are spaced at a distance d_x from each other, and the distance $\underline{\rho}' = x'\underline{x}_0 = nd_x\underline{x}_0$ so that the exponential in (C.10) becomes $\exp(jknd_xu)$. Further, we can assume that the element radiation pattern is the same for all of the array elements and hence (C.10) becomes:

$$\underline{E}_a(u,v) = \underline{g}(u,v) \sum_{-(N-1)/2}^{(N-1)/2} |a_n| e^{j\psi_n} \exp(jknd_xu) = F_a(u) \cdot \underline{g}(u,v) \tag{C.11}$$

which shows that the linear array radiation pattern can be expressed as the product of the vector element pattern $g(\underline{u},\underline{v})$ (assumed the same for all elements in the array) and a scalar algebraic term $F_a(u)$. This term can be defined as the *array factor* when the peak value is normalized to a unity value. If the phases ψ_n of the element excitations are assumed to be $\psi_n = 0$, the peak value of $F_a(u)$ is reached for $u = 0$.

An important consideration can be drawn from (C.11) by considering that if the phases ψ_n of the element excitations are assumed to be $\psi_n = -knu_0$ then the argument of the exponential becomes $jknd_x(u - u_0)$ and the array factor becomes $F_a(u - u_0)$. That is, the pattern peak is shifted to $u = u_0$ and the radiation pattern plotted in the u coordinates does not change its shape (i.e., beamwidth and sidelobe level) for that shift.

An important special case occurs when all excitation amplitudes are the same and can be assumed to be $|a_n| = 1$ for all n. In such a case the array factor has an explicit expression, by recognizing that the summation is in effect the summation of the terms of a geometrical series of ratio $\exp(jkd_x u)$, which gives

$$F_a(u) = \frac{\sin\left[\frac{N\pi d_x}{\lambda} u\right]}{N\sin\left[\frac{\pi d_x}{\lambda} u\right]} \tag{C.12}$$

The array factor of (C.12) at the wavelength λ for a regular linear array of 16 elements spaced as $d_x = 0.5\lambda$ and uniformly excited is shown in Figures C.6(a) and (b), respectively, for the beam pointing at $u(0°,0°)$ and $u_0(60°,0°)$.

From Figure C.6(a) we can determine that the –3-dB beamwidth θ_B at broadside ($\theta = 0°, \phi = 0°$) is $\theta_{B0} = 0.866\lambda/N\ dx$, whereas when scanned to $u_0(\theta_{\text{scan}}°,0°)$ it becomes $\theta_{B,\text{scan}} = \theta_{B0}/\cos\theta_{\text{scan}}$; that is, it widens with the scan angle.

From the preceding expression of the –3-dB beamwidth, we can see that as the frequency changes the pattern broadens at lower frequencies and narrows at higher frequencies as shown in Figure C.7.

Similarly to that shown for aperture illumination, the uniform excitation of the elements results in the highest gain for the array but it provides a first sidelobe of only –13.2 dB with respect to the peak , with the other sidelobes decaying asymptotically as $1/(n+1/2)$ for large n.

A further consideration relevant to the shift of the array beam to the direction u_0 by impressing on each element a phase shift $\psi_n = -knu_0$ is that a change of the frequency from $f_0 = c/\lambda_0$ to $f_0+\Delta f = c/(\lambda_0 - \Delta\lambda)$ with no change in the element phase shifters setting provides a further shift of the beam-pointing from u_0 to $u = u_0 f_0/(f_0+\Delta f)$; the beam is squinting toward the broadside if $\Delta f > 0$ and away from broadside if $\Delta f < 0$. It can be shown that a 3-dB loss in signal reception is incurred when the frequency shift Δf is such that:

$$\frac{\Delta f}{f_0} = \frac{\sin(\theta_{B,scan})}{u_0} \tag{C.13}$$

This equation indicates a limit in the phased array system bandwidth that is more stringent the larger the number of elements in the array (i.e., the narrower is its beamwidth). The above limit is overcome if the phase impressed to each element by the excitation is provided through time-delay lines (whose phase is invariant with the frequency). The beam squint due to frequency variation of 5% is shown in Figure C.8 for a 16-element linear array pointing at (30°,0°) (element spacing 0.75λ).

From (C.12) note that when scanning the beam to u_0 at the increase of the width of the spacing d_x with respect to the wavelength λ (i.e., an increase of the ratio d_x/λ), the scanned beam is accompanied by the appearance of a new lobe, called a *grating lobe*, in a direction opposite to the scanning direction, as shown in Figure C.9.

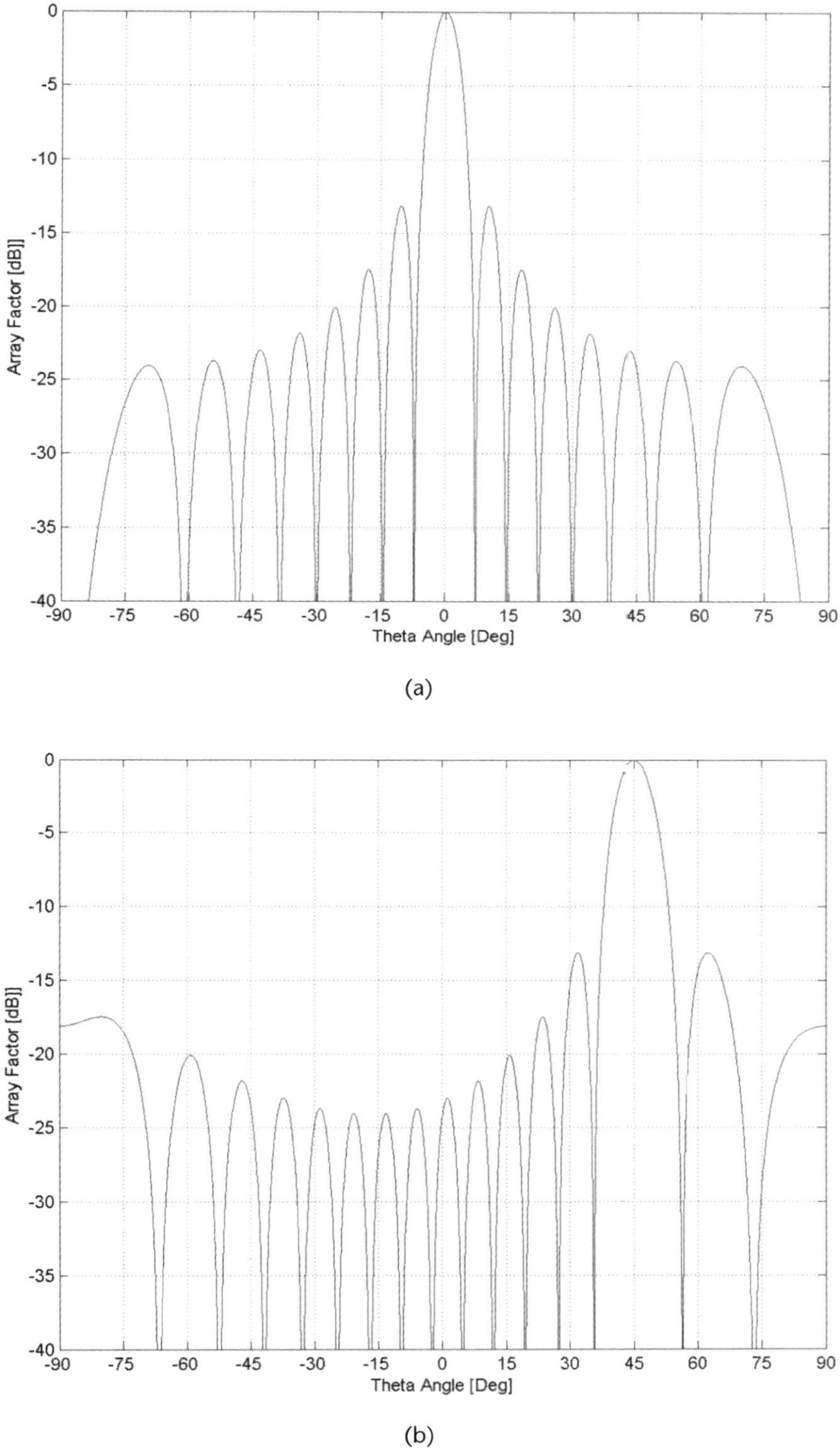

Figure C.6 Radiation patterns of a 16-element linear array with spacing 0.5λ: (a) pointing at (0°,0°) and (b) pointing at (60°,0°).

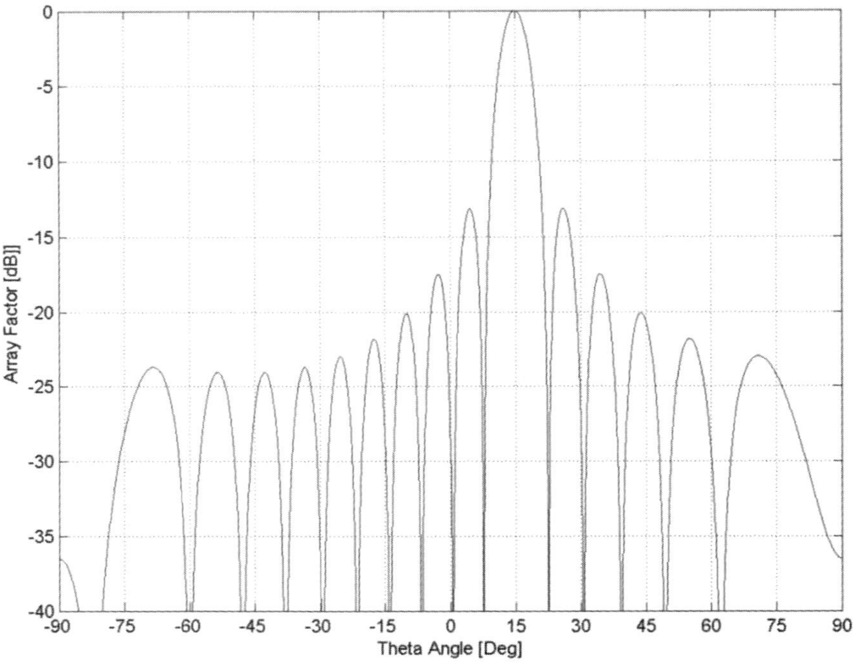

(a)

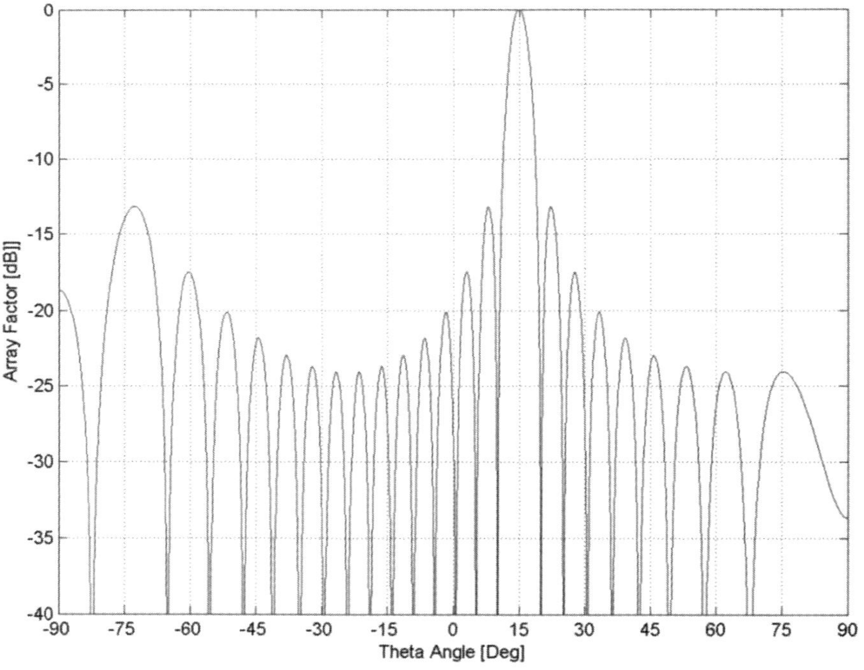

(b)

Figure C.7 Radiation pattern of a 16-element linear array pointed at (10°,0°): (a) at frequency $f_0(dx/\lambda = 0.5)$ and (b) at frequency $f_1(dx/\lambda = 0.75)$.

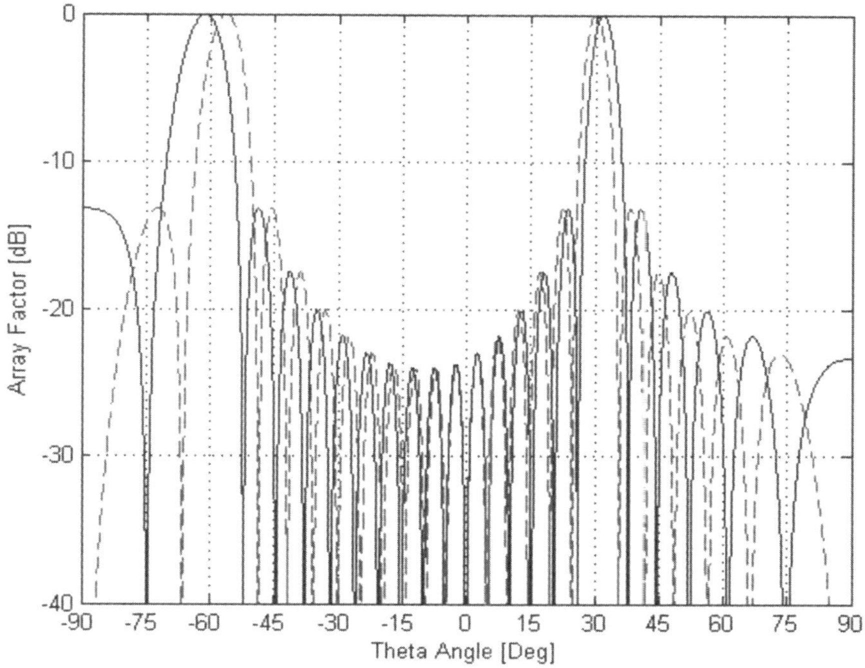

Figure C.8 Beam squint due to a frequency variation of 5% ($f_1 = 0.95f_0$) for a 16-element linear array pointing at (30°,0°) (element spacing 0.75λ).

The grating lobe conditions for a linear array at wavelength λ can be determined from the exponential of the terms in the summation of the scanned array factor $F_a(u - u_0)$, which is $jknd_x(u - u_0)$.

Due to its periodicity the exponential has a zero value for every direction u_g that satisfies the equation:

$$\frac{2\pi d_x}{\lambda}(u_g - u_0) = 2\pi g \text{ for } g = 0, \pm 1, \pm 2, \ldots$$
$$i.e., \quad u_g = u_0 \frac{\lambda}{\lambda_0} + g\frac{\lambda}{d_x} \tag{C.14}$$

having taken into account also the element impressed phase variations with frequency in the array bandwidth. For the case of an element phase impressed through time-delay lines, the above equation simplifies as

$$u_g = u_0 + g\frac{\lambda}{d_x} \tag{C.15}$$

Equation (C.15) defines a set of beams displaced from the main scanned beam at u_0 by the distance $g\lambda/d_x$. The grating lobes that appear in the "real space" $|u_g| \leq 1$ are real radiation beams, whereas the ones that appear in the "imaginary space" $|u_g| \geq 1$, are not radiating energy.

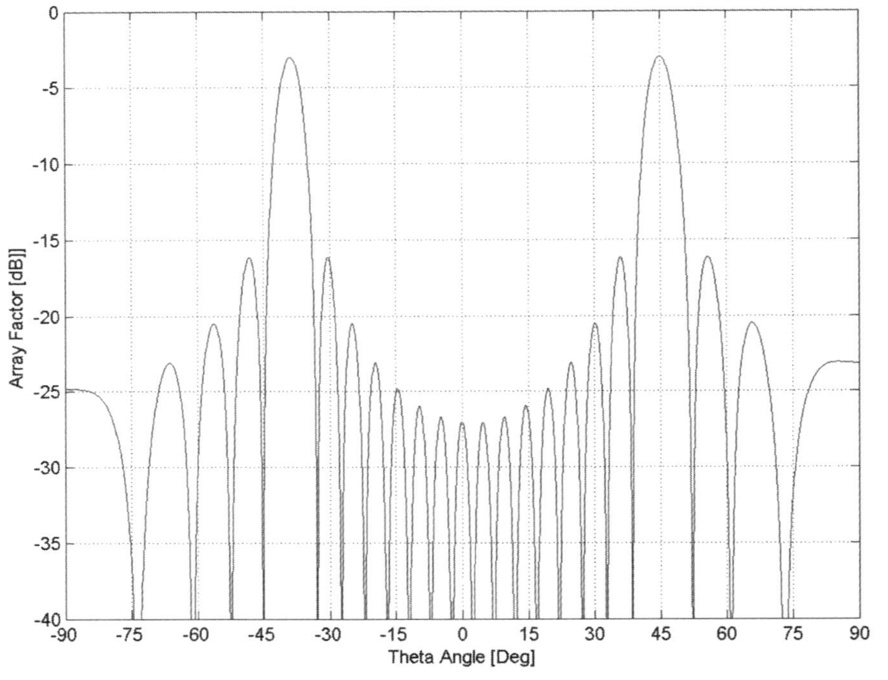

Figure C.9 Appearance of a grating lobe at (–40°,0°) when the beam of a 16-element linear array (spacing 0.75λ) is scanned to (45°,0°).

The condition of avoiding grating lobes in real space while scanning the beam to the maximum required direction $u_{\max}$ is derived from (C.15) as follows and sets a limit to the interelement spacing:

$$\frac{d_x}{\lambda} \le \frac{1}{u_g + u_{\max}} = \frac{1}{1 + \sin\theta_{\max,scan}} \tag{C.16}$$

Note that small arrays, as commonly used in the EW domain, might have a broad enough grating lobe beam when scanning that, even if the peak is in the imaginary space, they can still have a contribution in the real space. Therefore, for these relatively small arrays it is useful to select the spacing d_x such that the grating lobe is at least one-half beamwidth, $\theta_B/2$, beyond the imaginary space boundary:

$$\frac{d_x}{\lambda} = \frac{1}{1 + \sin\theta_{\max,scan} + \sin(\theta_B / 2)} \tag{C.17}$$

For a planar array the real space boundary is a circle (as shown in Figure C.10) since:

$$\sin\theta = \sqrt{u^2 + v^2} \le 1 \tag{C.18}$$

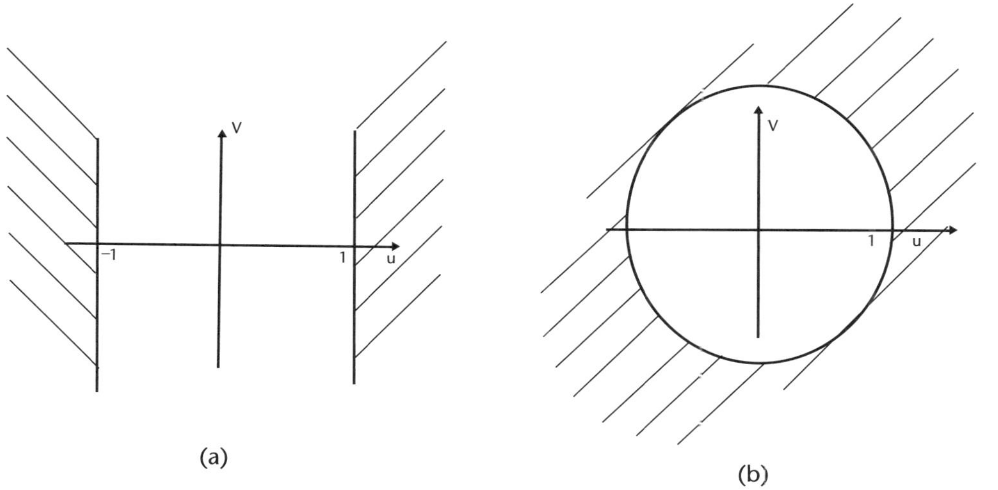

Figure C.10 Boundaries of the real space for a (a) linear array and (b) planar array.

Two types of lattices are primarily used for planar arrays: the *rectangular lattice* (with two spacings d_x and d) and the *triangular lattice* (both with two spacings d_x and d_y) as shown in Figure C.11.

The expression for the grating lobe locations of an array with rectangular lattice are derived from (C.14) above and are:

$$\begin{aligned} v_h &= \frac{\lambda}{\lambda_0} v_0 + h\frac{\lambda}{d_y} \quad \text{for } h = 0, \pm 1, \pm 2, \ldots \\ u_g &= \frac{\lambda}{\lambda_0} u_0 + g\frac{\lambda}{d_x} \quad \text{for } g = 0, \pm 1, \pm 2, \ldots \end{aligned} \tag{C.19}$$

These grating lobes are shown in Figures C.12 and C.13, respectively, for a square lattice and an equilateral triangular lattice, with a color scale that represents the relative amplitudes in decibels below the main beam peak. For the rectangular lattice

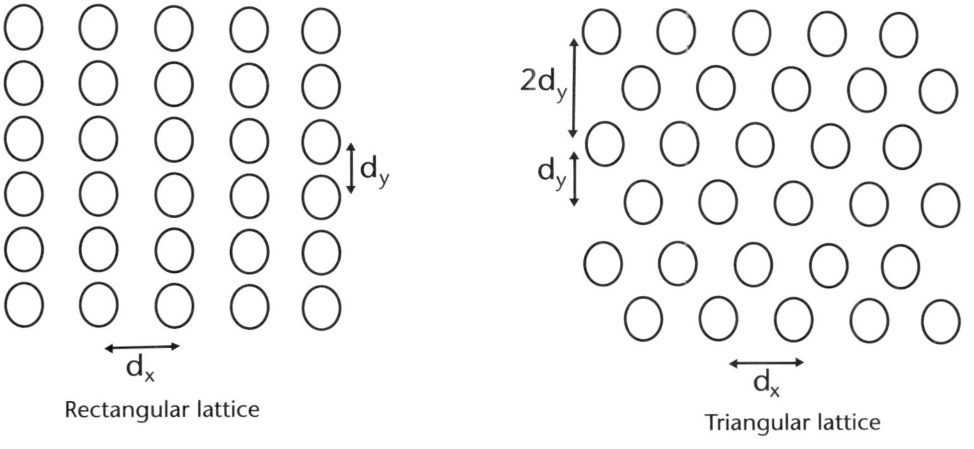

Figure C.11 Typical lattices for a planar phased array.

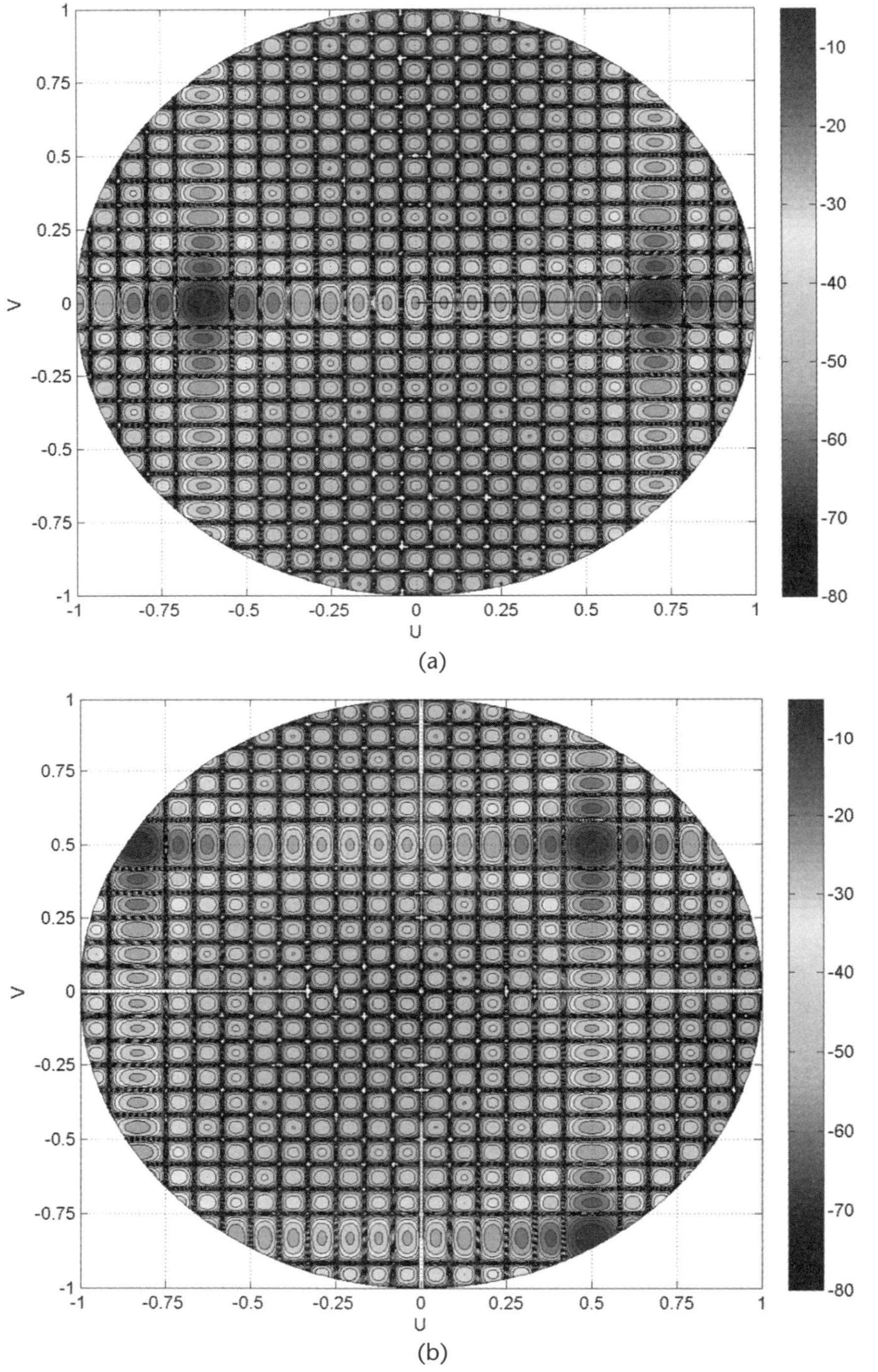

Figure C.12 Grating lobes in array factors of a 16- × 16-element planar array (square lattice with spacing $dx = dy = 0.75\lambda$): (a) main beam scanned at (45°,0°) and (b) main beam scanned at (45°,45°).

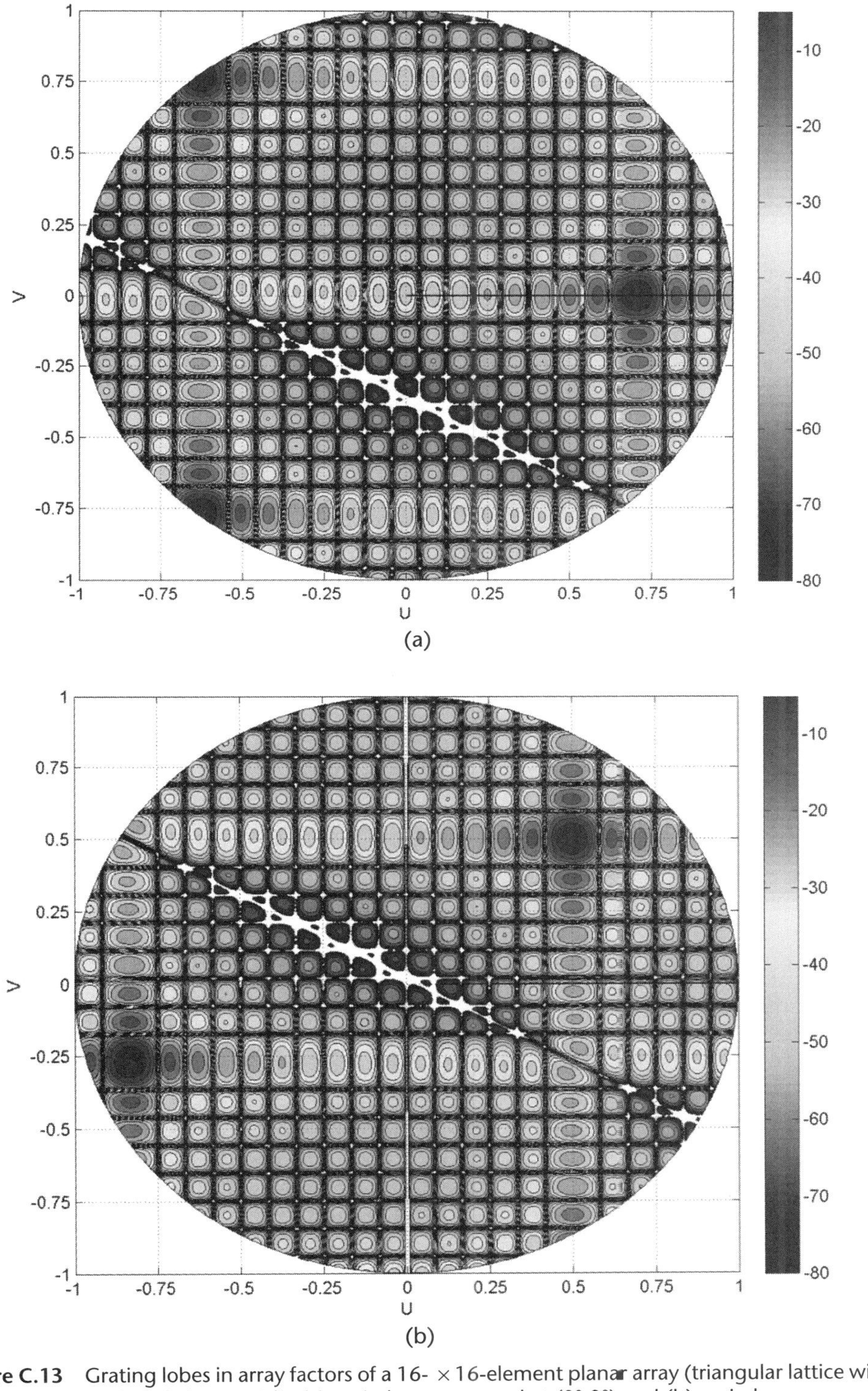

Figure C.13 Grating lobes in array factors of a 16- × 16-element planar array (triangular lattice with spacing $dx = 0.75\lambda$ and $dy = 0.65\,\lambda$): (a) main beam scanned at (0°,0°) and (b) main beam scanned at (45°,45°).

array the nearest grating lobes appear along the principal planes [i.e., along $(u,0)$ or $(0,v)$], so the condition to avoid them in the real space is the same as per (C.16) for both d_x and d_y, as shown in Figure C.14.

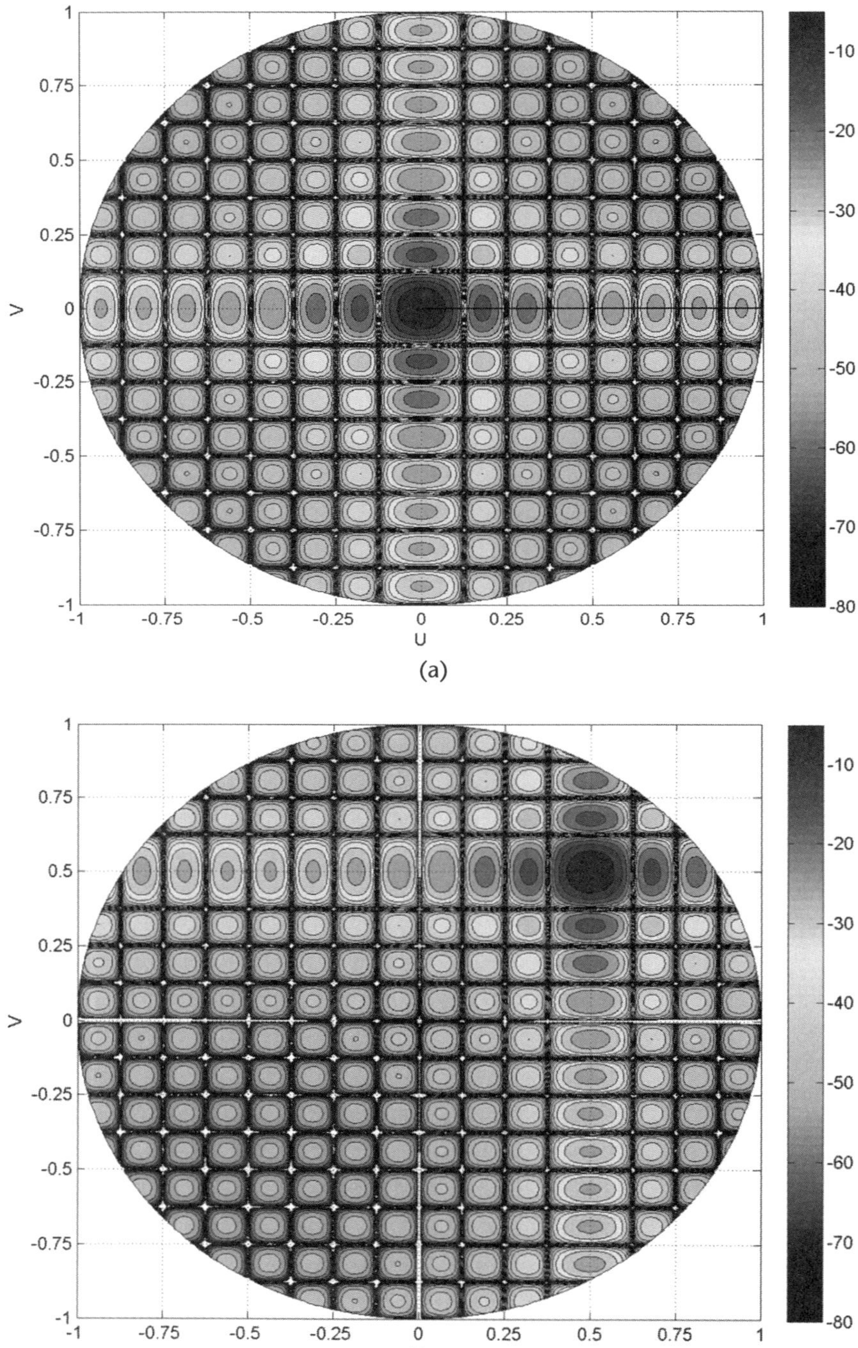

(a)

(b)

Figure C.14 No grating lobes in array factors of a 16- × 16-element planar array (square lattice with spacing $dx = dy = 0.5\lambda$): (a) main beam scanned at (0°,0°) and (b) main beam scanned at (45°,45°).

The triangular lattice allows for elements to be saved with respect to the square lattice in a planar array of given area size. This can be noted from Figure C.12 for the case of an equilateral triangular lattice where within an area $A = d_x \cdot 2d_y$ we count six elements for the rectangular lattice and five elements for the triangular lattice (i.e., a spare of 16.6% in the number of elements).

References

[1] Jasik, H., "Fundamentals of Antennas," in R. Johnson (Ed.), *Antenna Engineering Handbook*, New York: McGraw-Hill, 1993.

[2] Skolnik, M., *Introduction to Radar Systems*, New York: McGraw-Hill, 1962.

[3] Cooley, M., and Davis, D., "Reflector Antennas," Chap. 12 in M. Skolnik (Ed.), *Radar Handbook*, 3rd ed., New York: McGraw-Hill, 2008.

Appendix D
Analog Modulation Methods

D.1 Amplitude Modulation

The AM technique is shown in Figure D.1 where the carrier $c(t) = \cos 2\pi fct$ is modulated by the signal $m(t)$. The spectrum of finite bandwidth Δf is shown in Figure D.1(a) at baseband and in Figure D.1(b) when translated by the carrier.

The shape of the spectrum can be easily deducted by considering a modulating signal constituted by a single tone $m(t) = \cos 2\pi f_m t$ with $f_m < f_c$. In this case the output signal of Figure D.1 is

$$y(t) = \cos 2\pi f_c t \cdot \cos 2\pi f_m t = \frac{1}{2}\cos 2\pi (f_c + f_m)t + \frac{1}{2}\cos 2\pi (f_c - f_m)t$$

Because the AM is linear, the bandwidth of the spectrum reproduces the one at baseband. The modulation method reported in Figure D.1(c) is relevant to the case of double-sideband (DSB) AM. Further AM methods perform (1) suppression of the carrier using a method called DSB-SC (suppressed carrier) and (2) suppression of one sideband, using a method called single sideband. In particular, if the suppressed bandwidth is indicated by a 1 in the figure, it is termed upper sideband; if indicated by a 2, it is termed lower sideband. One of the bands can be suppressed as each of the sidebands contains the same information, thus increasing the efficiency of the modulation method.

D.2 Angle Modulation

The instantaneous phase of the modulated signal is $\varphi(t) = 2\pi f_c t + \phi(t)$ and its instantaneous frequency is

$$f_{inst} = \frac{1}{2\pi}\frac{d\phi}{dt} = f_c + \frac{1}{2\pi}\frac{d\varphi}{dt} \tag{D.1}$$

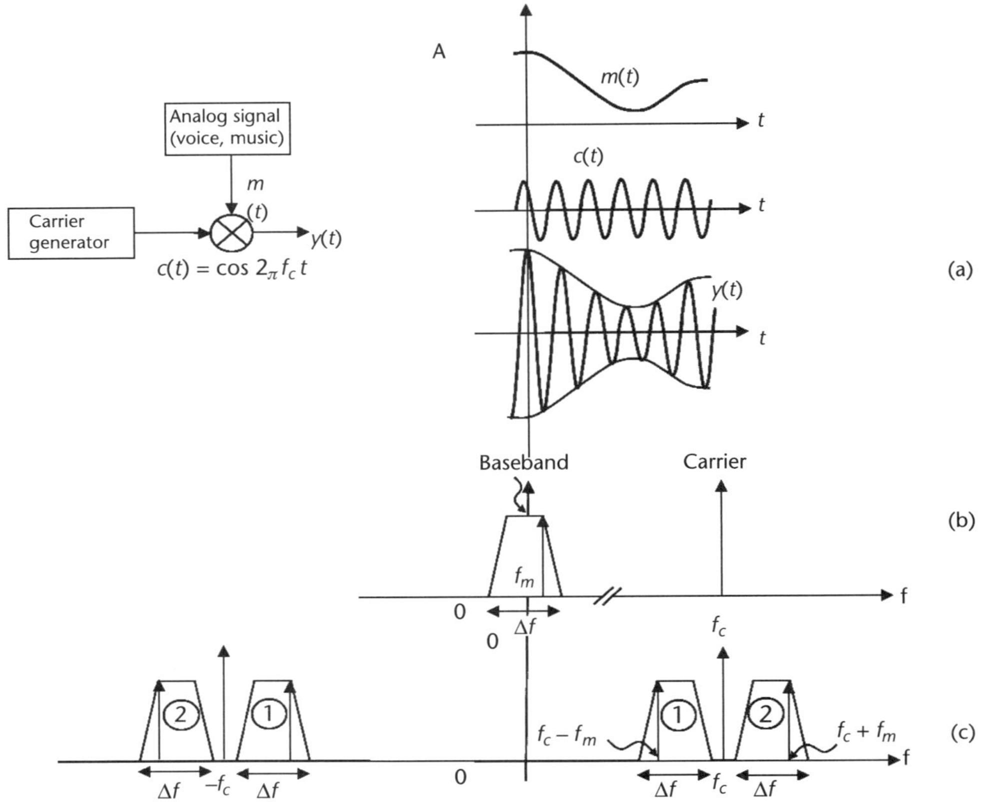

Figure D.1 Time representation of AM signals: (a) modulating signal spectrum (b) at baseband and (c) translated by the carrier.

For PM we have where k_p is the PM constant (radians/volt) and thus the modulated carrier signal is

$$s(t) = A\cos(2\pi f_c t + k_p m(t))$$

For FM we have

$$\varphi(t) = k_f \int_{-\infty}^{t} m(\tau) \cdot d\tau$$

where k_f is the FM constant (radians/second/volt) and thus the modulated carrier signal is

$$s(t) = A\cos(2\pi f_c t + k_f \int_{-\infty}^{t} m(\tau) \cdot d\tau$$

The shape of the FM spectrum is quite complicated. A simple example is shown in Figure D.2 by considering the modulation signal $m(t)$ to be a single tone $m(t) = M \cos 2\pi f_m t$ with $f_m << f_c$. In this case the modulated signal is

$$y(t) = A\cos(2\pi f_c t + \beta \sin(2\pi f_m t)) \tag{D.2}$$

where $\beta = k_f \dfrac{M}{2\pi f_m}$ is called the *modulation index* and is the ratio of the maximum frequency deviation of the carrier caused by the modulating signal to the maximum frequency of the modulating signal. Its spectrum is shown in Figure D.2.

Equation (D.2) can be expressed for small β as

$$s(t) = A\cos 2\pi f_c t - \frac{A\beta}{2}\cos\left[2\pi(f_c - f_m)t\right]\frac{A\beta}{2}\cos\left[2\pi(f_c + f_m)t\cdot\right] \tag{D.3}$$

and it is reported in Figure D.2. When $\beta < 1$, the modulation is called narrowband FM, and the bandwidth of the signal is about $2f_m$.

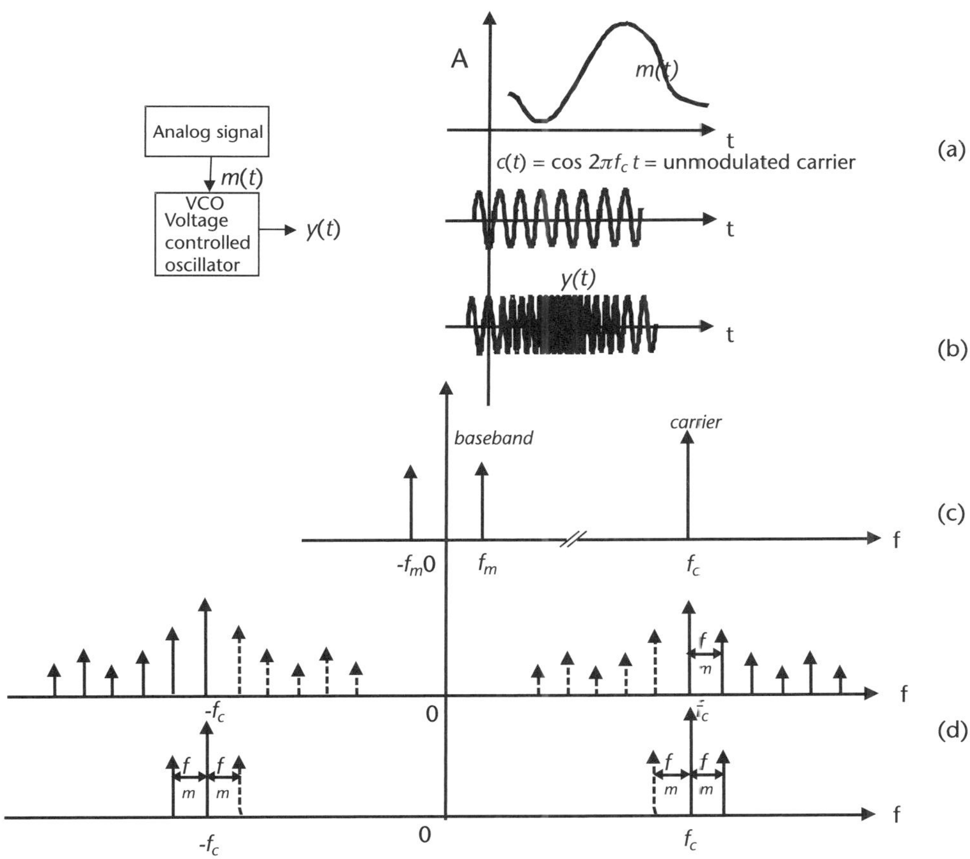

Figure D.2 Spectrum of FM modulation [1].

Because FM is a nonlinear modulation, the determination of its bandwidth is not simple. Carson [1] developed an approximation, called Carson's rule, which states that the bandwidth is

$$\begin{aligned} B_{IF} &= 2(\beta+1)f_m \text{ for large } \beta(\beta << 1) \\ B_{IF} &= 2(\beta+2)f_m \text{ for } 2<\beta<10 \end{aligned} \tag{D.4}$$

An effective and simple FM modulator is constituted by a VCO, in which the output frequency of oscillation is controlled by the input voltage $m(t)$ (provided that is slowly changing with respect to the oscillator frequency, which in our case is $f_m << f_c$).

The derivation of the PM spectrum can be conducted in the same manner as that for the FM spectrum. We consider the modulation signal $m(t)$ to be a single tone $m(t)= M\cos 2\pi f_m t$ with $f_m << f_c$. The same results apply. In particular Carson's rule is the same as (D.4) above (with the exception that now β = PM modulation index = $\Delta f/f_{PM}$, where f_{PM} is the largest modulating frequency).

D.3 Quadrature Modulation

Two pieces of data, A and B, normally associated with digital communications, can be transmitted at the same time if they are modulating the carrier frequency with orthogonal or quadrature waveforms, such as and . The quadrature modulation method is depicted in Figure D.3.

The transmitted signal is processed in the upper branch of the receiver. The relevant integrator, which integrates over a number p of periods $T = 1/f_c$, produces a signal $y_1(t)$ at the output:

$$\begin{aligned} y_1(t) &= \int_0^{pT} r(t)\sin 2\pi f_c t\cdot dt = \int_0^{pT} A\sin^2 2\pi f_c t\cdot dt + \int_0^{pT} B\cos 2\pi f_c t\cdot \sin 2\pi f_c t\cdot dt = \\ & A\int_0^{pT}\left(\frac{1-\cos 4\pi f_c t}{2}\right)\cdot dt + B\int_0^{pT}\frac{\sin 4\pi f_c t}{2}\cdot dt = \frac{A(pT)}{2} = kA \end{aligned} \tag{D.5}$$

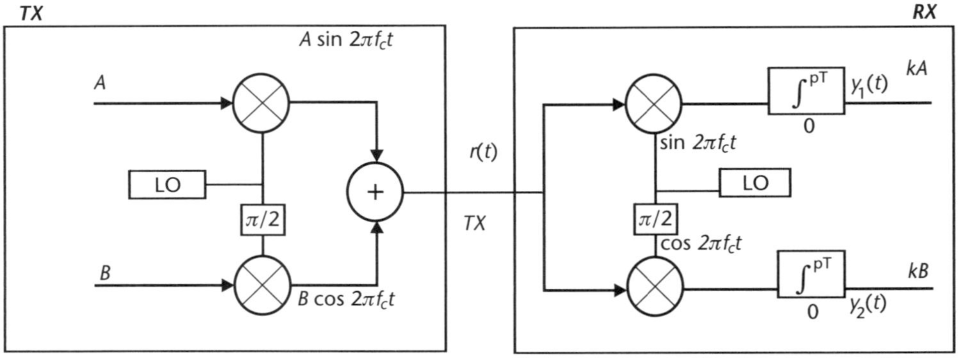

Figure D.3 Quadrature modulation method [1].

Following the same calculations as above, it can be shown that the output of the lower branch of the receiver is $y_2(t) = kB$.

The quadrature method shown in Figure D.3 requires that the transmitter and the receiver LOs be synchronized in order to be coherent in phase with one another and that the integration be performed in exactly a period pT. If that is not the case, the outputs can be different from kA and kB. This is why the quadrature modulation method is utilized only in digital communications; because the expected output from the upper receiver can be either kA or $-kA$ and it is easier to assign the more likely expected value is the output is different fromkA.

Reference

[1] Poisel, R., *Introduction to Communication EW Systems,* Norwood, MA: Artech House, 2002.

Appendix E
Evaluation of BER Increase for Noise and CW Tone Jamming in Communication Systems Employing BFSK Modulation

The most common form of frequency modulation is *binary frequency shift key* (BFSK) modulation, where two tones f_1 and f_2 are used to transmit, respectively, a binary symbol "0" or "1." By using 2^n frequencies (MFSK), it is possible to transmit n binary bits per frequency. The FSK method allows noncoherent detection techniques in the receiver, which allows for the use of oscillators that do not need to maintain phase coherency over large frequency changes as is required by MFSK modulation or FFH communication. Those oscillators are, however, capable of maintaining coherency within a hop of a SFH communication. The schematic block diagram of a MFSK communication transmitter is shown in Figure E.1.

To evaluate the effect of jamming on BFSK communication, let us consider the mechanics of superheterodyne detection. The receiver detector structure for noncoherent BFSK detection is shown in Figure E.2, where the channel level can be recognized as the identical structure used for the radar noncoherent detection analyzed in Appendix A.

The signals $x_i(t)$ after the two bandpass filters, centered, respectively, at $f_1 = f_c - \delta_f$ and $f_2 = f_c + \delta_f$, are ideally expressed as follows, where the indexes i and k are used so that the two channels can be considered generic:

$$x_i(t) = s_i(t) + j_i(t) + n_i(t) \tag{E.1}$$

and similarly for $x_k(t)$, where

$s_i(t) = A\cos(2\pi f_i t)$ with $i = 1$ or 2, is the received signal with amplitude A proportional to the transmitted power;

$n_i(t)$ and $n_k(t)$ are narrowband Gaussian noise;

$j_i(t)$, $j_k(t)$ are the filtered output of the jamming waveform $j(t)$ at the entrance of the receiver.

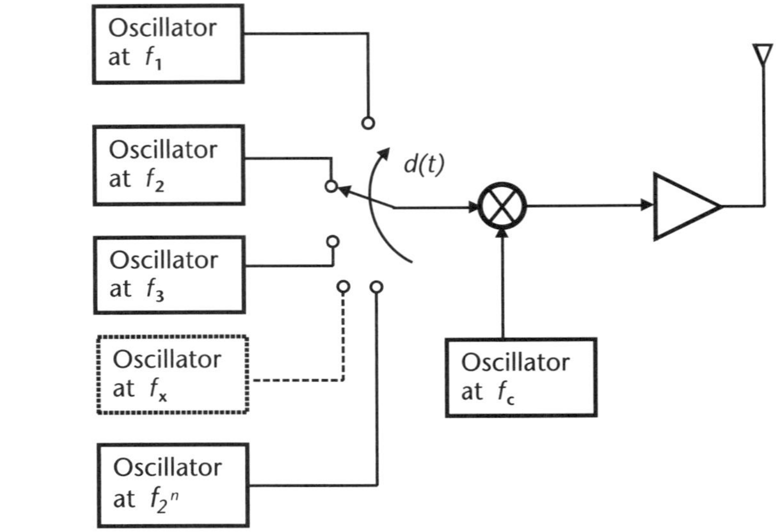

Figure E.1 Schematic block diagram of MFSK signal generation.

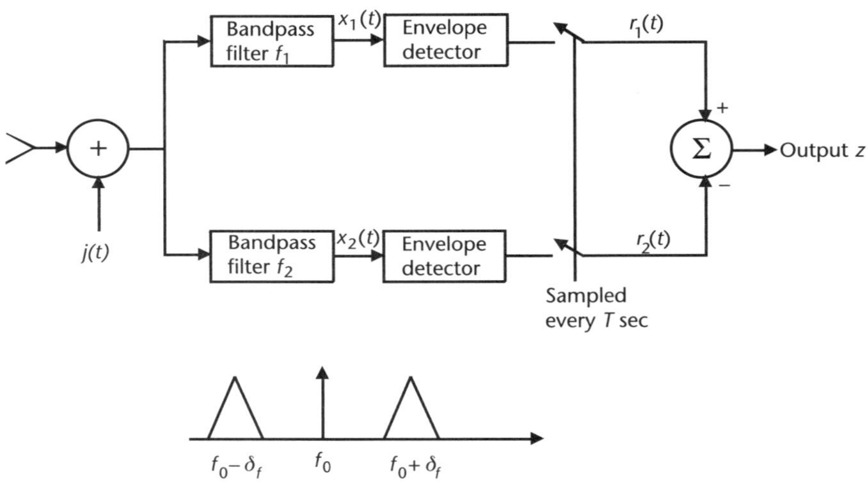

Figure E.2 BFSK noncoherent receiver detector.

The envelope detector outputs $r_1(t)$ and $r_2(t)$ are the ideal envelopes of the narrowband input $x_1(t)$ and $x_2(t)$, respectively. The output symbol $z(pT)$ of the receiver is based on the larger envelope sample between the two channel outputs at each signaling interval T.

By expressing the narrowband noise and jamming processes in terms of quadrature components, we get

$$\int_0^\infty dx \cdot x \cdot \exp\left[-\frac{x^2+a^2}{2}\right] \cdot I_0(ax) = 1 \tag{E.2}$$

where n_{ci} and n_{si} are statistically independent, zero-mean Gaussian processes with average power N. Because there is no bandwidth overlap between the two passband filters, then n_{ci} and n_{ck} (and similarly n_{si} and n_{sk}) are also statistically independent processes. With these assumptions let us assume that the symbol "1" is transmitted with frequency f_2. The probability density functions (pdfs) of $r_1(t)$ and $r_2(t)$, in the absence of jamming, can be expressed, respectively, with the Rayleigh and Rician distributions, having let:

$$x = \sqrt{2} \cdot r_2/\sigma_n \tag{E.3}$$

$$a = A/\left(\sqrt{2}\sigma_n\right) \tag{E.4}$$

where σ_n is the mean-square value of the noise (i.e., the noise power N) at the output of each bandpass filter, and $I_0(x)$is the modified Bessel function of the first kind and order zero and recognizing that:

$$A^2/\left(2\sigma_n^2\right) = SNR \tag{E.5}$$

and

$$\int_0^\infty dx \cdot x \cdot \exp\left[-\frac{x^2 + a^2}{2}\right] \cdot I_0(ax) = 1$$

An error will occur if $r_1 > r_2$ and this event has the following probability:

$$p(r_1 > r_2 | 1sent) = p(error|1sent) = \frac{1}{2}\exp\left[-\frac{SNR}{2}\right] \tag{E.6}$$

Similarly if the transmitted symbol was a "0," the same result would be obtained:

$$p(r_2 > r_1 | 0sent) = p(r_1 > r_2 | 1sent) = \frac{1}{2}\exp\left[-\frac{SNR}{2}\right]$$

The total error probability is

$$p_{error} = p_0 \cdot p(error|0 \cdot sent) + p_1 \cdot p(error|1 \cdot sent) = \frac{1}{2}\exp\left[-\frac{SNR}{2}\right] \tag{E.7}$$

assuming that the a priori probabilities of sending a symbol are p_0 and p_1, respectively, and that p_0 and p_1 are the same and equal to ½ because there should be no reason to transmit a preferred symbol. Similar expressions can be derived for

MFSK and are reported in Chapter 3 of [1]. The BER of the MFSK communication is shown in Figure E.3.

In the presence of jamming two possibilities have to be considered: (1) The jammer places an extra noise of power J in each of the two channels or (2) the jammer places its extra noise of power J in one channel only. In the first case the probability of error is easily evaluated by considering in (E.7) the total noise power $N + J$ and, hence,

$$p_{error}(both \cdot bands \cdot jammed) = \frac{1}{2}\exp\left[-\frac{S}{2(N+J)}\right] \tag{E.8}$$

In the second case the noise power in the two channels is different; in one we have only N and in the other $N + J$. With some mathematical passages similar to the earlier ones and with the same assumption about the a priori probabilities of sending the two symbols ,we have:

$$\begin{aligned} p_{error} &= p_0 \cdot p(error|\, jam \cdot in \cdot complementarychannel) + \\ &p_1 \cdot p(error|\, jam \cdot in \cdot signalchannel) = \\ &\frac{1}{2}\cdot\frac{N+J}{2N+J}\exp\left[-\frac{S}{2N+J}\right] + \frac{1}{2}\cdot\frac{N}{2N+J}\exp\left[-\frac{S}{2N+J}\right] = \frac{1}{2}\exp\left[-\frac{S}{2N+J}\right] \end{aligned} \tag{E.9}$$

In the case of one CW tone jamming expressed as:

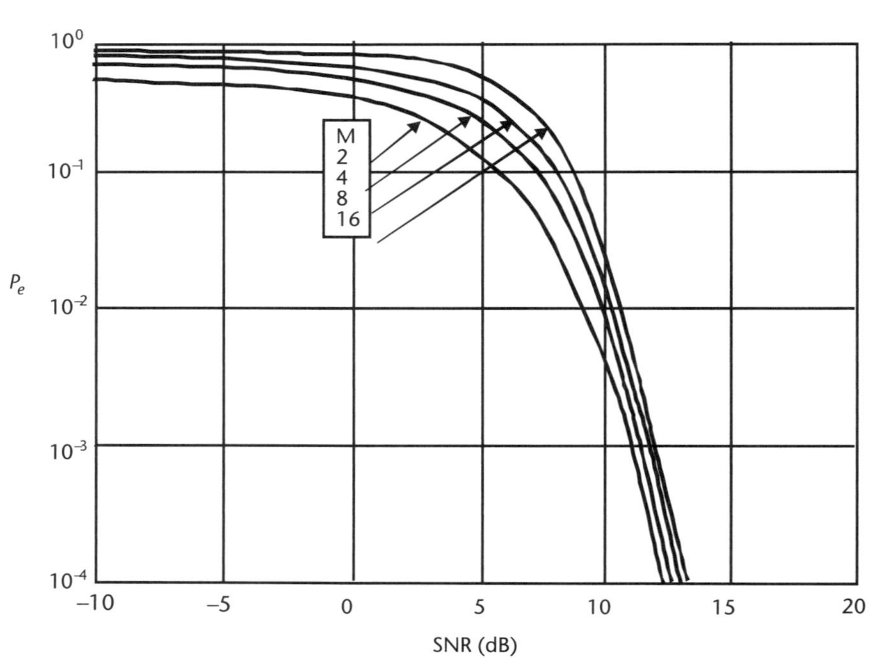

Figure E.3 BER of MFSK communication in the presence of noise only. (From: [1, Chap. 3].)

$$j(t) = \sqrt{J}\cos(2\pi f_i t + \phi)$$

the tone may have at the same frequency of the bandpass channel a random phase ϕ that is uniformly distributed in $[0.2\pi]$. Reference [1] shows that this phase provides an important multiplicative contribution (because it is independent of the other factors) to the probability of error. This is especially true if it is in the same data channel, because it may reinforce the signal rather than produce an error.

The probability of error, with both pdfs now being Rician, can be expressed [2] in terms of the Q-function, which is defined as:

$$Q(x, y) = \int_y^\infty du \cdot u \exp\left[-\frac{u^2 + x^2}{2}\right] I_0(ux)$$

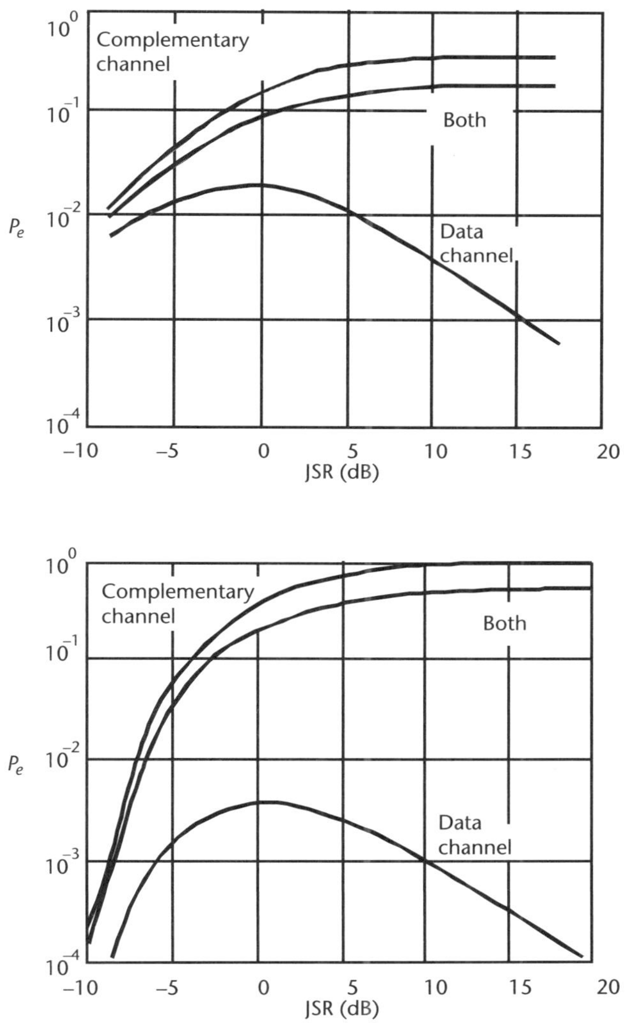

Figure E.4 BER of BFSK communication vs. JSR (dB) with parameter SNR = 10 and 20 dB. (From: [1].)

$$p(error|one.band.subjected.to.CW.tone) = \frac{1}{2}Q\left(\sqrt{\frac{JS}{N^2}}\right)$$

$$p(error|both.bands.subjected.to.CW.tones)$$
$$= \frac{1}{2} - \frac{1}{2\pi}\int_0^{\pi} d\phi \cdot \left[Q(h(\phi), \sqrt{J/N}) - Q\left[(\sqrt{J/N}, h(\phi)\right]\right]$$

where

$$h(\phi) = \left[S/N + 2\cos\phi \cdot \sqrt{JS/N^2} + J/N\right]^{1/2}$$

The BER of the BFSK communication in the presence of jamming is shown in Figure E.4.

References

[1] Poisel, R., *Introduction to Communication EW Systems,* Norwood, MA: Artech House, 2002.

[2] Pettit, R. H., *ECM and ECCM Techniques for Digital Communication Systems,* Belmont, CA: Lifetime Learning Publications, 1982.

List of Acronyms

A	Amplitude
A/C	Aircraft
AA	Antialiasing
AAA	Antiaircraft Artillery
AAM	Air-to-Air Missiles
ACM	Amplitude-Comparison Monopulse
ADC	Analoge-to-Digital Conversion
ADT	Automatic Detection and Tracking
AEF	Active Emitter File
AESA	Active Electronically Scanning Arrays
AEW	Airborne Early Warning
AFE	Antenna Front End
AGC	Automatic Gain Control
AGL	Agility
AI	Airbone Intercept Radar
AM	Amplitude modulation
AMTI	Automatic Moving Target Indicator
APA	Active Phased Array Antenna
ARM	Antiradiation Missiles
ARQ	Automatic Repeat Request
ASIC	Application Specific Integrated Circuit
ASK	Amplitude Shift Key
ASM	Air-to-Surface Missile
ASP	Antenna Scan Period
ASPJ	Airborne Self-Protection Jammers
AST	Antenna Scan Type
BBN	Broadband or Barrage Noise
BDA	Battle Damage Assessment
BER	Bit Error Rate
BFSK	Binary Frequency Shift Key
BOC	Binary Offset Carrier
BPF	Bandpass Filter
BPSK	Biphase Shift Key
BST	Burst
BW	Bandwidth

C2	Command and Control
CA	Cell-Averaging
CCD	Charge-Coupled-Detectors
CDMA	Code Division Multiple Access
C-ECM	Communication ECM
CEP	Circular Error Probability Area
C-ESM	Communications ESM
CFAR	Constant False Alarm Rate
CFD	Chaff and Flares Dispensers
CGCE	Constant Gain Cross-Eye
CHILL	Chaff Illumination
CHR	Channelised Receiver
CIC	Cascaded Integrated Comb
CIWS	Close-In Weapon Systems
CLOS	Command-to-Line of Sight
CM	Countermeasure
CNR	Combat Net Radios
COD	Coded
COMINT	Communications Intelligence
COMSEC	Communications Security
CORDIC	Coordinate Rotator Digital Computer
COSRO	Conical Scan on Receive Only
CPCE	Constant Power Cross-Eye
CPI	Coherent Processing Intervals
CPX	Complex
CRB	Cramer-Rao Bound
CRLB	Cramer -Rao Lower Bound
CSJ	Close Support Jammer
CVR	Crystal Video Receiver
CW	Continuous Wave
DAC	Digital-to-Analog Converter
DD	Differential Doppler
DDC	Digital Down Converter
DDS	Direct Digital Synthesis
DECM	Deception ECM
DEMUX	Demultiplexer
DEW	Direct Energy Weapons
DFD	Doppler Frequency Difference
DFT	Discrete Fourier Transform
DIAL	Differential Absorption Laser
DINA	Direct Noise Amplification
DIRCM	Directional IR Countermeasure
DLS	Decoy Launching System
DME	Distance Measuring Equipment
DMT	Discrete Multitone
DOA	Direction of Arrival
DR	Dynamic Range
DRFM	Digital RF Memories

DRx	Digital Receiver
DS	Direct Sequence
DSB	Double Sideband
DSP	Digital Signal Processing
EA	Electronic Attack
ECCM	Electronic Counter-Countermeasure
ECM	Electronic Countermeasures
EEP	Elliptical Error Probability Area
EJ or ESJ	Escort Support Jammer
E-L	Early-Late
ELINT	Electronic Intelligence
EM	Electromagnetic
EMCON	Electromagnetic Emissions Control
EMP	ElectromagneticNuclear Pulses
EMS	Electromagnetic Spectrum
ENOB	Effective Number of Bit
EO	Electro-Optical
EOB	Electronic Order of Battle
EP	Electronic Protection
EPI	Extended Phase Interferometry
EPL	ELINT Parameters Limits
ERP	Effective Radiated Power
ES	Electronic Warfare Support
ESM	Electronic Warfare Support Measures
EW	Electronic Warfare
EW	Early Warning Radar
F	Frequency
FAM	FFT Accumulation Method
FCR	Fire Control Radar
FCS	Fire Control System
FDMA	Frequency Division Multiple Access
FDOA	Frequency Difference on Arrival
FEC	Forward Error Correction
FFH	Fast Frequency Hop
FFT	Fast Fourier Transform
FH	Frequency Hopping
FIR	Finite Impulse Response
FIX	Fixed
FLIR	Forward Looking Infrared
FM	Frequency Modulation
FMCW	Frequency Modulated Continuous Wave
FOV	Field-of-View
FPA	Focal Plane Array
FPGA	Field Programmable Gate Array
FSK	Frequency Shift Key
GaAs	Gallium Arsenide
GCA	Ground-Controlled Approach
GCD	Greatest Common Divisor

GEO	Geostationary Earth Orbital
GNSS	Global Navigation Satellite Systems
GPS	Global Positioning System
GSM	Global System Mobile
HADF	High Accuracy Direction Finding
HD-TV	High Definition Television
HMI	Human-Machine Interface
HOP	Hopping
HPA	High Power Output Amplifier
HPRF	High PRF
HS	High-Speed
HVPS	High Voltage Power Supply
IADS	Integrated Air Defense System
ICW	Interrupted Continuous Wave
IDR	Instantaneous Dynamic Range
IED	Improvised Explosive Devices
IF	Intermediate Frequency
IFM	Instantaneous Frequency Measurement
IFOV	Instantaneous Field of View
IIR	Infinite Impulse Response
IPL	Inverse Passive Location
IR	Infrared
IRST	Infrared Search and Track
ISAR	Inverse SAR
ISTAR	Intelligence Gathering, Surveillance, Target Acquisition and Reconnaissance
JCR	Jammer Cancellation Ratio
JIT	Jitter
JPEG	Joint Pictures Expert Group
JSR	Jammer-to-Signal power Ratio
LACE	Low Altitude Cross-Eye
LB	Large Band
LCM	Least Common Multiple
LEO	Low Earth Orbital
LE-TDOA	Leading Edge TDOA
LFM	Linear Frequency Modulation
LNA	Low Noise Amplifier
LO	Local Oscillator
LOB	Line of Bearing
LORAN	Long-Range Navigation
LORO	Lobe on Receive Only
LOS	Line of Sight
LPF	Lowpass Filter
LPI	Low Probability of Intercept
LPIR	Low Probability of Intercept Radar
LPRF	Low PRF
LSB	Lower Sideband
LSM	Least-Squares Method

LUT	Look-Up Table
MAC	Multiplier Accumulator
MANPADS	Man Portable Air Defense Systems
MAW	Missile Approach Warner
MC	Memory Controller
MFAR	Multifunctional Phased Array Radars
MFSK	Multiple Frequency Shift Key
MI	Missile Illuminator
MIDS	Military Information Distribution System
MLE	Maximum Likelihood Estimation
MOP	Modulation on Pulse
MPEG	Motion Pictures Expert Group
MPM	Microwave Power Modules
MPRF	Medium PRF
MR(MRAD)	Milliradian
MRT	Minimum Resolvable Temperature
MSK	Active Missile Seeker
MTBF	Mean Time Between Failure
MTD	Moving Target Detection
MTF	Modulation Transfer Function
MTI	Moving Target Indicator
MTR	Missile Tracking Radar
MTV	Magnesium-Teflon-Viton
MUSIC	Multiple Signal Classification Algotithm
MWI	Moving Window Integrator
MWS	Missile Warning System
NAVSTAR	Navigation by Satellite Timing And Ranging
NB	Narrowband
NBC	Nuclear Biological Chemical
NBN	Narrowband noise
NCS	Net Control Station
NEMP	Nuclear EM Pulse
NEP	Equivalent Power
NET	Noise Equivalent Temperature
NIMD	Noncoherent Integration of N Maximum Detection
NLFM	Nonlinear Frequency Modulation
NN	Neural Network
OFDM	Orthogonal Frequency Division Multiplexing
OKPSK	Offset QPSK
OPG	Optical Parametric Generation
OPSEC	Operation Security
OTH	Over-the-Horizon
PACF	Periodic Autocorrelation Function
PAF	Periodic Ambiguity Function
PAR	Precision Approach Radars
PBN	Partial Bandnoise
PBR	Passive Bistatic Radar
PCL	Passive Coherent Location

PCM	Phase-Comparison Monopulse
PCM	Pulse Code Modulation
PCR	Pulse Compression Ratio
PD	Pulse Doppler
PDF	Probability Density Function
PDM	Pulse Descriptor Message
PDW	Pulse Descriptor Word
PET	Passive Emitter Tracker
PF	Position Fix
PGM	Precision Guided Munitions
PGR	Processing Gain of a Radar
PLL	Phase Lock Loop
PM	Phase Modulation
PN	Pseudo-Noise
POI	Probability of Interception
PRF	Pulse Repetitition Frequency
PRI	Pulse Repetition Interval
PRN	Pseudorandom Noise
PSK	Phase Shift Key
PSYOPS	Psychological Operations
PTOA	Predicted Time-of-Arrival
PW	Pulse Width
QAM	Quadrature Amplitude Modulation
QM	Quadrature Modulation Method
QMF	Quadrature Mirror Filter
QPSK	Quadrature PSK
RAF	Range Advantage Factor
RATG	Ram Air Turbine Generator
RCIED	Radio-Controlled IED
RCS	Radar Cross Section
RF	Radio Frequency
RFM	Range-Filter Map
RGPO	Range-Gate Pull-Off
RHCP	Right-Hand Circular Polarization
RRC	Radar Resolution Cell
RWR	Radar Warning Receiver
S/H	Sample and Hold
S/W	Software
SAM	Surface-to-Air Missile
SAR	Synthetic Aperture Radar
SATCOM	Satellite Communication Systems
SAW	Surface Acoustic Wave
SC- TDOA	Signal Correlation TDOA
SCR	Signal-to-Clutter Ratio
SEAD	Suppression of Enemy Air Defense
SEI	Specific Emitter Identification

SFDR	Spurious Free Dynamic Range
SFH	Slow Frequency Hopping
SH	Superheterodyne
SHR	Superhet Receiver
SIGINT	Signal Intelligence
SIJ	Stand-In Jammers
SLB	Sidelobe Blanker
SLC	Sidelobe-Canceller
SLD	Sliding
SLL	Sidelobe Level
SNR	Signal-to-Noise Ratio
SOJ	Stand-Off Jammer
SPJ	Self-Protection Jammer
SPS	Self-Protection System
SS	Spread-Spectrum
SSA	Solid State High Gain Amplifier
SSB	Single Sideband
SSCA	Strip Spectral Correlation Analysis
SSM	Surface-to-Surface Missile
SSW	Swept-Square-Wave
STAP	Space-Time Adaptive Processing
STG	Staggered
STT	Single Tracking Target
SWT	Swept
T/R	Transmit/Receive
TA	Target Acquisition Radar
TD	Towed Decoy
TDL	Tactical Data Link
TDMA	Time Division Multiple Access
TDOA	Time Difference Of Arrival
TERCOM	Terrain Contour Matching
TEWA	Threat Evaluation And Weapon Assignment
TOA	Time of Arrival
TOI	Time of Interception
TOT	Time-on-Target
TR	Transform Receivers
TTR	Target Tracking Radar
TVM	Track Via Missile
TWS	Track While Scan
TWT	Traveling Wave Tube
UAS	Unmanned Aerial Systems
UCA	Uniform Circular Array
UMOP	Unwanted MOP
USB	Upper Sideband
VCO	Voltage Controlled Oscillator
VGPO	Velocity Gate Pull-Off

VOR	VHF Omni-directional Radio-range
VRGPO	Velocity and Range-Gate Pull-Off
WB	Wideband
WGN	White Gaussian Noise
WO	Wide Open
WVD	Wigner-Ville Distribution

About the Author

Dr. Andrea De Martino graduated (Laurea) in nuclear engineering (electronic track) in 1970 and began his career as a researcher in the field of large systems theory and automatic control at the Electronic Engineering Institute of the University of Rome (1970–1971), where he then achieved a Ph.D. in automatic control systems in 1977. After completion of military duty as lieutenant in the Italian Air Force (1971–1972) he joined Selenia S.p.A—Rome/Pomezia, from 1972 to 1985,where he was involved in radar systems holding positions from radar system designer to chief engineer, ranging from surveillance and tracking radars to multifunction phased arrays radars, and he was then appointed as head of the land and shipboard EW systems department.

In 1985 he joined Elettronica S.p.A–Rome where he covered a number of department head positions ranging from new EW products, microwave to EFA-DASS (now Praetorian) equipment, and since 1999 he has held the position of chief technical officer. He was also appointed as technical director of the Euro-DASS Consortium, which comprises Selex Galileo (UK), Indra (Sp) and EADS (Ger), keeping that position from 2002 to 2009.

Dr. De Martino is holder of several patents, and author of many technical papers on radar and EW equipment techniques and technologies presented at various classified/unclassified conferences and appearing in specialized magazines.

Dr De Martino can be reached at: Elettronica S.p.A.—Via Tiburtina Valeria Km 13.600–00131 Rome, Italy; tel.: +39 06 4154370; e-mail: andrea.demartino@elt.it.

Index

D

E

Q

R

T

Recent Titles in the Artech House Radar Series

David K. Barton, Series Editor

Adaptive Antennas and Phased Arrays for Radar and Communications, Alan J. Fenn

Advanced Techniques for Digital Receivers, Phillip E. Pace

Advances in Direction-of-Arrival Estimation, Sathish Chandran, editor

Airborne Pulsed Doppler Radar, Second Edition, Guy V. Morris and Linda Harkness, editors

Bayesian Multiple Target Tracking, Lawrence D. Stone, Carl A. Barlow, and Thomas L. Corwin

Beyond the Kalman Filter: Particle Filters for Tracking Applications, Branko Ristic, Sanjeev Arulampalam, and Neil Gordon

Cognitive Radar: The Knowledge-Aided Fully Adaptive Approach, Joseph R. Guerci

Computer Simulation of Aerial Target Radar Scattering, Recognition, Detection, and Tracking, Yakov D. Shirman, editor

Design and Analysis of Modern Tracking Systems, Samuel Blackman and Robert Popoli

Detecting and Classifying Low Probability of Intercept Radar, Second Edition, Phillip E. Pace

Digital Techniques for Wideband Receivers, Second Edition, James Tsui

Electronic Intelligence: The Analysis of Radar Signals, Second Edition, Richard G. Wiley

Electronic Warfare in the Information Age, D. Curtis Schleher

Electronic Warfare Target Location Methods, Second Edition, Richard A. Poisel

ELINT: The Interception and Analysis of Radar Signals, Richard G. Wiley

EW 101: A First Course in Electronic Warfare, David Adamy

EW 102: A Second Course in Electronic Warfare, David L. Adamy

EW 103: Tactical Battlefield Communications Electronic Warfare, David Adamy

Fourier Transforms in Radar and Signal Processing,Second Edition, David Brandwood

Fundamentals of Electronic Warfare, Sergei A. Vakin, Lev N. Shustov, and Robert H. Dunwell

Fundamentals of Short-Range FM Radar, Igor V. Komarov and Sergey M. Smolskiy

Handbook of Computer Simulation in Radio Engineering, Communications, and Radar, Sergey A. Leonov and Alexander I. Leonov

High-Resolution Radar, Second Edition, Donald R. Wehner

Introduction to Electronic Defense Systems, Second Edition, Filippo Neri

Introduction to Electronic Warfare, D. Curtis Schleher

Introduction to Electronic Warfare Modeling and Simulation, David L. Adamy

Introduction to RF Equipment and System Design, Pekka Eskelinen

Introduction to Modern EW Systems, Andrea De Martino

The Micro-Doppler Effect in Radar, Victor C. Chen

Microwave Radar: Imaging and Advanced Concepts, Roger J. Sullivan

Millimeter-Wave Radar Targets and Clutter, Gennadiy P. Kulemin

Modern Radar Systems, Second Edition, Hamish Meikle

Modern Radar System Analysis, David K. Barton

Modern Radar System Analysis Software and User's Manual, Version 3.0, David K. Barton

Monopulse Principles and Techniques, Second Edition, Samuel M. Sherman and David K. Barton

MTI and Pulsed Doppler Radar with MATLAB®, Second Edition, D. Curtis Schleher

Multitarget-Multisensor Tracking: Applications and Advances Volume III, Yaakov Bar-Shalom and William Dale Blair, editors

Principles of High-Resolution Radar, August W. Rihaczek

Principles of Radar and Sonar Signal Processing, François Le Chevalier

Radar Cross Section, Second Edition, Eugene F. Knott et al.

Radar Evaluation Handbook, David K. Barton et al.

Radar Meteorology, Henri Sauvageot

Radar Reflectivity of Land and Sea, Third Edition, Maurice W. Long

Radar Resolution and Complex-Image Analysis, August W. Rihaczek and Stephen J. Hershkowitz

Radar Signal Processing and Adaptive Systems, Ramon Nitzberg

Radar System Analysis, Design, and Simulation, Eyung W. Kang

Radar System Analysis and Modeling, David K. Barton

Radar System Performance Modeling, Second Edition, G. Richard Curry

Radar Technology Encyclopedia, David K. Barton and Sergey A. Leonov, editors

Radio Wave Propagation Fundamentals, Artem Saakian

Range-Doppler Radar Imaging and Motion Compensation, Jae Sok Son, et al.

Robotic Navigation and Mapping with Radar, Martin Adams, John Mullane, Ebi Jose, Ba-Ngu Vo

Signal Detection and Estimation, Second Edition, Mourad Barkat

Space-Time Adaptive Processing for Radar, J. R. Guerci

Special Design Topics in Digital Wideband Receivers, James Tsui

Theory and Practice of Radar Target Identification, August W. Rihaczek and Stephen J. Hershkowitz

Time-Frequency Transforms for Radar Imaging and Signal Analysis, Victor C. Chen and Hao Ling

For further information on these and other Artech House titles, including previously considered out-of-print books now available through our In-Print-Forever® (IPF®) program, contact:

Artech House
685 Canton Street
Norwood, MA 02062
Phone: 781-769-9750
Fax: 781-769-6334
e-mail: artech@artechhouse.com

Artech House
16 Sussex Street
London SW1V HRW UK
Phone: +44 (0)20 7596-8750
Fax: +44 (0)20 7630-0166
e-mail: artech-uk@artechhouse com

Find us on the World Wide Web at: www.artechhouse.com